Lecture Notes in Computer Science 8836

Commenced Publication in 1973
Founding and Former Series Editors:
Gerhard Goos, Juris Hartmanis, and Jan van Leeuwen

Chu Kiong Loo   Keem Siah Yap
Kok Wai Wong Andrew Teoh
Kaizhu Huang (Eds.)

Neural Information Processing

21st International Conference, ICONIP 2014
Kuching, Malaysia, November 3-6, 2014
Proceedings, Part III

 Springer

Volume Editors

Chu Kiong Loo
University of Malaya, Kuala Lumpur, Malaysia
E-mail: ckloo.um@um.edu.my

Keem Siah Yap
Universiti Tenaga Nasional, Selangor, Malaysia
E-mail: yapkeem@uniten.edu.my

Kok Wai Wong
Murdoch University, Murdoch, WA, Australia
E-mail: k.wong@murdoch.edu.au

Andrew Teoh
Yonsei University, Seoul, South Korea
E-mail: bjteoh@yonsei.ac.kr

Kaizhu Huang
Xi'an Jiaotong-Liverpool University, Suzhou, China
E-mail: kaizhu.huang@xjtlu.edu.cn

ISSN 0302-9743　　　　　　　　　　e-ISSN 1611-3349
ISBN 978-3-319-12642-5　　　　　　　e-ISBN 978-3-319-12643-2
DOI 10.1007/978-3-319-12643-2
Springer Cham Heidelberg New York Dordrecht London

Library of Congress Control Number: 2014951688

LNCS Sublibrary: SL 1 – Theoretical Computer Science and General Issues

Typesetting: Camera-ready by author, data conversion by Scientific Publishing Services, Chennai, India

Printed on acid-free paper

Springer is part of Springer Science+Business Media (www.springer.com)

Preface

This volume is part of the three-volume proceedings of the 21st International Conference on Neural Information Processing (ICONIP 2014), which was held in Kuching, Malaysia, during November 3–6, 2014. The ICONIP is an annual conference of the Asia Pacific Neural Network Assembly (APNNA). This series of ICONIP conferences has been held annually since 1994 in Seoul and has become one of the leading international conferences in the area of neural networks.

ICONIP 2014 received a total of 375 submissions by scholars from 47 countries/regions across six continents. Based on a rigorous peer-review process where each submission was evaluated by at least two qualified reviewers, a total of 231 high-quality papers were selected for publication in the reputable series of *Lecture Notes in Computer Science* (LNCS). The selected papers cover major topics of theoretical research, empirical study, and applications of neural information processing research. ICONIP 2014 also featured a pre-conference event, namely, the Cybersecurity Data Mining Competition and Workshop (CDMC 2014) which was held in Kuala Lumpur. Nine papers from CDMC 2014 were selected for a Special Session of the conference proceedings.

In addition to the contributed papers, the ICONIP 2014 technical program included a keynote speech by Shun-Ichi Amari (RIKEN Brain Science Institute, Japan), two plenary speeches by Jacek Zurada (University of Louisville, USA) and Jürgen Schmidhuber (Istituto Dalle Molle di Studi sull'Intelligenza Artificiale, Switzerland). This conference also featured seven invited speakers, i.e., Akira Hirose (The University of Tokyo, Japan), Nikola Kasabov (Auckland University of Technology, New Zealand), Soo-Young Lee (KAIST, Korea), Derong Liu (Chinese Academy of Sciences, China; University of Illinois, USA), Kay Chen Tan (National University of Singapore), Jun Wang (The Chinese University of Hong Kong), and Zhi-Hua Zhou (Nanjing University, China).

We would like to sincerely thank Honorary Chair Shun-ichi Amari, Mohd Amin Jalaludin, the members of the Advisory Committee, the APNNA Governing Board for their guidance, the members of the Organizing Committee for all their great efforts and time in organizing such an event. We would also like to take this opportunity to express our deepest gratitude to all the technical committee members for their professional review that guaranteed high quality papers.

We would also like to thank Springer for publishing the proceedings in the prestigious LNCS series. Finally, we would like to thank all the speakers, authors,

Workshop and Tutorial Chairs

Chen Change Loy	Chinese University of Hong Kong, SAR China
Ying Wah Teh	University of Malaya, Malaysia
Saeed Reza	University of Malaya, Malaysia
Tutut Harewan	University of Malaya, Malaysia

Special Session Chairs

Thian Song Ong	Multimedia University, Malaysia
Siti Nurul Huda Sheikh Abdullah	Universiti Kebangsaan Malaysia, Malaysia

Financial Chair

Ching Seong Tan	Multimedia University, Malaysia

Sponsorship Chairs

Manjeevan Seera	University of Malaya, Malaysia
John See	Multimedia University, Malaysia
Aamir Saeed Malik	Universiti Teknologi Petronas, Malaysia

Publicity Chairs

Siong Hoe Lau	Multimedia University, Malaysia
Khairul Salleh Mohamed Sahari	Universiti Tenaga Nasional, Malaysia

Asia Liaison Chairs

ShenShen Gu	Shanghai University, China

Europe Liaison Chair

Wlodzislaw Duch	Nicolaus Copernicus University, Poland

America Liaison Chair

James T. Lo	University of Maryland, USA

Advisory Committee

Lakhmi Jain, Australia
David Gao, Australia
BaoLiang Lu, China
Ying Tan, China
Jin Xu, China.
Irwin King, Hong Kong, SAR China
Jun Wang, Hong Kong, SAR China
P. Balasubramaniam, India
Kunihiko Fukushima, Japan
Shiro Usui, Japan
Minho Lee, Korea
Muhammad Leo Michael Toyad
 Abdullah, Malaysia
Mustafa Abdul Rahman, Malaysia
Narayanan Kulathuramaiyer,
 Malaysia
David Ngo, Malaysia

Siti Salwah Salim, Malaysia
Wan Ahmad Tajuddin Wan Abdullah,
 Malaysia
Wan Hashim Wan Ibrahim, Malaysia
Dennis Wong, Malaysia
Nik Kasabov, New Zealand
Arnulfo P. Azcarraga, Phillipines
Wlodzislaw Duch, Poland
Tingwen Huang, Qatar
Meng Joo Err, Singapore
Xie Ming, Singapore
Lipo Wang, Singapore
Jonathan H. Chan, Thailand
Ron Sun, USA
De-Liang Wang, USA
De-Shuang Huang, China

Technical Committee

Ahmad Termimi Ab Ghani
Mark Abernethy
Adel Al-Jumaily
Leila Aliouane
Cesare Alippi
Ognjen Arandjelovic
Sabri Arik
Mian M. Awais
Emili Balaguer-Ballester
Valentina Emilia Balas
Tao Ban
Sang-Woo Ban
Younès Bennani
Asim Bhatti
Janos Botzheim
Salim Bouzerdoum
Ivo Bukovsky
Jinde Cao
Jiang-Tao Cao
Chee Seng Chan
Long Cheng
Girija Chetty

Andrew Chiou
Pei-Ling Chiu
Sung-Bae Cho
Todsanai Chumwatana
Pau-Choo Chung
Jose Alfredo Ferreira Costa
Justin Dauwels
Mingcong Deng
M.L. Dennis Wong
Hongli Dong
Hiroshi Dozono
El-Sayed M. El-Alfy
Zhouyu Fu
David Gao
Tom Gedeon
Vik Tor Goh
Nistor Grozavu
Ping Guo
Masafumi Hagiwara
Osman Hassab Elgawi
Shan He
Haibo He

Sven Hellbach
Jer Lang Hong
Jinglu Hu
Xiaolin Hu
Kaizhu Huang
Amir Hussain
Kazushi Ikeda
Piyasak Jeatrakul
Sungmoon Jeong
Yaochu Jin
Zsolt Csaba Johanyák
Youki Kadobayashi
Hiroshi Kage
Joarder Kamruzzaman
Shin'Ichiro Kanoh
Nikola Kasabov
Rhee Man Kil
Kyung-Joong Kim
Kyung-Hwan Kim
Daeeun Kim
Laszlo T. Koczy
Markus Koskela
Szilveszter Kovacs
Naoyuki Kubota
Takio Kurita
Olcay Kursun
James Kwok
Sungoh Kwon
Weng Kin Lai
Siong Hoe Lau
Yun Li Lee
Minho Lee
Nung Kion Lee
Chin Poo Lee
Vincent Lemaire
L. Leng
Yee Tak Leung
Bin Li
Yangming Li
Ming Li
Xiaofeng Liao
Meng-Hui Lim
C.P. Lim
Chee Peng Lim
Kim Chuan Lim

Hsuan-Tien Lin
Huo Chong Ling
Derong Liu
Zhi-Yong Liu
Chu Kiong Loo
Wenlian Lu
Zhiwu Lu
Bao-Liang Lu
Shuangge Ma
Mufti Mahmud
Kenichiro Miura
Hiroyuki Nakahara
Kiyohisa Natsume
Vinh Nguyen
Tohru Nitta
Yusuke Nojima
Anto Satriyo Nugroho
Takenori Obo
Toshiaki Omori
Takashi Omori
Thian Song Ong
Sid-Ali Ouadfeul
Seiichi Ozawa
Worapat Paireekreng
Paul Pang
Ying Han Pang
Shaoning Pang
Hyung-Min Park
Shri Rai
Mallipeddi Rammohan
Alexander Rast
Jinchang Ren
Mehdi Roopaei
Ko Sakai
Yasuomi Sato
Naoyuki Sato
Shunji Satoh
Manjeevan Seera
Subana Shanmuganathan
Bo Shen
Yang Shi
Tomohiro Shibata
Hayaru Shouno
Jennie Si
Jungsuk Song

Table of Contents – Part III

Signal and Image Processing

The 2014 Cybersecurity Data Mining Competition and Workshop (CDMC2014)

Intelligent Systems for Supporting Decision-Making Processes: Theories and Applications

Neuroengineering and Neuralcomputing

Cognitive Robotics

Security in Signal Processing and Machine Learning

Learning Systems for Social Network and Web Mining

Real Time Crowd Counting with Human Detection and Human Tracking

Xinjian Zhang and Liqing Zhang

Key Laboratory of Shanghai Education Commission for Intelligent Interaction and Cognitive
Engineering,
Department of Computer Science and Engineering,
Shanghai Jiao Tong University, China
{zha,lqzhang}@sjtu.edu.cn

Abstract. Real-time crowd counting is of many potential applications, such as
surveillance, crowd flow control in subway. In this paper, we propose a fast and
novel method for estimating the number of people in crowded surveillance
scenes. This method is able to count people in real time and is robust to
changes of illumination and background. The combined rectangle features and
cascade of boosted classifier are employed to train a multi-scale head-shoulder
detector. The detector can detect human in every frame with a high accuracy.
Then human tracking is used to track the detected people and remove duplicates
in successive frames. Experiments on a real-world video show the proposed
method can give an accurate estimation in real time.

Keywords: Crowd Counting, Human Detection, Human Tracking.

1 Introduction

Real-time crowd counting in videos becomes more and more important for public
area monitoring for the purpose of safety and security. The goal of crowd counting is
to estimate the number of people passing through a given line or a given area. It has
many valuable real-world applications, such as controlling the number of people in
the venues, estimating the people flow in the subway station, counting people entering
and exiting. There are still many challenges to be solved in this task. First, in crowded
scenes, the occlusion between people is serious. Second, the resolution of video in
surveillance camera is relatively low, detailed information is lost. In real-word places,
such as in subway stations and libraries, we find that most monitors are above the
front of people's heads. This is because at this position, monitors can capture faces,
dresses and other characteristics of pedestrians passing through. So in our scenario,
we assume that the cameras are installed at a high place, facing the crowd flow
direction.

Most previous approaches can be divided into two categories. The first type is
based on the counting-by-regression framework, where extracted features are directly
regressed to the number of people. The second type is mainly based on multi-target
human detection.

C.K. Loo et al. (Eds.): ICONIP 2014, Part III, LNCS 8836, pp. 1–8, 2014.

The counting-by-regression methods extract low level features, such as the foreground pixels, HOG features, and such local features are transformed to the number of people using regression. Cong et al. [1] used 1-D flow velocity estimation to extract dynamic mosaic. Then they did regression between the features of dynamic mosaic (pixels number, edge pixels number, width of the mosaic) and number of people. Ma and Chan [2] used local HOG features as the low level feature and did regression between the meaningful local descriptors and the number of the people. These methods only used low level features, so most of them can run in real time. But these systems can't be large-scale deployed because extracting these low level features is highly dependent on the prior knowledge of the background and the position of cameras. When the scene changes, these low level features will change a lot and the pre-trained regression function will fail to work.

Multi-target human detection can be used to count people in crowd scenes. Lin et al. [3] used Harr wavelet transform to extract the area with head-like contour, then used Support Vector Machine to determine whether these areas are heads or not. Their method can count people in a single image. Li et al [4] used a foreground segmentation algorithm and a HOG based head-shoulder detection algorithm to estimate the number of people in images. But these methods can't work in real time; the heavy computational cost limits their applications to real-world problems.

In recent years, there are great progresses in object detection. It is possible to implement robust pedestrian detection in real time. In this paper, we propose a novel real-time crowd counting method. Fig. 1 shows the block diagram of the whole system. We use combined rectangle features and cascade of boosted classifier to train a head-shoulder detector. The detector can detect human in every frame with a high accuracy and then human tracking is used to track the detected people. The number of detected pedestrians is equal to the number of the tracks.

The rest of this paper is organized as follows: Section 2 introduces the training process of the head-shoulder detector; Section 3 proposes how to remove duplicate detected pedestrians by human tracking, followed by experimental results and discussions in Section 4. Finally conclusions are given in section 5.

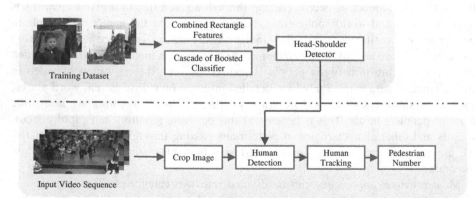

Fig. 1. Framework of the Proposed Crowd Counting System

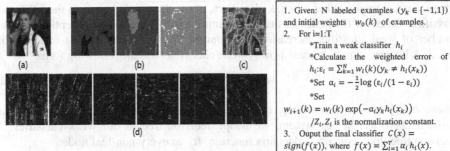

1. Given: N labeled examples $(y_k \in \{-1,1\})$ and initial weights $w_0(k)$ of examples.
2. For i=1:T
   *Train a weak classifier h_i
   *Calculate the weighted error of
   $h_i{:}\varepsilon_i = \sum_{k=1}^{N} w_i(k)(y_k \neq h_i(x_k))$
   *Set $\alpha_i = -\frac{1}{2}\log(\varepsilon_i/(1-\varepsilon_i))$
   *Set
   $w_{i+1}(k) = w_i(k)\exp(-\alpha_i y_k h_i(x_k))$
   $/Z_i, Z_i$ is the normalization constant.
3. Ouput the final classifier $C(x) = sign(f(x))$, where $f(x) = \sum_{i=1}^{T} \alpha_i h_i(x)$.

Fig. 2. All channels, (a) is the original image, (b) is the RGB channels, (c) is gradient magnitude, (d) is the six orientations quantized gradient

Fig. 3. Training Processing of Adaboost

2   Head-Shoulder Detection

Human detection is one of key issues in computer vision. In real world applications, people concern both accuracy and speed. In our system, we use the combined rectangle features and cascade of boosted classifier to realize real time human detection.

2.1   Combined Rectangle Feature

A rectangle feature is the sum of values in a given rectangle and given channel. Dollar et al. [5] augmented channels using linear and non-linear transform. There are total ten channels used in this paper, as shown in Fig. 2. Integral image is an image representation, allowing fast rectangle feature extraction. It was first proposed by Viola and Jones [6] for face detection. A rectangle feature can be computed by only 3 operations with integral channels:

$$r(x_s, y_s, x_e, y_e) = I(x_e, y_e) - I(x_s-1, y_e) - I(x_e, y_s-1) + I(x_s-1, y_s-1) \quad (1)$$

where I is one integral channel of the image. A combined rectangle feature is the linear combination of several rectangle features:

$$R_{combined} = \sum_{i=1}^{K} w_i r_i, \quad \sum_{i=1}^{K} w_i = 1 \quad (2)$$

where K is the number of rectangles. K=1, 2, 3, 4 is used in this paper. The position of rectangles, channel index and weights are all random sampled.

2.2   Cascade of Boosted Classifier

Adaboost is a fast approach to train a classifier when given a lot of simple features. It's widely used in object detection [5, 6, 7]. The training error of Adaboost is bounded by:

$$\varepsilon_{train} \leq \prod_{i=1}^{T} 2\sqrt{\varepsilon_i(1-\varepsilon_i)} \quad (3)$$

where T is the total number of weak classifiers, ε_i is the training error of the i-th weak classifier. Equation (3) indicates that the training error decreases exponentially as the number of weak classifier increasing. The training process of Adaboost is shown in Fig. 3. A boosted classifier consists of N weak classifiers has the following form:

$$C(x) = \sum_{i=1}^{N} \alpha_i h_i(x) \qquad (4)$$

where $h_i(x)$ is a weak classifier, α_i is the weight of the classifier, $C(x)$ is the final boosted classifier. In our system, we adopt decision tree as our weak classifier. A decision tree h_{tree} contains a decision function h_j at every non-leaf node:

$$h_j = p_j \times sign(x(k_j) - t_j) \qquad (5)$$

Where $p_j \in \{-1, 1\}$ is the polarity. When training the decision tree, we find the best feature $(x(k_j))$ and threshold (t_j) to minimize the error. We cascade several boosted classifiers to get our final classifier, and refer to [6] for details. Given a trained cascade of classifiers, the false positive rate F and detection rate D are:

$$F = \prod_{i=1}^{N} F_i \text{ and } D = \prod_{i=1}^{N} D_i \qquad (6)$$

where F_i and D_i is the false positive rate and detection rate for one layer.

On one hand, when one people can't be detected at one frame, it can still be detected out in other frames. On the other hand, it is difficult to remove the object that is falsely detected as a people. Therefore in our system, we use a low detection rate. Meanwhile, the false positive is very low.

3   Human Tracking Based Number Estimation

In this section, we describe how the detected pedestrians, obtained in individual frames, are converted into the number of pedestrians in the videos. We adopt the tracking method proposed in [8], for its speed and accuracy. We randomly sample many positions around the current position. All the samples are compressed by the same sparse measurement matrix. Then the tracking task is converted to a classification problem. We deal it with Native Bayesian Classifier. The sample with the maximal classifier response will be the position of the pedestrian in the next frame.

As shown in Fig. 4, in every frame, we hold a pool of candidates and track each candidate. In the next frame, we get the tracking results. At the same time, the head-shoulder detector will detect new pedestrians in this frame. We calculate the overlap between the new detected pedestrian and the tracking results. If the overlap between them exceeds a predefined threshold, they will be considered as the same person; we will use the detected location as the candidate location, and this will avoid deviation caused by tracking. Otherwise the detected result will be recognized as a new person. After this merging process, we get the candidates in the second frame. The overlap threshold will be determined by experiments. It's defined as (7):

$$R = \frac{S_1 \cap S_2}{S_1 \cup S_2} \qquad (7)$$

We also add the following constraints to the tracking process. They will make the tracking process faster and more accurate:

- The velocity of the pedestrians can't be too fast;
- The size of the detected pedestrians are stationary;
- Pedestrians can only walk towards the camera.

This process will carry on in each frame. When the tracks touch the bottom boundary, or the left and right boundary, we think the pedestrian will disappear from the interested area in the next frame and will not track it anymore. This tracking process establishes matches between head-shoulders crossing frames. By counting the number of tracks, we get the number of pedestrians in the video.

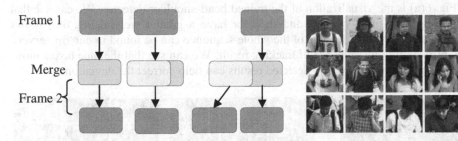

Fig. 4. Red boxes are candidate pedestrians; green boxes are tracking results; yellow boxes are detection results

Fig. 5. Some Typical Positive Examples

4 Experimental Results

4.1 Dataset

Dataset for Training Head-Shoulder Detector. We collect 985 positive examples from 423images. 336 of them are from the well-known INRIA set and the others are cropped from our recorded videos. When training, all these examples will be normalized to 48*48 pixels. For the negative examples, 564 of them are from the INRIA negative set, 297 are cropped from our recorded videos without head-shoulder. Five thousand negative examples will be sampled from the negative images when training. Some typical examples of header-shoulder are shown in Fig. 5.

Dataset for Testing. We test our algorithm on a real video taken in the subway station. There is a stationary camera mounted at the subway tunnels, above the front of pedestrian's head. The orientation of the camera is about 30 degree. There are 1000 frames in total and the frame rate is 10 fps, which is available at our ftp server[1]. The crowd density in the tunnels ranges from sparse to very crowd.

[1] `ftp://zha:public@public.sjtu.edu.cn/TestSequence/`

4.2   Human Detection Results

We drop the pixels far away from the camera and too close to the camera. The detection is performed on the cropped images (320*126 pixels). We labeled 200 successive frames. Following PASCAL VOC evaluation protocol [9], a detected bounding box is considered correct if and only if it overlaps more than 50% with one ground truth bounding box, otherwise it will be considered as a false positive.

In the 200 labeled frames, there are total 499 ground truth boxes. Our detector detect out 379 of them and the number of false positive boxes is 5. The detection rate is about 76.0% and false positive per image is 2.5%. But when one pedestrian is not detected in one frame, it can still be detected out in other frames. We count identical pedestrians in the 200 frames, there are 69 pedestrians in total, and 66 of them are detected out at least once. The detection rate is 95.7%.

Fig. 6(a) is the visualization of the trained head-shoulder detector. We can see that pixels at the position of head and shoulder have a greater weight. Fig. 6(b) shows some detection results. Results of the whole sequence can be found in our ftp server[2]. Fig. 6(c) shows one sequence of tracking result. We can see that the blue boxes move with people tightly and newly detected results can help correct the deviation.

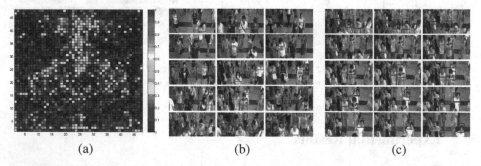

(a)   (b)   (c)

Fig. 6. (a) Visualize spatial support of the trained detector, (b) Examples of detection results for different levels of person density. The size of the bounding boxes shows the size of detected pedestrians, (c) An example tracking sequences. The blue boxes are the tracking results; the red boxes are the detecting results in current frame.

4.3   Crowd Counting Results

Because every pedestrian goes through the monitored area takes about 10 frames and our method can't guarantee the pedestrian coming first being counted first, we divide all the frames into bins and count number in each bin. There are ten frames in each bin. In each bin, we count the number of ground truth, number of predicted, number of correctly predicted and number of false positive. We also compare our method with a regression method. The regression method only counts pedestrians passing through the bottom of this area, so the total counted number is less than our method. It does regression between the number of foreground pixels and number of pedestrian. We use the following criterion to evaluate our system:

---

[2]   ftp://zha:public@public.sjtu.edu.cn/DetectResult.avi/

$$MSE = \frac{\sum_{All\ bins} (\#of\ ground\ truth\ -\ \#of\ predicted)^2}{total\ \#\ of\ bins} \tag{8}$$

$$Absolute\ Error = \frac{\sum_{All\ bins} |\#of\ ground\ truth\ -\ \#of\ predicted|}{total\ \#\ of\ bins} \tag{9}$$

$$FFPB = \frac{\sum_{All\ bins} |\#of\ false\ positive|}{total\ \#\ of\ bins} \tag{10}$$

The MSE and Absolute Error of the predicted number and correctly predicted number are calculated respectively. Some results are shown in Table 1 and the FFPB is 0.38. The accumulated counting results are shown in Fig. 7. A video of result can be found on our ftp server[3].

Table 1. The Results of Pedestrian Counting

	Total Ground Truth Number	Predicted Number	MSE	Absolute Error
Our Method (Total)	439	421	1.28	0.72
Our Method (Correct)	439	383	1.06	0.56
Linear Regression	380	320	1.38	0.87

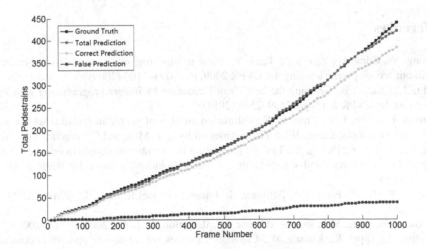

Fig. 7. Total Number of Pedestrians

From the experimental results, we can see that our method achieves a high accuracy and the performance is better than regression method. Our methods can also adapt to other scenes more conveniently and locate the position of pedestrians. This means our method can give a more detail description of the crowd.

---

[3] ftp://zha:public@public.sjtu.edu.cn/CountRes.avi

4.4   Speed Measurement

All the code is written by Matlab. Our computer is equipped with Intel i7-3770 CPU@3.4GHZ, 8G memory. All speed results are given on the cropped images (320*126 pixels), averaged over 1000 frames of the video sequence. The detection speed achieves 77Hz and the final counting system achieves 45Hz. This means our method is applicable to the real-time applications.

5   Conclusion

In this paper, we propose a novel method to estimate the number of pedestrians in crowded scenes. We use the simple combined rectangle features and cascade of boosted classifier to implement real time human detection. The head-shoulder detector can detect pedestrians in every frame with a high accuracy. Then human tracking is used to track the detected pedestrians and remove duplicates. This method can not only count pedestrians in crowd scenes, but can also locate pedestrians in videos, which means it can do more things besides crowd counting. Experiments on a real word video show our system is able to achieve satisfactory performance in real time.

Acknowledgement. The work was supported by the national natural science foundation of China (Grant No. 91120305, 61272251).

References

1. Cong, Y., Gong, H., Zhu, S.C., Tang, Y.: Flow mosaicking: real-time pedestrian counting without scene-specific learning. In: CVPR 2009, pp. 1093–1100 (2009)
2. Ma, Z., Chan, A.B.: Crossing the line: Crowd counting by integer programming with local features. In: CVPR 2013, pp. 2539–2546 (2013)
3. Lin, S.-F., Chen, J.-Y., Chao, H.-X.: Estimation number of people in crowded scenes using perspective transformation. IEEE Transactions on System, Man, and Cybernetics 31 (2001)
4. Li, M., Zhang, Z., Huang, K., Tan, T.: Estimating the number of people in crowded scenes by MID based foreground segmentation and head-shoulder detection. In: ICPR 2008, pp. 1–4 (2008)
5. Dollár, P., Tu, Z., Perona, P., Belongie, S.: Integral channel features. In: BMVC 2009, pp. 1–11 (2009)
6. Viola, P.A., Jones, M.J.: Robust real-time face detection. In: ICCV 2001, p. 747 (2001)
7. Dollár, P., Appel, R., Kienzle, W.: Crosstalk cascades for frame-rate pedestrian detection. In: Fitzgibbon, A., Lazebnik, S., Perona, P., Sato, Y., Schmid, C. (eds.) ECCV 2012, Part II. LNCS, vol. 7573, pp. 645–659. Springer, Heidelberg (2012)
8. Zhang, K., Zhang, L., Yang, M.-H.: Real-Time Compressive Tracking. In: Fitzgibbon, A., Lazebnik, S., Perona, P., Sato, Y., Schmid, C. (eds.) ECCV 2012, Part III. LNCS, vol. 7574, pp. 864–877. Springer, Heidelberg (2012)
9. Everingham, M., Gool, L.V., Williams, C.K.I., Winn, J., Zisserman, A.: The PASCAL visual object classes challenge (VOC 2008) results (2008)

A New Method for Removing Random-Valued Impulse Noise

Qiyu Jin[1], Li Bai[2], Jie Yang[1], Ion Grama[3], and Quansheng Liu[3]

[1] Institute of Image Processing and Pattern Recognition, Shanghai Jiao Tong University, No. 800 Dongchuan Road, Minhang District, Shanghai 200240, China
[2] School of Computer Science, University of Nottingham UK
[3] Laboratoire de Mathmatiques de Bretagne Atlantique, UMR 6205, Université de Bretagne-Sud, Campus de Tohaninic, BP 573, 56017 Vannes, France
jinqiyu@sjtu.edu.cn

Abstract. A new algorithm for removing random-valued impulse noise is proposed. We use a standardized version of the Rank Ordered Absolute Differences statistic of Garnett et al. [1] to attribute weights to noisy pixels. These weights are then incorporated into the Optimal Weights Filter approach from [2,3] to construct a new filter. Simulation results show that our method performs significantly better than a number of existing techniques.

Keywords: random-valued impulse noise, denoising, Optimal Weights Filter, Non-Local Means, Rank Ordered Absolute Difference.

1 Introduction

Random-valued impulse noise can be systematically introduced into digital images during acquisition and transmission [4]. Impulse noise is characterized by replacing a portion of an images pixel values with random values, leaving the remainder unchanged. In most applications, denoising is fundamental to subsequent image processing operations, such as edge detection, image segmentation, object recognition, etc. The goal of denoising is to effectively remove noise from an image while keeping its features intact. To this end, a variety of techniques have been proposed to remove impulse noise.

Recently, an edge-preserving regularization method has been proposed to remove impulse noise [5]. It uses a nonsmooth data-fitting term along with edge-preserving regularization. In order to improve this variational method in removing impulse noise, a two-stage method was proposed in [6] and [7]. It is efficient in dealing with high noise ratio, e.g., ratio as high as 90% for salt-and-pepper impulse noise and 50% for random-valued impulse noise. Its performance is impaired by the inaccuracy of the noise detector in the first phase. In order to find a better noise detector, especially for the random-valued impulse noise, Garnett et al. [1] introduced a new local image statistic called ROAD to identify the impulse noisy pixels. The result is a trilateral filter, which performs well for removing impulse noise. However, when the noise level is high, it blues the

C.K. Loo et al. (Eds.): ICONIP 2014, Part III, LNCS 8836, pp. 9–16, 2014.

images significantly. Dong et al. [8] amplified the differences between noisy pixels and noise-free pixels in ROAD, by introducing a new statistic called ROLD, so that the noise detection becomes more accurate. When the random-valued impulse noise ratio is as high as 60%, they still can remove most of the noise while preserving image details.

The ROAD statistic is efficient but turns out to be sensible to the proportion of the impulse noise, so that it is difficult to determine the parameters of the concerned filters. In this paper, we propose a standardized version of the ROAD statistic, called SROAD, to provide a more stable filter for which the determination of the parameters is simpler. We define new impulsive weights to magnify the difference of SROAD values between noisy pixels and noise free pixels. We then propose an efficient filter that combines the SROAD impulse noise detector with the optimal weights algorithm from [2,3] for removing random-valued impulse noise. Extensive experimental results show that our method performs significantly better than many known techniques.

2   Optimal Weights Filter for Random-Valued Impulse Noise

2.1   Impulse Noise Model

An image containing random-valued impulse noise can be described as follows:

$$y(x) = \begin{cases} u(x), \text{ with probability } 1 - p; \\ n(x), \text{ with probability } p, \end{cases} \tag{1}$$

where $u(x)$, $x \in \mathbf{I} = \{1, 2, \cdots, M\} \times \{1, 2, \cdots, N\}$, is the original image, $n(x)$, $x \in \mathbf{I}$, are independent random variables uniformly distributed in $[s_{\min}, s_{\max}]$, $s_{\min}$ and $s_{\max}$ being respectively the lowest and the highest pixel luminance values within the dynamic range, and p denotes the proportion of noisy pixels. The goal is to recover the original image $u(x)$, $x \in \mathbf{I}$, from the observed image $y(x)$, $x \in \mathbf{I}$.

2.2   Standardized Rank Ordered Absolute Differences

The ROAD (Rank Ordered Absolute Differences) statistic introduced by Garnett et al. [1] is known to be efficient in removing impulse noise. However this statistic is too sensitive to the proportion p of noisy points. We find that the operability of the ROAD statistic can be improved by its standardization.

For any pixel $x_0 \in \mathbf{I}$, we define the square window of pixels (whose center is excluded) of size $(2d + 1) \times (2d + 1)$:

$$\Omega_{x_0,d}^0 = \{x : 0 < \|x - x_0\|_\infty \le d\}, \tag{2}$$

where d is a positive integer and $\| \cdot \|_\infty$ denotes the supremum norm: $\|y\|_\infty = \max\{|y_1|, |y_2|\}$ for $y = (y_1, y_2)$. We define the SROAD statistic by

$$\mathrm{SROAD}(x_0) = \frac{1}{K} \sum_{i=1}^{K} r_i(x_0), x_0 \in \mathbf{I}, \tag{3}$$

where $r_i(x_0)$ is the i-th smallest term in the set $\{|y(x) - y(x_0)| : x \in \Omega^0_{x_0,d}\}$ and $2 \le K < \text{card } \Omega^0_{x_0,d}$. Without the coefficient it is just the ROAD statistic introduced by Garnett et al. [1]. The factor $1/K$ makes the statistic less sensible to variations of the level of the noise p and to the choice of K.

We then define the impulsive weight as follows:

$$J(x, H) = e^{-\frac{(\text{SROAD}(x)-b)^2_+}{2H^2}}, \tag{4}$$

where b is the hard threshold of SROAD values, H is a parameter and $(\cdot)_+$ is the positive part function: $(y)_+ = \max\{y, 0\}$. The impulsive weights measure the degree of contamination of a given pixel. With these weights we are able to construct a filter which is stable to the variations of the impulse noise levels: the constructed filter can remove most of the noise while preserving image details even when the impulse noise ratio is as high as 60%. Furthermore, we do not need the computationally expensive joint impulsivity weights introduced in Garnett et al. [1].

2.3 Construction of Optimal Weights Impulse Noise Filter

Now, we adapt the Optimal Weights Filter [2,3] to treat random-valued impulse noise. For any pixel $x_0 \in \mathbf{I}$ and a given $h \in \mathbb{N}_+$, the square window of pixels

$$\mathbf{U}_{x_0,h} = \{x \in \mathbf{I} : \|x - x_0\|_\infty \le h\} \tag{5}$$

will be called *search window* at x_0. The size of the square search window $\mathbf{U}_{x_0,h}$ is the positive integer number $D = (2h+1)^2 = \text{card } \mathbf{U}_{x_0,h}$. For any pixel $x \in \mathbf{U}_{x_0,h}$ and a given $\eta \in \mathbb{N}_+$, a second square window of pixels $\mathbf{V}_{x,\eta} = \mathbf{U}_{x,\eta}$ will be called for short a *similarity patch* at x in order to be distinguished from the search window $\mathbf{U}_{x_0,h}$. The size of the similarity patch $\mathbf{V}_{x,\eta}$ is the positive integer $S = (2\eta+1)^2 = \text{card } \mathbf{V}_{x,\eta}$. The vector $\mathbf{Y}_{x,\eta} = (y(x))_{x \in \mathbf{V}_{x,\eta}}$ formed by the values of the observed noisy image at pixels in the patch $\mathbf{V}_{x,\eta}$ will be called simply *data patch* at $x \in \mathbf{U}_{x_0,h}$.

Consider the weighted patch distance

$$\|\mathbf{Y}_{x_0,\eta} - \mathbf{Y}_{x,\eta}\|_{J,\kappa} =$$

$$\sqrt{\frac{\sum\limits_{x' \in \mathbf{V}_{x_0,\eta}} \kappa(x')J(x', H)J(T_x x', H)(y(T_x x') - y(x'))^2}{\sum\limits_{x' \in \mathbf{V}_{x_0,\eta}} \kappa(x')J(x', H)J(T_x x', H)}},$$

where T_x is the translation map defined by $T_x y = y + x - x_0$, and κ is the smoothing kernel defined by

$$\kappa(x) = \sum_{k=\max(1,j)}^{h} \frac{1}{(2k+1)^2} \tag{6}$$

if $\|x - x_0\|_\infty = j$ for some $j \in \{0, 1, \cdots, h\}$ and $x \in \mathbf{U}_{x_0, \eta}$. Introduce the impulse detection distance by

$$\widehat{\rho}_{J,\kappa,x_0}(x) = \left(\|\mathbf{Y}_{x_0,\eta} - \mathbf{Y}_{x,\eta}\|_{J,\kappa} - \mu \right)_+, \tag{7}$$

where μ is parameter which controls the robustness of the estimate.

We define our new filter, called *Optimal Weights Impulse Noise Filter* (OW-INF) by

$$\widetilde{u}_h(x_0) = \frac{\displaystyle\sum_{x \in \mathbf{U}_{x_0,h}} J(x, H_2) K_{\mathrm{tr}}\left(\frac{\widehat{\rho}_{J,\kappa,x_0}(x)}{\widehat{a}_J}\right) y(x)}{\displaystyle\sum_{x \in \mathbf{U}_{x_0,h}} J(x, H_2) K_{\mathrm{tr}}\left(\frac{\widehat{\rho}_{J,\kappa,x_0}(x)}{\widehat{a}_J}\right)}, \tag{8}$$

where the bandwidth $\widehat{a}_J > 0$ can be calculated as in Remark 1 of [2] (cf. the algorithm below), H_2 is a parameter and K_{tr} is the triangular kernel:

$$K_{\mathrm{tr}}(t) = (1 - t)_+.$$

Notice that H and H_2 may take different values. The weights defined by the triangular kernel appears as optimal in [2,3].

To give some insights on the filter (8), note that the function $J(x, H_2)$ acts as a filter on the points contaminated by the impulse noise. In fact, if x is an impulse noisy point, then $J(x, H_2) \approx 0$. So, in the new filter, the basic idea is to apply the Optimal Weights Filter [2] by giving nearly 0 weights to impulse noisy points. The computational algorithm is as follows.

– **Algorithm :**   Optimal Weights Impulse Noise Filter
– **Step 1**
   For each $x \in \mathbf{I}$ compute:
   $\mathrm{ROADG}(x) = \frac{1}{K} \sum_{i=1}^{K} r_i(x)$
   $J(x, H) = \exp\left(-\frac{(\mathrm{ROADG}(x) - b)_+^2}{H^2}\right)$
   $J(x, H_2) = \exp\left(-\frac{(\mathrm{ROADG}(x) - b)_+^2}{H_2^2}\right)$
– **Step 2** Repeat for each $x_0 \in \mathbf{I}$
   if $\mathrm{ROADG}(x) = 0$
   $\widetilde{u}_h(x_0) = u(x_0)$.
   else
   a) compute $\{\widehat{\rho}_{J,\kappa,x_0}(x) : x \in \mathbf{U}_{x_0,h}\}$ by (7);
   b) compute the bandwidth $\widehat{a}$ *at* x_0 :
   reorder $\{\widehat{\rho}_{J,\kappa,x_0}(x) : x \in \mathbf{U}_{x_0,h}\}$ as an increasing sequence, say
   $\widehat{\rho}_{J,\kappa,x_0}(x_1) \leq \widehat{\rho}_{J,\kappa,x_0}(x_2) \leq \cdots \leq \widehat{\rho}_{J,\kappa,x_0}(x_M)$
   loop from $k = 1$ to M
   if $\sum_{i=1}^{k} \widehat{\rho}_{J,\kappa,x_0}(x_i) > 0$

if $\dfrac{\sigma^2 + \sum\limits_{i=1}^{k} \widehat{\rho}_{J,\kappa,x_0}^2(x_i)}{\sum\limits_{i=1}^{k} \widehat{\rho}_{J,\kappa,x_0}(x_i)} \geq \widehat{\rho}_{J,\kappa,x_0}(x_k)$

then $\widehat{a} = \dfrac{\sigma^2 + \sum\limits_{i=1}^{k} \widehat{\rho}_{J,\kappa,x_0}^2(x_i)}{\sum\limits_{i=1}^{k} \widehat{\rho}_{J,\kappa,x_0}(x_i)}$

else quit loop

else continue loop

end loop;

c) compute the estimated weights: for $i = 1, ..., M$,

$\widehat{w}(x_i) = \dfrac{J(x_i, H_2) K_{tr}\left(\frac{\widehat{\rho}_{x_0}(x_i)}{\widehat{a}}\right)}{\sum_{j=1}^{M} J(x_j, H_2) K_{tr}\left(\frac{\widehat{\rho}_{x_0}(x_j)}{\widehat{a}}\right)}$;

d) compute the filter $\widetilde{u}_h$ at x_0 :

$\widetilde{u}_h(x_0) = \sum_{i=1}^{M} \widehat{w}(x_i) y(x_i)$.

To avoid the undesirable border effects in our simulations, we mirror the image outside the image limits. In more detail, we extend the image outside the image limits symmetrically with respect to the border. At the corners, the image is extended symmetrically with respect to the corner pixels.

Here the parameter σ acts as a smoothing factor for the restored image. The larger the value of σ, the more smooth the denoised image is. When computing SROAD values, we follow approximately the rules:

$$d = [4p + 1] \quad \text{and} \tag{9}$$
$$K = (2d + 1)^2 \times \min(0.5, -p/4 + 0.55). \tag{10}$$

For example when the noise ratio is 60% we use 7×7 windows and $K = 19$; when the noise ratio is 40% we use 5×5 windows and $K = 10$; when the noise ratio is 20% we use 3×3 windows and $K = 4$. The other parameters are chosen as follows:

$$S = (2[10p + 7] + 1)^2, \quad D = (2[4p + 1] + 1)^2,$$
$$H = 5 + \frac{30}{1 + 20p} \quad \text{and} \quad H_2 = 27 - 20p.$$

We point out that the values of parameters or the coefficients in the above formulae can vary within certain range. The dependence of the filter on the values of H and H_2 in a neighborhood of the suggested value given above is not very noticeable.

3   Simulation

In this section, the proposed algorithm is evaluated and compared with several other existing techniques for removing random-valued impulse noise. Extensive experiments are conducted on four standard 512×512 , 8-bit gray-level images

Table 1. Comparison of restoration in PSNR(db) for images corrupted with random-valued impulse noise

Images	Baboon			Bridge			Lena			Pentagon		
p%	20%	40%	60%	20%	40%	60%	20%	40%	60%	20%	40%	60%
Method	PSNR	PSNR	PSNR	PSNR	PSNR	PSNR	PSNR	PSNR	PSNR	PSNR	PSNR	PSNR
MF [9]	22.52	20.65	19.36	25.04	22.17	19.36	32.37	27.64	21.58	28.29	25.16	23.41
SS-I [10]	22.46	21.35	19.42	25.90	22.85	19.04	33.43	27.75	20.61	28.28	26.43	23.85
ACWM[11]	24.17	21.58	19.56	27.08	23.23	19.27	36.07	28.79	21.19	30.23	26.84	23.50
PWMAD[12]	23.78	21.56	19.68	26.90	23.83	20.83	36.50	31.41	24.30	30.11	27.33	24.46
IMF[13]	24.18	21.41	19.08	27.05	23.88	19.74	36.90	30.25	22.96	30.42	26.93	23.72
TriF [1]	24.18	21.60	19.52	27.60	24.01	20.84	36.70	31.12	26.08	30.33	27.14	24.60
ACWM-EPR [7]	23.97	21.62	19.87	27.31	24.60	20.89	36.57	32.21	24.62	30.03	27.35	24.59
ROLD-EPR [8]	24.49	21.92	20.38	**27.86**	24.79	**22.59**	37.45	32.76	29.03	30.73	27.73	25.70
FWNLM [14]	23.45	21.71	20.45	26.82	24.23	22.23	34.95	32.12	28.03	30.26	27.48	25.48
OWINF	**25.01**	**22.41**	**20.46**	**27.86**	**24.91**	22.49	**37.56**	**33.07**	**29.05**	**31.18**	**28.19**	**25.78**

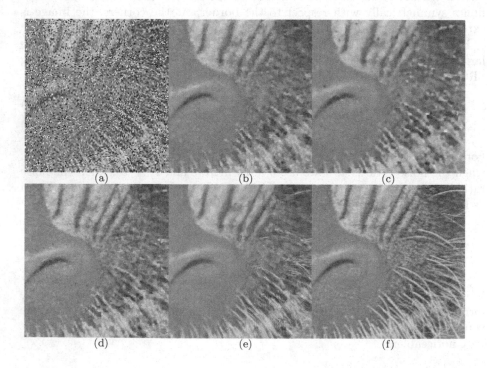

Fig. 1. Results of different methods in restoring 40% corrupted images "Baboon": (a)the noisy image; (b) results after the ACWM-EPR method [7]; (c) results after the ROAD-trilateral filter [1]; (d) results after the ROLD-EPR method [8]; (e) results after our OWINF; (f) the original image

with different features, including "Baboon," "Bridge", "Lena", and "Pentagon". Our experiments are done in the same way as in [8] in order to produce comparable results. The authors of [8] kindly provided us with their set of noisy images, restored images and PNSR values[1].

3.1 PSNR Comparison

We first concentrate on directly comparable and quantitative measures of image restoration. In particular, we evaluate the performance by using the peak signal-to-noise ratio (PSNR) [15]. If u is the original image and $\widetilde{u}$ is a restored image of u, the PSNR of $\widetilde{u}$ is given by

$$PSNR = 20\log_{10}\frac{255}{\sqrt{MSE}},$$

$$MSE = \frac{1}{\operatorname{card}\mathbf{I}}\sum_{x\in\mathbf{I}}(u(x) - \widetilde{u}(x))^2.$$

Larger PSNR values signify better restoration.

In Table 1, we list the best PSNR values from all considered methods for the four images with $p \in \{20\%, 40\%, 60\%\}$. The best values are marked in bold. From Table 1, it is clear that OWINF proposed in this paper provides significant improvement over all other algorithms for the images "Baboon", "Lena" and "Pentagon". For the image "Bridge", ROLD-EPR and our algorithm all provide satisfactory denoising performance.

3.2 Visual Quality

Our main goal was to ensure that our approach provides improved denoising and visually pleasing results. To compare the results subjectively, we enlarge portion of the images restored by some methods listed in Table 1. Fig. 1 shows the results in restoring 40% corrupted images of "Baboon". In the images restored by ACWM-EPR [7] and ROAD-trilateral filter [1], we can see that there are still some loss of details in the hair around the mouth of the baboon. The visual qualities of images restored by ROLD-EPR [8] are improved obviously, but we can still find a few noise around the nose of baboon. Our restored images are quite good: they not only retain the abundance of image details, but also keep the continuity of the details.

4 Conclusions

In this paper, we use a standardized version of ROAD statistic [1] to define new impulsive weights in order to measure the degree of the contamination of a pixel by a random impulse noise. Then we combine it with the Optimal Weights Filter from [2,3] to get a new filter for removing impulse noise. Simulation results show that our method is competitive compared with a number of existing methods both quantitatively and visually.

---

[1] All of them are in www.math.cuhk.edu.hk/~rchan/paper/dcx/

Acknowledgements. This research is partly supported by NSFC, China (No: 6127325831100672, 61375048,) China Postdoctoral Science Foundation (No. 2014M551412).

References

1. Garnett, R., Huegerich, T., Chui, C., He, W.: A universal noise removal algorithm with an impulse detector. IEEE Transactions on Image Processing 14(11), 1747–1754 (2005)
2. Jin, Q., Grama, I., Liu, Q.: Removing gaussian noise by optimization of weights in non-local means. arXiv:1109.5640 (2011)
3. Jin, Q., Grama, I., Liu, Q.: A new poisson noise filter based on weights optimization. Journal of Scientific Computing, 1–26 (2012)
4. Gonzalez, R.C., Woods, R.E.: Digital Image Processing. Prentice-Hall, Englewood Cliffs (2002)
5. Nikolova, M.: A variational approach to remove outliers and impulse noise. Journal of Mathematical Imaging and Vision 20(1-2), 99–120 (2004)
6. Chan, R.H., Ho, C.-W., Nikolova, M.: Salt-and-pepper noise removal by median-type noise detectors and detail-preserving regularization. IEEE Transactions on Image Processing 14(10), 1479–1485 (2005)
7. Chan, R.H., Hu, C., Nikolova, M.: An iterative procedure for removing random-valued impulse noise. IEEE Signal Processing Letters 11(12), 921–924 (2004)
8. Dong, Y., Chan, R.H., Xu, S.: A detection statistic for random-valued impulse noise. IEEE Transactions on Image Processing 16(4), 1112–1120 (2007)
9. Pratt, W.: Median filtering. Image Process. Inst., Univ. Southern California, Los Angeles (1975)
10. Sun, T., Neuvo, Y.: Detail-preserving median based filters in image processing. Pattern Recognition Letters 15(4), 341–347 (1994)
11. Chen, T., Wu, H.R.: Adaptive impulse detection using center-weighted median filters. IEEE Signal Processing Letters 8(1), 1–3 (2001)
12. Crnojevic, V., Senk, V., Trpovski, Z.: Advanced impulse detection based on pixel-wise mad. IEEE Signal Processing Letters 11(7), 589–592 (2004)
13. Wenbin, L.: A new efficient impulse detection algorithm for the removal of impulse noise. IEICE Transactions on Fundamentals of Electronics, Communications and Computer Sciences 88(10), 2579–2586 (2005)
14. Wu, J., Tang, C.: Random-valued impulse noise removal using fuzzy weighted non-local means. Signal, Image and Video Processing 8(2), 349–355 (2014)
15. Bovik, A.C.: Handbook of image and video processing. Access Online via Elsevier (2010)

CTR Prediction for DSP with Improved Cube Factorization Model from Historical Bidding Log

Lili Shan, Lei Lin, Di Shao, and Xiaolong Wang

School of Computer Science and Technology, Harbin Institute of Technology, Harbin, China
{shanll,linl,dshao,wangxl}@insun.hit.edu.cn

Abstract. In the real-time bidding (RTB) display advertising ecosystem, de-
mand-side-platforms (DSPs) buy ad impressions through real-time auction or
bidding from ad exchanges for advertisers. Receiving a bid request, DSP needs
predict the click through rate (CTR) for ads and determine whether to bid and
calculates the bid price according to the CTR estimated. In this paper, we
address CTR estimation in DSP as a recommendation issue. Due to the complicated trilateral interactions among users, ads and publishers (web pages), conventional matrix factorization does not perform well. Adopting ideas from
high-order singular value decomposition (HOSVD), we extend two dimensional
matrix factorization model to three dimensional cube factorization containing
users, ads and publishers, and propose an improved cube factorization model to
address it. We evaluate its performance over a real-world advertising dataset
and the results demonstrate that the improved cube factorization model outperforms the matrix factorization.

Keywords: click through rate estimation, demand-side platform, real time bidding, cube factorization model.

1 Introduction

In the real time bidding (RTB) display advertising ecosystem, ad exchanges aggregate
ad impressions from multiple publishers and send them to several demand-side platforms (DSPs) via real time auction. Receiving a bid request, each DSP needs to use
bidding algorithms to determine whether to bid the ad impression and search for an
optimal bid price for each impression. This bid price must be not higher than the expected cost-per-impression(eCPM) which is equal to the click-through-rate (CTR) for
the impression multiplied by the cost-per-click (CPC), or the conversion rate (CVR)
multiplied by the cost-per-action (CPA) [1,2]. If a CPC or CPA goal is fixed in ad-
vance, the eCPM directly depends on how well the CTR or CVR can be estimated.
Due to the difficulty of tracking the conversion actions of audience, CPC is nowadays
prevalent cost model how advertiser pays DSP. Therefore, we mainly focus on the
approach of CTR estimation for ad impressions used in DSPs.

We consider CTR prediction problem as a recommendation problem that ads need
to be recommended for users. Regularized matrix factorization models are known
generate high quality rating predictions for recommender systems [3,4]. However,

C.K. Loo et al. (Eds.): ICONIP 2014, Part III, LNCS 8836, pp. 17–24, 2014.

matrix factorization models the bilateral interaction between users and items, that is to say, though three types of attributes are available respectively according to a user, an ad and a publisher in our problem, they would be divided into two groups on *x-axis* and *y-axis* respectively [4]. Without loss of generality, it is supposed that one group is composed of all attributes from a user and the other group is composed of those from an ad or a publisher. So the interactions between the ad and the publisher inside the second group will not to be learned. However the CTR estimation in RTB is a tripartite interaction among pages, users and ads. Because the topic of the page p clicked by the user u reveals the user's intention, whether the user u will click the ad a on the page p is influenced not only by how well the content of the ad a conform the preference of the user u but also by what extent the product in the ad a matches the topic of the page p. For example, if the page p is a web page containing certain expertise, the ads recommending professional publications with respect to that expertise may be more appropriate to be impressed than game ads. In order to learn this tripartite interaction, we propose an improved cube factorization model based on high-order Singular Value Decomposition(HOSVD) to extends two dimensional matrix factorization model containing users and ads to three dimensional cube factorization covering users, ads and publishers. This model learns the interaction among users, ads and pages and outperforms the matrix factorization in addressing CTR estimation in our experiments.

The contribution of this paper is two-fold:

- We address the issue of click-through rate prediction for DSP by introducing improved cube factorization model based on high-order SVD. Our model shows its superior performance in handling sparse data than matrix factorization. Furthermore, it also presents better scalability than matrix factorization.
- We conduct various experiments on large-scale real-world bidding log data to evaluate our model and algorithm. Our results show that our improved cube factorization model is a highly promising direction for CTR estimation for DSP.

2   Related Work

There are a number of published studies on click-through rate prediction for search advertising or web search [5,6,7]. Due to many new challenges different from previous application situations, such as more seriously sparse data, more types of ad slot and more complex possibility of user action etc., it is hard to directly apply these approaches to solve our problem. B. Kanagal etc. [8] propose a novel focused matrix factorization model which learns users' preferences towards the specific campaign products for audience selection in display advertising. However, similar to recommendation, only the relevance of the user preference and the ad campaign need to be considered in audience selection. While for real-time CTR prediction in RTB, the user, the ad and the context of publishing are all factors deservedly taken into account. Jinlong Wu etc. [9] present cube factorization model(CF) based on high-order SVD to transform click-through rate prediction problem in personalization web search into rating prediction issue, and verify its performance on artificial datasets. Due to different context of RTB from web search mentioned above, cube factorization model is hardly applied to directly handle click-through rate prediction problem for DSP. Therefore, we perform adaptation on the cube factorization model in order to make it applicable for our problem and improve its performance on real-world datasets.

Few published literature adopts the method related to factorization models to handle our problem. Therefore, we adopt the similar method to the solution proposed by Tianqi Chen etc. [4] as the baseline. Tianqi Chen etc. [4] combine feature-based factorization models and achieve first place in Track1 of KDD Cup 2012.

3 CTR Prediction for DSP with Improved Cube Factorization Model

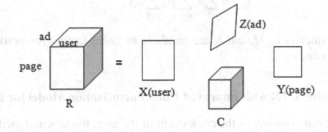

Fig. 1. Third-Order HOSVD

3.1 Problem Setup and Formulation

A bid request that an ad exchange sends to a DSP is denoted by $bid = \{user: u, page: p\}$ which indicates that the user u clicks the page p. The DSP has an ads set $A = \{a_1, a_2, ... , a_n\}$ whose member needs to be displayed. These data are all aggregated in the DSP side when a bid request arrived. The goal of the DSP bidding algorithm is to determine which ad in A has the highest probability of being clicked by the user u on the page p. The random variable X is used as the notation of click event outcome, and X equals to 1 if the user u click the ad a_k on page p, -1 if not. Mathematically, it is formulated as:

$$a^* = \underset{k=1,...,n}{\arg\max}\ prob(X = 1|u, p, a_k) \tag{1}$$

In which, a^* is the most optimal ad for bid. A bid price will be calculated according to the CTR estimation of a^* and submitted to the ad exchange for bidding.

We regard the problem as a cube complement issue with users, publishers and ads on x, y and z axes respectively. Accordingly, the value of the element (x, y, z) represents the quantity $prob(X = 1|u, p, a)$. Our objective is to estimate different quantities $prob$ according to different triples (u, p, a).

3.2   High-order SVD Based Factor Model

Lathauwer etc. [10] extend two-dimensional matrix Singular Value Decomposition (SVD) to high-order tensor and obtain high-order SVD (HOSVD). According to HOSVD, if given $N = 3$, then a tensor R can be factored into $R = C \times_1 X \times_2 Y \times_3 Z$ (Fig. 1). Matrix factorization closely related to SVD is used to solve matrix filling problem. Similarly, we apply third-order SVD-based factor model to address three-dimensional cube complement problem. Specifically, it is supposed that *x-axis, y-axis and z-axis* of R represent users, pages and ads respectively(Fig. 1), then according to HOSVD, the value of element(x, y, z) indicates the probability $\hat{r}_{upa}$ of the user u will click the ad a on the page p and can be estimated by the equation as follow:

$$\hat{r}_{upa} = \sum_{l=1}^{L} u_l \sum_{m=1}^{M} p_m \sum_{n=1}^{N} a_n c_{lmn} \tag{2}$$

Where, parameters L, M, and N are numbers of latent factors respectively corresponding to u, p and a.

3.3   Estimating CTR with Improved Cube Factorization Model for DSP

In order to alleviate sparsity of the click event of the user, the tags assigned to the user are incorporate into our model to represent the user. Since tags are composed of three different types of user attributes including gender, personal follows and purchase behaviors, it is good not to normalize the coefficients of tags in our experiments. Details are shown in equation (3) where $T(u)$ is the set of tags the user u has.

$$u = \sum_{t \in T(u)} x_t \tag{3}$$

One benefit of the factorization model is its flexibility in dealing with various data aspects. However, much of the observed variation in click events is due to effects associated with users, publishers or ads, known as biases or intercepts, independent of any interactions [9]. For example, some users show higher tendency in clicking ads than others, and some ads also receive more clicks than others. Therefore, a first-order approximation of the bias involved in r_{upa} is as follows:

$$b = b_u + b_p + b_a \tag{4}$$

$$b_u = \sum_{t \in T(u)} b_t \quad or \quad b_u = \frac{\sum_{t \in T(u)} b_t}{|T(u)|} \tag{5}$$

The notation b denotes the bias involved in r_{upa}. The notations b_u, b_a and b_p involved in b indicate the observed deviations of the user u, the ad a and the publisher p respectively. We examine two ways to combine tag bias as user bias. Details are shown in equation (5).

For the sake of enhancing efficiency for training, we borrow the way how matrix factorization approach is developed from SVD via absorbing the singular value matrix and deal with the tensor C in the similar means. Our experimental results show that extremely similar performance is achieved whether or not the tensor C is considered. Final estimation formulation with bias extend is as follows:

$$\hat{r}_{upa} = \sum_{l=1}^{L}\left(\sum_{t \in T(u)} x_{tl}\right)\sum_{m=1}^{M} p_m \sum_{n=1}^{N} a_n + b_u + b_p + b_a \tag{6}$$

The parameters u, p, a, b_u, b_p, b_a are learned by minimizing the squared error function of train set as follows:

$$L = \frac{1}{2}\sum_{(u,p,a)\in S}\left[\left(r_{upa} - \hat{r}_{upa}\right)^2 + \lambda\left(\sum_{t \in T(u)}\|x_t\|^2 + \|p\|^2 + \|a\|^2 + b_u^2 + b_p^2 + b_a^2\right)\right] \tag{7}$$

Where, S is the set of samples for training, λ is a regularization coefficient.

4   Experimental Evaluation

4.1   Experimental Setup

To evaluate our proposed models, we use the second season log data of Global bidding algorithm competition released by the DSP company iPinYou[1] recently. The train set contains bidding, impression, click, and conversion logs collected from eighteen advertising campaigns during seven days. A set of bidding logs from the following three days is used for offline testing purpose. We split the training dataset into two parts according to the impression date and use the last two days' data as validation set. There are totally 14,758,859 impression records and 11,117 click records in the whole dataset which contains 18 advertising campaigns, 74 ad creatives, 12,456,794 users, 45 user tags, 28505 domains. In order to alleviate the sparseness of the click sample in train set, we choose the combination of the tags the user u possesses to represent the user. The domain of the web page is used to characterize the publisher p. Finally, the ad campaign is selected to stand for the ad a.

To verify the effectiveness of our approach, we use feature-based matrix factorization approach [3], [4] as the baseline, which estimates the CTR as equation (8) and is denoted by the notation FMF. In equation (8), $T(u)$ is the set of tags the user u possesses, and the sum vector of latent factors of the tags the user u have is adopted to describe the user u.

$$\hat{r}_{u,a,p} = \left(\sum_{t \in T(u)} x_t\right)(a + p) + \frac{\sum_{t \in T(u)} b_t}{|T(u)|} + b_a + b_p \tag{8}$$

---

[1] http://contest.ipinyou.com/

We employ the area under the ROC curve (AUC) to compare our model with the baseline. AUC is a commonly used metric for testing the quality of CTR prediction.

To update the model, we conduct stochastic gradient descent training. The number of iterations for SGD inference process is set as 100. We also design experiments to select the appropriate number of the factors.

4.2   Experimental Results and Discussions

Impact of the Number of Factors.

Firstly, we investigate the influence of the number of factors L, M and N on our models, because the time cost of the training algorithm is directly associated with these parameters. Fig. 2 presents the results where horizontal axis is the training set size from one day to five days. As shown in the figure, our model shows relatively stable performance on diverse number of factors from 2 to 10. The number 4 in the horizontal axis means: $L=M=N=4$. In inference algorithm, the time complexity of calculating e_{upa} or updating u, p and a is $O(|S|LMN)$. That is to say, an appropriate choice of this parameter such as 4 can achieve an optimal balance between better prediction quality and less training time.

Coincidentally, when the number of factors is 4, the performance of the comparing algorithm (FMF) is also optimized. So, unless otherwise specified, four factors are used in the following experiments.

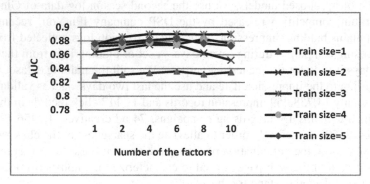

Fig. 2. Impact of the Number of Factors

Combination Approach of Tag Bias for Users.

Table 1. Impact of Combination Approach of Bias for Tags

Train Size(days)	Averaging	Summing
1	0.8336	0.7689
2	0.8746	0.7807
3	0.89	0.7773
4	0.8784	0.74
5	0.8815	0.7298

Since the tags are combined to express the user, the bias according to the tags also needs to be combined. We examine two combination solutions: averaging and summing. The results are shown in table 1 and the former outperforms the later over diverse size of datasets. That means the sum of the bias of all tags may be so strong and causes overfitting. Unless otherwise specified, we use the averaging solution to represent user bias in the following experiments.

Prediction Quality.
Finally, we compare the AUC quality of our improved cube factorization model to the baseline system. Fig.3 shows the results of different methods on the data set. Our model not only presents better robustness on diverse size of train dataset than the baseline but also outperforms the baseline approach. This figure demonstrates that it is more reasonable to consider the CTR estimation for DSP as a cube complement problem than two-dimensional matrix fill problem. Because our model properly considers the information interaction among three types of objects: users, ads and publishers. Therefore, the improved cube factorization model shows its superior ability in addressing this problem.

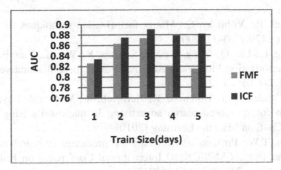

Fig. 3. Prediction Quality of Different Model

In addition, almost all experimental results show that the estimations based on the last three days' history behavior data of users are the most effective results. This reveals that short-term interests of users conduct a greater impact on our problem than long-term interests.

5   Conclusions

In this paper, we focus on the CTR prediction problem in RTB for DSP. We propose an improved cube factorization model to address this issue. In order to alleviate the sparsity of positive samples, the tags of the audience are combined to characterize users' preference. Biases respectively according to users, ads and publishers are also added to the final estimation to model the first-order approximations. For efficiency's sake, the tensor C is absorbed by other parameters without loss of effectiveness. Compared to matrix factorization, our model has superior ability in modeling the interaction among three-dimensional objects: users, ads and publishers, and also outperforms the matrix factorization in real-world data set. Furthermore, our model also shows relatively stable performance both on diverse number of factors and on different size of train set.

Acknowledgments. This work is supported by the National Natural Science Foundation of China (61100094 & 61300114), Specialized Research Fund for the Doctoral Program of Higher Education (20132302120047), and Key Basic Research Foundation of Shenzhen (JC201005260118A).

References

1. Lee, K., Orten, B.B., Dasdan, A., et al.: Estimating conversion rate in display advertising from past performance data: U.S. Patent Application 13/584, 545[P] (2012)
2. Chen, Y., Berkhin, P., Anderson, B., Devanur, N.R.: Real-time bidding algorithms for performance-based display ad allocation. In: Proceedings of the 17th ACM SIGKDD International Conference on Knowledge Discovery and Data Mining, pp. 1307–1315. ACM (August 2011)
3. Koren, Y., Bell, R., Volinsky, C.: Matrix factorization techniques for recommender systems. Computer 42(8), 30–37 (2009)
4. Chen, T., Tang, L., Liu, Q., Yang, D., Xie, S., Cao, X., Wu, C., Yao, E., Liu, Z., Jiang, Z.: Combining Factorization Model and Additive Forest for Collaborative Followee Recommendation, KDD CUP (2012)
5. Graepel, T., Candela, J.Q., Borchert, T., Herbrich, R.: Web-scale bayesian click-through rate prediction for sponsored search advertising in microsoft's Bing search engine. In: International Conf. on Machine Learning (2010)
6. Chen, Y., Yan, T.W.: Position-normalized click prediction in search advertising. In: Proceedings of the 18th ACM SIGKDD International Conference on Knowledge Discovery and Data Mining, pp. 795–803. ACM (2012)
7. Wang, T., Bian, J., Liu, S., Zhang, Y., Liu, T.-Y.: Psychological advertising: exploring user psychology for click prediction in sponsored search. In: Proceedings of the 19th ACM SIGKDD International Conference on Knowledge Discovery and Data Mining, pp. 563–571. ACM (2013)
8. Kanagal, B., Ahmed, A., Pandey, S., Josifovski, V., Garcia-Pueyo, L., Yuan, J.: Focused matrix factorization for audience selection in display advertising. In: Data Engineering (ICDE), pp. 386–397 (April 2013)
9. Wu, J.: Collaborative Filtering On the Netix Prize Dataset, Peking University doctoral dissertation, pp. 87–104 (May 2010)
10. Lathauwer, L.D., Moor, B.D., Vandewalle, J.: A multilinear singular value decomposition. SIAM J. Matrix Anal. Appl. 21(4), 1253–1278 (2000)

Image Retrieval Using a Novel Color Similarity Measurement and Neural Networks

Cheng Yang and Xiaodong Gu[*]

Department of Electronic Engineering, Fudan University, Shanghai 200433, China
xdgu@fudan.edu.cn, guxiaodong@263.net

Abstract. Automatic feature extraction combined with proper similarity measurement plays an important role in Content-based Image Retrieval(CBIR). This paper introduces a new similarity measurement named Weighted Main Colors First (WMCF), derived from three conditions to approximate human perception, to improve retrieval performance in CBIR. Meanwhile, the texture feature (Ptex) for CBIR is extracted by using unit-linking Pulse Coupled Neural Network (PCNN). This PCNN-based texture feature consists of a series of image gradient entropy values. Experimental results show that Ptex distinguishes different textures very well, and that WMCF has better performance than Comparing Histogram by Clustering (CHIC) and Optimal Color Composition Distance (OCCD) with much lower time complexity. Compared with Fixed Cardinality (FC), Block Difference of Inverse Probabilities (BDIP) and Normalized Moment of Inertia (Nmi), our approach makes 7% improvement and obtains a better ANMRR (Average Normalized Modified Retrieval Rank).

Keywords: Content-based image retrieval (CBIR), Pulse coupled neural network (PCNN), Similarity measurement, Weighted main colors first (WMCF).

1   Introduction

Effective content-based image retrieval on large image databases requires robust feature extraction and the corresponding similarity measurement [1]. Color feature is effective for color image retrieval because of their robustness to noise, degradation, size change and rotation. Meanwhile, texture feature plays an important role in image classification. The combination of the color and the texture features accords with the human perception of different color patterns and shows an excellent performance on color image retrieval [2]. Besides feature extraction, effective image retrieval should also take into account similarity measurement.

The simplest and usual color feature is color histogram [3]. Combining with Euclidean similarity measurement, it provides a solution to content-based image retrieval. Based on it, many color feature extraction methods are derived to obtain better performance, such as color coherence vector, color moments [4] and color correlograms. For texture features, the early work includes Haralick's co-occurrence matrix, Tamura

---

[*] Corresponding author.

C.K. Loo et al. (Eds.): ICONIP 2014, Part III, LNCS 8836, pp. 25–32, 2014.

texture and wavelet moments. Meanwhile, more effective similarity measurements and metric learning have also been developed for image retrieval. For instance, histogram intersection [3] is more effective for histogram representation than traditional L2 distance. These simple methods have two disadvantages. The first is that they usually cannot obtain satisfying retrieval results on large image database because their corresponding similarity measurements donot conform to the human perception well. Second, these methods usually require huge data storage, which limits database expansion. In 1999, inspired by the probability theorem that a random vector can be characterized by its statistical moments, literature[6] used binary quaternion-moment-preserving (BQMP) threshold technique to extract color features, being an adaptive color quantization method based on color distribution. In 2010 Chen et al. proposed Fixed Cardinality (FC) to extract color features according to the image color distribution [5]. They also proposed a clustering based similarity measurement Comparing Histogram by Clustering (CHIC) [5]. FC with CHIC achieves better performance than traditional histogram and color moments.

Pulse Coupled Neural Network (PCNN) was proposed based on the experimental observations of synchronous pulse burst in the cat visual cortex [7]. In this paper, we use unit-linking PCNN [8] to extract texture features and then use BQMP technique to extract color features. Meanwhile, a new color similarity measurement based on several human perception conditions is proposed to enhance the overall performance. Compared with other similarity measurements, such as CHIC and Optimal Color Composition Distance (OCCD) [9], our similarity measurement gives a superior performance on retrieval precision and efficiency. Experimental results also show our approach can improve the retrieval performance greatly with less time complexity.

2   Texture Feature and Color Feature Extraction

2.1   Unit-Linking PCNN Based Texture Feature Extraction

We use the unit-linking Pulse Coupled Neural Network (PCNN), a simplified version of the PCNN, to extract features for image retrieval. Unit-linking PCNN is composed of three parts (the receptive field, the modulation field and the pulse generator). In image retrieval, one to one correspondence exists between one pixel and a neuron. One neuron receives the external input signal from one channel (F channel) and connects with other neurons in its neighboring field by the other channel(L channel). The neuron combines the responses of two channels to produce the internal activity, and delivers it to the pulse generator. The pulse generator compares the internal activity with the decaying threshold to produce the output signal. Some literatures give details of this model, such as [8].

Using unit-linking PCNN to extract texture features, the input signal of the unit-linking PCNN is the gradient image including the contrast and edge distribution information of the original image. The gradient image is produced by the vector gradient method of Lee and Cok [10]. There are two methods for PCNN to extract texture features. One is the image entropy and the other is the time signature, the former of which is used in our approach. Unit-linking PCNN uses the gradient image to get a series of image entropy values of the gradient images, which contain the texture information of images. During each iteration, the feature of each binary image is calculated by image entropy. Our experiment results show image entropy achieve a better performance than time signature on texture representation. The entropy sequence forms the texture feature vector, called unit-linking PCNN based texture feature (Ptex) in this paper. For texture similarity measurement, Euclidean distance has proved to be an effective one in our experiment.

2.2   Color Feature Extraction

Literature [6] uses the quaternion to represent a color and propose a moment-preserving threshold technique called Binary Quaternion Moment-Preserving (BQMP). A quaternion is an extension of complex number, being denoted as $\hat{q} = q_0 + q_1 \cdot i + q_2 \cdot j + q_3 \cdot k$. Quaternion can represent 4-D vectors. A color image can be represented by 3-D vectors (R, G, B). Therefore, these vectors can be represented by quaternion with $q_1=R, q_2=G, q_3=B$ and $q_0=0$. The principle of BQMP in a quaternion dataset is to divide the dataset into two parts and each part is represented by a quaternion, i.e. $\hat{z}_0$ and $\hat{z}_1$, with keeping the first three moments of the resultant two-level dataset invariant. Pei and Cheng proved that this moment-preserving division can get similar performance as optimum Bayesian classifier [6]. The FC algorithm in [5] is based on the BQMP threshold technique. The input in this algorithm is a quaternion dataset which will be split iteratively. During each iteration, FC finds a splitable cluster whose variance is maximal and then uses the BQMP threshold technique to split the cluster into two new clusters. The iteration continues until enough clusters have been gotten. After getting enough clusters, we record each cluster's representative quaternion and percentage, equal to the main colors of the image and their corresponding percentages. At last, the color information of the image is stored as a set of color-percentage pairs, namely $\{(C_1,p_1), (C_2,p_2)...(C_N,p_N)\}$, where C_i is a vector representing color and p_i is the corresponding percentage. Our method is based on this BQMP technique.

3   Our Weighted Main Colors First Distance

The BQMP color feature is a set of color-percentage pairs. Because of BQMP adaption to color distribution, these extracted colors are not fixed like color quantization. Therefore, we cannot use Euclidean distance or histogram intersection for this feature. Here, we propose a new color similarity measurement, Weighted Main Colors First distance (WMCF), which is derived from three conditions.

   Although images with the same semantic meaning may contain different colors, similar colors still bring the same semantic meaning in most cases, for instance, sunset images with dominant red and sea images with dominant blue. Therefore, comparing images with proper color similarity measurement that approximates human perception influences the final retrieval results greatly. In this paper, we think the color feature and its similarity measurement should meet the following three conditions.

Firstly, a color image can be described by a few colors, which originates from the fact that our perception can only perceive several colors in an image. This condition assures that the number of our extracted colors can be much fewer than FC. Secondly, colors with more percentage, which we call main colors, contribute more to the image semantic meaning. Thirdly, two images are identical only if similar colors are with similar percentages. Although all cases do not meet the second and the third conditions, these two ones are quite reasonable based on our experience. Next, we will design a similarity measurement that meets these conditions.

Assume we have a query image and a target image. After color feature extraction, a set of color-percentage pairs from query image is obtained. According to the second condition, we first sort them based on each color's percentage so that we can get information of different color's contribution to the image. Because the color with largest percentage contributes most to the image, we give this color a priority to find the most similar color of the target image. After search, this color from the query image forms a corresponding pair with the newly obtained color from the target image. This process continues until all the colors in the query image find their correspondences in the target image. Fig.1 illustrates this process.

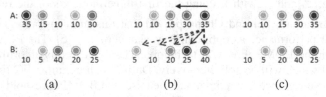

Fig. 1. Illustration of searching for corresponding colors. (a) Two original color sets,(b) Colors sorted and search starting; (c) Final corresponding colors.

For each corresponding colors, we calculate two variables according to the third condition, one is the color difference (Cdiff). $Cdiff = \left\| \vec{C_1} - \vec{C_2} \right\|_2$,and the other is the percentage difference (Vdiff), $Vdiff = |p_1 - p_2|$, where C is a vector that represents color and p is its percentage. According to the third condition that two images are supposed to be identical only if similar colors have similar percentages, which is a logical AND relation between color difference and percentage difference. We also take it into account, which colors with more percentage have more influence to the image. Thus, we propose $dist(A,B) = \sum_i p_i \cdot Cdiff \cdot (1 + Vdiff)^4$ to calculate the distance between two images A and B. In this equation, adding 1 to the Vdiff avoids such an extreme circumstance that two very different colors (large Cdiff) with equal percentage (Vdiff equals 0) may equal to the 0 distance, which is irrational. Because human color perception with different percentages is not linear, we add nonlinear mechanism to this equation and we find the power curve gets the best performance for colors in the range from 0 to 255 and percentages in the range from 0 to 1. Note that this equation is from the view of A, which means searching for corresponding colors is for the colors in image A. p_i is the color percentage in image A.

The similarity measurement described above may lead to asymmetry, i.e. dist(A,B)≠dist(B,A). We consider the larger one between dist(A,B) and dist(B,A) as the last similarity. The reason is that larger distance means smaller similarity and from our experience we have the lower risk to compare two object's resemblance with choosing the smaller similarity measurement. It also accords with the fuzzy set theory. Finally, our Weighted Main Colors First (WMCF) measure is defined as $WMCF(A,B) = \max\{dist(A,B), dist(B,A)\}$ .Using our proposed similarity measurement, we need much fewer color-percentage pairs than FC, which leads to reduce storage and time complexity

4   Experimental Results and Discussion

4.1   Image Databases and Performance Measures

Corel Database(DB) composed of 1000 color images with 10 classes are used to evaluate WMCF measurement. Our Brodatz-based texture database including 1776 texture images, is used to evaluate the retrieval performance of textures using unit-linking PCNN. We use the precision and the recall measures to evaluate retrieval performance. Suppose that a query q, a set of images $S(q)$ relevant to the query q, and a set of retrieved images $A(q)$ are given. Recall $R(q)$ and precision $P(q)$ are given as $R(q) = |A(q) \cap S(q)| / S(q)$ and $P(q) = |A(q) \cap S(q)| / A(q)$, where the operator $| \cdot |$ returns the size of a set. In our experiments, each image in a test DB is chosen as a query and the others in the DB became target images for such a query. Besides precision and recall, the ANMRR (Average Normalized Modified Retrieval Rank) used in entire MPEG-7 color core experiments is also adopted. It is defined as,

$$ANMRR = \frac{1}{NQ} \sum_{q=1}^{NQ} \frac{\frac{1}{NG(q)} \sum_{k=1}^{NG(q)} Rank*(k) - 0.5 \times [1 + NG(q)]}{1.25 \times K(q) - 0.5 \times [1 + NG(q)]}, \text{ where } Rank*(k) = \begin{cases} Rank(k) & Rank(k) \le K(q) \\ 1.25K & Rank(k) > K(q) \end{cases}$$

$NG(q)$ is the size of the images relevant to a query image q in the database. $Rank(k)$ is the rank of these images sorted. NQ is number of query images, and $K(q)$ specifies the 'relevant ranks' for each query. $K(q) = \min(4NG(q), 2GTM)$,where GTM is the maximum of $NG(q)$ for all queries. ANMRR not only evaluates the precision of a retrieval method but also takes into account the rank of the returned images, which means relevant images with high ranks are better results than those with low ranks. This accords with the requirement of image retrieval. A lower ANMRR value means more accurate retrieval performance.

4.2   Experimental Results

Fig.2(a) shows four precision-recall curves (P-R curves) of the results of unit-linking PCNN 'Entropy', unit-linking PCNN 'Timesig', traditional PCNN 'Tp' [7] and 'Gabor' feature[11] on our Brodatz-based texture database. Fig.2(a) illustrates that both two features produced by unit-linking PCNN, namely 'Entropy' and 'Timesig', perform a bit better than 'Gabor' when the recall is not high. 'Gabor' outperforms unit-linking PCNN when we want to retrieve more relevant images, but the difference is acceptable. Unit-linking PCNN performs better than traditional PCNN model (Tp). It is clear that the entropy produced by unit-linking PCNN works better than the time signature. It shows that unit-linking PCNN is competent for texture feature extraction and image entropy is better than time signature to represent textures.

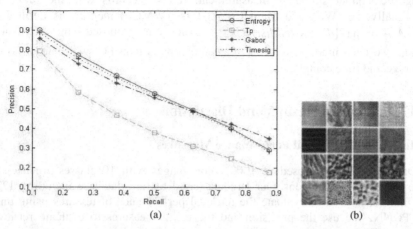

(a)   (b)

Fig. 2. (a) P-R curves of different methods on Brodatz-based texture database, (b) Examples from Brodatz-based texture database

To evaluate our WMCF measurement, we compare it with the Comparing Histograms by Clustering (CHIC) in [5] and Optimal Color Composition Distance (OCCD) in [9]. Both two methods are competent for measuring distance between sets of color-percentage pairs. CHIC is a measure based on clustering and calculates the intra-cluster divergence. OCCD can find an optimal mapping between those sets. For CHIC, the parameter Td is set as 30 and number of colors is chosen as 32 as the paper set. The minimal unit of OCCD is set as 1%. Both OCCD and our WMCF use 15 main colors based on the experiments. Fig.3(a) illustrates that the proposed WMCF measure performs better than CHIC and OCCD when the recall is not high. This is because we take into account the different contributions of main colors and the other two methods treat these colors as an ensemble to measure the distance. In high recall, our WMCF gets similar performance as the other two. In practice, in most cases one just requires a few relevant images rather than all, thus the superior performance of our WMCF at the low recall can lead to a more effective retrieval. It also can be found that OCCD performs a little better than CHIC. Although OCCD can find an optimal mapping, it needs to quantize the dataset firstly, which may introduce quantization error. However, our WMCF measure does not need quantization so that it avoids the quantization error. As to time complexity, WMCF takes advantage over the OCCD and CHIC. For CHIC, the time complexity is $O(n_1^2 \log n_1)$, where n_1 is the sum of lengths of two quaternion features. OCCD has the time complexity $O(n_2^3)$ and WMCF's time complexity is just $O(n_2^2)$, where n_2 is the number of main colors. On Corel database, retrieval time of WMCF, OCCD, CHIC are 2.48s, 21.03s, and more than 100s respectively.

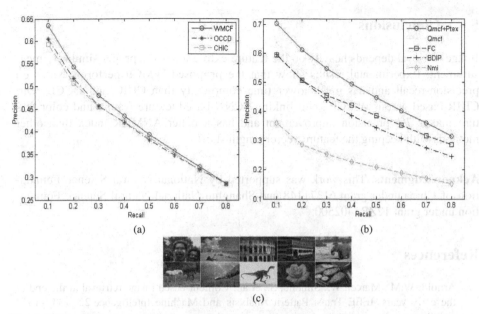

Fig. 3. (a)The P-R curves of three methods on Corel database, (b) P-R curves of different methods on Corel database, (c)Examples from Corel database

Table 1. ANMRR indexes of retrieval methods on Corel database

Method	BDIP	NMI	FC	Qmcf	Qmcf+Ptex
ANMRR	0.5379	0.6501	0.4986	0.4801	0.4417

WMCF measurement based quaternion color feature is called Qmcf in this paper. Meanwhile, Qmcf is combined with the unit-linking PCNN based (entropy) texture features (Ptex) for retrieval. Our approach is compared with FC with CHIC measure[5], Block Difference of Inverse Probabilities (BDIP)[12], and PCNN-Normalized Moment of Inertia (Nmi) algorithm. Fig. 3(b) shows that our Qmcf does better than other methods when recall is not high. When we add texture information into color features, i.e. Qmcf+Ptex, the performance improved greatly with maximal 10% gain on precision, which shows that color feature combined with texture feature can perform more effective retrieval. We also note that the color feature based method FC gets more robust performance than texture feature based BDIP method, which shows that color is more important than texture in color image retrieval. Because the distance of texture feature is obtained by traditional L_2 distance that does not cost much time, the time complexity analysis aforementioned is still valid here and Qmcf+Ptex can perform a quick retrieval on Corel database. Table 1 shows ANMRR indexes of those methods. The smaller the ANMRR, the better the retrieval performance is. Qmcf+Ptex has the smallest ANMRR index which means it achieves the best performance. It indicates that our method can return the relevant images at the front locations, which accords with the requirement of image retrieval.

5   Conclusions

Image retrieval depends heavily on the feature extraction and a proper similarity measurement. Experimental results show that the proposed WMCF performs better on precision-recall and has much lower time complexity than CHIC and OCCD. Our CBIR based on our WMCF , unit-linking PCNN based texture feature and color feature makes 7% precision improvement and has a better ANMRR index than other methods with keeping the feature vector length short.

Acknowledgments. This work was supported by National Natural Science Foundation of China under grant 61371148 and Shanghai National Natural Science Foundation under grant 12ZR1402500.

References

1. Arnold, W.M., Marcel, W., Simone, S., et al.: Content-based image retrieval at the end of the early years. IEEE Trans. Pattern Analysis and Machine Intelligence 22, 1349–1379 (2000)
2. Mojsilovic, A., Kovacevic, J., Hu, J., et al.: Matching and retrieval based on the vocabulary and grammar of color patterns. IEEE Trans. Image Processing 9, 38–54 (2000)
3. Swain, M., Ballard, D.: Color indexing. Int. J. Comput. Vis. 7, 11–32 (1991)
4. Paschos, G., Radev, I., Prabakar, N.: Image content-based retrieval using chromaticity moments. IEEE Trans. Knowledge and Data Engineering. 15, 1069–1072 (2003)
5. Chen, W., Liu, W., Chen, M.: Adaptive color feature extraction based on image color distributions. IEEE Trans. Image Processing 19, 2005–2016 (2010)
6. Pei, S., Cheng, C.: Color image processing by using binary quaternion moment-preserving thresholding technique. IEEE Trans. Image Processing 8, 614–628 (1999)
7. Johnson, J.L., Padgett, M.L.: PCNN models and applications. IEEE Trans. on Neural Networks 10, 480–498 (1999)
8. Gu, X.D.: Feature extraction using unit-linking pulse coupled neural network and its applications. Neural Process. Lett. 27, 25–41 (2007)
9. Mojsilovic, A., Hu, J., Soljanin, E.: Extraction of perceptually important colors and similarity measurement for image matching, retrieval, and analysis. IEEE Trans. Image Processing 11, 1238–1248 (2002)
10. Lee, H., Cok, D.: Detecting boundaries in a vector field. IEEE Trans. Signal Process. 39, 1181–1194 (1991)
11. Manjunath, B., Ma, W.: Texture features for browsing and retrieval of image data. IEEE Trans. Pattern Analysis and Machine Intelligence 18, 837–842 (1996)
12. Chun, Y.D., Seo, S.Y., Kim, N.C.: Image retrieval using BDIP and BVLC moments. IEEE Trans. Circuits and Systems for Video Technology 13, 951–957 (2003)

Bottom-Up Visual Saliency Using Binary Spectrum of Walsh-Hadamard Transform

Ying Yu[1], Jie Lin[2], and Jian Yang[1]

[1] School of Information Science and Engineering, Yunnan University, Kunming 650091, China
[2] Department of Information Management, Yunnan Normal University, Kunming 650500, China
yuying.mail@163.com, linjie@ynnu.edu.cn, nxryang@126.com

Abstract. Detection of visual saliency is valuable for applications like robot navigation, adaptive image compression, and object recognition. In this paper, we propose a fast frequency domain visual saliency method by use of the binary spectrum of Walsh-Hadamard transform (WHT). The method achieves saliency detection by simply exploiting the WHT components of the scene under view. Unlike space domain-based approaches, our method performs the cortical center-surround suppression in frequency domain and thus has implicit biological plausibility. By virtue of simplicity and speed of the WHT, the proposed method is very simple and fast in computation, and outperforms existing state-of-the-art saliency detection methods, when evaluated by using the capability of eye fixation prediction.

Keywords: Visual attention, Saliency detection, Walsh-Hadamard transform.

1 Introduction

Visual saliency refers to the perceptual quality that makes an object or location stand out or pop out relative to its neighbors and thereby attract visual attention. Typically, visual attention is either driven by fast, pre-attentive, bottom-up visual saliency, or controlled by slow, task-dependent, top-down cues [1].

This paper is primarily concerned with the automatic detection of bottom-up visual saliency, which has already attracted intensive investigations in the area of computer vision in relation to robotics, cognitive science and neuroscience. One of the most influential computational models of bottom-up saliency detection was proposed by Itti et al. [2], which is designed conforming to the neural architecture of the human early visual system and thereby has biological plausibility. Itti et al.'s model has been shown to be successful in detecting salient objects and predicting human fixations. However, the model is ad-hoc designed and suffers from over-parameterization.

Some recent works addressed the question of "what attracts human visual attention" in an information theoretic way, and proposed a series of attention models based on information theory. These models based on information theory include the attention model based on information maximization [3], the graph-based visual saliency approach [4], and the discriminant center-surround approach [5]. While these

C.K. Loo et al. (Eds.): ICONIP 2014, Part III, LNCS 8836, pp. 33–41, 2014.
© Springer International Publishing Switzerland 2014

information theory-based models show better performance in saliency detection than Itti et al.'s model, they are more computationally expensive for some real-world systems.

Another kind of saliency models are implemented in the frequency domain, which are not at all biologically motivated, but they have fast computational speed and good consistency with psychophysics. These frequency domain models include the so-called spectral residual approach [6], and the approach using phase spectrum of quaternion Fourier transform [7]. Later works proposed by Yu et al. [8][9] asserted that visual saliency can be describes in terms of spatial correlation in the visual space, and that saliency information can be generated within a simple normalization process for principal component analysis (PCA) coefficients of the scene under view. Yu et al.'s saliency model has neurobiological plausibilities because the principal components of natural scenes can be obtained by using a Hebbian-based neural network.

In this paper, we propose a bottom-up visual saliency method based on the Walsh-Hadamard transform (WHT). Our saliency method simply projects the whole image into the WHT space and utilizes the signs of the WHT components to compute the saliency information of the visual space. This significantly reduces computations because unlike all spatial domain approaches, our method does not need to decompose the input image into numerous feature maps separated in orientation and scale, and then compute saliency at every spatial location of every feature map. Such a computation process may be quick for the massively parallel connections of the human visual pathway, but is comparatively slow for computer processors. The WHT [10][11] is perhaps the most well-known of the non-sinusoidal orthogonal transforms, which has gained prominence in various digital signal processing applications, since it can essentially be computed using additions and subtractions only. Consequently its hardware implementation is also simpler. The proposed saliency method is referred to as binary spectrum of Walsh-Hadamard transform (BWHT) in this paper. As compared to other frequency domain approaches, our method is simpler and faster in computation, and requires fewer storage spaces.

The remainder of this paper is organized as follows. Section 2 describes the proposed method of bottom-up visual saliency as well as its neurobiological plausibility. Section 3 presents the experiments and quantifies the consistency of our saliency method with eye fixation data. Finally, conclusions are given in Section 4.

2   Proposed Method

In this section, we begin by providing an interpretation of bottom-up visual saliency, and then propose a saliency detection method based on the WHT. We will explain how our proposed method relates to visual saliency.

2.1 Visual Saliency

Li [12] hypothesized that the primary visual cortex (V1) creates a bottom-up saliency map of the visual space and the contextual influence is necessary for saliency computation. For example, a red flower is salient in a context of green leaves. Each neuron in V1 is tuned to a particular visual feature such as color and orientation. The dominant contextual influence in V1 is the so-called "iso-feature suppression", i.e., nearby neurons tuned to similar features are linked by intra-cortical inhibitory connections [13]. Besides Li's hypothesis, a number of recent studies (e.g., [3][5][14][15]) have attempted to describe visual saliency in terms of surprise, interest, innovation, self-information and center-surround discrimination. These studies provided a general idea that higher information entropy accounts for higher saliency.

Our visual environment is highly structured and thereby much information redundancy exists in the visual input. It has been shown that the dominant redundancy of our visual input arises from second order input statistics and that the human visual system is capable of reducing such redundancy of visual sensory data [16]. Yu et al. [9] found that visual saliency can be described in terms of statistical correlation in the visual space, and employed the PCA projection vectors to capture the second order correlated components among image pixels. They have attempted to suppress highly correlated image components and meanwhile highlight salient image regions by normalizing the PCA coefficients of the input image. Following Yu et al.'s interpretations of visual saliency, in the next subsection we use the WHT to capture highly correlated components in visual space and suppress them so as to highlight salient visual features.

2.2 Saliency Map

It has been noted that like the PCA for natural images, the WHT components reflect global features in the visual space, and the redundancy reflected in the second-order correlations between pixels can be captured by the WHT components of the image [10][11]. According to such an interpretation of visual saliency in the previous subsection, image regions with high spatial correlation with its surroundings can be suppressed through a normalization operation upon the WHT components. As a result, salient locations can be relatively highlighted.

As compared to the PCA for natural images, the WHT is much simpler and faster, and has many fast algorithms for its computation. Moreover, a 2-dimensional WHT is separately performed in row and column, and therefore its computational complexity is significantly lower than a PCA transformation.

We start by considering a gray-scale image X. According to previous analysis, we first conduct a 2-dimensional WHT on the image. Next, we normalize the WHT components by setting all positive coefficients to a value of 1 and all negative coefficients to a value of -1. This 2-dimensional orthogonal transformation followed by a normalization operation can be easily formulated as

$$B = \text{sign}(\text{WHT}(X)), \tag{1}$$

where "WHT($\cdot$)" denotes a 2-dimensional Walsh-Hadamard transform, and the notation "sign($\cdot$)" is a signum function. The matrix B is referred to as binary spectrum of Walsh-Hadamard transform (BWHT) in this paper. It retains only the sign of each WHT component, discarding the amplitude information across the entire frequency spectrum. Note that B is expressed in binary codes (i.e., 1s and -1s) and thereby is very compact, with a single bit per component. The signum function, which normalizes the WHT coefficients, suppresses highly correlated components in the visual space and thereby accomplishes the computation of visual saliency in the WHT domain.

To recover the saliency information in the visual space, we conduct an inverse WHT on the binary spectrum B, which is formulated as

$$F = \text{abs}(\text{IWHT}(B)), \tag{2}$$

where "IWHT($\cdot$)" denotes the corresponding inverse Walsh-Hadamard transform, and the notation "abs($\cdot$)" is an absolute value function. Normally, the obtained matrix F, which carries the saliency information, is post-processed by convolution with a Gaussian filter for smoothing. This operation can be formulated as

$$S = G * F^2, \tag{3}$$

where G is a 2-dimensional Gaussian kernel, and S is the corresponding saliency map of the input image X. Note that F is squared for visibility.

It is worth stating that we resize the image to a width of 64px and keep its aspect ratio before computing the saliency map. This spatial scale is chosen according to the heuristics of other frequency domain approaches (e.g., [6][7][9]).

In the human visual pathway, the color space of natural images is decomposed into well decorrelated channels. The RGB color space is highly correlated, but an LAB color space transformation results in well decorrelated color channels for natural color images. In addition, the transformation is perceptually uniform, and it produces three biologically plausible channels: a luminance channel, a red-green opponent channel and a blue-yellow opponent channel.

The complete BWHT algorithm from input image to final saliency map is given as follows.

1. Perform an LAB color space transformation
2. Resize the image to a suitable scale
3. Perform a Walsh-Hadamard transform for each color channel and calculate the binary spectrum of all WHT components using equation (1)
4. Obtain the saliency maps of each color channel using equation (2)
5. Take the spatial maximum across the saliency maps of all color channels to obtain the final saliency map
6. Post-process the saliency map by convolution with a Gaussian filter for smoothing and visibility as formulated in equation (3)

For recombination, we take the maximum value, as argued by Li and Dayan [13], at each pixel location of the corresponding saliency maps instead of spatial summation used by most models. The complete flow of the proposed method is illustrated in Fig. 1. The input image is initially decomposed into three biologically motivated channels: a luminance channel and two color opponent channels. Each of the three channels is then subjected to a Walsh-Hadamard transformation. Then, the binary spectrum of WHT is obtained by taking the signs of the WHT components of each channel. Afterward, the binary spectrum of each channel is subjected to an inverse Walsh-Hadamard transformation so that the saliency map of each channel is generated. Finally, a final saliency map is obtained by taking the spatial maximum value across all three saliency maps. Note that the saliency map is a topographically arranged map that represents visual saliency of a corresponding visual scene. The objects or locations with high saliency values may stand out or pop out relative to their surroundings, and thus attract our visual attention. From Fig. 1, it can be seen that the salient objects are the mountain tents, which pop out from the background.

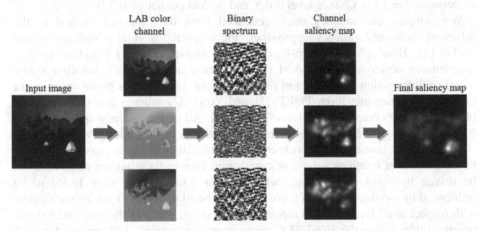

Fig. 1. An illustration of the BWHT method from input image to final saliency map

3   Experimental Validation

In this section, we present the experiments and quantify the consistency of our saliency method with eye fixation data. We compare our method to six popular state-of-the-art saliency approaches in literature by providing an objective evaluation as well as the visual comparison of all saliency maps.

To validate the saliency maps generated by our method, we use the data set of 120 color images from an urban environment and corresponding human eye-fixation data from 20 subjects provided by Bruce and Tsotsos [3]. These color images consist of indoor and outdoor scenes, of which some have very salient items, and others have no particular regions of interest. In order to quantify the consistency of a particular saliency map with a set of fixations of the image, we employ an objective evaluation metric that is referred to as receiver operating characteristic (ROC) area under the curve (AUC). Note that a number of published papers employed ROC-AUC score to evaluate a saliency map's ability to predict human eye fixations.

Following Tatler et al.'s approach [17], we compute the ROC-AUC score conforming to the following procedure. For one image, the positive point set is composed of the fixated locations from all subjects on that image, whereas the negative point set is composed of the non-fixated locations of the image. Each saliency map is binarized by a particular threshold and thereby considered as a binary classifier. At a particular threshold level, a binary saliency map can be divided into the target (white) region and the background (black) region. The true positive rate (TPR) is the proportion of the positive points that fall in the target region of the binary saliency map. The false positive rate (FPR) can be calculated in the same way by using the negative point set. Varying the threshold yields an ROC curve of TPRs versus FPRs, of which the area beneath provides a good measure of the capability of the saliency map to accurately predict where human eye fixations occurred on an image. Since the AUC is a portion of the area of the unit square, its value will always be between 0 and 1.0. Chance level is 0.5, and perfect prediction is 1.0.

We compare our saliency maps generated from the proposed method to the following published saliency approaches: the original Itti et al.'s saliency model (ITTI) [2], Harel et al.'s graph-based visual saliency (GBVS) [4], Gao et al.'s discriminant center-surround model (DISC), Bruce and Tsotsos's attention model based on information maximization (AIM) [3], Guo and Zhang's phase spectrum of quaternion Fourier transform (PQFT) [7], and Yu et al.'s saliency approach based on pulsed principal component analysis (PPCA) [9]. All of the saliency approaches are based on the original Matlab implementations available on the author's websites.

An important note about these experiments is that the ROC-AUC score is sensitive to the number of fixations we use in calculation. Former fixations are more likely to be driven by bottom-up manner, whereas later fixations are more likely to be influenced by top-down cues [17]. We calculate the ROC-AUC scores for each image with respect to all fixations, and repeat the process but use only the first two fixation points. Table 1 lists the ROC-AUC score averaged over all 120 images for each saliency method. As expected, the ROC-AUC scores with only the first two fixations are higher than those with all fixations. It can be seen that in both tests our BWHT method has the best capability for predicting eye fixations.

Table 1. The ROC-AUC performance of all seven methods

Method	BWHT	PPCA	PQFT	AIM	DISC	GBVS	ITTI
All fixations	**0.7792**	0.7766	0.7751	0.7706	0.7605	0.7127	0.7062
First 2 fixations	**0.7983**	0.7907	0.7846	0.7777	0.7683	0.7267	0.7182

Fig. 2 gives the saliency maps for 6 sample images from the image data set, which provides a qualitative comparison of all saliency methods. A fixation density map, generated for each image by convolution of the fixation map for all subjects with a Gaussian filter, serves as ground truth. Analysing the qualitative results, we can see that BWHT shows more resemblance to the ground truth. The regions highlighted by our proposed saliency method overlap to a surprisingly large extent with those image regions looked at by humans in free viewing. In addition, high contrast straight edges are suppressed to a much great extent using frequency domain approaches. Good performance with respect to color pop-out is also observed with BWHT compared to the other approaches.

We also record the computational time cost per image in a standard desktop computing environment. Table 2 shows each method's Matlab runtime measurements averaged over the data set. It can be noticed that, not only is BWHT the most predictive of fixations, it also runs faster than all competitors in our tests of computational performance. Note that three frequency domain methods (i.e., BWHT, PPCA and PQFT) are significantly faster than others. This is due to their small number of channels and calculations compared to other saliency methods. PPCA

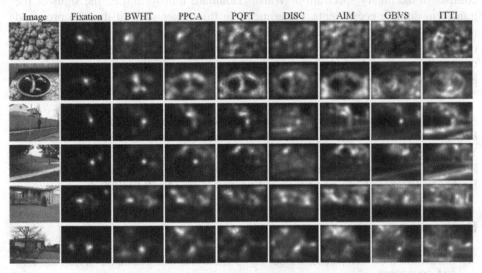

Fig. 2. Qualitative analysis of results for the Bruce data set

employs the PCA transform and has a computational complexity of $O(N^2)$, where N denotes the total number of pixels of the image. PQFT uses the fast Fourier transform has a computational complexity of $O(N\log N)$. Compared to PPCA and PQFT, the computation of BWHT is mainly comprised of the Walsh-Hadamard transform that can essentially be computed using additions and subtractions only. In computational mathematics, the fast Walsh-Hadamard transform requires only $N\log N$ additions or subtractions and thereby its hardware implementation can be much simpler. Compared to the BWHT, which uses only three color channels at a single spatial scale, ITTI and GBVS rely on seven feature channels and multiple spatial scales; AIM uses 25 filters of 1,323 dimensions. Although these approaches can be accelerated with efficient C implementations, the computational complexity of the BWHT is lower, as suggested by the Matlab runtimes. All seven saliency approaches are implemented in the Matlab R2012a environment on such a computer platform as Intel 3.3 GHz CPU with 8 GB of memory.

Table 2. Computational time cost per image for all seven methods

Method	BWHT	PPCA	PQFT	AIM	DISC	GBVS	ITTI
Time (s)	**0.0018**	0.2337	0.0151	5.0766	1.3778	2.5957	1.1842

4   Conclusions

This paper aims to find a bottom-up visual saliency method based on the Walsh-Hadamard transform. We manifested that the saliency information of an image consists in the binary spectrum of Walsh-Hadamard transform, i.e., the signs of the transform domain coefficients. Experiments in this paper showed that the proposed method is simple and efficient in saliency detection, and outperforms existing state-of-the-art saliency detection approaches. The potentials of our method lies in real-time and interdisciplinary applications focused on computer vision in relation to psychology, robotics and neuroscience.

Acknowledgments. This research was supported by the National Natural Science Foundation of China (Grant No. 61263048), by the Scientific Research Foundation of Yunnan Provincial Department of Education (2012Y277), by the Scientific Research Project of Yunnan University (2011YB21), and by the Young and Middle-Aged Backbone Teachers' Cultivation Plan of Yunnan University (XT412003).

References

1. Itti, L., Koch, C.: Computational modeling of visual attention. Nature Rev. Neurosci. 2(3), 194–203 (2001)
2. Itti, L., Koch, C., Niebur, E.: A model of saliency-based visual attention for rapid scene analysis. IEEE Trans. Patt. Anal. and Mach. Intell. 20(11), 1254–1259 (1998)
3. Bruce, N.D., Tsotsos, J.K.: Saliency, attention, and visual search: An information theoretic approach. Journal of Vision 9(3), 1–24 (2009)
4. Harel, J., Koch, C., Perona, P.: Graph-based visual saliency. In: Proc. NIPS (2006)
5. Gao, D., Mahadevan, V., Vasconcelos, N.: On the plausibility of the discriminant center-surround hypothesis for visual saliency. Journal of Vision 8(7), 1–18 (2008)
6. Hou, X., Zhang, L.: Saliency detection: A spectral residual approach. In: Proc. CVPR (2007)
7. Guo, C., Zhang, L.: A novel multiresolution spatiotemporal saliency detection model and its applications in image and video compression. IEEE Trans. Image Process 19(1), 185–198 (2010)
8. Yu, Y., Wang, B., Zhang, L.: Hebbian-based neural networks for bottom-up visual attention systems. In: Proc. ICONIP (2009)
9. Yu, Y., Wang, B., Zhang, L.: Hebbian-based neural networks for bottom-up visual attention and its applications to ship detection in SAR images. Neurocomputing 74(11), 2008–2017 (2011)
10. Ahmed, N., Rao, K.R.: Walsh-Hadamard transform. In: Orthogonal Transforms for Digital Signal Processing, pp. 99–152. Springer, Heidelberg (1975)

11. Kunz, H.O.: On the equivalence between one-dimensional discrete Walsh-Hadamard and multidimensional discrete Fourier transforms. IEEE Trans.Comput. C-28(3), 267–268 (1979)
12. Li, Z.: A saliency map in primary visual cortex. Trends Cognit. Sci. 6(1), 9–16 (2002)
13. Li, Z., Dayan, P.: Pre-attentive visual selection. Neural Network 19(9), 1437–1439 (2006)
14. Itti, L., Baldi, P.: Bayesian surprise attracts human attention.In: Proc. NIPS (2005)
15. Zhang, L., Tong, M.H., Marks, T.K., Shan, H., Cottrell, G.W.: SUN: A Bayesian framework for saliency using natural statistics. Journal of Vision 8(7), 1–20 (2008)
16. Barlow, H.B.: Possible principles underlying the transformation of sensory messages. Sensory Communication, 217–234 (1961)
17. Tatler, B.W., Baddeley, R.J., Gilchrist, I.D.: Visual correlates of fixation selection: effects of scale and time. Vision Research 45(5), 643–659 (2005)

Sonification for EEG Frequency Spectrum and EEG-Based Emotion Features

Yuxi Zhang, Yifeng Huang, Junwei Yue, and Liqing Zhang*

Key Laboratory of Shanghai Education Commission for Intelligent Interaction and Cognitive Engineering, Department of Computer Science and Engineering, Shanghai Jiao Tong University, Shanghai 200240, China
{snow_sword@,hyf042@,sonicmisora@,lqzhang@}sjtu.edu.cn

Abstract. Sonification is the use of representations of data through sound to convey information. It is particularly meaningful if the data are involved in time. This paper present a hybrid sonification method and aims to directly expressed the emotion hidden in the EEG signal through sound. The hybrid method mainly consists of two parts: (1) Frequency Mapping Representation (FMR) and (2) Emotion Feature Representation (EFR).

1 Introduction

Electroencephalogram (EEG) signal is the recording of electrical activity along the human scalp, which contains the information of human brain states, and furthermore reflects our minds. There have been many works on recognizing human brain states based on EEG signals, including motor imagery[1], emotion classification[2] and some other aspects. These works have made some achievements on recognizing human brain states based on EEG signal. However, these works mainly focused on the classification of the EEG signal, rather than the representation of the EEG signal.

These years, the representation of the big data has an increasing significance, because it can intuitively display the data and help people easily find the information hidden behind the data. Sonification, which is a kind of representation method, is an approach of representing data through acoustic sound. There have been many techniques in sonification. Auditory Icons generate the sound by selecting one in a set of sound pieces according to a classification process[3]. Audification maps the data directly to the audible domain[4]. In Parameter Mapping, data are mapped into the parameters in a sound synthesis algorithm[5]. *T. Hermann* et al. presented a Model Based Sonification, which maps the data to the elements in a "virtual physics" to generate sounds[6].

Compared to visualization, which is very popular and well-developed, sonification is more intuitive when representing signals, for acoustic sound is essentially a kind of time-involved signal as well. Therefore, applying sonification to EEG signal representation is quite meaningful. *Gerold Baier et al.* presented a sonification way based on multi-channel EEG signals[7]. *Dan wu et al.* built a sonification representation for rapid-eye movement sleep (REM) and slow-wave sleep (SWS) signals[8]. However,

---

* Corresponding author.

C.K. Loo et al. (Eds.): ICONIP 2014, Part III, LNCS 8836, pp. 42–49, 2014.

most of these papers did not build a relationship between the frequency in EEG signals and that in the music. This paper will focus on the sonification way based on the frequency domain of EEG signals. In addition, based on *Duan's* work on emotion classification, we will put the emotion features and classification results into the music so that listeners can feel the positive or the negative emotion directly from the sound.

2   Methodology

In this paper, a hybrid sonification method is presented to represent emotion features based on EEG signals. This hybrid method mainly consists of two parts: (1) Frequency Mapping Representation (FMR) based on the raw EEG data and (2) Emotion Feature Representation (EFR) based on the emotion features. Both of these two parts are conducted in the time-frequency domain. Therefore, we firstly use Shot-Time Fourier Transform (STFT) to obtain the time-frequency feature from the original signal in time domain before these two methods.

When both of these two sounds are generated, in order to represent the raw EEG data and the emotion feature at the same time, we will mix FMR sound and EMR sound together, while FMR sound is regarded as the background with a low volume and EMR sound as the theme with a high volume.

2.1   Frequency Mapping Representation

This section contains two main parts: (1) build the mapping from original EEG frequency to target audible frequency and (2) build the mapping from preprocessed amplitude to target euphonious amplitude. Once these two mappings are done, we just need to use invert fourier transforms to obtain the original signal.

Frequency Part. The frequency of EEG signal ranges from 0 Hz to 50 Hz, which is almost inaudible. Therefore, we need to map the frequency to the audible area (20 Hz - 20 kHz). However, the frequency close to 20 kHz would make the audio too harsh to listen, while on the other hand, most people are insensible of the frequency close to 20 Hz. Therefore, we need to take a subset $[f_{low}, f_{high}]$ of the audible area as the target domain.

In addition, according to Fechner's law, the pitch and its corresponding frequency follow the exponential relationship, which is shown in Equation 1

$$f = f_{base} \times 2^{\frac{p-p_{base}}{12}} \tag{1}$$

where $f_{base} = 440$ and $p_{base} = 69$ in MIDI standard.

In order to make the sound more uniformly distributed in the audible area, it is better to first map the original frequency to the pitch and then map the pitch to the target frequency, instead of directly mapping the original frequency to target frequency. Based on this idea, we can figure out the pitch area $[p_{low}, p_{high}]$ corresponding to the frequency area $[f_{low}, f_{high}]$. Consequently, we map the original frequency area $[0, 50]$ to the pitch area $[p_{low}, p_{high}]$ linearly, so the mapping function is

$$p = \lceil k \cdot f_{origin} + b \rceil \tag{2}$$

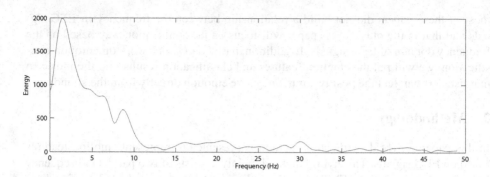

Fig. 1. Raw frequency spectrum in a certain time window

where

$$b = p_{low}$$

$$k = \frac{p_{high} - p_{low}}{50}$$

Note that we use ceiling function to guarantee the pitch value is an integer, which makes the sound more musical.

Combine Equation 1 and Equation 2, we obtain the final mapping function.

$$f_{target} = f_{base} \times 2^{\frac{\lceil k \cdot f_{origin} + b \rceil - p_{base}}{12}} \qquad (3)$$

Amplitude Part. In EEG signal, as Figure 1 shows, the energy in low frequency part is always much larger than which in high frequency part. If we directly map the amplitude to the target frequency domain, the audio we hear is always too low. To solve this problem, given the time-frequency spectrum $y(t, f)$, we firstly need to normalize the amplitude as Equation 4 shows, which makes the energy in different frequency part comparable.

$$\widetilde{y}(t, f) = \frac{y(t, f) - \mu(f)}{\sigma(f)} \qquad (4)$$

where $|\mu(f)|$ and $|\sigma(f)|$ are the mean amplitude and the standard deviation in frequency f respectively. Note that $|\widetilde{y}(t, f)|$ may be negative, while the amplitude is always positive, so we shift $|\widetilde{y}(t, f)|$ to a positive area by minus the minimum value as Equation 5 shows.

$$\widetilde{y}'(t, f) = \widetilde{y}(t, f) - \min_{t', f'}\{\widetilde{y}(t', f')\} \qquad (5)$$

The continuous line in Figure 2 shows the normalized amplitude in a certain time window. From the figure we can find that almost every frequency has its own component, which would make the sound too noisy. To make the sound more musical and euphonious, here we choose the peak values and valley values, representing the local maximum amplitude and local minimum amplitude respectively, and to ignore the other

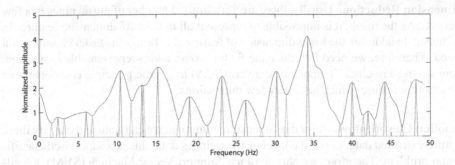

Fig. 2. Normalized frequency spectrum in a certain time window

frequency components. Notice that the normalized amplitude represent the relationship between the amplitude at this time and the average amplitude of all the time, and the peak values and valley values are really meaningful, for they are the local maximum deviation from the average.

Based on the peak-valley detection, the sound wave can be constructed through invert-fourier transform. Here we divide each time window into two parts. The first part is the peak-based sound wave while the second is the valley-based sound wave. The first part is much stronger than the second part so that people can distinguish the peak-based pieces from valley-based pieces.

2.2 Emotion Feature Representation

Besides generating audio from the raw EEG signals, we also want to put the emotion features into our audio. In other words, we are going to play the emotion features in the music form, so that we can tell the differences among the features through the audio. In order to achieve this goal, here we have four steps:

1. Extract the emotion features from the raw EEG signals;
2. Reduce the feature dimension to obtain the principle components;
3. Classify the emotion from the reduced feature of EEG signals;
4. Generate the music with both reduced features and classification results.

Feature Extraction. In neuroscience, EEG signals are often divided into 5 frequency bands: δ (1 - 3Hz), θ (4 - 7Hz), α (8 - 13Hz), β (14 - 30Hz) and γ (31 - 50Hz). Different frequency bands have different biological meanings and play different roles in brain state recognition. Based on the previous STFT transform, we adopt this partition and take the average energy in each band as the emotion features in each time window.

In addition, we take the channels into consideration. According to *Duan*'s work [2], differential asymmetry (DASM) features have a good performance on emotion classification. Therefore, for each frequency bands, we calculate the differences in each hemisphere asymmetry electrode pairs. Denote M as the number of hemisphere asymmetry electrode pairs, then we have a $5 \times M$ dimension feature in total within each time window.

Dimension Reduction. Usually, there are very limited number of music notes in a few seconds. As the result, it is impossible to represent all the $5 \times M$ dimension features in the music. In addition, the lower dimension of feature also brings the faster classification speed. Therefore, we need to reduce the feature to an audio-representable dimension. Here we use Principal Component Analysis (PCA) to get the principle components of the emotion feature, which have very few dimensions.

Emotion Classification. After the dimension reduction, we can train and classify these feature-extracted data. Given the label of the training data, this is a supervised classification problem. Therefore, we choose to use Support Vector Machine (SVM), a well-developed algorithm as our classification algorithm.

Music Generation. There are two parts in music generation: note generation and chord generation. Because notes and chords in the music are highly related to the emotion expressed by this music, we will not only generate the music from the emotion features to make the features audible, but also from the classification results to express the emotion directly through the music.

Denote classification confidence $p \in [0, 1]$ and emotion feature $\mathbf{x} \in \mathbf{R}^d$, where d is the dimension of the reduced feature. We are going to design *note generation function* $\mathbf{M}(\mathbf{x}, p) : (\mathbf{R}^d, [0, 1]) \rightarrow \text{NOTE}^d$ and *chord generation function* $\mathbf{C}(\mathbf{x}, p) : (\mathbf{R}^d, [0, 1]) \rightarrow \text{CHORD}$, where NOTE and CHORD are the sets of notes and chords respectively.

The main idea of our algorithm consists of two parts: (1) mapping the values of the features to the pitches of the note and (2) mapping the result *happy* to a major tune and a harmony chord, and meanwhile mapping *sad* to a minor tune and a disharmony chord.

For note generation, we can predesign a scale of major tune and a scale of minor tune with m ascending notes. Then we just need to convert the value of features to the index of the scale sequence. Here we apply sigmoid function in the value-to-index function, for it has a limited range and is almost linear around 0. The note generation function is shown as Equation 6.

$$note_i = \mathbf{M}_{(i)}(x_i, p) = \begin{cases} MajorScale[index(x_i)], \ p \geq 0.5 \\ MinorScale[index(x_i)], \ p < 0.5 \end{cases} \quad (6)$$

where

$$index(\mathbf{x}_i) = \lceil \frac{m}{1 + e^{-\alpha(\mathbf{x}_i - \bar{x})}} \rceil, \ \alpha \text{ is a constant} \quad (7)$$

$$\bar{x} = \frac{1}{d \times T} \sum_{i=1}^{d} \sum_{t=1}^{T} x_i^{(t)}, \ T \text{ is the total time length} \quad (8)$$

For chord generation, the case may be a little complicated. Because the chord in the music is basically related to the note occurring most frequently, we firstly need to find the *mode* among the notes generated in note generation part, and then take this note as basic note and build k types of chords representing emotion from saddest to happiest. Therefore, we need to predesign k chords for m notes. Denote *ChordSet(x, j)* as the predesigned chord with base note x and type j, then the chord generation function is shown as Equation 9.

$$chord = C(\mathbf{x}_i, p) = ChordSet(\widetilde{note}, \lceil p \times k \rceil) \qquad (9)$$

where

$$\widetilde{note} = mode\{\mathbf{M}(\mathbf{x}, p)\} \qquad (10)$$

3 Experiment

3.1 Data Acquisition

Equipment. A 62-channel electrode cap was used to collect the EEG signals in our experiment, while ESI NeuroScan System was used to recording the data with sample rate 200 Hz synchronously.

Subjects. Three men and three women participated in the EEG signals acquisition experiments. They were aged between 22 and 24, and were all in good condition during the experiment. All of them were informed of the harmlessness of the equipment.

Stimuli. Each volunteer watched twelve movie clips. Six clips expressed positive emotion and the other six clips expressed negative emotion. Each movie lasted for about four minutes long. All the movies were in English.

3.2 Frequency Mapping Representation

In FMR part, we take window size of STFT transform as 1 second, which means we generate the audio second by second. In addition, according to the relationship between pitch and frequency, we take $[f_{low}, f_{high}] = [65, 1976]$ as our target domain, corresponding to the pitch area from C2 to B6, because the pitches in these areas are more euphonious. Consequently, according to MIDI standard, we figure out that the pitch area $[p_{low}, p_{high}] = [36, 95]$.

3.3 Emotion Feature Representation

In EFR part, we take window size of STFT transform as 2 seconds, for the emotion cannot vary too much within 2 seconds. In feature extraction, we take 12 hemisphere asymmetry electrode pairs: Fp1-Fp2, F3-F4, F7-F8, FT7-FT8, FC3-FC4, T7-T8, P7-P8, C3-C4, TP7-TP8, CP3-CP4, P3-P4, O1-O2. Therefore, we have $5 \times 12 = 60$ dimensions in total. In dimension reduction, we reduce the dimension to 16, based on which the classification still has a good result. In classification, we take 8 clips of EEG data as training data (4 happy and 4 sad) and 4 clips (2 happy and 2 sad) as testing data.

In music generation part, we predesign a C Major Scale and a c Harmonic Minor Scale with 10 notes as the first two lines in Figure 3. In addition, we also predesign a set of six chords for each notes. The third line in Figure 3 shows a set of chords for note C4.

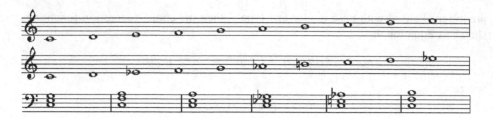

Fig. 3. The first line represents the C Major Scale. The second line represents the c Harmonic Minor Scale. The third line shows a set of chords with base note C3 (corresponding to note C4).

Table 1. Emotion classification Accuracy of six subjects

Subject	1	2	3	4	5	6
Accuracy (%)	72.08	76.17	81.70	55.36	78.94	73.44

4 Results

Table 1 shows the emotion classification results. According to the classification results, we generate 6 music pieces corresponding to the testing EEG signals. The music pieces can be found in the supporting material. In order to evaluate our sonification results, we take an evaluation test. We take 12 music episodes of 30 seconds from the 6 music pieces, 6 corresponding to positive emotion and 6 corresponding to negative emotion. Ten volunteers participated in the tests. All the volunteers have normal music appreciation abilities. Each volunteer was required to listen to all of the 12 music episodes and give a rating from 1 to 5 to each episode, where 1 stands for most positive and 5 stands for most negative.

Table 2 shows the rating results for twelve music episodes. We can find that the rating results have a positive correlation with the classification accuracy. For example, we have a terrible classification result on Music 4, and correspondingly two close ratings for the sounds, while sounds based on other good classification results have distinct differences. Therefore we can say that our sonification methods properly represent the emotion features and the classification results.

5 Supporting Materials

All the sound files can be found at http://bcmi.sjtu.edu.cn/~zhangyuxi/

Acknowledgments. The work was supported by the national natural science foundation of China (GrantNo. 91120305, 61272251). We thank *Ruonan Duan* for EEG data supply and *Ye Liu* for discussions and paper guidance. In addition, we also acknowledge the music theory support from *Yang Zhang*. Finally, we acknowledge the support from *Center for Brain-Like Computing and Machine Intelligence* (BCMI) laboratory.

Table 2. Evaluation results from five raters

Music	Episode	Rater 1	Rater 2	Rater 3	Rater 4	Rater 5	Average
1	1	1	2	2	2	2	1.8
	2	5	2	3	5	4	3.8
2	3	1	1	3	4	3	2.4
	4	4	2	3	3	4	3.2
3	5	2	2	3	2	2	2.2
	6	5	3	4	2	4	3.6
4	7	3	4	5	2	3	3.4
	8	3	3	5	2	3	3.2
5	9	1	1	3	1	1	1.4
	10	5	2	4	3	3	3.4
6	11	1	2	3	1	2	1.8
	12	5	3	3	3	3	3.4

References

1. Zhang, H., Liang, J., Liu, Y., Wang, H., Zhang, L.: An iterative method for classifying stroke subjects motor imagery eeg data in the bci-fes rehabilitation training system. In: Sun, F., Hu, D., Liu, H. (eds.) Foundations and Practical Applications of Cognitive Systems and Information Processing. AISC, vol. 215, pp. 363–373. Springer, Heidelberg (2014)
2. Duan, R.N., Zhu, J.Y., Lu, B.L.: Differential entropy feature for eeg-based emotion classification. In: 2013 6th International IEEE/EMBS Conference on Neural Engineering (NER), pp. 81–84. IEEE (2013)
3. Gaver, W.W.: Using and creating auditory icons (1994)
4. Kramer, G.: Auditory display: Sonification, audification, and auditory interfaces. Addison-Wesley, Reading (1994)
5. Scaletti, C.: Sound synthesis algorithms for auditory data representations. In: Santa Fe Institude Studies in the Sciences of Complexity-Proceedings, vol. 18, p. 223. Addison-Wesley Publishing Co. (1994)
6. Hermann, T., Ritter, H.: Listen to your data: Model-based sonification for data analysis. Advances in Intelligent Computing and Multimedia Systems 8, 189–194 (1999)
7. Baier, G., Hermann, T., Stephani, U.: Multi-channel sonification of human eeg. In: Proceedings of the 13th International Conference on Auditory Display (2007)
8. Wu, D., Li, C.Y., Yao, D.Z.: Scale-free music of the brain. PloS One 4(6), e5915 (2009)

Forecasting Crowd State in Video
by an Improved Lattice Boltzmann Model

Ye Tao, Peng Liu, Wei Zhao, and XiangLong Tang[1]

Pattern Recognition Research Center, School of Computer Science and Technology,
Harbin Institute of Technology, Harbin 150001, China

Abstract. Fluid methods have been introduced to analysis of crowd movements in videos recent years. Among these methods, Lattice Boltzmann model has been widely used as a quite convenient tool. Moreover, the Lattice Boltzmann model describes crowd movement as fluid, and the particles of the fluid flow randomly. Therefore, it is very difficult for the model to simulate the crowds purpose drive. In this study, a lattice Boltzmann based model added with a traction force term, which represents the crowds purpose drive toward the exit, is proposed. The model input is optical flow velocity field. Less error in the velocity fields computing and the capability in forecasting the crowd state is obtained.

Keywords: Video analysis, crowd state forecasting Lattice Boltzmann model traction force.

1 Introduction

Due to increasing populations and higher mobility, mass events, such as sports events, festivals, or concerts, attract growing numbers of attendees, and thus security measures are becoming more and more important. Nevertheless, despite adequate precautions even video surveillance are adopted, deadly stampedes and crowd disasters still occur rather frequently[12][11]. Experimental studies and simulations on video data are conducted widely. A system in predicting abnormal state of masses in real-time is presented in this paper. By optical flow, the system avoids the need for detection and tracking of individual pedestrians, which is a tough task due to the inappropriate camera viewpoints and the occlusions occurred in the large number of people. An improved physical model is then introduced to simulate walking crowd and predict crowd states automatically.

To understand human behaviour and improve existing physical models, experimental studies are conducted by researches. Parameters such as crowd density, speed, flow, and crowd pressure[14] are determined either manually[13] or by means of digital image processing[7,6]. They usually do not adopt real data except experimental data. To avoid occlusions and to facilitate automatic video analysis, video-based experiments are typically carried out using top-view cameras. Former researches often detect and track individuals, but holistic approaches that make use of optical flow features are proposed recently. Among physical models, microscopic models pay much attention to the details, well the

C.K. Loo et al. (Eds.): ICONIP 2014, Part III, LNCS 8836, pp. 50–57, 2014.

macroscopic ones focus on the global property of the crowd states. Both of them can not describe the effect caused by human will and environment barriers. Mesoscopic models, as an efficiently alternative computational technique to Navier-Stokes solvers[5], has attained wide popularity in simulating various fluid flow problems. The lattice Boltzmann model (LBM)[2] has recently been introduced as a new computational tool in fluid dynamics and systems governed by related partial differential equation (PDE). Meanwhile, it has been an alternative for computational fluid dynamics (CFD). The LBM originates from a Boolean fluid model, which is originally developed to overcome certain drawbacks of LGA, known as the lattice gas automata, such as the presence of statistical noise and lack of Galilean invariance in modelling fluid based upon kinetic theory[10].

Based on the kinetic theory, the LBM studies the dynamics of fictitious particles by using the density distribution functions. It contains collision and streaming sub-processes, both of which can be solved directly in calculation step. The macroscopic variables, such as density and momentum, can be calculated by the distribution functions. The advantages of LBM are simplicity, easy implementation, explicit calculation and intrinsic parallel nature[8]. Two problems of the LBM are considered in this paper. First, the velocities at the exit might be negative. Second, it is not sensitive to human velocities.

In the following section, LBM is introduced in detail. In Section 3, A traction force representing individuals proceed willing is added to LBM. The output velocity after adding force term is deduced. In Section 4, an experiment in predicting crowd states is presented. The results show that improved LBM could avoid negative-velocity near the exit and provide a good capability in predicting the abnormal crowd states. Conclusion is given at last.

2   Lattice Boltzmann Model

LBM originated from lattice gas cellular automata (LGA) initially proposed by Frisch, et al[9]. They have shown that the Navier-Stokes(N-S) equations can be derived in a suitable macroscopic limit from the particle distribution function representing streaming and collision of fluid particles. Suppose the particles are distributed over two-dimensional square lattices, spreading and colliding on them. In this study, the D2Q9 model, which has two dimensions and 9 directions, is selected, and the structure of a lattice is shown in Fig.1.

Fig.1(a) is the D2Q9 Spatial configuration of LBM, the spreading and collision of particles in the lattice are shown in Fig.1(b) and Fig.1(c) . f_α is the particle distribution function in the direction of α. $f_\alpha(X, t)$ denote the particle state at time t and at position $X = (x, y)$. Different from the N-S equations, the LBM studies the evolution of distribution functions. The governing formula is[4]

$$f_\alpha(X + e_\alpha \Delta t, t + \Delta t) - f_\alpha(X, t) = -\frac{f_\alpha(X, t) - f_\alpha^{eq}(X, t)}{\tau} \qquad (1)$$

where f_α^{eq} is its corresponding equilibrium distribution function in α direction;τ is the single relaxation parameter; e_α is the particle velocity. Note $\Omega(f_\alpha)$ as

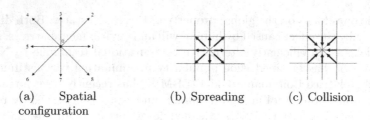

(a)   Spatial configuration

(b) Spreading

(c) Collision

Fig. 1. The structure of a lattice

collision operator. It is exactly the right side of Eq.(1)

$$\Omega(f_\alpha) = -\frac{f_\alpha(X,t) - f_\alpha^{eq}(X,t)}{\tau} \tag{2}$$

In former works, the equilibrium distribution function is expressed as[3]

$$f_\alpha^{eq}(X,t) = \rho\omega_\alpha \left[1 + \frac{e_\alpha \cdot u}{c_s^2} + \frac{(e_\alpha \cdot u)^2 - c_s^2 |u|^2}{2c_s^4} \right] \tag{3}$$

where ω_α are constants; c_s is the sound of speed. The constant value of ω_α and c_s may change in different lattice structure. u is the optical flow velocity vector of an image pixel at time t at position $X = (x, y)$. The velocity set is given by

$$e_{\alpha} = \begin{cases} 0, & \alpha = 0 \\ (\cos((\alpha - 1)\pi/4), \sin((\alpha - 1)\pi/4))c, & \alpha = 1, 3, 5, 7 \\ \sqrt{2}(\cos((\alpha - 1)\pi/4), \sin((\alpha - 1)\pi/4))c, & \alpha = 2, 4, 6, 8 \end{cases} \tag{4}$$

where $c = \Delta x/\Delta t$; Δx is the lattice gap. In general, $c = 1$ implies $\Delta x = \Delta t$. In D2Q9 model, $\omega_0 = 4/9$, $\omega_1 = \omega_3 = \omega_5 = \omega_7 = 1/9$, and $\omega_2 = \omega_4 = \omega_6 = \omega_8 = 1/36$; $c_s = c/\sqrt{3}$. From the conservation laws of mass and momentum, the macroscopic density ρ and fluid velocity u are calculated in terms of distribution functions. They are

$$\rho = \sum_\alpha f_\alpha, \rho u = \sum_\alpha e_\alpha f_\alpha \tag{5}$$

3   Lattice Boltzmann Model with Traction Force

The discrete form of distribution functions can be given as follows

$$f_\alpha(X + e_\alpha \Delta t, t + \Delta t) - f_\alpha(X, t) = -\frac{f_\alpha(X,t) - f_\alpha^{eq}(X,t)}{\tau} + \Delta t F_\alpha(X, t) \tag{6}$$

where, $-\frac{f_\alpha(X,t) - f_\alpha^{eq}(X,t)}{\tau}$ is the linear equation of collision term $\Omega(f_\alpha)$. The continuous Boltzmann equation can be given as follow

$$\frac{\partial f}{\partial t} + \xi \cdot \nabla f + a \cdot \nabla_\xi f = \Omega(f) \tag{7}$$

where f is the particle distribution function, $\boldsymbol{\xi}$ and $\boldsymbol{a}$ are, respectively, velocity and acceleration of a certain particle. The force term $\boldsymbol{a} \cdot \nabla_{\boldsymbol{\xi}} f$ is unknown, but it can be written into an expansion of $\boldsymbol{\xi}$ as follows:

$$\boldsymbol{a} \cdot \nabla_{\boldsymbol{\xi}} f = \rho \omega(\boldsymbol{\xi})[c^{(0)} + c_\alpha^{(1)} \xi_\alpha + c_{\alpha\beta}^{(2)} \xi_\alpha \xi_\beta + \cdots] \tag{8}$$

the first few coefficients $c_{i_1,i_2\cdots i_n}^{(n)}$ can be easily obtained by the following moment constraints:

$$\int \boldsymbol{a} \cdot \nabla_{\boldsymbol{\xi}} f d\boldsymbol{\xi} = 0 \tag{9}$$

$$\int \boldsymbol{\xi} \boldsymbol{a} \cdot \nabla_{\boldsymbol{\xi}} f d\boldsymbol{\xi} = -\rho \boldsymbol{a} \tag{10}$$

$$\int \xi_i \xi_j \boldsymbol{a} \cdot \nabla_{\boldsymbol{\xi}} f d\boldsymbol{\xi} = -\rho(a_i u_j + a_j u_i) \tag{11}$$

Therefore, ignore the order of $O(u)$ and $O(\xi^2)$, we have

$$\boldsymbol{a} \cdot \nabla_{\boldsymbol{\xi}} f = -3\rho\omega(\boldsymbol{\xi}) \frac{1}{c_s^2} [(\boldsymbol{\xi} - \boldsymbol{u}) + \frac{3}{c_s^2}(\boldsymbol{\xi} \cdot \boldsymbol{u}) \cdot \boldsymbol{\xi}] \cdot \boldsymbol{a} \tag{12}$$

It should be stressed that every term in Eq.(6) must be treated equally to maintain the same order of accuracy. Specifically, the expansion of the force term must be of second order in $\boldsymbol{\xi}$ and of first order in $\boldsymbol{u}$, in order to be consistent with the expansion of the equilibrium distribution function given by Eq.(3). Following the same discretization procedure for the equilibrium distribution function, the traction force is obtained.

$$F_\alpha = -3\rho\omega_\alpha[\frac{1}{c_s^2}(\boldsymbol{e}_\alpha - \boldsymbol{u}) + 3\frac{(\boldsymbol{e}_\alpha \cdot \boldsymbol{u}) \cdot \boldsymbol{e}_\alpha}{c_s^4}] \cdot \boldsymbol{a} \tag{13}$$

The above force term also satisfies the discrete counterpart of Eq.(9)(10)(11). If only the Eq.(9)(10) are satisfied, and meanwhile the Eq.(11)is replaced by $\sum_\alpha e_{\alpha,i} e_{\alpha,j} F_\alpha = 0$ in the discrete case, then the force term reduces to $F_\alpha = -3\rho\omega_\alpha \frac{e_\alpha}{c_s^2} \cdot \boldsymbol{a}$. This is the way of force term often used in Eq.(6). It is the discrete form of traction force term. $\boldsymbol{e}_\alpha$ is the particle velocity. $\boldsymbol{a}$ is traction acceleration, which can be replaced by acceleration of gravity. The magnitude of traction force is the same as gravity, but its direction points to exit of the site. Therefore, the force term also can be given as $-3\rho\omega_\alpha \frac{e_\alpha}{c_s^2} \cdot \boldsymbol{g}$.

Procedure of the proposed algorithm is sum up in **Algorithm 1**

4   Experimental Results and Analysis

In this section, a few examples are presented to illustrate the new models prediction performance by using data set PETS2009 [1].

1-87th frames velocity fields are used as the input of the LBM to predicts the 88-140th frames velocity fields by Mengs method [8] and by the proposed LBM

Algorithm 1. Lattice Boltzmann Model with traction force algorithm

Input:

$I_{m \times n} = 1$, I is the image space

$t = 1$, iteration control variable

N, the Maximum Iterations constant

$u(X, t)$ optical flow fields matrices

Output:

1. **while** $t \leq N$ **do**

2.   COLLISION STEP:

      Update the collision state parameter of all lattices by Eq.(6)

3.   STREAMING STEP:

      Update the STREAMING state parameter of all lattices by $f_\alpha(X + e_\alpha \Delta t, t + \Delta t) = f_\alpha(X, t + \Delta t)$.

4.   MACROSCOPIC VARIABLES:

      Compute and save the variables ρ and u via Eq.(5)

5. $t + +$

6. **end while**

with traction force. Fig.2(a) shows the velocity field of 114th frame by Mengs method, where, there are lots of negative velocities at the exit. The first order or the second order vortex of fluid dynamic phenomenon at corners of the exit leads to this phenomenon. Well, the real crowd state should not evolve like that. The velocity field of the 114th frame predicted by the proposed method is showed in Fig.2(b). It can be found that the crowds near the exit are trend to get out of it. It is accordance with the real case. The method adds traction force term, reflecting the marching trend, into the crowd states evolution. An increment of distribution function f_α is produced because of the effect of traction force. This increment could avoid the negative velocity of Mengs model effectively.

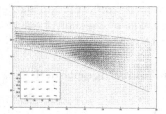
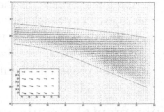

(a) the velocity field of 114th frame by Mengs method, inset shows a zoomed in version of the exit region

(b) the velocity field of the same frame by this paper, inset shows a zoomed in version of the exit region

Fig. 2. The velocity fields in the exit of the scene

The forecast velocity error is calculated by Eq.(14) and is shown in Fig.3.

$$\varepsilon(x,y) = \frac{1}{M \times N} \sum_{i=1}^{M} \sum_{j=1}^{N} (v_{i,j}(x,y) - u_{i,j}(x,y))^2 \qquad (14)$$

where, the size of the frame is $M \times N$; the predict velocity $v(x,y)$ is calculated by these two models, respectively, and $u(x,y)$ is the real velocity fields in 1-140th frame.

The curves in Fig.3 shows the velocity error surface of the Mengs method and the proposed method, where the error decreases when the iteration going further, and increases with the frame increases. Fig.3(a)shows the error surface drawn by Mengs model. Its error increases to maximum during frame 80th to 100th because the crowd begin to run in these frames. The error begins to decrease while the crowd state is steady. The model is not good when crowd velocity changes rapidly. Fig.3(b) shows the error surface calculated by our model. The error surface is reduced obviously. The added force term makes the model more sensitive to the change of fluid velocity.

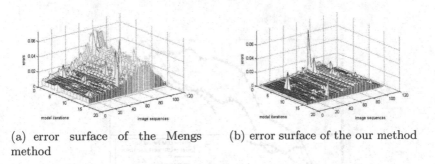

(a) error surface of the Mengs method

(b) error surface of the our method

Fig. 3. A comparison of error surface against the Mengs method

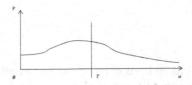

Fig. 4. Histograms of predict velocity amplitude

The LBM with traction force can predict the velocity fields of many frames after the current frames. Every frames velocity magnitude distribution trend can be predicted by histograms where u is velocity and y is the statistic of a certain velocity. For the specialties of scene, this experiment set a empirical permit maximum velocity $T_1 = 0.20$. This value is shown as a vertical line in histograms, see Fig.4. The area of the right side of the vertical line in histogram is calculated and if the area is greater than S_1, called empirical permit maximum area, then the crowd state is regarded as abnormal. The threshold parameters T_1 and S_1 can be chosen by the feature of different scenes and safety requirement.

Fig.5 shows a comparison of the two models, the vertical coordinate is the area value. The crowd state is regarded as abnormal when $T_1 = 0.20$ and the area is greater than S_1, assume $S_1 = 2000$, in 107th frame in the real image frames, see Fig.5. The proposed model predicts the abnormal situation in the 96th frame, while the Mengs model detects the situation in the 115th frame, at least 21 frames advantage are gotten by us. The experiment shows that the LBM with traction force is able to predict the abnormal situation. Particles in the LBM without traction force lost their speed because boundary conditions exist and particles collisions occur. This is the reason why the detect delay happens in Fig.5. The adding of the traction force fills the loss of the velocity and makes the LBM with traction force have the capability to predict abnormal situations. Hereby, the LBM with traction force is more sensitive to the variation of the velocity, consequently a good forecast performance is obtained.

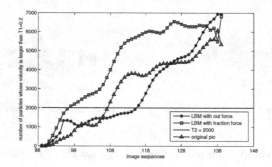

Fig. 5. Contrasting figure of prediction gotten by the two methods

5   Conclusion

Although plenty of works have been carried out in detecting crowd states from a video, it is now still difficult to forecast the crowd state. This study proposed a forecast model by adding a traction force term in the Boltzmann equation after analysing the features of crowd movement. The force term could reduce the negative velocities in the exit region, and make up for the velocity and momentum loss caused by particle collisions. All these could make the model describe and predict the velocity fields more accurately and, therefore, have the ability in crowd state predicting.

References

1. http://www.cvg.rdg.ac.uk/PETS2009/a.html
2. Ammar, A.: Lattice boltzmann method for polymer kinetic theory. Journal of Non-Newtonian Fluid Mechanics 165(19), 1082–1092 (2010)
3. Eifi, B.: Discrete lattice effects on the forcing term in the lattice boltzmann method. Physical Review E Phys Rev. E 65, 046308 (2002)
4. He, X., Luo, L.S.: A priori derivation of the lattice boltzmann equation. Physical Review E 55(6), R6333 (1997)
5. Higuera, F., Succi, S., Benzi, R.: Lattice gas dynamics with enhanced collisions. EPL (Europhysics Letters) 9(4), 345 (1989)
6. Johansson, A., Helbing, D., Al-Abideen, H.Z., Al-Bosta, S.: From crowd dynamics to crowd safety: A video-based analysis. Advances in Complex Systems 11(04), 497–527 (2008)
7. Liu, X., Song, W., Zhang, J.: Extraction and quantitative analysis of microscopic evacuation characteristics based on digital image processing. Physica A: Statistical Mechanics and its Applications 388(13), 2717–2726 (2009)
8. Meng, J., Qian, Y., Li, X., Dai, S.: Lattice boltzmann model for traffic flow. Physical Review E 77(3), 036108 (2008)
9. Pomeau, B.H.Y., Frisch, U.: Lattice-gas automata for the navier-stokes equation. Phys. Rev. Lett. 56(14), 1505 (1986)
10. Qian, Y.H., d'Humières, D., Lallemand, P.: Lattice bgk models for navier-stokes equation. EPL (Europhysics Letters) 17(6), 479 (1992)
11. Rodriguez, M., Ali, S., Kanade, T.: Tracking in unstructured crowded scenes. In: 2009 IEEE 12th International Conference on Computer Vision, pp. 1389–1396. IEEE (2009)
12. Rodriguez, M., Laptev, I., Sivic, J., Audibert, J.Y.: Density-aware person detection and tracking in crowds. In: 2011 IEEE International Conference on Computer Vision (ICCV), pp. 2423–2430. IEEE (2011)
13. Seyfried, A., Steffen, B., Klingsch, W., Boltes, M.: The fundamental diagram of pedestrian movement revisited. Journal of Statistical Mechanics: Theory and Experiment 2005(10), P10002 (2005)
14. Steffen, B., Seyfried, A.: Methods for measuring pedestrian density, flow, speed and direction with minimal scatter. Physica A: Statistical Mechanics and Its Applications 389(9), 1902–1910 (2010)

Properties of Multiobjective Robust Controller Using Difference Signals and Multiple Competitive Associative Nets in Control of Linear Systems

Weicheng Huang, Yuki Ishiguma, and Shuichi Kurogi

Kyushu Institute of Technology, Tobata, Kitakyushu, Fukuoka 804-8550, Japan
{ko@kurolab.,ishiguma@kurolab.,kuro@}cntl.kyutech.ac.jp
http://kurolab.cntl.kyutech.ac.jp/

Abstract. Recently, we have developed multiobjective robust controller using difference signals of nonlinear plant for multiple CAN2s to learn and approximate Jacobian matrices of the nonlinear dynamics. Here, the CAN2 is an artificial neural net for learning efficient piecewise linear approximation of nonlinear function. So far, by means of numerical experiments, we have shown that the controller is capable of coping with the change of plant parameter values as well as the change of control objective by means of switching multiple CAN2s. However, the controller have not been analyzed enough. This paper clarifies several properties of the controller by means of examining the control of linear plants.

Keywords: Multiobjective robust control, Switching of multiple CAN2s, Difference signals, Generalized predictive control, Jacobian matrix of nonlinear dynamics.

1 Introduction

Recently, we have developed multiobjective robust controller using difference signals of nonlinear plant to be controlled and multiple CAN2s (competitive associative nets) [1,2,3]. Here, the CAN2 is an artificial neural net introduced for learning efficient piecewise linear approximation of nonlinear function by means of competitive and associative schemes [5,6,7]. Thus, a CAN2 is capable of leaning piecewise Jacobian matrices of nonlinear dynamics of a plant by means of feeding difference signals of the plant to the CAN2. In [1], we have constructed a robust controller using multiple CAN2s to learn to approximate the plant dynamics for several parameter values. In [2], we have focused on a multiobjective robust control, where we consider two conflicting control objectives for a nonlinear crane system: one is to reduce settling time and the other is to reduce overshoot. Our method enables the controller to flexibly cope with those objectives by means of switching two sets of CAN2s for reducing settling time and overshoot, respectively. In [3], we have tried to improve the control performance by means of replacing single CAN2s by bagging CAN2s and shown several properties of the controller. From the point of view of multiobjective control [8], the settling time is reduced by tuning the number of units of the CAN2s, while the overshoot on average is reduced by bagging CAN2s replacing single CAN2s and an overshoot for the plant with certain parameter values is reduced by an augmentation of bagging CAN2s.

C.K. Loo et al. (Eds.): ICONIP 2014, Part III, LNCS 8836, pp. 58–67, 2014.

However, these properties as well as other properties of the controller have not been analyzed enough so far. In order to examine the controller, we analyze the controller by means of applying it to simple linear plants. In the next section, we show the method to control nonlinear and linear plant. In Sect. 3, we examine the method by means of numerical experiments applied to linear plants involving changeable parameter values.

2   Multiobjective Robust Controller Using Difference Signals and CAN2s

2.1   Plant Model Using Difference Signals

Suppose a plant to be controlled at a discrete time $j = 1, 2, \cdots$ has the input $u_j^{[\mathrm{p}]}$ and the output $y_j^{[\mathrm{p}]}$. Here, the superscript "[p]" indicates the variable related to the plant for distinguishing the position of the load, (x, y), shown below. Furthermore, suppose that the dynamics of the plant is given by

$$y_j^{[\mathrm{p}]} = f(x_j^{[\mathrm{p}]}) + d_j^{[\mathrm{p}]} , \tag{1}$$

where $f(\cdot)$ is a nonlinear function which may change slowly in time and $d_j^{[\mathrm{p}]}$ represents zero-mean noise with the variance σ_d^2. The input vector $x_j^{[\mathrm{p}]}$ consists of the input and output sequences of the plant as $x_j^{[\mathrm{p}]} \triangleq \left(y_{j-1}^{[\mathrm{p}]}, \cdots, y_{j-k_y}^{[\mathrm{p}]}, u_{j-1}^{[\mathrm{p}]}, \cdots, u_{j-k_u}^{[\mathrm{p}]} \right)^T$, where k_y and k_u are the numbers of the elements, and the dimension of $x_j^{[\mathrm{p}]}$ is given by $k = k_y + k_u$. Then, for the difference signals $\Delta y_j^{[\mathrm{p}]} \triangleq y_j^{[\mathrm{p}]} - y_{j-1}^{[\mathrm{p}]}$, $\Delta u_j^{[\mathrm{p}]} \triangleq u_j^{[\mathrm{p}]} - u_{j-1}^{[\mathrm{p}]}$, and $\Delta x_j^{[\mathrm{p}]} \triangleq x_j^{[\mathrm{p}]} - x_{j-1}^{[\mathrm{p}]}$, we have the relationship $\Delta y_j^{[\mathrm{p}]} \simeq f_x \Delta x_j^{[\mathrm{p}]}$ for small $\| \Delta x_j^{[\mathrm{p}]} \|$, where $f_x = \partial f(x)/\partial x \big|_{x = x_{j-1}^{[\mathrm{p}]}}$ indicates the Jacobian matrix (row vector). If f_x does not change for a while after the time j, then we can predict $\Delta y_{j+l}^{[\mathrm{p}]}$ by

$$\widehat{\Delta y}_{j+l}^{[\mathrm{p}]} = f_x \widehat{\Delta x}_{j+l}^{[\mathrm{p}]} \tag{2}$$

for $l = 1, 2, \cdots$, recursively. Here, $\widehat{\Delta x}_{j+l}^{[\mathrm{p}]} = (\widetilde{\Delta y}_{j+l-1}^{[\mathrm{p}]}, \cdots, \widetilde{\Delta y}_{j+l-k_y}^{[\mathrm{p}]}, \widetilde{\Delta u}_{j+l-1}^{[\mathrm{p}]}, \cdots, \widetilde{\Delta u}_{j+l-k_u}^{[\mathrm{p}]})^T$, and the elements are given by

$$\widetilde{\Delta y}_{j+m}^{[\mathrm{p}]} = \begin{cases} \Delta y_{j+m}^{[\mathrm{p}]} & \text{for } m < 1 \\ \widehat{\Delta y}_{j+m}^{[\mathrm{p}]} & \text{for } m \geq 1 \end{cases} \quad \text{and} \quad \widetilde{\Delta u}_{j+m}^{[\mathrm{p}]} = \begin{cases} \Delta u_{j+m}^{[\mathrm{p}]} & \text{for } m < 0 \\ \widehat{\Delta u}_{j+m}^{[\mathrm{p}]} & \text{for } m \geq 0. \end{cases} \tag{3}$$

Here, $\widehat{\Delta u}_{j+m}^{[\mathrm{p}]}$ $(m \geq 0)$ is the predictive input (see Sect. 2.3). Then, we have the prediction of the plant output from the predictive difference signals as

$$\widehat{y}_{j+l}^{[\mathrm{p}]} = y_j^{[\mathrm{p}]} + \sum_{m=1}^{l} \widehat{\Delta y}_{j+m}^{[\mathrm{p}]}. \tag{4}$$

For linear plants, the plant function in (1) and the Jacobian matrix in (2) are modified as $f(x_j^{[\mathrm{p}]}) = A x_j^{[\mathrm{p}]}$ and $f_x = A$, where $A \in \mathbb{R}^{1 \times k}$ is constant.

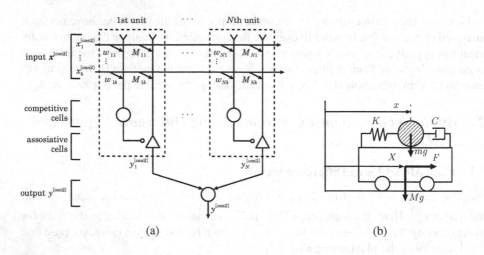

Fig. 1. Schematic diagram of (a) CAN2 and (b) a linear plant model of a car and the load

2.2   CAN2 for Learning and Identifying Nonlinear and Linear Plants

A CAN2 has N units. The ith unit has a weight vector $\boldsymbol{w}_i \triangleq (w_{i1}, \cdots, w_{ik})^T \in$ $\mathbb{R}^{k \times 1}$ and an associative matrix (row vector) $\boldsymbol{M}_i \triangleq (M_{i1}, \cdots, M_{ik}) \in \mathbb{R}^{1 \times k}$ for $i \in I = \{1, 2, \cdots, N\}$ (see Fig. 1(a)). For a given dataset $D^{[n]} = \{(\Delta \boldsymbol{x}_j^{[p]}, \Delta y_j^{[p]}) \mid j = 1, 2, \cdots, n\}$ obtained from the plant to be controlled, we train a CAN2 by feeding the input and output pair of the CAN2 as $(\boldsymbol{x}^{[\mathrm{can2}]}, y^{[\mathrm{can2}]}) = (\Delta \boldsymbol{x}_j^{[p]}, \Delta y_j^{[p]})$. We employ an efficient batch learning method shown in [10]. Then, for an input vector $\Delta \boldsymbol{x}_j^{[p]}$, the CAN2 after the learning predicts the output $\Delta y_j^{[p]} = f_{\boldsymbol{x}} \Delta \boldsymbol{x}_j^{[p]}$ by

$$\widehat{\Delta y}_j^{[\mathrm{p}]} = \boldsymbol{M}_c \Delta \boldsymbol{x}_j^{[\mathrm{p}]}, \tag{5}$$

where c denotes the index of the unit selected by

$$c = \operatorname*{argmin}_{i \in I} \| \Delta \boldsymbol{x}_j^{[\mathrm{p}]} - \boldsymbol{w}_i \|^2. \tag{6}$$

Here, we have assumed the following conjecture shown in [2,3]. Namely, $\boldsymbol{M}_c \simeq f_{\boldsymbol{x}}$ may not be identified via $\Delta \boldsymbol{x}_j^{[p]}$ because $f_{\boldsymbol{x}}$ is not the function of $\Delta \boldsymbol{x}_j^{[p]}$ generally. However, an enlarged vector $\Delta \boldsymbol{z}_j^{[p]} = (\Delta y_{j-1}^{[p]}, \cdots, \Delta y_{j-k_y'}^{[p]}, \Delta u_{j-1}^{[p]}, \cdots, \Delta u_{j-k_u'}^{[p]})$ for $k_y' = k + k_y$ and $k_u' = k + k_u$ enables a Jacobian matrix $f_{\boldsymbol{z}} = \partial f / \partial \boldsymbol{z}$ to be a function of $\Delta \boldsymbol{z}_j^{[p]}$ when the elements in $\Delta \boldsymbol{z}_j^{[p]}$ vary sufficiently and the plant parameter does not change for a while. Thus, the above method with $\Delta \boldsymbol{x}_j^{[p]}$ in (6) replaced by an enlarged $\Delta \boldsymbol{z}_j^{[p]}$ is supposed to select an appropriate cth unit in the situation of multiobjective and robust control assuming the change of both plant parameters and control objectives. However, this conjecture is hard to be verified because Jacobian matrix for a certain duration of time involves approximation error in general, thus k_y and k_u to identify the Jacobian matrix depend on the approximation error allowable for the control.

On the other hand, the above conjecture does not seem to be applied to a linear plant with $f(x_j^{[p]}) = Ax_j^{[p]}$ because there is only one Jacobian matrix $f_x = A$. Thus, the CAN2 with multiple units after the learning of the plant dynamics is considered to have the following associative matrix for the ith unit as

$$M_i = A + \delta A_i \tag{7}$$

where δA_i denotes approximation error of A. The cth unit with the error δA_c is selected by (6) whose weight vector w_c is near to the current difference input vector $\Delta x_j^{[p]}$ (or enlarged $\Delta z_j^{[p]}$) consisting of the trained difference trajectory, i.e. $\Delta y_{j-l}^{[p]}$ and $\Delta u_{j-l}^{[p]}$ for $l = 1, 2, \cdots$. This interpretation of erroneous associative matrices can be also applied to the control of nonlinear plants, and seems more plausible than the above conjecture for nonlinear plants if the present controller also works for linear plants.

2.3 GPC Using Difference Signals

The GPC (Generalized Predictive Control) is an efficient method for obtaining the predictive input $\widehat{u}_j^{[p]}$ which minimizes the following control performance index [9]:

$$J = \sum_{l=1}^{N_y} \left(r_{j+l}^{[p]} - \widehat{y}_{j+l}^{[p]} \right)^2 + \lambda_u \sum_{l=1}^{N_u} \left(\widehat{\Delta u}_{j+l-1}^{[p]} \right)^2, \tag{8}$$

where $r_{j+l}^{[p]}$ and $\widehat{y}_{j+l}^{[p]}$ are desired output and predictive output, respectively. The parameters N_y, N_u and λ_u are constants to be designed for the control performance. We obtain $\widehat{u}_j^{[p]}$ by means of the GPC method as follows: at a discrete time j, use CAN2 to predict $\Delta y_{j+l}^{[p]}$ by (2) and then $\widehat{y}_{j+l}^{[p]}$ by (4). Then, owing to the linearity of these equations, the above performance index is written as

$$J = \|r^{[p]} - G\Delta u^{[p]} - \overline{y}^{[p]}\|^2 + \lambda_u\|\widehat{\Delta u}\|^2 \tag{9}$$

where $r^{[p]} = \left(r_{j+1}^{[p]}, \cdots, r_{j+N_y}^{[p]} \right)^T$ and $\widehat{\Delta u}^{[p]} = \left(\widehat{\Delta u}_j^{[p]}, \cdots, \widehat{\Delta u}_{j+N_u-1}^{[p]} \right)^T$. Furthermore, $\overline{y}^{[p]} = \left(\overline{y}_{j+1}^{[p]}, \cdots, \overline{y}_{j+N_y}^{[p]} \right)^T$ and $\overline{y}_{j+l}^{[p]}$ is the natural response $\widehat{y}_{j+l}^{[p]}$ of the system (1) for the null incremental input $\widehat{\Delta u}_{j+l}^{[p]} = 0$ for $l \geq 0$. Here, we actually have $\overline{y}_{j+l}^{[p]} = y_j^{[p]} + \sum_{m=1}^{l} \overline{\Delta y}_{j+m}^{[p]}$ from (4), where $\overline{\Delta y}_{j+l}^{[p]}$ denotes the natural response of the difference system of (2) with f_x replaced by M_c. The ith column and the jth row of the matrix G is given by $G_{ij} = g_{i-j+N_1}$, where g_l for $l = \cdots, -2, -1, 0, 1, 2, \cdots$ is the unit step response $y_{j+l}^{[p]}$ of (4) for $\widehat{y}_{j+l}^{[p]} = \widehat{u}_{j+l}^{[p]} = 0$ ($l < 0$) and $\widehat{u}_{j+l}^{[p]} = 1 (l \geq 0)$. It is easy to derive that the unit response g_l of (4) is obtained as the impulse response of (2). Then, we have $\widehat{\Delta u}^{[p]}$ which minimizes J by $\widehat{\Delta u}^{[p]} = (G^T G + \lambda_u I)^{-1} G^T (r^{[p]} - \overline{y}^{[p]})$, and then we have $\widehat{u}_j^{[p]} = u_{j-1}^{[p]} + \widehat{\Delta u}_j^{[p]}$.

2.4   Control and Training Iterations

We execute iterations of the following phases to obtain the training data for CAN2s respectively and train the CAN2s.

(i) **Control Phase:** Control by a default control schedule at the first iteration, and by the GPC using the CAN2s obtained by the previous training phase otherwise.
(ii) **Training Phase:** Train the CAN2s with the dataset $D^{[n]} = \{(\Delta x_j^{[p]}, \Delta y_j^{[p]} | j = 1, 2, \cdots, n)\}$ obtained in the control phase.

The control performance at an iteration depends on the CAN2 obtained at the previous iterations. So, for the actual control of the plant, we use the best CAN2s obtained through a number of iterations as shown below.

2.5   Switching Multiple CAN2s For Multiobjective Robust Control

To cope with the change of plant parameters and the change of control objective, we employ the following method to switch CAN2s for each control objective O_l ($l = 1, 2, \cdots$). Let $\mathrm{CAN2}_{O_l}^{[\theta_s]}$ denote the best CAN2 from the point of view of O_l obtained for the plant with parameter θ_s ($s \in S = \{1, 2, \cdots, |S|\}$) through the above control and training iterations.

Step 1: At each discrete time j in the control phase, obtain $M_c^{[s]}$ ($= M_c$ in (6)) for all $\mathrm{CAN2}_{O_l}^{[\theta_s]}$ ($s \in S$).

Step 2: Select the s^*th CAN2, or $\mathrm{CAN2}_{O_l}^{[\theta_{s^*}]}$, which provides the minimum MSE (mean square prediction error) for the recent N_e predictions, or

$$s^* = \operatorname*{argmin}_{s \in S} \frac{1}{N_e} \sum_{l=0}^{N_e-1} \left\| \Delta y_{j-l}^{[p]} - \widehat{\Delta y}_{j-l}^{[p][s]}) \right\|^2, \tag{10}$$

where $\widehat{\Delta y}_{j-l}^{[p][s]} = M_c^{[s]} \Delta x_{j-l}^{[p]}$ (see (5)) denotes the prediction by $\mathrm{CAN2}_{O_l}^{[\theta_s]}$.

3   Numerical Experiments Using Linear Plant Model

3.1   A Car and Load System

We consider a linear model plant of a car and the load shown in Fig. 1(b). This model is derived from the overhead traveling crane system [3] by means of replacing the nonlinear crane by a load (mass) with a spring and a damper. From the figure, we have the motion equations given by

$$m\ddot{x} = -K(x - X) - C(\dot{x} - \dot{X}) \tag{11}$$

$$M\ddot{X} = F + K(x - X) \tag{12}$$

where x and X are the positions of the load and the car, respectively, m and M are the weights of the load and the car, respectively, K the spring constant, C the damping coefficient, and F is the driving force of the car. From the above equations, we have the following state-space representation for the state $x = (x, \dot{x}, X, \dot{X})^T$,

$$\dot{x} = \begin{bmatrix} 0 & 1 & 0 & 0 \\ -\frac{K}{m} & -\frac{C}{m} & \frac{K}{m} & \frac{C}{m} \\ 0 & 0 & 0 & 1 \\ \frac{K}{M} & 0 & -\frac{K}{M} & 0 \end{bmatrix} x + \begin{bmatrix} 0 \\ 0 \\ 0 \\ \frac{1}{M} \end{bmatrix} F \tag{13}$$

3.2 Parameter Settings

Suppose that the controller has to move the load on the car from $x = 0$ to the destination position $x_d = 5\text{m}$ by means of operating F. We obtain discrete signals by $u_j^{[p]} = F(jT_v)$ and $y_j^{[p]} = x(jT_v)$ with (virtual) sampling period $T_v = 0.5\text{s}$. Here, we use virtual sampling method shown in [4], where the discrete model is obtained with T_v (virtual sampling period) while the observation and operation are executed with shorter actual sampling period $T_a = 0.01\text{s}$. We have used $N_y = 20$, $N_u = 1$ and $\lambda_u = 0.01$ for the GPC. and $N_e = 8$ samples for (10).

The parameters of the plant are set as follows; th weight of the car $M = 100\text{kg}$, the spring constant $K = 15 \text{ kg/s}^2$, the damping coefficient $C = 10 \text{ kg/s}$, and the maximum driving force $F_{\max} = 10\text{N}$. To achieve the robustness to the load weight for $m = 10, 15, 20, \cdots, 100$ [kg], we train the CAN2s with PLANT$^{[\theta_s]}$ for the load weight $\theta_s = m = 10, 40, 70, 100$ [kg] and $s = 1, 2, 3, 4$, respectively, where PLANT$^{[\theta]}$ indicate the plant with the parameter θ. Let CAN2$_{\text{OS}}^{[\theta_s]}$ and CAN2$_{\text{ST}}^{[\theta_s]}$ denote the best CAN2s which have achieved smallest overshoot and settling time, respectively, through 10 control and training iterations. Here, at each iteration, we train the CAN2 with the control dataset of two recent iterations, i.e. the current and the previous ones, because the number of obtained data becomes huge and time consuming as the number of iterations increases and the control performance does not seem improved even if we use all data. In order to uniquely select the CAN2, the overshoot x_{OS} and the settling time t_{ST} are ordered by $x_{\text{OS}} + \epsilon t_{\text{ST}}$ and $t_{\text{ST}} + \epsilon x_{\text{OS}}$, respectively, with small $\epsilon = 10^{-2}$. We have used the set of CAN2s, or CAN2$_{\text{OS}}^{[\theta_S]} = \{\text{CAN2}_{\text{OS}}^{[\theta_s]} | s \in S\}$ and CAN2$_{\text{ST}}^{[\theta_S]} = \{\text{CAN2}_{\text{ST}}^{[\theta_s]} | s \in S\}$ for the switching controller explained in Sect. 2.5, where $S = \{1, 2, 3, 4\}$.

3.3 Results and Analysis

Result Using CAN2s with Single Units. First, we have examined the controller using true linear models and CAN2s with single units ($N = 1$). We use the input vector $\Delta x_j^{[p]}$ with $k_y = 4$ and $k_u = 1$, which is not enlarged $\Delta z_j^{[p]}$ but has the original minimum dimension of the true dynamics. From the experimental result shown in Table 1, we can see that the controller using true model has achieved increasing settling time t_{ST} and overshoot x_{OS} with the increase of the load weight $m = 10, 40, 70, 100$ [kg] for θ_i ($i = 1, 2, 3, 4$). This is because the present controller uses the performance index J to be minimized shown in (9) with fixed control parameter values ($N_y = 20$, $N_u = 1$ and $\lambda_u = 0.01$). The conventional GPC has to tune the control parameters for minimizing t_{ST} and x_{OS} for the plants with different parameter values, while the present

64 W. Huang, Y. Ishiguma, and S. Kurogi

Table 1. Experimental result of settling time t_{ST} [s] and overshoot x_{OS} [mm] obtained by the controller using true linear models $PLANT^{[\theta_i]}$ and trained $CAN2^{[\theta_i]}$ with single units. The ith raw from the top shows the result of the control of $PLANT^{[\theta_i]}$ with $\theta_i = m = 10, 40, 70, 100$ [kg] for $i = 1, 2, 3, 4$, respectively.

	t_{ST}	x_{OS}		t_{ST}	x_{OS}		t_{ST}	x_{OS}
$PLANT^{[\theta_1]}$	33.8	0	$CAN2^{[\theta_1]}_{ST}$	32.4	186	$CAN2^{[\theta_1]}_{OS}$	32.7	170
$PLANT^{[\theta_2]}$	35.1	15	$CAN2^{[\theta_2]}_{ST}$	22.5	9	$CAN2^{[\theta_2]}_{OS}$	24.2	0
$PLANT^{[\theta_3]}$	41.5	86	$CAN2^{[\theta_3]}_{ST}$	15.9	63	$CAN2^{[\theta_3]}_{OS}$	27.5	9
$PLANT^{[\theta_4]}$	48.5	141	$CAN2^{[\theta_4]}_{ST}$	29.9	44	$CAN2^{[\theta_4]}_{OS}$	30.1	43

controller shows different performances by using different CAN2s for minimizing t_{ST} and x_{OS}, respectively, as shown in Table 1. The difference of the performance is supposed to be obtained from the training datasets for the CAN2s derived from control trajectories which may involve degenerations and/or fluctuations through control and training iterations. However, the performance in Table 1 is not so good as to apply it to the switching control using multiple CAN2s, e.g. $CAN2^{[\theta_1]}_{OS}$ could not have achieved overshoot less than $x_{OS} = 170$[mm] for the plant θ_1 with $m = 10$ [kg].

Result Using CAN2s with Multiple Units. In order to improve the control performance, we use CAN2s with multiple units. Here, note that the CAN2s with multiple units involve erroneous models as shown in (7). However, the batch learning algorithm of the CAN2 (see [10]) tries to reduce the total approximation error for a given training dataset by means of using the condition called asymptotic optimality to equalize the approximation errors for all units of the CAN2. Thus, we may expect that the error of the associative matrix in (7), $\delta A_i = M_i - A$, does not grow so much for all units and provides a variety of allowable control performances.

A statistical result of settling time t_{ST} and overshoot x_{OS} obtained by the controllers using multiple units is shown in Table 2, and four examples of time course of the input F, the output X and x for the best and the worst control result using multiple CAN2s are shown in Fig. 2. We can see that the best control for reducing settling time (top left) and overshoot (lower right) are reasonable, while the worst control for reducing settling time (top right) and overshoot (bottom right) are not so bad from their objectives.

From Table 2, we can see that the mean, max and std of settling time achieved by the controller using multiple $CAN2^{[\theta_s]}_{ST}$ are smaller than those by the controller using single $CAN2^{[\theta_s]}_{ST}$ for $s = 1, 2, 3, 4$. Incidentally, the controller using $CAN2^{[\theta_2]}_{OS}$ has achieved smaller mean, min and std of settling time, but $CAN2^{[\theta_2]}_{OS}$ is the CAN2 having achieved the minimum overshoot for θ_2 and we cannot find out any reason for this good performance in settling time.

On the other hand, the controller using multiple $CAN2^{[\theta_s]}_{OS}$ could not achieved smaller performance than the controller using single $CAN2^{[\theta_3]}_{OS}$. It seems that this is owing that $CAN2^{[\theta_s]}_{OS}$ involves $CAN2^{[\theta_2]}_{OS}$ which has a big mean overshoot 49.4[mm]. In our previous study [3], we have shown a method of augmentation of CAN2s to reduce plant-parameter-specific overshoots, and we apply the method as follows. First, we examined

Table 2. Statistical summary of the performance obtained by the controller using CAN2s with multiple units for the control of test plants with $m = 10, 15, 20, \cdots, 100$ [kg]. The columns of "trained θ_i" indicate the result by the controller applied to the training plants θ_i with $m = 10, 40, 70, 100$ [kg] for $i = 1, 2, 3, 4$, respectively. The columns of "mean", "min", "max" and "std" for "settling time" and "overshoot" indicate the minimum, maximum and standard deviation of the control result for all test plants. We denote $CAN2_{OS}^{[\theta_{S'}]} = CAN2_{OS}^{[70kg]} \cup CAN2_{OS}^{[97kg]}$ and $CAN2_{OS}^{[\theta_{S''}]} = CAN2_{OS}^{[70kg]} \cup CAN2_{OS}^{[90kg]} \cup CAN2_{OS}^{[97kg]}$. The boldface figures indicate the best (smallest) result in each block, while the italicface figures show the result not corresponding the control objective of the CAN2 shown on the leftmost column.

CAN2 used for the controller	settling time t_{ST} [s]					overshoot x_{OS} [mm]				
	trained θ_i	mean	min	max	std	trained θ_i	mean	min	max	std
$CAN2_{ST}^{[\theta_1]}$	19.6	26.06	**19.6**	35.4	5.91	*98*	*97.3*	*55.0*	*142.0*	*28.5*
$CAN2_{ST}^{[\theta_2]}$	20.5	26.18	20.3	35.5	5.93	*59*	*108.3*	*51.0*	*172.0*	*43.9*
$CAN2_{ST}^{[\theta_3]}$	25.1	27.36	23.1	35.1	3.55	*44*	*75.6*	*34.0*	*162.0*	*41.6*
$CAN2_{ST}^{[\theta_4]}$	28.6	30.68	25.6	39.0	3.74	*94*	***25.9***	***0.0***	***103.0***	***36.1***
$CAN2_{ST}^{[\theta_S]}$	—	**25.29**	22.2	**34.9**	**3.38**	—	*52.1*	*11.0*	*136.0*	*42.1*
$CAN2_{OS}^{[\theta_1]}$	*32.9*	*31.89*	*27.8*	*35.5*	*2.50*	0	11.4	**0.0**	66.0	20.1
$CAN2_{OS}^{[\theta_2]}$	*21.2*	*24.76*	**15.9**	*33.8*	*3.82*	0	49.4	**0.0**	264.0	72.6
$CAN2_{OS}^{[\theta_3]}$	*38.8*	*39.89*	*36.8*	*44.2*	***1.89***	0	3.2	**0.0**	28.0	7.4
$CAN2_{OS}^{[\theta_4]}$	*74.9*	*65.25*	*59.6*	*78.9*	*5.57*	0	6.8	**0.0**	38.0	11.5
$CAN2_{OS}^{[\theta_S]}$	—	*37.11*	*31.9*	*45.5*	*4.04*	—	6.6	**0.0**	35.0	11.9
$CAN2_{OS}^{[\theta_{S'}]}$	—	*41.61*	*36.6*	*44.6*	*2.08*	—	3.1	**0.0**	59.0	13.2
$CAN2_{OS}^{[\theta_{S''}]}$	—	*38.55*	*35.3*	*43.0*	*2.11*	—	**0.0**	**0.0**	**0.0**	**0.0**

the overshoot obtained by the controller using $CAN2_{OS}^{[\theta_3]} = CAN2_{OS}^{[70kg]}$, and it has the overshoot 3, 11, 18 and 28 [mm] for the plant with $m = 85, 90, 95$ and 100 [kg], respectively, and 0 [mm] for other test plants. Therefore, we next examined the controller using multiple $CAN2_{OS}^{[70kg]} \cup CAN2_{OS}^{[97kg]}$, and have an overshoot 59[mm] for the plant with $m = 100$ [kg] and 0[mm] for other test plants. Finally, we executed trial and error, and we have multiple $CAN2_{OS}^{[\theta_{S''}]} = CAN2_{OS}^{[70kg]} \cup CAN2_{OS}^{[90kg]} \cup CAN2_{OS}^{[97kg]}$ which has no overshoot for all test plants as shown in Table 2.

4   Conclusion

We have examined the multiobjective robust controller using difference signals and multiple CAN2s by means of applying it to linear model plants. From the result of numerical experiments as well as theoretical analysis, the following properties are obtained. (1) The dimension of the input vector to select the associative matrix of the CAN2 to approximate the Jacobian matrix of the plant to be controlled does not have to be enlarged, which may reject the conjecture shown in [2,3] that the enlargement is necessary for the present method. (2) The present controller using fixed GPC parameters provides various control performances by means of involving errors of associative

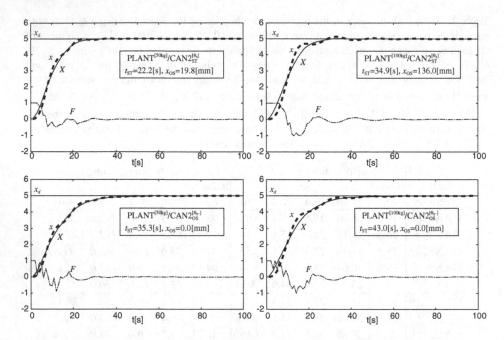

Fig. 2. Examples of time course of x[m], X[m] and F[10N]. Among the control of all test plants, the results of the smallest and biggest settling time by multiple $\mathrm{CAN2}_{\mathrm{ST}}^{[\theta_S]}$ are shown on the top left and right, respectively, and those of the smallest and biggest settling time without overshoot ($x_{\mathrm{OS}} = 0$[mm]) by $\mathrm{CAN2}_{\mathrm{OS}}^{[\theta_{S''}]}$ are shown on the bottom left and right, respectively.

matrices of CAN2s to learn Jacobian matrices, which enables the controller to be multi-objective robust controller by means of switching CAN2s. (3) Plant-parameter-specific overshoots can be reduced by the augmentation of CAN2s, which has also been shown possible for nonlinear plants [3].

Acknowledgement. This work was supported by JSPS KAKENHI Grant Number 24500276.

References

1. Kurogi, S., Yuno, H., Nishida, T., Huang, W.: Robust control of nonlinear system using difference signals and multiple competitive associative nets. In: Lu, B.-L., Zhang, L., Kwok, J. (eds.) ICONIP 2011, Part III. LNCS, vol. 7064, pp. 9–17. Springer, Heidelberg (2011)
2. Huang, W., Kurogi, S., Nishida, T.: Robust controller for flexible specifications using difference signals and competitive associative nets. In: Huang, T., Zeng, Z., Li, C., Leung, C.S. (eds.) ICONIP 2012, Part V. LNCS, vol. 7667, pp. 50–58. Springer, Heidelberg (2012)
3. Huang, W., Kurogi, S., Nishida, T.: Performance Improvement Via Bagging Competitive Associative Nets for Multiobjective Robust Controller Using Difference Signals. In: Lee, M., Hirose, A., Hou, Z.-G., Kil, R.M. (eds.) ICONIP 2013. LNCS, vol. 8226, pp. 319–327. Springer, Heidelberg (2013)

4. Kurogi, S., Nishida, T., Sakamoto, T., Itoh, K., Mimata, M.: A simplified competitive associative net and a model-switching predictive controller for temperature control of chemical solutions. In: Proc. of ICONIP 2000, pp. 791–796 (2000)
5. Kurogi, S., Ren, S.: Competitive associative network for function approximation and control of plants. In: Proc. NOLTA 1997, pp. 775–778 (1997)
6. Kohonen, T.: Associative Memory. Springer (1977)
7. Ahalt, A.C., Krishnamurthy, A.K., Chen, P., Melton, D.E.: Competitive learning algorithms for vector quantization. Neural Networks 3, 277–290 (1990)
8. Deb, K.: Multi-objective Optimization Using Evolutionary Algorithms. John Wiley & Sons (2009)
9. Clarki, D.W., Mohtadi, C.: Properties of generalized predictive control. Automatica 25(6), 859–875 (1989)
10. Kurogi, S., Sawa, M., Ueno, T., Fuchikawa, Y.: A batch learning method for competitive associative net and its application to function approximation. In: Proc. of SCI 2004, vol. 5, pp. 24–28 (2004)

Calibrating Independent Component Analysis with Laplacian Reference for Real-Time EEG Artifact Removal

Hussein A. Abbass[1,2]

[1] School of Engineering and Information Technology, The University of New South Wales, ADFA, Canberra, ACT 2600, Australia
{h.abbass@adfa.edu.au}
[2] Department of Electrical and Computer Engineering, National University of Singapore, 4 Engineering Drive, 117583, Singapore

Abstract. Independent Component Analysis (ICA) has emerged as a necessary preprocessing step when analyzing Electroencephalographic (EEG) data. While many studies reported on the use of ICA for EEG, most of these studies rely on visual inspection of the signal to detect those components that need to be removed from the signal. Little has been done on how to process EEG data in real-time, autonomously, and independent of a human expert inspecting the data. A few attempts have been made in the literature to design standard procedures on the processing of EEG data in real-time environments. To enable standardization to occur, the work and discussion of this paper focus on understanding the impact of different preprocessing steps on the performance of ICA. A proposed cut-off threshold for ICA is demonstrated to produce reliable and sound processing when compared to a Laplacian reference system. A methodology for real-time processing that is simple and efficient is being suggested.

Keywords: Electroencephalography, Independent Component Analysis, EEG Preprocessing.

1 Introduction

Independent Component Analysis (ICA) has been used widely for removing artifacts. A carefully designed experiment [1] using Magnetoencephalographic (MEG) data demonstrated that ICA can detect and isolate eye movement, eye blinking, cardiac, myographic, and respiratory artifacts. The data in this study was bandpass filtered at 0.03-90Hz for MEG, and 0.1-100Hz for Vertical and Horizontal electrooculography (EOG), and Electrocardiogram (ECG). It was then digitally low-pass filtered with a cutoff frequency of 45Hz. A second study [2] demonstrated that ICA can isolate ocular artifact. The author claimed that it is better to use ICA to isolate this type of artifact than measuring the artifact using an EOG then subtracting it from the EEG signal. The latter can remove proper EEG data as well.

These studies rely on visual inspection of the components to determine which of them contains an artifact, there are two main problems that still remain unsolved. Firstly,

C.K. Loo et al. (Eds.): ICONIP 2014, Part III, LNCS 8836, pp. 68–75, 2014.

how to remove artifact automatically? This is especially critical in real-time adaptive automation and augmented cognition applications [3,4], where the EEG signals need to be analyzed autonomously without any interference from human experts. A few attempts has been made to answer this question in [5,6], where the issue of automatic removal of artifact was addressed while the issue of speed for real-time application was not studied. Secondly, there has been no theoretical proof that those independent components capturing the artifacts mentioned in previous studies [1,2] do not also capture legitimate EEG information. In fact, the claim presented in [2] that ICA is better in isolating ocular artifact than simply subtracting the EOG signal from the corresponding EEG was only substantiated with a synthetic example. The example was too simple; therefore, was not representative as we will see in the remainder of this paper.

In addition to the above challenges, previous work would normally rely on a synthetic data that starts with n sources and generates n equal number of mixed signals. In EEG, this is almost never the case. It is common knowledge, for example, that an electrode would be sensing many sources and artifacts simultaneously. While we are not attempting to do localization of sources, which is a different problem all together, it is important to consider the fact that the number of signals/channels will always be smaller than the number of sources within the EEG domain. In this paper, we study the impact of this assumption on the performance of ICA.

The primary aim of this paper is to establish a heuristic that can guide the process of cleaning EEG in real-time operations. Common electromyogram (EMG) artifact removal techniques are first discussed in Section 2, followed by the methodology in Section 3, results in Section 4 and conclusion in Section 5.

2   Common EMG Removal Techniques

2.1   Independent Component Analysis

Over two decades of research on ICA, alternatively known as a technique for the source separation problem, have passed, while the technique is still finding more and more applications. In the problem of source separation, one can imagine multiple people talking in a cocktail party. Signals obtained from distributed independent microphones will be a mixture of all voices and background noise. The source separation problem attempts to find a linear transformation from the collected mixed signals to the original sources. If x, c, and n are random vectors representing the mixed signals, independent components, and a noise source, respectively, and M is a linear transformation matrix, the problem can be formulated mathematically as follows: $x = Mc + n$.

Recovering the original components, c, is not possible because of the noise term [7]. Instead, the model can be rewritten as: $x = As$, where s denotes a source. Assuming a number of realizations of the vector x in the form of a matrix X, the objective is to find the mixing matrix, A, to recover the sources (ie. Components), S as follows: $X = AS$.

Once A is calculated, one can find the separating matrix, W, to estimate the sources, $\hat{S} = W^T X$. Since we can only measure the mixed signal X, we have many more unknowns than known variables. Fortunately, it turns out that we only need to make the following two assumptions to be able to solve this system of linear equations [8]:

1. The components are statistically independent; that is, $Prob(c_i, c_j) = Prob(c_i) * Prob(c_j)$, where $Prob(c_i, c_j)$ denotes the joint probability distribution between any two different components i and j.
2. All independent components have non-gaussian distribution; although, if only one of them has a gaussian distribution, the rest can still be recovered.

The fast ICA algorithm (fastica) [8] is an efficient algorithm for estimating the independent components. The algorithm used in fastica performs two preprocessing steps by default. First, each x is centered on zero by subtracting the mean. Second, the vector x is whitened by transforming it into a new vector that is white. In essence, this guarantees that the inputs to the independent component algorithm are uncorrelated. ICA does not take the order of the data into account, it works on a random vector and is not designed to take the time-ordered signal information into consideration.

2.2 Referencing Techniques

Signals in the brain are measured by their electric potential difference volts. The EEG data can be referenced in many different ways, including ear-lobe (unipolar), or re-referencing to one of the EEG electrodes (bipolar). However, the latter case would eliminate one channel from the data; thus reducing the number of signals available for analysis. While bipolar measurements are common in classical neuro-feedback studies, there seems to be no advantage in using this type of referencing in quantitative EEG studies.

Two most common referencing methods are the Common Average Reference (CAR) and the Laplacian filter. Theoretically, CAR works best with many electrodes. However, McFarland et.al. [9] demonstrated that CAR and Laplacian filters were highly correlated when a 19-electrode according to the 10-20 standard measurement system is used.

In CAR, all EEG readings at each time step are averaged. All electrodes are referenced to this common average by subtracting it from all electrodes. It is calculated using the equation

$$x_i^{CAR} = x_i - \sum_j \left(\frac{x_j}{n} \right)$$

where x_i representing the electrical potential difference between the electrode and the ear (or otherwise) reference, and x_i^{CAR} representing the signal filtered with CAR.

The Laplacian filter relies on finite differences to approximate the second derivative of the instantaneous spatial voltage distribution. In essence, a Laplacian filter is local, while a CAR filter is global. The equation of Laplacian is:

$$x_i^{LaP} = x_i - \sum_{j \in N(i)} \left(\frac{x_j}{d_{ij} \times \sum_{k \in N(i)} \left(\frac{1}{d_{ik}} \right)} \right)$$

where, $N(i)$ is the set of electrodes in the neighborhood of electrode i, and d_{ik} represents the distance between electrodes i and k. For equal distances, and letting $|N(i)|$ denoting the neighborhood size of electrode i, the formulae is reduced to

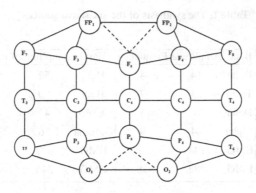

Fig. 1. Laplacian referencing network

$$x_i^{LaP} = x_i - \sum_{j \in N(i)} \left(\frac{x_j}{|N(i)|} \right)$$

The neighborhood structure used in this paper is presented in Figure 1. The dotted lines from F_z to FP_1 and FP_2 are used in the absence of FP_z. In this case, FP_1 and FP_2 are averaged to approximate the value for FP_z used in the filtering process of F_z. The same holds in the case of the dotted lines from P_z to O_1 and O_2 in the absence of O_z. The distances used in this paper are 6cm.

3   Methodology

A synthetic data set is designed for the analysis in this paper. We first assume 6 sources of signals. Two of the sources are assumed to be EMG operating at high frequency and overlapping with Beta and Gamma bands. We intentionally do not use a low-pass filter in a preprocessing step because the overlap with the Gamma band would remove EEG signals during filtering. Each source operates with a mixture of two frequencies representative of classic EEG bands as shown in Table 1.

The sampling rate is 256Hz; while it is greater than the required Nyquist frequency, we needed to standardize it in our work to have comparable results. We sample 2 minutes of data. We assume that the fifth source was activated in the last 250ms of every second, and the sixth source was activated in the last 500ms of every second; thus creating aperiodic EMG interference. All other sources were periodic and were activated from time 0. This setup was formulated to mimic situations we encountered in real EEG signals.

Pearson correlation coefficient was calculated between all sources. It was close to zero ($< \pm 0.06$) in all cases; indicating that the sources are at least uncorrelated.

The six sources are mixed into four signals, x_1, x_2, x_3, x_4, as follows:

$$x_1 = s_1 + 0.9 * s_5$$

$$x_2 = s_2 + 0.9 * s_6$$

Table 1. The synthesis of the six signal sources

Source ID	Band	Amplitude	Frequency	Band	Amplitude	Frequency
s_1	Delta	14	4	Beta	52	22
s_2	Theta	23	7	Beta	70	19
s_3	Delta	16	5	Alpha	43	11
s_4	Alpha	44	9	Gamma	56	47
s_5	EMG	144	31	EMG	337	51
s_6	EMG	282	28	EMG	246	49

$$x_3 = s_3 + s_5$$
$$x_4 = s_4 + s_6$$

The two odd-numbered mixed signals share the same EMG source with different weights, while the even-numbered mixed signals share a different EMG source. This is to mimic a left and right hemisphere electrode positions accompanied with left and right local muscle movements.

All signals and sources are preprocessed to have zero mean and unit variance to eliminate differences in magnitude and facilitate the ICA calculations. Pearson correlation coefficient between the original sources and mixtures and the estimated independent components is shown in Table 2.

Table 2. Pearson correlation coefficient between the original sources and mixtures and the estimated independent components

	s_1	s_2	s_3	s_4	s_5	s_6	x_1	x_2	x_3	x_4
$\hat{s}_1$	-0.3	0.2	-0.4	0.2	-0.7	0.4	-0.7	0.4	-0.8	0.5
$\hat{s}_2$	0.2	0.2	0.3	0.5	0.4	0.7	0.4	0.6	0.4	0.8
$\hat{s}_3$	0	0.8	0	-0.6	0.1	0.1	0	0.7	0.1	-0.4
$\hat{s}_4$	-0.8	0	0.6	0	-0.1	0	-0.6	0	0.4	0

We notice that the estimated components 1 and 2 are highly correlated with all mixtures. They are also highly correlated with the fifth and sixth sources representing the EMG. This suggests that, if the artifact impacts multiple signals, the associated components would be highly correlated with all impacted signals. All estimated components are visualized in Figure 2, where it is clearly shown that indeed the first and second estimated components are contaminated with the artifact.

As the correlation coefficient can be positive or negative, the sum of squared correlations between a component and mixtures is a good indicator for artifacts. If this sum is

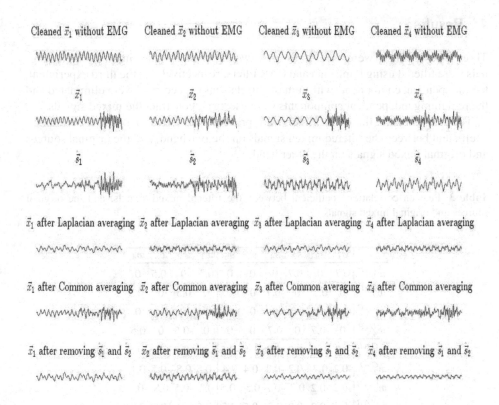

Fig. 2. Original and correct signals

greater than 1.0-1.5, there is a high probability that this component should be removed as an artifact. The logic behind this is as follows. An independent component is unlikely to have a perfect correlation close to either ends of ±1 with an EEG signal given that the close proximity of electrodes generate an overlap in the captured signals. Also, noise and artifact sources would make it almost impossible for an independent component to be perfectly correlated with a signal.

Therefore, it is reasonable to expect that the highest correlation will be around 0.9 or less; in which case, it is unlikely that this component will be highly correlated with a second EEG signal except when both signals share an artifact. Squaring the previous correlation coefficient would reduce the value down to 0.8. For the sum to exceed 1.0-1.5, the squared correlations from all other electrodes should sum to a value greater than 0.2-0.7. This requires either relatively high correlations with some other signals or a close to a uniform medium level correlation with the rest of the signals. The adjustment of this threshold would depend on the number of electrodes used. As the number of electrodes increases, the number of electrodes with high correlation with an independent component increases because of the closer proximity of electrodes. Therefore, this threshold is safer to be chosen higher than 1.0 as the number of electrodes increases.

4 Results

Three types of filters were used. In the first two experiments, the simulated mixed signals were filtered using Laplacian and CAR filters, respectively. In the third experiment, the independent components with sum of correlations exceeding 1 were eliminated and the remaining independent components were used to reconstruct the mixed signals.

The results of the three experiments are presented in terms of Pearson correlation coefficient between the filtered mixed signals on the one hand, and the original sources and original mixed signals on the other hand.

Table 3. Pearson correlation coefficient between the filtered mixed signals, and the original sources and original mixed signals

	s_1	s_2	s_3	s_4	s_5	s_6	x_1	x_2	x_3	x_4
x_1^{LaP}	0.7	0	-0.7	0	0	0	0.5	0	-0.5	0
x_2^{LaP}	0	0.7	0	-0.7	0	0	0	0.5	0	-0.5
x_3^{LaP}	-0.7	0	0.7	0	0	0	-0.5	0	0.5	0
x_4^{LaP}	0	-0.7	0	0.7	0	0	0	-0.5	0	0.5
x_1^{CAR}	0.7	-0.2	-0.2	-0.2	0.4	-0.5	0.8	-0.5	0.2	-0.5
x_2^{CAR}	-0.2	0.7	-0.2	-0.2	-0.4	0.4	-0.5	0.8	-0.5	0.1
x_3^{CAR}	-0.2	-0.2	0.7	-0.2	0.5	-0.5	0.2	-0.5	0.8	-0.5
x_4^{CAR}	-0.2	-0.2	-0.2	0.7	-0.5	0.5	-0.5	0.1	-0.5	0.8
x_1^{IC}	0.8	0.1	-0.6	0	0.1	0	0.6	0.1	-0.4	0
x_2^{IC}	0	0.8	0	-0.6	0.1	0.1	0.1	0.7	0.1	-0.4
x_3^{IC}	-0.8	0.1	0.6	-0.1	-0.1	0	-0.6	0.1	0.4	-0.1
x_4^{IC}	0	-0.8	0	0.6	-0.1	-0.1	0	-0.7	-0.1	0.4

The Laplacian filter has the best performance since it has a zero correlation with the two EMG sources: s_5 and s_6. CAR has the worst performance, which is expected since the number of mixed signals is few and the current was not calculated over a closed surface for CAR to work. However, CAR plays a secondary role in this study given the small number of signals, where it can be seen as a Laplacian over a larger area.

The important piece of result is that the heuristic used for determining those IC to be excluded from the signal worked perfectly well. While there remains a small correlation between the filtered mixed signal and the two EMG sources, the correlation between each filtered mixed signal and its corresponding original signal is higher than that with other original signals.

In other words, the averaging occurred with Laplacian, while eliminated the EMG signal, it simply distributed equally the EEG signal information of each mixed signal across other mixed signals in its neighborhood. Therefore, x_1^{LaP} is correlated with

x_1 positively and equally negatively with x_3. ICA, on the other hand, maintained maximum correlation with the original signal, where the highest positive correlation between x_1^{IC} and all original mixed signals occurs with x_1.

5 Conclusion

It is shown that the impact of artifacts from EMG signals can be eliminated or significantly reduced using independent component analysis as well as Laplacian filter. The latter eliminated the impact completely but generated signals that are equally correlated with its source and non-sources. Moreover, there is a large assumption that the neighborhood for Laplacian is chosen properly. The filter does not produce the intended result as demonstrated with the results of the common average reference, which in our chosen small example can be seen as a Laplacian with a larger than appropriate neighborhood. In practice, such a perfect neighborhood for Laplacian is neither known or possible.

Independent component analysis, on the other hand, is more robust in separating EMG sources. While a small residual may remain, it is minimum and the filtered data is maximally correlated with the original mixed signal.

Space constraints did not allow the presentation of results on real EEG data. However, the findings are similar in the sense that, independent component analysis maintained the original information efficiently. For future work, we are developing more studies to validate the threshold used in this work to eliminate those independent components that are suspects of containing artifacts.

References

1. Vigário, R., Jousmäki, V., Hämäläninen, M., Hari, R., Oja, E.: Independent component analysis for identification of artifacts in magnetoencephalographic recordings. Advances in Neural Information Processing Systems, 229–235 (1998)
2. Vigário, R.: Extraction of ocular artefacts from EEG using independent component analysis. Electroencephalography and Clinical Neurophysiology 103(3), 395–404 (1997)
3. Abbass, H., Tang, J., Amin, R., Ellejmi, M., Kirby, S.: Augmented cognition using real-time EEG-based adaptive strategies for air traffic control. In: International Annual Meeting of the Human Factors and Ergonomic Society, HFES. SAGE (2014)
4. Abbass, H., Tang, J., Amin, R., Ellejmi, M., Kirby, S.: The computational air traffic control brain: Computational red teaming and big data for real-time seamless brain-traffic integration. Journal of Air Traffic Control 56(2), 10–17 (2014)
5. Mognon, A., Jovicich, J., Bruzzone, L., Buiatti, M.: Adjust: An automatic eeg artifact detector based on the joint use of spatial and temporal features. Psychophysiology 48(2), 229–240 (2011)
6. Nolan, H., Whelan, R., Reilly, R.: Faster: Fully automated statistical thresholding for eeg artifact rejection. Journal of Neuroscience Methods 192(1), 152–162 (2010)
7. Comon, P.: Independent component analysis, a new concept? Signal Processing 36(3), 287–314 (1994)
8. Hyvärinen, A., Oja, E.: Independent component analysis: algorithms and applications. Neural Networks 13(4), 411–430 (2000)
9. McFarland, D.J., McCane, L.M., David, S.V., Wolpaw, J.R.: Spatial filter selection for eeg-based communication. Electroencephalography and Clinical Neurophysiology 103(3), 386–394 (1997)

Unsupervised Segmentation Using Cluster Ensembles

Wei Zhang[1], Jie Yang[1,*], Wenjing Jia[2], Nikola Kasabov[3],
Zhenhong Jia[4], and Lei Zhou[1]

[1] Institute of Image Processing and Pattern Recognition, Shanghai Jiao Tong University, China
[2] School of Computing and Communications, University of Technology, Sydney, Australia
[3] Knowledge Engineering and Discovery Research Institute, Auckland University of
Technology Auckland, New Zealand
[4] School of Information Science and Engineering, Xinjiang University, Urumqi, China
{zhangwei666,jieyang,311821-8.15}@sjtu.edu.cn,
Wenjing.Jia@uts.edu.au, nkasabov@aut.ac.nz, jzhh@xju.edu.cn

Abstract. We propose a novel framework for automatic image segmentation. In this approach, a mixture of several over-segmentation methods are used to produce superpixels and then aggregation is achieved using a cluster ensemble method. Generated by different existing segmentation algorithms, superpixels can describe the manifold patterns of a natural image such as color space, smoothness and texture. We use them as the initial superpixels. Grouping cues which affect the performance of segmentation can also be captured. After the over-segmentation, the simultaneous collection of superpixels is expected to achieve synergistic effects and ensure the accuracy of the segmentation. For this purpose, cluster ensemble methods are used to process the initial segmentation results and produce the final result. Our method achieves significantly better performance on the Berkeley Segmentation Database compared to state-of-the-art techniques.

Keywords: segmentation, superpixels, cluster ensembles, LDAPPA, multi-label.

1 Introduction

Image segmentation is a fundamental computer vision problem, which has wide applications in computer vision area.

It is challenging to segment images accurately and efficiently. In the past few years, many methods have been developed to address this problem. For example, Comaniciu and Meer's Mean Shift [6] is a general nonparametric technique, which is proposed for the analysis of a complex multimodal feature space and to delineate arbitrarily shaped clusters in it. Felzenszwalb and Huttenlocher's FH [4] makes greedy decisions and produces segmentations that satisfy global properties. Shi and Malik's Ncuts [7] measures both the total dissimilarity between different groups as well as the total similarity within the groups. Arbelaez and Fowlkes's OWT- UCM [8]

---

* Corresponding author.

C.K. Loo et al. (Eds.): ICONIP 2014, Part III, LNCS 8836, pp. 76–84, 2014.

consists of generic machinery for transforming the output of any contour detector into a hierarchical region tree. Hoiem, Efros and Hebert's algorithm [5] re-estimates boundary strength as the segmentation progresses, which is based on the agglomerative merging.

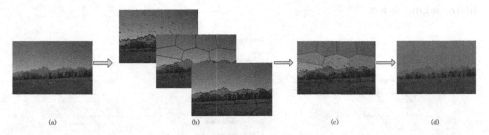

Fig. 1. System overview. Given an input image (a), we divide it into superpixels (b) using three segmentation methods respectively, i.e., FH [4], Mean Shift [6], and Gabor-texture [18]. Then, synthesizing these tree resultant images, we get the co-segmentation result (c). Finally, we get the output image (d) through cluster ensembles.

In order to enhance the reliability of over-segmentation, especially for images with complex background, and to improve the quality of segmentation, we propose an automatic image segmentation framework. It integrates three segmentation methods, i.e., Mean Shift [6], Gabor-Texture [18], and FH [4] to obtain the initial superpixels and multiple labels in those methods. Then, we use the cluster ensemble method Labels and Distances based Affinity Propagation Partitioning Algorithm (referred to as LDAPPA) to get the synergistic effect of segmentation and improve the accuracy of the segmentation.

The main contributions of this paper are summarized as follows:

1. We show that cluster ensemble can be highly powerful on image segmentation. Through cluster ensemble, we can get a good understanding of the complementary relationship between different segmentation methods, which is especially important in feature extraction.
2. Compared to state-of-the-art techniques, we have achieved competitive results on the Berkeley Segmentation Database. The performance is quantified using four criteria: PRI, BDE, VoI and GCE. Using the proposed framework, PRI rises from 0.7735 to 0.8059, and BDE reduces from 13.3087 to 12.0407. The definitions of those criteria will be given in Section 4.

2   Segmentation Using Cluster Ensemble

As shown in Fig. 1, our approach is composed of three parts: over segmentation, multiple labeling and cluster ensemble. The over segmentation and multiple labeling share the same segmentation results. In this section, we will illustrate each of them.

2.1   Over Segmentation

A standard image preprocessing step in many segmentation algorithms is to partition an input image into a set of superpixels [10], which are referred to as "perceptually meaningful atomic regions". In most existing works, superpixels are exploited to initialize segmentation.

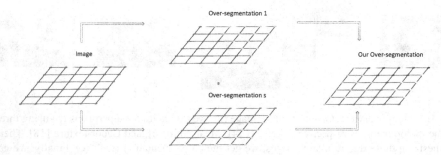

Fig. 2. Our over-segmentation model. In this paper, over-segmentation number s equals 3.

According to the definition of superpixel, being meaningful and atomic are two important properties. In this paper, we consider superpixels as atomic so that the superpixels, treated as basic unit elements, can be labeled by different segmentation methods. As a matter of fact, the results of other segmentation algorithms can be treated as superpixels as well. Those segmentation results satisfy the definition of the superpixel, which are both meaningful and atomic on the local information. For the integrity of this paper, the segmentation methods we used are briefly introduced in the next sub-section.

The resultant superpixels can describe the manifold patterns of a natural image. In our framework, we propose to combine the results of those segmentation methods to get superpixels. At first, we exploit superpixels with N segmentation methods. Formally, we denote B^i as the boundaries of the segmentation result of input image $I(i = 1,2, \cdots s)$. Let B be the boundaries of over-segmentation, where $B = B^1 \cup B^2 \cup \cdots \cup B^s$. According to the over-segmentation boundaries B, the input image I is partitioned into superpixels. Denote S as the set of superpixels of I, where $S = \{s_j\}_{j=1}^n$. n is the number of superpixels of image I. By now, we have got the basic unit elements used in the rest of this paper. The model of our over-segmentation method is shown in Fig. 2.

2.2   Multiple Labeling

The superpixels, generated by different over-segmentation algorithms, can describe the manifold patterns of a natural image such as color space, smoothness and texture. According to those features, images can be labeled diversely. The universality of all of the features can differentiate an object from its surroundings. The segmentation methods that we used are briefly illustrated as follows, i.e., Felzenszwalb and Huttenlocher's FH approach [4], Comaniciu and Meer's Mean Shift based segmentation [6], and Gabor-texture [18].

FH Segmentation

According to the FH segmentation approach, $G = (V, E)$ is an undirected graph with vertices $u \in V$ corresponding to the set of elements to be segmented, and edges $(v_i, v_j) \in E$ corresponding to pairs of neighboring vertices. Each edge $(v_i, v_j) \in E$ has a corresponding weight $w(v_i, v_j)$, which is a non-negative measure of the dissimilarity between two neighboring elements v_i and v_j. The weight of an edge is some dissimilarity measure between the two pixels connected by that edge (e.g., the difference in intensity, color, motion, location or some other features). In FH, they use the absolute intensity difference between pixels as edge weight, i.e.,

$$w(v_i, v_j) = |I(p_i) - I(p_j)|. \tag{1}$$

Segmentation Using Mean Shift

For a color input image I, geographic coordinates (g_x, g_y) and color tion(r, g, b) constitute a 5-D space. Let $K(x)$ denote a kernel function that indicates
how much x contributes to the estimation of the mean. Then, the sample mean m at x with kernel K is given by:

$$m(x) = \frac{\sum_{i=1}^{n} K(x - x_i)x_i}{\sum_{i=1}^{n} K(x - x_i)}. \tag{2}$$

The difference $m(x) - x$ is called mean shift. The main idea of Mean Shift algorithm is to iteratively move data points to their mean, which means that in each iteration move $m(x)$ to x until $(x) \approx x$.

Texture Segmentation Using Gabor Filters

Texture segmentation is the process of partitioning an image into regions based on their texture [18]. To extract the texture feature of an image, Gabor filters are used. A Gabor function in the spatial domain is a sinusoidal modulated Gaussian. For a 2-D Gaussian curve with a spread of σ_x and σ_y in x and y directions and a modulating frequency of u_0, the real impulse response and frequency response of the filter are given as:

$$h(x, y) = \frac{1}{2\pi\sigma_x\sigma_y} \exp\left\{-\frac{1}{2}\left[\frac{x^2}{\sigma_x^2} + \frac{y^2}{\sigma_y^2}\right]\right\} \cos(2\pi u_0 x). \tag{3}$$

$$H(u, v) = exp\{-2\pi^2[\sigma_x^2(u - u_0) + \sigma_y^2 v^2]\} + exp\{-2\pi^2[\sigma_x^2(u + u_0) + \sigma_y^2 v^2]\}. \tag{4}$$

2.3   Cluster Ensembles

Cluster ensembles address the problem of combining multiple 'base clusters' of the same set of objects into a single consolidated cluster. Cluster ensembles have emerged as a much more powerful method than single clustering. Besides, it also improves both the robustness and the stability of unsupervised classification solutions [17].

In the previous subsection, we summarized the three segmentation techniques used in our approach. They produce both over-segmentation results and the labels of super-pixels. In this subsection, we will use this information as the input of the clustering ensembles.

For classical cluster ensemble algorithms such as CSPA, HGPA and MCLA [19], attention was focused on the resultant labels of different clustering methods. However, in image segmentation, the locations of pixels and superpixels are also important. In our cluster ensemble method, we combine the labels and the locations of superpixels.

Algorithm 1. Labels and Distances based Affinity Propagation Partitioning Algorithm

Input: The labels set, the superpixels set S and the location set $Locations$.
Output: A partition of S.

1: According to Eq. (5), compute τ_{ij} , $i \neq j, i, j = 1,2 \cdots, n$.
2: Calculate each $r(i, j)$ and $a(i, j)$ until a clustering center is found or the number of iterations is out of range
3: Re-label all the superpixels.

Similarity between superpixels can be estimated by the number of clusters shared by two superpixels and their locations. Formally, let us denote $X = \{x_1, x_2, x_3, \cdots, x_n\}$ as the n input data points, where $X = I$ in the case of segmentation, and apply different clustering algorithms to X. Denote the clustering result of S_i obtained using algorithm k as l_i^k, $L = \{l_i^k\}_{i=1,2\cdots,n}^{k=1,2\cdots,s}$ and $Distances = \{d_{ij}\}_{i,j=1,2\cdots,n}$, $\min_{i \neq j}(d_{ij}) = 1$. In this paper d_{ij} is defined as the minimum Euclidean distance between two superpixels. Using consensus function Γ to process $S_1, S_2, S_3, \cdots, S_n$, the similarity of superpixels, denoted by a matrix Γ, contains the values:

$$\Gamma = [\tau_{ij}] = \tau(S_i, S_j) = \begin{cases} -\sum_{k=1}^{s}\left\|(l_i^k - l_j^k)\right\| \times d_{ij} & \text{if } i \neq j \\ \sum_{i=1}^{n}\sum_{j=1,j\neq i}^{n}\tau_{ij} /n \times (n-1) & \text{if } i = j \end{cases} \tag{5}$$

According to the similarity matrix Γ, we apply the affinity propagation (AP) [20] to converge the superpixels and get the final labels. The responsibility $r(i, k)$ and availability $a(i, k)$ are iterated using the following rules:

$$r(i, j) \leftarrow s(i, j) - \max_{j' \ s.t. \ j' \neq j} \{a(i, j') + s(i, j')\}. \tag{6}$$

$$a(i, j) \leftarrow \min\{0, r(j, j) + \sum_{i' \ s.t.i' \neq \{i,j\}} \max\{0, r(i', j)\}\}. \tag{7}$$

We summarize our algorithm of cluster ensembles in Algorithm 1, which we call Labels and Distances based Affinity Propagation Partitioning Algorithm (LDAPPA) since it combines both the results of other segmentation methods and the locations of the superpixels. In our experiments, the performance of LDAPPA was better than CSPA, HGPA and MCLA.

3   Experimental Results

In this section, we evaluate the performance of our method on the Berkeley Database [16], which consists of 300 natural images of diverse scene categories (In the Berkeley Database, on average, five ground truths are available per image). We illustrate some of our results compared to other methods in Fig. 3. To get a better understanding of the performance of our method, we use four criteria to quantify the performance of different segmentation algorithms: Probabilistic Rand Index (PRI) [12], Variation of Information (VoI) [13], Global Consistency Error (GCE) [14] and Boundary Displacement Error (BDE) [15].

Table 1. Quantitative comparison of our algorithm with other segmentation methods over Berkeley database. The three best results are highlighted in bold for each criterion.

Methods	PRI	VoI	GCE	BDE
Region-Growing	0.7522	8.2724	**0.0486**	14.1395
FH(knn)	**0.7718**	**2.9423**	0.1773	14.5734
FH(adjacent)	**0.7735**	3.4037	0.1362	14.4551
Ncuts	0.6832	4.8527	0.3303	17.1833
Mean Shift	0.7476	7.3141	**0.0690**	14.2260
Gabor-texture	0.7078	**3.1157**	0.2668	**13.3087**
Best of CSPA, HGPA and MCLA	0.7467	4.0949	0.1857	**13.5845**
Our Method	**0.8058**	4.9186	**0.0943**	**12.0407**

According to the four criteria above, a segmentation result is deemed as better if its PRI is larger and the other three performance measures are smaller.

We compare the average scores of our approach and the nine benchmark algorithms, i.e., Region-Growing, FH [4], Ncuts [7], Mean Shift [6], Gabor-texture [18] and cluster ensemble methods CSPA, HGPA and MCLA. The scores are shown in Table 1, with the three best results highlighted in bold for each criterion. And the performance of our method can be further improved by introducing new segmentation results. As shown in Table. 1, our method has achieved a significantly better performance on the Berkeley Segmentation Database compared to state-of-the-art techniques, where PRI rises from 0.7735 to 0.8059, and BDE reduces from 13.3087 to 12.0407. We can see that our method ranks first in PRI and BDE by a large margin compared to other methods, and third in GCE.

Some segmentation examples can be visualized in Fig. 3 (j). Fig. 3 (i) illustrates the performance of our method with different combinations of other segmentation results. With the use of complementary segmentation methods, our co-segmentation result preserve the object details, which results in the reduction of BDE. Besides, cluster ensembles ensure the increase of PRI.

4 Conclusions

In this paper, we propose an automatic image segmentation framework. Firstly, we use over-segmentation to process the input images, so that our algorithm starts with superpixels rather than pixels. Then the segmentation methods we used can capture the various features of images. With those initial segmentation results we can get the synergistic effect and improve the accuracy of the segmentation. Our method has achieved significantly better performance on the Berkeley Segmentation Database compared to state-of-the-art techniques. For future work, we will explore how to incorporate high-level segmentation methods into the proposed segmentation method.

Fig. 3. Visual comparison of our algorithm with other segmentation methods. (a) Test images, (b) Ground Truth, (c) Region Growth, (d) FH, (e) Mean Shift, (f) Ncuts, (g) Gabor-texture, (h) Best of CSPA, HGPA and MCLA, and our method in (i) and (j) using different parameters, respectively.

Acknowledgements. This research is partly supported by NSFC, China (No: 61273258, 61105001), Ph.D. Programs Foundation of Ministry of Education of China (No.20120073110018).

References

1. Zhou, L., Gong, C., Li, Y., Qiao, Y., Yang, J., Kasabov, N.: Salient Object Segmentation Based on Automatic Labeling. In: Lee, M., Hirose, A., Hou, Z.-G., Kil, R.M. (eds.) ICONIP 2013, Part III. LNCS, vol. 8228, pp. 584–591. Springer, Heidelberg (2013)
2. Kim, T.H., Lee, K.M., Lee, S.U.: Learning full pairwise affinities for spectral segmentation. IEEE Transactions on Pattern Analysis and Machine Intelligence 35(7), 1690–1703 (2013)
3. Ren, Z., Shakhnarovich, G.: Image segmentation by cascaded region agglomeration. In: 2013 IEEE Conference on Computer Vision and Pattern Recognition (CVPR), pp. 2011–2018. IEEE (June 2013)
4. Felzenszwalb, P.F., Huttenlocher, D.P.: Efficient graph-based image segmentation. International Journal of Computer Vision 59(2), 167–181 (2004)
5. Hoiem, D., Efros, A.A., Hebert, M.: Recovering occlusion boundaries from an image. International Journal of Computer Vision 91(3), 328–346 (2011)
6. Comaniciu, D., Meer, P.: Mean shift: A robust approach toward feature space analysis. IEEE Transactions on Pattern Analysis and Machine Intelligence 24(5), 603–619 (2002)
7. Shi, J., Malik, J.: Normalized cuts and image segmentation. IEEE Transactions on Pattern Analysis and Machine Intelligence 22(8), 888–905 (2000)
8. Arbelaez, P., Maire, M., Fowlkes, C., Malik, J.: Contour detection and hierarchical image segmentation. IEEE Transactions on Pattern Analysis and Machine Intelligence 33(5), 898–916 (2011)
9. Fern, X.Z., Brodley, C.E.: Solving cluster ensemble problems by bipartite graph partitioning. In: Proceedings of the Twenty-First International Conference on Machine Learning, p. 36. ACM (July 2004)
10. Ren, X., Malik, J.: Learning a classification model for segmentation. In: Proceedings of the Ninth IEEE International Conference on Computer Vision, pp. 10–17. IEEE (October 2003)
11. Van den Bergh, M., Boix, X., Roig, G., de Capitani, B., Van Gool, L.: SEEDS: Superpixels extracted via energy-driven sampling. In: Fitzgibbon, A., Lazebnik, S., Perona, P., Sato, Y., Schmid, C. (eds.) ECCV 2012, Part VII. LNCS, vol. 7578, pp. 13–26. Springer, Heidelberg (2012)
12. Unnikrishnan, R., Pantofaru, C., Hebert, M.: Toward objective evaluation of image segmentation algorithms. IEEE Transactions on Pattern Analysis and Machine Intelligence 29(6), 929–944 (2007)
13. Meilă, M.: Comparing clusterings: an axiomatic view. In: Proceedings of the 22nd International Conference on Machine Learning, pp. 577–584. ACM (August 2005)
14. Martin, D., Fowlkes, C., Tal, D., Malik, J.: A database of human segmented natural images and its application to evaluating segmentation algorithms and measuring ecological statistics. In: Proceedings of the Eighth IEEE International Conference on Computer Vision, ICCV 2001, vol. 2, pp. 416–423. IEEE (2001)
15. Freixenet, J., Muñoz, X., Raba, D., Martí, J., Cufí, X.: Yet another survey on image segmentation: Region and boundary information integration. In: Heyden, A., Sparr, G., Nielsen, M., Johansen, P. (eds.) ECCV 2002, Part III. LNCS, vol. 2352, pp. 408–422. Springer, Heidelberg (2002)

16. Maire, M., Arbeláez, P., Fowlkes, C., Malik, J.: Using contours to detect and localize junctions in natural images. In: IEEE Conference on Computer Vision and Pattern Recognition, CVPR 2008, pp. 1–8. IEEE (June 2008)
17. Topchy, A.P., Jain, A.K., Punch, W.F.: A Mixture Model for Clustering Ensembles. In: SDM (April 2004)
18. Hammouda, K., Jernigan, E.: Texture segmentation using gabor filters. Center for Intelligent Machines, McGill University, Canada (2000)
19. Strehl, A., Ghosh, J.: Cluster ensembles—a knowledge reuse framework for combining multiple partitions. The Journal of Machine Learning Research 3, 583–617 (2003)
20. Frey, B.J., Dueck, D.: Clustering by passing messages between data points. Science 315(5814), 972–976 (2007)

Similar-Video Retrieval via Learned Exemplars and Time-Warped Alignment

Teruki Horie, Masafumi Moriwaki*, Ryota Yokote**, Shota Ninomiya,
Akihiro Shikano, and Yasuo Matsuyama***

Waseda University, Department of Computer Science and Engineering,
Tokyo, 169-8555, Japan
{t.horie,masa.m,rrryokote,nino225,a.shikano,yasuo}@wiz.cs.waseda.ac.jp
http://www.wiz.cs.waseda.ac.jp

Abstract. New learning algorithms and systems for retrieving similar
videos are presented. Each query is a video itself. For each video, a set
of exemplars is machine-learned by new algorithms. Two methods were
tried. The first and main one is the time-bound affinity propagation. The
second is the harmonic competition which approximates the first. In the
similar-video retrieval, the number of exemplar frames is variable according to the length and contents of videos. Therefore, each exemplar possesses responsible frames. By considering this property, we give a novel
similarity measure which contains the Levenshtein distance (L-distance)
as its special case. This new measure, the M-distance, is applicable to
both of global and local alignments for exemplars. Experimental results
in view of precision-recall curves show creditable scores in the region of
interest.

Keywords: Similar-video retrieval, exemplar, time-bound affinity propagation, M-distance, numerical label.

1 Introduction

Machine learning or computational intelligence has discovered its own new values
in the age of big data. Today, various types of unstructured data are continually
accumulated. A typical case can be found in videos. Advent of smart phones
made users produce and upload their own videos to the Web easily. However,
most of them are structured poorly. This hinders users from utilizing rich hidden
resources. The tendency is worse than the era of the static image retrieval [1] [2].
Reflecting this situation, efforts have been made on content-based approaches to
the video retrieval as is surveyed in [3]. In this paper, we give new machine learning methods for automatic exemplar extraction and novel similarity measures,
as well as their applications to similar-video retrieval.

---

* Currently with NS Solutions Corporation, Tokyo, Japan.
** Currently with Donuts, Co., Tokyo, Japan.
*** This work was supported by the Grant-in-Aid for Scientific Research 26330286, and
Waseda University Special Research Projects 2013A-6248, 2013B-094, 2014K-6129.

C.K. Loo et al. (Eds.): ICONIP 2014, Part III, LNCS 8836, pp. 85–94, 2014.

The organization of this paper is as follows. In Section 2, we give a novel learning algorithm called time-bound affinity propagation (TBAP). This has the frame-order awareness which cannot be realized by the original affinity propagation (AP) of [4]. This is an unsupervised learning algorithm which finds representative frames in a video. In Section 3, we give a class of new similarity measures with time-warping which includes the Levenshtein distance (L-distance) [5], the Needleman-Wunsch algorithm [6] and the Smith-Waterman algorithm [7] as its special cases. Such a new measure, the M-distance, is an important part of this paper's similar-video retrieval. In Section 4, a test set of videos is designed. We will prepare a data set which depends on subject's sensibility as less as possible. In Section 5, we give a class of alternative learning methods to the time-bound affinity propagation. That is based on the harmonic competition [8]. Section 6 gives concluding remarks.

2   Problem Description

2.1   Exemplars Reflecting Time Information

Let $\{x_t\}_{t=1}^n$ be a given time series. x_t can be any vector. In this paper, this is a feature vector series in terms of the color structure descriptor (CSD) of MPEG-7. The CSD is a patch-based histogram.

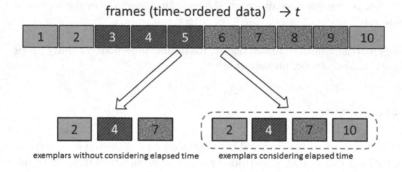

Fig. 1. Exemplars reflecting time information

Figure 1 illustrates a time series of frames $\{x_t\}_{t=1}^{10}$. Let this conceptual video have three similar frames of $\{1, 2, 10\}$, $\{3, 4, 5\}$ and $\{6, 7, 8, 9\}$. If time information or frame ordering were not considered, a learning system would choose only frames $\{2, 4, 7\}$ as exemplars (Fig. 1 bottom left). But, this is not appropriate as a label for the video retrieval. Rather, we want to have a learning algorithm to find $\{2, 4, 7, 10\}$ as exemplars (Fig. 1 bottom right). In this case, the exemplar set also gives responsible frames or dominant neighbors. For instance, we have $\{(1, 2, 0), (1, 4, 1), (1, 7, 2), (0, 10, 0)\}$. In the next section, we will give a new learning algorithm to obtain an order-aware exemplar set like Fig. 1 bottom right.

2.2 Time-Bound Affinity Propagation

Before going to the algorithm for the exemplar frame learning, it is necessary to have a right understanding about the following items.

(a) The plain affinity propagation algorithm can produce only the case of Fig. 1 bottom left. We need to obtain an algorithm which is aware of the frame ordering.

(b) Since we want to obtain exemplars, i.e., existing frames, the affinity propagation was set as a basic tool. However, the harmonic competition [8] which is a generalization of the vector quantization will also be applicable to the order-aware exemplar extraction.

In this section, we focus on item (a). Item (b) will be discussed in Section 5.

Our method to find order-aware exemplars is as follows.

[**Intra-Video Processing: Time-Bound Affinity Propagation (TBAP)**]

Step 1: A time series of feature vectors of video frames $\{x_t\}_{t=1}^n$ is given. Each vector is a normalized CSD histogram whose summation is unity. A similarity measure $s(x_i, x_j) \overset{\text{def}}{=} s(i, j)$ is given. Here, $s(k, i) > s(k, j)$ holds if and only if x_i is more similar to x_k than x_j. In our experiments, we will use

$$s(i, j) = \bar{D} - \|x_i - y_j\|. \tag{1}$$

Here, $\bar{D}$ is a constant which can be chosen by users.[1] Other design parameters appearing in subsequent steps are set here. A convergence criterion is also specified here.

Step 2: Prepare a matrix $A = [a_{ij}]$ whose initial value is a zero matrix O. This is called the availability.

Step 3: Pick up x_i, x_j, and x_k by considering their temporal ordering in a window.

Window length: Set a sliding window of length $2w - 1$.

Windowing (Order awareness property 1):
The following computation of the responsibility matrix and the availability matrix is computed for

$$i: \ 1 \leq i \leq n,$$
$$j: \ 1 \leq j \leq n \ \text{constrained by} \ i - w < j < i + w. \tag{2}$$

Responsibility matrix update (Order awareness property 2):
$R = [r_{ij}]$, which is symmetric, is updated by

$$\rho_{ij} := s(i, j) - \max_{k: \ k \neq j} \{a_{ik} + s(i, k)\}, \tag{3}$$

$$r_{ij} := (1 - \lambda)\rho_{ij} + \lambda r_{ij}. \tag{4}$$

---

[1] The choice of $\bar{D}$ does not affect the result of this TBAP which is an intra-video processing. Therefore, it can be set zero here. However, for the inter-video comparison based on the similarity, $\bar{D}$ becomes an important design parameter for users. Its default value will be discussed in Section 3.1.

Availability matrix update (Order awareness property 3):
The availability matrix A is updated by

$$\alpha_{ii} := \sum_{k \neq i} \max\{r_{ki}, 0\}, \tag{5}$$

$$\alpha_{ij} := \min\{0, \ r_{jj} + \sum_{k: \ k \neq i, \, k \neq j} \max\{r_{kj}, 0\}\}, \ (i \neq j), \tag{6}$$

$$a_{ij} := (1 - \lambda)\alpha_{ij} + \lambda a_{ij} \ \ (\text{including } i = j). \tag{7}$$

The design parameter $\lambda \in (0, 1)$ is a dumping factor.

Step 4 (Order awareness property 4): If a convergence criterion is not satisfied for $a_{ij} + r_{ij}$, then Step 3 is repeated. If the convergence is met,

$$\underset{j: \ i-w<j<i+w \ \text{for} \ 1 \leq i \leq n}{\arg\max} \ \{a_{ij} + r_{ij}\} \tag{8}$$

is adopted as an exemplar index. The final exemplar set is determined by collecting such indices.

Theoretical Consideration: The affinity propagation (AP) [4] was theoretically derived from the maximization of a similarity measure with a constraint of the node labeling in view of message passing. The labeling stands for the identification of exemplars. In this paper, however, each node has a sequence index. That is, the node is a frame of a video. Therefore, we added a constraint on the message passing so that the set of nodes is a time series. This is our TBAP algorithm.

3   Distance Measure and Similarity Comparison

3.1   Data Normalization and Distance Measure

The similarity measure $s(i, j)$ can be any as long as it leads to the convergence of the algorithm. We found that the form of equation (1) gives the convergence of the algorithm if λ is chosen appropriately. The bias $\bar{D}$ in Equation (1) has effects on inter-video comparison appearing in later sections.

Since each vector x_t is normalized to have only nonnegative elements whose summation makes unity, possible choices of $\bar{D}$ are as follows.

(a) The average of all possible data distances: This is acceptable only if the data size is small.
(b) A fixed choice of $\sqrt{2}$, $\sqrt{(d-2)/(2d)}$, or $\sqrt{1 - (1/d)}$: Here, d is the dimension of x_t which resides in a simplex. Note that $\sqrt{2}$ is the edge length of this simplex. $\sqrt{(d-2)/(2d)} \approx 1/\sqrt{2}$ is the radius of the interior sphere. $\sqrt{1 - (1/d)} \approx 1$ is the radius of the exterior sphere. In experiments, we will use the exterior radius with $d = 768$ which is our size of CSD bins by the HSV expression of the color space (Hue-Saturation-Value).

3.2 Similarity Comparison 1: M-distance for Global Alignment

Here, we give a method to compare two different videos with different exemplar sets. Although we use the terminology of videos, the method is applicable to any time series with exemplars. Our method generalizes the Levenshitein distance (L-distance) [5] of discrete text processing, and the Needleman-Wunsch algorithm (NW algorithm)[6] for the global alignment in bioinformatics. After the name of the L-distance, the similarity comparison below will be called M-distance for the global alignment.[2]

[Global Alignment and Retrieval]

Step 1: For the video v_A, sets of exemplars $\{e_i^A\}$ and accompanied dominance lengths by the relevance $\{E_i^A\}$, $(i = 1, 2, \cdots)$ are given. For the video v_B, similar sets are given. The similarity measure (1) is chosen here.

Step 2: Fill a global alignment table, and then backtrack a path by the following dynamic programming procedure.

(2-1) A gap penalty g is chosen as a design parameter.

(2-2) Make a table.

Fill the $\{i = 0\}$-th row by $(0, -gE_1^B, -g\sum_{j=1}^{2} E_j^B, -g\sum_{j=1}^{3} E_j^B, \cdots)$.

Fill the $\{j = 0\}$-th column by $(0, -gE_1^A, -g\sum_{i=1}^{2} E_i^A, -g\sum_{i=1}^{3} E_i^A, \cdots)$.

(2-3) Starting from the position of $(i, j) = (1, 1)$, fill elements by

$$f(i, j) = \max$$
$$\{f(i-1, j) - gE_i^A, f(i-1, j-1) + r(i,j)s(i,j), f(i, j-1) - gE_j^B\}. \quad (9)$$

To a cell which gave the maximum, an arrow is directed as a pointer. Here, $r(i, j)$ is a weight which reflects the exemplar dominance. We will use $r(i, j) = (E_i^A + E_j^B)/2$ in experiments. If we backtrack from the bottom right element of the value f_{last}, the path gives a global alignment.

Step 3: The similarity between v^A and v^B is computed by

$$u(A, B) = h(f_{\text{last}})/w(\sum_i E_i^A, \sum_j E_j^B). \quad (10)$$

Here, $h(x)$ is a monotone increasing function. w is an averaging function. The simplest one is an arithmetic mean.

3.3 Similarity Comparison 2: M-distance for Local Alignment

If video lengths are considerably different, the global alignment might deviate from human sensibility. In such a case, we use a local alignment which can compare the most similar parts. The following algorithm generalizes the Smith-Waterman algorithm for the local alignment [7] in bioinformatics. Only the difference from the global alignment is described.

[Local Alignment and Retrieval]

---

[2] The M-distances of Section 3.2 and Section 3.3 are due to the last and second authors, Matsuyama and Moriwaki.

Step 1: This step is the same as the global alignment.

Step 2: Fill a local alignment table and backtrack a path by the following dynamic programming procedure.

(2-1) A gap penalty g is chosen as a design parameter.

(2-2) Make a table. Fill elements in the $\{i = 0\}$-th row and $\{j = 0\}$-th column all by zero. Starting from the position of $(i, j) = (1, 1)$, fill elements by

$$f(i, j) = \max$$
$$\{0, \ f(i-1, j) - gE_i^A, f(i-1, j-1) + r(i, j)s(i, j), f(i, j-1) - gE_j^B\}. \ (11)$$

To a cell which gives a non-zero maximum, an arrow is directed as a pointer. We backtrack from the position of the largest value $f_{\max}$. This path gives a local alignment.

Step 3: This step is the same as the global alignment.

4   Experiments

4.1   Data Preparation

Since end users of the retrieval are human, the final similarity judgment strongly depends on their sensibility. Therefore, it is desirable that the similarity judgment depends on subjects as less as possible. But, such a simple set would be too easy to judge even by plain machines. Therefore, we designed a data set so that the following is satisfied.

(a) Source video data are totally unlabeled.

(b) Precision and recall on their retrieval can be judged mechanically.

(c) Each video possesses temporal changes of concepts so that the time-dependent property of Fig. 1 bottom right can be identified.

Fig. 2 and Fig. 3 illustrate the generation procedure of the source data.

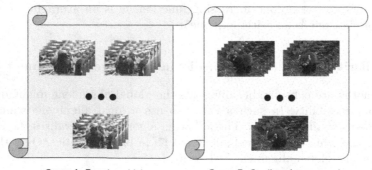

Group A: Peeping chick   Group B: Cautious lesser panda

Fig. 2. Groups of videos made from NHK Creative Library

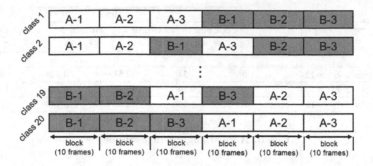

Fig. 3. Twenty classes of videos

Fig. 2 illustrates two groups of videos extracted from the NHK Creative Library which is royalty free [9]. One is a scene of a peeping chick (group A). The other is that of a cautious lesser panda (group B). From these groups, 20 classes of video films were generated as is illustrated in Fig. 3. Each class has 21 videos. Thus, there are 420 test videos, all of which are different each other.

4.2 Alignment Example

On the prepared video set, we tried the time-bound affinity propagation of Section 2.2 to find exemplars accompanied with responsible frames. This is an important step to give a *numerical annotation* to each video. It is equivalent to give a structure to the unorganized video set of Fig. 3 in terms of numerical tags. Such tags can be computed either on-line or off-line.

The M-distance is computed by the procedures of Section 3.2 or Section 3.3. Fig. 4 shows a global alignment by $\bar{D} = 1/\sqrt{1 - (1/d)}$, $d = 3 \times 256 = 768$, and $g = 0.05$. The backtracking starts from the last element. Fig. 5 illustrates a local alignment by the same $\bar{D}$ and g. Here, the backtracking starts from the largest value. This gave a local matching of Video A to a segment of Video B.

			Video B			
$\downarrow i$	$\xrightarrow{\quad} j$		exemplar 1 $E^B_1 = 8$	exemplar 2 $E^B_2 = 12$	exemplar 3 $E^B_3 = 10$	exemplar 4 $E^B_4 = 7$
			0.0 $\leftarrow$	-0.4 $\leftarrow$	-1.0 $\leftarrow$	-1.5 $\leftarrow$ -1.85
Video A	exemplar 1 $E^A_1 = 12$ $\uparrow$		-0.6 $\nwarrow$	9.54 $\nwarrow$	10.13 $\leftarrow$	9.63 $\leftarrow$ 9.28
	exemplar 2 $E^A_2 = 7$ $\uparrow$		-0.95 $\uparrow$	9.19 $\nwarrow$	17.81 $\nwarrow$	17.88 $\leftarrow$ 17.53
	exemplar 3 $E^A_3 = 10$ $\uparrow$		-1.45 $\uparrow$	8.69 $\nwarrow$	18.68 $\nwarrow$	26.95 $\leftarrow$ 26.60

Fig. 4. A global alignment example

4.3 Evaluation by Precision Recall Curves

Since the video data set was designed deliberately, a mechanical judgment of the precision (11-point interpolated precision) and recall is possible. Numerical

			Video B			
$\longrightarrow$ j $\downarrow$ i			exemplar 1 $E^B_1 = 8$	exemplar 2 $E^B_2 = 12$	exemplar 3 $E^B_3 = 10$	exemplar 4 $E^B_4 = 7$
		0	0	0	0	0
exemplar 1	$E^A_1 = 12$	0 ↖	9.54 ↖	10.53 ←	10.03 ←	9.68
exemplar 2	$E^A_2 = 7$	0 ↑	9.19 ↖	17.81 ↖	18.28 ←	17.93
exemplar 3	$E^A_3 = 10$	0 ↖	8.83 ↖	18.68 ↖	26.95 ↖	26.66

(Video A label at left of rows)

Fig. 5. A local alignment example

values are computed as follows.

$$\text{recall} = |\text{correct videos found}| \, / \, N_{\text{same\_class}} \qquad (12)$$

$$\text{precision} = |\text{correct videos found}| \, / \, |\text{top rank videos to be checked}| \qquad (13)$$

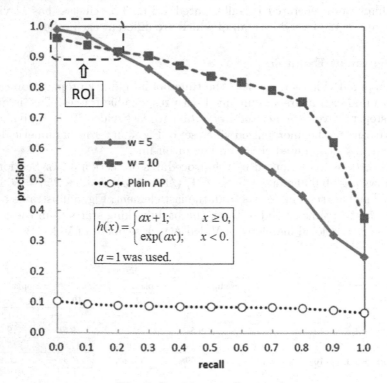

Fig. 6. Precision-recall curves

Fig. 6 illustrates the precision recall curves. We can find that, in its region of interest (recall $\leq 20\%$), the precision is very satisfactory since correct videos are almost always included. On the other hand, the plain AP is unsatisfactory since the time-bound property of Section 2.2 is not considered.

5 Alternative Methods

Here, we consider possibilities of other learning algorithms with the help of Table 1 which summarizes characteristics of exemplar finding methods. This table suggests that a competitive learning approach is possible as a version of the harmonic competition [8].

Table 1. Comparison of methods

Method	Elements	# of exemplars	Mode
time-bound AP	exemplar	variable	successive
harmonic competition	mean vector → exemplar	pre-specified	batch

Step 1: Data set $\{f(x_t, t)\}_{t=1}^n$ and the number of clusters are given.
Step 2: Iterations for learning are conducted until the convergence is met.
Step 3: The nearest frame to each centroid is regarded as an exemplar.

Comparison with TBAP: We conducted a set of preliminary experiments by $f(x_t, t) = [x_t, \alpha t]^T$ with $\alpha = 0.015$. Its performance was slightly inferior to the result of Fig. 6. But, the learning speed of a single run was much faster than the affinity propagation family.

6 Concluding Remarks

We presented algorithms and systems for the similar-video retrieval. Since a video has a large size, the set of exemplars and their responsible frames are usually computed off-line. Therefore, they can be used as numerical labels for structure information on a big data set. The reverse direction, or the retrieval, is fast by computing the M-distance.

In [4], it is pointed out that finding the exemplars has a relationship to the labeling by a mechanism of the Hopfield network [10]. After the structural and algorithmic speedup became mature, configurations using such a strategy (e. g., [11]) could be used.

References

1. Katsumata, N., Matsuyama, Y.: Database Retrieval for Similar Images Using ICA and PCA Bases. Engineering Applications of Artificial Intelligence 18, 705–717 (2005)
2. Matsuyama Laboratory: Waseda Image Searchable Viewer (2006), http://www.wiz.cs.waseda.ac.jp/rim/wisvi-e.html
3. Hu, W., Xie, N., Li, L., Zeng, X., Maybank, S.: A Survey on Visual Content-Based Video Indexing and Retrieval. IEEE Trans. SMC 41, 797–819 (2011)

4. Frey, B.J., Dueck, D.: Clustering by Passing Messages between Data Points. Science 315(5814), 972–976 (2007)
5. Levenshtein, V.I.: Binary Codes Capable of Correcting Deletions, Insertions, and Reversals. Soviet Physics Doklady 10(8), 707–710 (1966)
6. Needleman, S.B., Wunsch, C.D.: A General Method Applicable to the Search for Similarities in the Amino Acid Sequence of Two Proteins. J. Mol. Bio. 48, 443–453 (1970)
7. Smith, T.F., Waterman, M.S.: Identification of Common Molecular Subsequences. J. Mol. Biol. 147, 195–197 (1981)
8. Matsuyama, Y.: Harmonic Competition: A Self-Organizing Multiple Criteria Optimization. IEEE Trans. Neural Networks 7, 652–668 (1996)
9. NHK creative library, http://www1.nhk.or.jp/creative/
10. Hopfield, J.J.: Neural Networks and Physical Systems with Emergent Collective Computational Abilities. Proc. National Academy of Science 79, 2554–2558 (1982)
11. Cheng, L., Hou, Z.-G., Tan, M.: Relaxation Labeling Using an Improved Hopfield Neural Network. Lecture Notes in Control and Information Sciences (345), 430–439 (2006)

Automatic Image Annotation Exploiting Textual and Visual Saliency

Yun Gu[1], Haoyang Xue[1], Jie Yang[1,*], and Zhenhong Jia[2]

[1] Institute of Image Processing and Pattern Recognition, Shanghai Jiao Tong University, Shanghai, China
[2] School of Information Science and Engineering, Xinjiang University, Urumqi, China
{geron762,xuehaoyangde,jieyang}@sjtu.edu.cn,jzhh@xju.edu.cn

Abstract. Automatic image annotation is an attractive service for users and administrators of online photo sharing websites. In this paper, we propose an image annotation approach exploiting visual and textual saliency. For textual saliency, a concept graph is firstly established based on the association between the labels. Then semantic communities and latent textual saliency are detected; For visual saliency, we adopt a dual-layer BoW (DL-BoW) model integrated with the local features and salient regions of the image. Experiments on NUS-WIDE dataset demonstrate that the proposed method outperforms other state-of-the-art approaches.

Keywords: Image Annotation, Visual Saliency, Textual Saliency.

1 Introduction

With the explosive growth of web images, image annotation has drawn wide attentions in recent years. Given an image, the goal of image annotation is to analyze its visual content and assign the labels to it. Numerous approaches have been proposed for automatic image annotation. Search-based methods like [5,6] and learning-based methods like [11,9] are demonstrated with good performance on state-of-art datasets. However, most of them focus on learning with pre-extracted features while some works are dealing with the visual representation.[2] learns the probability distribution of a semantic class from images with weakly labeled information. In [7], the images are coded with sparse features via over-segmenatation for label-to-region annotation. In this paper, we focus on a combined task which provides better visual representation and annotation performance simultaenously.

Evidence from visual cognition researchers demonstrates that people are usually attracted with the salient object standing out from the rest of the scene[13]. Then, the rest of the scene will be recognized via the its visual features and concept correlation with the salient object. It naturally leads to the adoption

---

* Corresponding author.

C.K. Loo et al. (Eds.): ICONIP 2014, Part III, LNCS 8836, pp. 95–102, 2014.
© Springer International Publishing Switzerland 2014

of visual saliency model for image annotation. However, the number of images with region-wise labels is quite limited. In most cases, we can only get the images with some tags. Although the salient region can be extracted by some saliency detection methods, the corresponding "salient" tag is not easy to obtain.

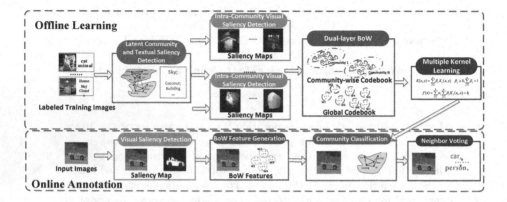

Fig. 1. The main framework of TVSA

In todays image annotation, the number of labels (i.e. concepts/tags) is quite large and label concurrence is pretty common. Intuitively, the non-salient objects,i.e. background scene, are likely to occur with the salient objects in various scenes. For instance, the tag "sky" may appear in urban views which is often associated with "road", etc. However, "sky" can also appear in outdoor scenes with "dog" and "trees",etc. Since these two scenes are quite different, we can infer that the label "sky" is an "background"(i.e. non-salient) tag. Therefore, the coherence of the label concurrence may reveal the textual saliency.

In this paper, a Textual-Visual Saliency based Annotation (TVSA) method is proposed for image annotation by learning training sample based on visual and textual saliency. Figure 1 illustrates our framework, which consists of two parts: offline learning and online annotation.

Offline Learning: Given the labeled training samples, a concept graph is firstly established by exploiting the association between the concepts. Then concept communities and latent textual saliency are detected from concept graph. In each community, the salient region of images are detected which is used for dual-layer Bag-of-Words (DL-BoW) generation. The community classifiers are trained with Multiple-Kernel SVM based on the local features (DL-BoW) and global features of training samples in each concept community.

Online Annotation: The DL-BoW feature is firstly generated for the un-labeled image. Then, corresponding community of the image is determined by the community classifier. Finally, neighbor-voting annotation is performed with training samples according to the result of community classification.

The rest of our paper is organized as follows: The main details of TVSA are described in Section 2; In Section 3, we evaluate the performance of TVSA with some other approaches. Finally, the conclusion is presented in Section 4.

2   Methodology

2.1   Textual Saliency Detection

The first step of TVSA is to construct a concept graph based on the tagged images. In this paper, we construct a directed-weighted graph $G = \{V, E\}$. The elements of vertex set V are tags from concept set $C = \{c_1, c_2, , c_m\}$. The concept c_i is connected with c_j by a directed edge e_{ij} if an image in training set is tagged with c_i and c_j at the same time. Let w_{ij} denote the weight of e_{ij} which implies the semantic correlation between two concepts and determined as follows:

$$w_{c_i, c_j} = P(c_j|c_i) = \frac{N(c_i, c_j)}{N(c_i)} \tag{1}$$

where $P(c_j|c_i)$ is the conditional probability of concept c_j given c_i, $N(c_i)$ stands for the number of images tagged with concept c_i in the image collection and $N(c_i, c_j)$ stands for the number of images tagged with concept c_i and c_j simultaneously.

Concepts which often appear in the same scene or have similar semantic characteristics are likely to be grouped into the same community. If an untagged sample is allocated to specific community, the concepts in this community are likely to be candidating labels for the image. In this paper, a fast unfolding algorithm [1] is applied to realize the latent community detection. It is proved a promising algorithm to generate proper communities under optimal time-complexity.

After latent community detection, each tag is assigned with the corresponding community. We define the correlation between tag (c_i) and community (COM_k) as follows:

$$Corr(c_i, COM_k) = \frac{1}{N_{COM_k}} \sum_{c_j \in COM_k} w_{c_j, c_i} = \frac{1}{N_{COM_k}} \sum_{c_j \in COM_k} \frac{N(c_i, c_j)}{N(c_i)} \tag{2}$$

where N_{COM_k} denotes the number of concepts in COM_k. The textual saliency of tag c_i assigned with COM_k is defined as:

$$Sal(c_i) = \frac{Corr(c_i, COM_k)}{\sum_{m=1}^{N_{COM}} Corr(c_i, COM_m)} \tag{3}$$

where N_{COM} denotes the number of communties. Sal_{c_i} indicates the intra-community correlatation and inter-community discrimination. With larger Sal_{c_i}, the tag c_i is likely to be asscociated only with COM_k,i.e. a salient tag. Given a textual saliency threshold T_{txt}, tags are divided into two sets with high saliency and low saliency respectively. Noted that we will assign the training samples with the corresponding community by voting on the number of salient tags.

2.2 Visual Saliency Detection

In each community, The visual saliency of a pixel refers to its relative attractiveness with respect to the whole image. To generate a saliency map for each image, a MATLAB implementation of Manifold Ranking-Based Visual Saliency[10] is applied to compute saliency values of pixels, with the values normalized to a range between 0 and 1. The higher the saliency value is, the more attractive an image pixel would be. As reported in [13], the salient portions often correspond to semantic objects in an image. Given a saliency value threshold T_{vis}, we can divide an image into two disjoint regions, one of high saliency and the other of low saliency. They will both be used to extract the visual words indicating the saliency-level.

2.3 Dual-Layer Bag of Salient Words

In our work, SIFT is adopted to extract the local features in training images. Firstly, we extract visual words according to region saliency in each community. Then, the global codebook is generated according to the community-wise codebook.

In the specific community,a $M \times N$ image I_k is featured with a saliency map $\{M_{k,m \times n}\}$, $m \leq M, n \leq N$ and n_k SIFT descriptors $\{D_{k,j}\}$, $j = 1...n_k$. We generate the intra-community codebook with the SIFT features and the corresponding value of the saliency map for high and low salient regions respectively. For instance, the distance between two SIFT descriptors $D_{k,i}$ and $D_{k,j}$ in salient region is defined as:

$$d_{i,j} = \|D_{k,i} - D_{k,j}\| \exp^{\frac{\|M_{k,i} - M_{k,j}\|}{\sigma}} \tag{4}$$

where $M_{k,i}$ is the saliency-level of the SIFT descriptor defined by the saliency map. The codebook can be generated by clustering based on the distance measured as Eq.4. However, for non-salient regions, we directly use $\|D_{k,i} - D_{kj}\|$ for similarity measurement since the saliency value are quite closed for them.As a result, the community-wise codebook consisting of visual words for salient and non-salient region is obtained.

Based on the community-wise codebook, we can obtain the global codebook by clustering the visual words from all communities for salient and non-salient regions. The DL-BoW features of image are generated according to the global codebook for salient and non-salient regions.

2.4 Community Classifier:Learning and Inference

We define the score of interpreting an image I with the corresponding community as :

$$F(I) = \Theta^T \Phi(I) = \theta^T \phi_{sal}(I) + \eta^T \phi_{unsal}(I) + \beta^T \omega(I) \tag{5}$$

In the following, we describe in detail each term in Eq.(5).

Bag-of-Salient-Words $\theta_{sal}^T \phi_{sal}(I)$: For an unlabeled image I, we can extract the local feature based on salient visual words. θ_i can be weight associated with the similarity between each training samples I_k and the unlabeled image. Therefore, we can parameterize this potential function as :

$$\theta^T \phi_{sal}(I) = \sum_{I_k \in I_{Com}} \theta_k K_{sal}(I, I_k) \tag{6}$$

where $K_{sal}(I, I_k)$ is a similarity function, I_{Com} denote the images in specific community.

Bag-of-non-salient-Words $\theta_{unsal}^T \phi_{unsal}(I)$: This potential function captures the similarity on non-salient words between each training samples I_k and the unlabeled image. As shown above, we can parameterize it as:

$$\eta^T \phi_{unsal}(I) = \sum_{I_k \in I_{Com}} \eta_k K_{unsal}(I, I_k) \tag{7}$$

Global features $\beta^T \omega(I)$: This part indicates how likely the image I assigned with this community based on global features of I. It is shown as:

$$\beta^T \omega(I) = \sum_{I_k \in I_{Com}} \beta_k K_{global}(I, I_k) \tag{8}$$

We learn our model in a multiple-kernel learning SVM framework. The multiple-kernel SVM model can be trained with adaptively-weighted combined kernels and each kernel is in accordance with a specific type of visual feature. The decision function is defined as follows:

$$\begin{aligned}
F(I) &= \sum_{I_k \in I_{Com}} \theta_i K_{sal}(I, I_k) + \eta_i K_{unsal}(I, I_k) + \beta_i K_{global}(I, I_k) \\
&= \sum_{I_k \in I_{Com}} \alpha_k \{ \frac{\theta_k}{\alpha_k} K_{sal}(I, I_k) + \frac{\eta_k}{\alpha_k} K_{unsal}(I, I_k) + \frac{\beta_k}{\alpha_k} K_{global}(I, I_k) \} \\
&= \sum_{I_k \in I_{Com}} \alpha_k \sum_m w_m K_m(I, I_k) = \sum_{I_k \in I_{Com}} \alpha_k K(I, I_k)
\end{aligned} \tag{9}$$

where $K(\cdot)$ is the combined kernel, $K_m(\cdot)$ is the sub-kernel of m_{th} visual feature and w_m is the weight for sub-kernel to be learnt. In order to get a sparse solution, we add the l_1 norm constraints and the learning problem is shown as follows:

$$\begin{aligned}
\min &\frac{1}{2} \|F\| + C \sum_{I_k \in I_{Com}} \xi_k \\
s.t. &F(I) = \sum_{I_k \in I_{Com}} \alpha_k K(I, I_k) \\
&K(I, I_k) = \sum_m w_m K_m(I, I_k), w_m \geq 0, \sum_m w_m = 1 \\
&\xi_k \geq 0, y_k F(I_k) \geq 1 - \xi_k
\end{aligned} \tag{10}$$

As reported in previous work, multiple-kernel SVM shows better performance than conventional SVM learnt with combined features. We solve this problem via SimpleMKL[8].

2.5 Labeling: Neighbor-Voting in Communities

The corresponding communities of an untagged image can be determined by the trained community classifiers. A naive KNN search is carried out to realize the initial annotation in each community based on the Euclidean distance between the visual features of the untagged image and the ones in the community. Noted that we will firstly tag the image with the salient tags. The non-salient tag is assigned based both on the correlation of salient tag and the visual feature. Let $r(I, r_{c_i}^{sal})$ denote the relevance between image I and salient tag c_i. $r(I, r_{c_i}^{sal})$ is determined by the K-nearest-neighbors measured with Bag-of-Salient-Words feature and global features:

$$r(I, r_{c_i}^{sal}) = \frac{1}{K}\{ \sum_{I_j \in \mathcal{N}_K^{sal}(I)} w_{sal} r(I_j, r_{c_i}^{sal}) + \sum_{I_j \in \mathcal{N}_K^{global}(I)} w_{global} r(I_j, r_{c_i}^{sal})\} \quad (11)$$

where w_{sal} and w_{global} are kernel weight obtained in (10); $\mathcal{N}_K^{sal}(I)$ is the K-nearest-neighbors measured with salient word feature; $\mathcal{N}_K^{global}(I)$ is the K-nearest-neighbors measured with global feature which can reduce the impact of false/miss salient regions. Similarly, replace "sal" with "unsal" in (11). The relevance between the unlabeled image and non-salient tags are determined as:

$$r(I, r_{c_i}^{unsal}) = \frac{1}{K}\{ \sum_{I_j \in \mathcal{N}_K^{unsal}(I)} w_{unsal} r(I_j, r_{c_i}^{unsal}) + \sum_{I_j \in \mathcal{N}_K^{global}(I)} w_{global} r(I_j, r_{c_i}^{unsal})\}$$

$$(12)$$

The final tagging information of the image is a combination of salient and non-salient tags.

3 Experiments

In this section, some experiments are conducted to evaluate the performance of the proposed method on NUS-WIDE[3] dataset which contains 27807 images in training parts and 27808 images in testing parts. All images are tagged with labels from 81 Ground Truth. The comparison between TVSA and state-of-the-art methods MLKNN[12], MLNB[11], RLVT[6], RANK[6],NBVT[5] and LCMKL[4] is also presented to show the proposed method progresses towards better performance. All of the experiments are executed on a PC with Intel 2.4GHz CPU and 10GB RAM on MATLAB.

For TVSA, we use [10] to extract saliency map and detect 500D BoW feature for salient and non-salient regions respectively. Global features including Color Moments(225D) and Color Histogram (64D) are also adopted as visual representation.For the baseline methods, a 1000D BoW feature and the global features mentioned above are deployed. The parameter settings for TVSA are listed as follows: The threshold of textual saliency (T_{txt}) is set to 0.4 while for the visual saliency (T_{vis}) is the mean-value of image's saliency map. The number of neighbours for neighbot-voting is 100. The scaling factor σ in Eq.4 is 10.

In this paper, Precision, Recall and F1-score are used to measure the performance of image annotation. For concept c_i, they are determined as follows:

$$Precision(c_i) = \frac{N_{corr}}{N_{tagged}}; Recall(c_i) = \frac{N_{corr}}{N_{all}}$$

$$F_1 - score(c_i) = 2\frac{Precision(c_i) \times Recall(c_i)}{Precision(c_i) + Recall(c_i)}$$

(13)

where N_{tagged} denotes the number of images tagged with a specific concept c_i, in testing part by image annotation, N_{corr} denotes the number of images tagged correctly according to the original tagging information and N_{all} denotes the number of images tagged with c_i in training part. For each concept, we can obtain Precison, Recall and F1-score respectively. The global performance is obtained via averaging over all concepts. To make fair comparisons, the top five relevant concepts of the image are selected for annotation. Table 1 shows the performance of image annotation on NUS-WIDE:

Table 1. The performance comparison on NUS-WIDE 81 tags

Method	MLKNN	MLNB	RLVT	RANK	NBVT	LCMKL	TVSA
Precision	0.122	0.110	0.192	0.181	0.127	0.237	0.263
Recall	0.210	0.302	0.186	0.187	0.177	0.233	0.282
F1-score	0.154	0.161	0.187	0.184	0.148	0.235	0.272

As shown in Table 1, we observe that the proposed method outperforms the compared method on Avg. Precsion, Avg. Recall and Avg. F1-score with the top five relevant tags.

Finally, we also discuss the selection of key paramters of TVSA including threshold of visual saliency(T_{txt}) and textual saliency (T_{vis}). For visual saliency, it is not appropriate to set a fixed threshold since the distribution of saliency map varies in different images. The mean-value of image's saliency map is a relative simple and good choise. For textual saliency, the threshold is seleted by cross-validation among $\{0.1, 0.2, ..., 0.9\}$. We found that $T_{txt} = 0.4$ achieves the best performance.

4 Conclusion

In this paper, a Textual-Visual Saliency based framework for image annotation is proposed. Our work integrates the textual saliency on labels and visual saliency on images. A concept graph is constructed which implies a dense sematic intra-community correlation of concepts. The dual-layer Bag-of-Words provide a good visual representatiopn based on local features and salienct regions. The robust multiple-kernel SVM is applied for community classification. Experiments on NUS-WIDE dataset demonstrate that the proposed method outperforms other state-of-the-art approaches.

Acknowledgments. This research is partly supported by NSFC, China (No: 6127325861105001) Ph.D. Programs Foundation of Ministry of Education of China (No.20120073110018).

References

1. Blondel, V.D., Guillaume, J.L., Lambiotte, R., Lefebvre, E.: Fast unfolding of communities in large networks. Journal of Statistical Mechanics: Theory and Experiment 2008(10), P10008 (2008)
2. Carneiro, G., Chan, A., Moreno, P., Vasconcelos, N.: Supervised learning of semantic classes for image annotation and retrieval. IEEE Transactions on Pattern Analysis and Machine Intelligence 29(3), 394–410 (2007)
3. Chua, T.S., Tang, J., Hong, R., Li, H., Luo, Z., Zheng, Y.: Nus-wide: A real-world web image database from national university of singapore. In: Proceedings of the ACM International Conference on Image and Video Retrieval, p. 48. ACM (2009)
4. Li, Q., Gu, Y., Qian, X.: Lcmkl: latent-community and multi-kernel learning based image annotation. In: Proceedings of the 22nd ACM International Conference on Information & Knowledge Management, pp. 1469–1472. ACM (2013)
5. Li, X., Snoek, C.G., Worring, M.: Learning social tag relevance by neighbor voting. IEEE Transactions on Multimedia 11(7), 1310–1322 (2009)
6. Liu, D., Hua, X.S., Yang, L., Wang, M., Zhang, H.J.: Tag ranking. In: Proceedings of the 18th International Conference on World Wide Web, pp. 351–360. ACM (2009)
7. Liu, X., Cheng, B., Yan, S., Tang, J., Chua, T.S., Jin, H.: Label to region by bilayer sparsity priors. In: Proceedings of the 17th ACM International Conference on Multimedia, pp. 115–124. ACM (2009)
8. Sonnenburg, S., Rätsch, G., Schäfer, C., Schölkopf, B.: Large scale multiple kernel learning. The Journal of Machine Learning Research 7, 1531–1565 (2006)
9. Yan, R., Natsev, A., Campbell, M.: A learning-based hybrid tagging and browsing approach for efficient manual image annotation. In: IEEE Conference on Computer Vision and Pattern Recognition, CVPR 2008, pp. 1–8. IEEE (2008)
10. Yang, C., Zhang, L., Lu, H., Ruan, X., Yang, M.H.: Saliency detection via graph-based manifold ranking. In: IEEE Conference on Computer Vision and Pattern Recognition, CVPR 2013, pp. 3166–3173 (2013)
11. Zhang, M.L., Peña, J.M., Robles, V.: Feature selection for multi-label naive bayes classification. Information Sciences 179(19), 3218–3229 (2009)
12. Zhang, M.L., Zhou, Z.H.: Ml-knn: A lazy learning approach to multi-label learning. Pattern Recognition 40(7), 2038–2048 (2007)
13. Zhu, G., Wang, Q., Yuan, Y.: Tag-saliency: Combining bottom-up and top-down information for saliency detection. Computer Vision and Image Understanding 118, 40–49 (2014)

Classification of Fish Ectoparasite Genus *Gyrodactylus* SEM Images Using ASM and Complex Network Model

Rozniza Ali[1,2], Bo Jiang[3], Mustafa Man[1], Amir Hussain[2], and Bin Luo[3]

[1] School of Informatic and Applied Mathematics, Universiti Malaysia Terengganu, Malaysia
{rozniza,mustafaman}@umt.edu.my
[2] Institute of Computing Science and Mathematics, University of Stirling, UK
{ali,ahu}@cs.stir.ac.uk
[3] Computer Science Department, Anhui University, China
{jiangbo,luobin}@ahu.edu.cn

Abstract. Active Shape Models and Complex Network method are applied to the attachment hooks of several species of *Gyrodactylus*, including the notifiable pathogen *G. salaris*, to classify each species to their true species type. ASM is used as a feature extraction tool to select information from hook images that can be used as input data into trained classifiers. Linear (*i.e.* LDA and K-NN) and non-linear (*i.e.* MLP and SVM) models are used to classify *Gyrodactylus* species. Species of *Gyrodactylus*, ectoparasitic monogenetic flukes of fish, are difficult to discriminate and identify on morphology alone and their speciation currently requires taxonomic expertise. The current exercise sets out to confidently classify species, which in this example includes a species which is notifiable pathogen of Atlantic salmon, to their true class with a high degree of accuracy. The results show that Multi-Layer Perceptron (MLP) is the best classifier for performing the initial classification of *Gyrodactylus* species, with an average of 98.36%. Using MLP classifier, only one species has been misallocated. It is essential, therefore, to employ a method that does not generate type I or type II misclassifications where *G. salaris* is concerned. In comparison, only K-NN classifier has managed to to achieve full classification on the *G. salaris*.

Keywords: *Gyrodactylus*, classification, Active Shape Model, Complex Network.

1 Introduction

There are over 440 described species of *Gyrodactylus* which are typically small (<1mm), ectoparasitic monogenetic flukes of fish [14]. While most species of *Gyrodactylus* are non-pathogenic, causing little harm to their hosts, other species like *Gyrodactylus salaris* Malmberg, 1957, which is an OIE (Office International des Epizooties) - listed pathogen of Atlantic salmon, has led to a catastrophic decimation in the size of the juvenile salmon population in over 40 Norwegian rivers [7].

Uncontrolled increases in the size of the parasite population on resident salmon populations have necessitated extreme measures such as the use of the biocide rotenone to kill-out entire river systems, to remove the entire fish population within a river and the parasite [7]. Given the impact that *G. salaris* has had in Norway and elsewhere in Scandinavia [1], many European states including the UK now have mandatory surveillance

C.K. Loo et al. (Eds.): ICONIP 2014, Part III, LNCS 8836, pp. 103–110, 2014.

programmes screening wild salmonid populations (*i.e.* brown trout, charr, grayling, Atlantic salmon etc) for the presence of notifiable pathogens including *G. salaris*. Current OIE methodologies for the identification of *G. salaris* from other species of *Gyrodactylus* that occur on salmonids require confirmation from both morphological and molecular approaches, which can be time consuming. If *G. salaris* specimens, however, are overlooked in a diagnostic sample or misclassified, the environmental and economic implications can be severe [20]. For this reason and because of the widely varying pathogenicity seen between closely related species, accurate pathogen identification is of paramount importance.

The discrimination of species from their congeners, however, is compounded by a limited number of morphological discrete characteristics which makes identification difficult. The task of morphological identification is, therefore, currently heavily reliant upon a limited number of domain experts available to analyse and determine species groups. This time can be dramatically reduced if the initial identification of *G. salaris* or *G. salaris*-like specimens by the morphology step can be improved and accelerated. In the event of a suspected outbreak, the demand for identification may significant exceed the available supply of suitable expertise and facilities. There is, therefore, a real need for the development of rapid, accurate, semi-automatic / automatic diagnostic tools that are able to confidently identify *G. salaris* in any population of specimens.

The aims of the current study were to explore the potential use of an Active Shape Model (ASM) combining with Complex Network method to extract features information from the attachment hooks of each species of *Gyrodactylus*. Given the small size of the marginal hook sickles (*i.e.* <7m), which are regarded as the most taxonomically informative morphological structure, this study will begin with an assessment of scanning electron microscope (SEM) images which give the best quality images. Given the subtle differences in the hook shape of each species, it is hoped that this approach moves towards the rapid automated classification of species with improved rates of correct classification over existing methods and negates the current laborious process of taking manual measurements which are used to assist experts in identifying species.

2   Specimen Preparation

Specimens of *Gyrodactylus* (*G. derjavinoides* n = 25; *G. salaris* n = 34; *G. truttae* n = 9) were removed from their respective salmonid hosts and fixed in 80% ethanol. Subsequently specimens were prepared for scanning electron microscopy (SEM) by transferring individual, distilled water rinsed, specimens onto 13 mm diameter round glass coverslips, where they had their posterior attachment organ excised using a scalpel and the attachment hooks released using a proteinase-K based digestion fluid (*i.e.* 100 μg/ml proteinase K, 75 mM Tris-HCl, pH 8, 10 mM EDTA, 5% SDS). Once the hooks were freed from enclosing tissue, the preparations were flushed with distilled water, air-dried, sputter-coated with gold and then examined and photographed using a JEOL JSM5200 scanning electron microscope operating at an accelerating voltage of 10 kV.

3 Segmentation and Feature Extraction

The application of the ASM method as a segmentation (landmark points) of tool and extracting features using Complex Network approach to the analysis of *Gyrodactylus* attachment hooks is presented in Fig. 2. Specimens of *Gyrodactylus* were picked from the skin and fins of salmonids and their attachment hooks released by proteolytic digestion. Images of the smallest hook structures, the marginal hook sickles which are the key to separating species and typically measure less than 0.007 mm in length, were captured using a scanning electron microscope. The images were pre-processed before being subjected to an Active Shape Model and Complex Network feature extraction step to define 110 landmarks and to fit the model to the training set of hook images. A Complex Network reduced the data to 49 variables which were used to train 4 classifiers (K-NN, LDA, MLP, SVM) and separate the three species of *Gyrodactylus* which includes the notifiable pathogen, *G. salaris*. Abbreviations: K-NN, K Nearest Neighbors; LDA, Linear Discriminant Analysis; MLP, Multi-Layer Perceptron; SVM, Support Vector Machine.

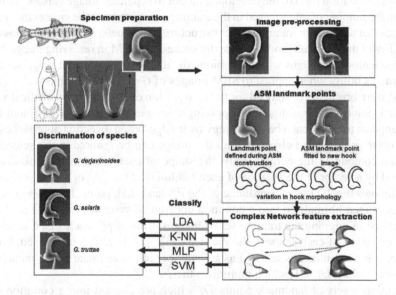

Fig. 1. The methodological approach used in the current study, the ASM were used as landmark points to segment the focus object, while the Complex Network were used to extract the informative features. Four classifiers were accessed and compared in species identification.

3.1 Existing Methods of Parasite Identification and Classification

A number of statistical classification based approaches applied to morphological data [20], [21], and molecular-based techniques targeting specific genomic regions [13], [19], have been developed to discriminate the pathogenic species, *G. salaris*, from other non-pathogenic species of *Gyrodactylus* that co-occur on salmonid hosts. While each technique is able to detect *G. salaris* within a population of specimens and to discriminate it from its congeners with high levels of correct classification, the techniques can be time consuming [20]. If image recognition software could be developed to extract key discriminatory features from the attachment hooks of each species, then it is anticipated that the identification process could be accelerated with equivalent or better rates of correct identification.

3.2 Landmark Points Using ASM

ASM is a feature extraction based technique that has been successfully applied in human face [9] and leaf [12] recognition, the screening of skin cancers [8], and, in the segmentation of lung radiographs [17] and of protozoan parasites from images captured with the light microscope [16], among a range of other studies. The ASM technique permits users to construct a general shape model which is subsequently applied to all images in order to landmark the image area for every given image, providing a pattern that encapsulates the variation seen across the range of shape images. The subsequent ability (classification rate) of the developed model to separate "image classes" is in part based on the number of images used in the training set - in theory, the greater the number of images that are used in training and constructing the models, the better the classification ability of the resultant model. Given the success of ASM in resolving image-based, shape recognition problems within the biomedical sphere, the current study set out to determine its utility when applied to SEM images of *Gyrodactylus* hooks.

ASM were originally developed for the recognition of landmarks on medical x-rays. Landmark points can be acquired by applying a sample template to a "problem area", which appears to represent a better strategy over edge-based detection approaches [18], as any noise or unwanted objects within the image can be ignored in the selection of the shape contour. In the current study, the shape of each attachment hook image is presented by a vector of the position of each landmark, $D = (d_1, e_1, ..., d_n, e_n)$, where $(d_i e_i)$ denotes the 2D image coordinate of the ith landmark point. The shape vector of the hook is then normalised into a common coordinate system. Procrustes analysis is then applied in aligning the training set of images. This aligns each shape so that the sum of distances of each shape to the mean $F = \sum |D_i - \bar{D}|^2$ is minimised. For this purpose, one hook image is selected as an example initial estimate of the mean shape and scaled so that $|\bar{D}| = 1$, which minimises the F.

Assuming s sets of landmark points D_i which are aligned into a common shape pattern for each species, if this distribution can be modelled, then new examples can be generated similar to those in the original training set s, and then these new shapes can be examined to decide whether they represent reasonable examples. In particular, $D = M(b)$ is used to generate new vectors, where b is a vector of parameters of the model. If the distribution parameters can be modelled, $p(b)$, these can then be limited such that the generated D's are similar to those in the training set. Similarly it should be possible to estimate $p(D)$ using the model.

Once the ASM model has been constructed, it is important to fit the defined model to a series of new input images to determine the parameters of the model that are the best descriptors of hook shape. ASM finds the most accurate parameters of the defined model for the new hook images. The ASM fitting attempts to "best fit" the defined model parameter to each image. Cootes *et al.* [10] explained that by adjusting each model parameter from the defined model will permit an extraction pattern of the image series to be created. During the model fitting, it measures newly introduced images and uses this model to correct the values of current parameters, leading to a better fit.

Once the shape of the images available, then the next step will be the feature extraction. The landmark point need to perform fist, where the SEM images of *Gyrodactylus* specimens contain tissue that difficult to distinguish from actual shape. The Complex Network perform feature extraction using the landmark points of information.

3.3 Extraction Features Using Complex Network

Recently, complex network based shape representation has been shown effectively and widely used in shape and image recognition and retrieval [5,4,6]. In general, this method consists of the following two steps.

(1) Shape representation with complex network model. First, N landmark (key) points should be extracted from the shape contour. Then, with these landmark points, we can construct a complex network $G =< V, E >$ as follows. Each landmark point is represented as a vertex in the network. For each pair of vertices, there is an edge with the corresponding weight w_ij representing the Euclidean distance between them. Therefore, the network can be represented by a $N \times N$ weight matrix W, normalized into interval $[0, 1]$ [5,4].

(2) Feature extraction. There are two main kinds of characteristic (measurements) that can be used to characterize topological connectivity of the complex network. One is the static statistic measurements, and the other is dynamic evolution [5,4]. The five static measurements used in this paper are the maximum degree, average degree, average joint degree, average shortest path length and entropy. Dynamic evolution is also an important characteristic for complex networks. In this paper, we use the evolution process proposed in the work [5,4]. Figure 1 shows the complex network representation and its dynamic evolution process.

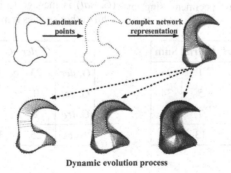

Fig. 2. Shape representation and the dynamic evolution process of complex network

4   Experimental Results

Although the attachment apparatus of *Gyrodactylus* consists of three main elements (*i.e* two larger centrally positioned anchors or hamuli; two connecting bars between the hamuli; and, 16 peripherally distributed marginal hooks), this study sets out to classify species based on features extracted from the sickles of the marginal hooks only. As the study is based on the analysis of biological structures, these require processing subsequent to capture in order to standardised the position and format of the image. Processing to standardise the orientation of the image is applied to reduce processing time and complexity during the training and construction of the ASM model. Then, the data were assessed using four methods of machine learning classifiers, namely are Linear Discriminant Analysis (LDA), K Nearest Neighbor (K-NN), Multi-layer Perceptron (MLP) and Support Vector Machine (SVM). For each approach, a 10-fold cross validation was used *i.e.* the data were divided into k (10) subsets, where k-1 subsets were used for training and the remaining subset used as the test set. This process was repeated 10 times using a different test set on each run and the average classification performance computed.

The K-NN classifier improved upon the classification of *G. salaris* specimens with all being correctly classified (Table 2), while two more species remain miclassified; such as *G. derjavinoides* specimens was misallocated as *G. salaris* and *G. truttae*. Also some of *G. truttae* that has been misclassified as *G. salaris*. Other classifier model LDA (Table 1) and SVM (Table 4) were also expremented. Among these two models, SVM has perform better than LDA, where using SVM classifier, *G. derjavinoides* has managed to have full classification. The MLP classifier, was able to correctly classify all specimens of *Gyrodactylus* to their true class, except for one specimens of *G. salaris* which were classified as *G. truttae* (Table 3). Comaparing to the other models, MLP has achieved highest classification rate at 98.38%. This is not surprising, since MLP is a well performance classifier in many field [11], [15].

Table 1. A confusion matrix showing the classification of *Gyrodactylus* specimen using an LDA classifier

	G. der	G. sal	G. tru	Sum
G. der	24	0	1	25
G. sal	1	28	5	34
G. tru	0	2	7	9
Sum	25	30	13	68

Table 2. Using the K-NN classifier, *G. salaris* (*G. sal*) is manage to have full classification, while other species remain misclassified

	G. der	G. sal	G. tru	Sum
G. der	23	1	1	25
G. sal	0	34	0	34
G. tru	2	1	6	9
Sum	25	36	7	68

This achievement is same as performance using ASM-PCA [3], and this performance is better than 25 point-to-point measurements manually extracted from light micrographs of 557 specimens (*i.e.* 92.59%) [2], this approach appears promising and now will be applied to hooks prepared for light microscopy hopefully with equal or better

Table 3. MLP classifer performs well with the correct classification *G. derjavinoides* (*G. der*) and *G. truttae* (*G. tru*)

	G. der	G. sal	G. tru	Sum
G. der	25	0	0	25
G. sal	0	33	1	34
G. tru	0	0	9	9
Sum	25	33	10	68

Table 4. Two specimens (*G. salaris* (*G. sal*) and *G. trutte* (*G. tru*)) are unable to achieve full classification using SVM classifier

	G. der	G. sal	G. tru	Sum
G. der	25	0	0	25
G. sal	0	32	2	34
G. tru	0	1	8	9
Sum	25	33	7	68

rates of correct classification. The ASM and Complex Network based approach applied to SEM images of the hook sickles of *Gyrodactylus* appears to out perform or equal other methods that have been tested to identify and discriminate this species with confidence. This study will continue and will explore the potential of using the ASM and Complex Network method in combination with multi-stage or ensemble classification techniques to improve upon the classification accuracy of each species using image taken with light microscope.

5   Conclusion

The current study set out to explore the utility of a novel ASM and Complex Network based approach in extracting and thus classifying species of *Gyrodactylus* which are ectoparasites of fish. ASM and Complex Network applied to 68 SEM images of the marginal hook sickle was able to overcome the limitation and difficulties in extracting feature information from the hooks. The best approach, which used a MLP method of classification, where only one species remain misclassified.

This work continues, exploring the more pertinent and realistic research problem of classifying specimens based on light microscope images which necessitates image pre-processing. In addition, this work will assess the performance of this method on larger datasets and will explore new methods based on an ensemble of classifiers, which have shown promising results, with the aims of providing a reliable model for the identification of species, including the pathogen *G. salaris*, by non-experts and fish health researchers.

Acknowledgements. Gratefully acknowledges the colloboration work between Stirling University, United Kingdom and Anhui University, China (Sino-UK Higher Education Research Partnership Funding Call). To the team of parasitology, Stirling University, thank you for allowing the use of *Gyrodactylus* dataset in this study.

References

1. Alenäs, I.: Gyrodactylus salaris på lax i svenska vattendrag och laxproblematiken på svenska västkysten. Vann 1, 135–142 (1998) (In Swedish)

2. Ali, R., Hussain, A., Bron, J.E., Shinn, A.P.: Multi-stage classification of Gyrodactylus species using machine learning and feature selection techniques. Intelligent Systems Design and Applications, 457–462 (2011)
3. Ali, R., Hussain, A., Bron, J.E., Shinn, A.P.: The use of asm feature extraction and machine learning for the discrimination of members of the fish ectoparasite genus Gyrodactylus. Neural Information Processing 7666, 457–462 (2011)
4. Backes, A.R., Bruno, O.M.: Shape classification using complex network and multi-scale fractal dimension. Pattern Recognition Letters 31(1), 45–51 (2010)
5. Backes, A.R., Casanova, D., Bruno, O.M.: A complex network-based approach for boundary shape analysis. Pattern Recognition 42(8), 54–67 (2009)
6. Backes, A.R., Martinez, A.S., Bruno, O.M.: Texture analysis using graphs generated by deterministic partially self-avoiding walks. Pattern Recognition 44(8), 1684–1689 (2011)
7. Bakke, T.A., Cable, J., Harris, P.D.: The biology of gyrodactylid monogeneans: the "Russian-doll killers". Advances in Parasitology 64, 161–376 (2007)
8. Blackledge, J.M., Dubovitskiy, A.: Object detection and classification with applications to skin cancer screening. Intelligent Systems 1(1), 34–45 (2008)
9. Choi, J., Chung, Y., Kim, K., Yoo, J.: Face recognition using energy probability in DCT domain, pp. 1549–1552. IEEE (2006)
10. Cootes, T.F., Edwards, G.J., Taylor, C.J.: Active appearance models. Pattern Analysis and Machine Intelligence 23(6), 681–685 (2001)
11. Cordella, L.P., Limongiello, A., Sansone, C.: Network intrusion detection by a multi-stage classification system. In: Roli, F., Kittler, J., Windeatt, T. (eds.) MCS 2004. LNCS, vol. 3077, pp. 324–333. Springer, Heidelberg (2004)
12. Du, J.X., Wang, X.F., Zhang, G.J.: Leaf shape based plant species recognition. Applied Mathematics and Computation 185(2), 883–893 (2007)
13. Hansen, H., Bachmann, L., Bakke, T.: Mitochondrial DNA variation of Gyrodactylus spp. (Monogenea, Gyrodactylidae) populations infecting Atlantic salmon, grayling, and rainbow trout in Norway and Sweden. Parasitology 33, 1471–1478 (2003)
14. Harris, P.D., Shinn, A.P., Cable, J., Bakke, T.A.: Nominal species of the genus Gyrodactylus v. Nordmann 1832 (Monogenea: Gyrodactylidae), with a list of principal host species. Systematic Parasitology 59, 1–27 (2004)
15. Kabir, M.F., Schmoldt, D.L., Araman, P.A., Schafer, M.E., Lee, S.M.: Classifying defects in pallet stringers by ultrasonic scanning. Wood and Fiber Science 34 (2003)
16. Lai, C.H., Yu, S.S., Tseng, H.Y., Tsai, M.H.: A protozoan parasite extraction scheme for digital microscope images. Computerized Medical Imaging and Graphics 34, 122–130 (2010)
17. Lee, J.S., Wu, H.H., Yuan, M.Z.: Lung segmentation for chest radiograph by using adaptive active shape models. In: Int. Conf. on Information Assurance and Security, pp. 383–386 (2009)
18. Maini, R., Aggarwal, H.: Study and comparison of various image edge detection techniques. Image Processing 3(1), 1–60 (2009)
19. Meinila, M., Kuusela, J., Zietara, M.: Brief report: Primers for amplifying 820 bp of highly polymorphic mitochondrial COI gene of Gyrodactylus salaris. Hereditas 137, 72–74 (2002)
20. Shinn, A.P., Collins, C., García-Vásquez, A., Snow, M., Paladini, G., Lindenstrøm, T., Longshaw, M., Matějusová, I., Stone, D.M., Turnbull, J.F., Picon-Camacho, S.M., Vázquez Rivera, C., Duguid, R.A., Mo, T.A., Hansen, H., Olstad, K., Cable, J., Harris, P.D., Kerr, R., Graham, D., Yoon, G.H., Buchmann, K., Raynard, R., Irving, S., Bron, J.E.: Multi-centre testing and validation of current protocols for Gyrodactylus salaris (Monogenea) identification. International Journal of Parasitology 40, 1455–1467 (2010)
21. Shinn, A.P., Hansen, H., Bachmann, L., Bakke, T.A.: The use of morphometric characters to discriminate specimens of laboratory-reared and wild populations of Gyrodactylus salaris and G. thymalli (monogenea). Folia Parasitologica 51, 239–252 (2004)

Linked Tucker2 Decomposition for Flexible Multi-block Data Analysis

Tatsuya Yokota[1] and Andrzej Cichocki[1,2]

[1] RIKEN Brain Science Institute, Saitama, Japan
{yokota,cia}@brain.riken.jp
[2] Systems Research Institute, Warsaw, Poland

Abstract. In this paper, we propose a new algorithm for a flexible group multi-way data analysis called the linked Tucker2 decomposition (LT2D). The LT2D can decompose given multiple tensors into common factor matrices, individual factor matrices, and core tensors, simultaneously. When we have a set of tensor data and want to estimate common components and/or individual characteristics of the data, this decomposition model is very useful. In order to develop an efficient algorithm for the LT2D, we imposed orthogonality constraints to factor matrices and applied alternating least squares (ALS) algorithm to the optimization criterion. We conducted some experiments to demonstrate the advantages and convergence properties of the proposed algorithm. Finally, we discuss potential applications of the proposed method.

Keywords: Tensor decompositions, Tucker2 decomposition, Group data analysis, Alternating Least Squares (ALS), common and individual components, Group component analysis.

1 Introduction

The tensor decomposition is an important technique for time series signal analysis, image analysis, neuroscience, psychological data analysis, chemometrics and other multi-way data processing [3]. Especially, analyzing a set of tensors is a quite important to study due to many applications. The tensor decomposition is known as one of the method to extract common factors from blocks of dataset. If dataset consists of a set of third order tensors (3D-tensors), the sizes of which are the same, such dataset can be regarded as a fourth order tensor (4D-tensor) via concatenation procedure. In this case, 4D-tensor decomposition can be applied to extract common factor matrices. For example, multi-way principal component analysis was applied to feature extraction for EEG classification [11]. However, when the set of tensors have different size in several modes, 4D-tensor decomposition can not be applied. In such a case, we have to consider more flexible decomposition model to analyze the group of data.

In this paper, we consider the set of 3D-tensors, the sizes of first and second modes of which are the same but the sizes of third mode of which are generally different. Population value decomposition (PVD) [4], which is one of the

C.K. Loo et al. (Eds.): ICONIP 2014, Part III, LNCS 8836, pp. 111–118, 2014.
© Springer International Publishing Switzerland 2014

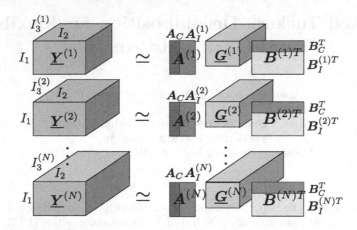

Fig. 1. Linked Tucker2 decomposition for 3rd-order tensors

analysis method to extract left and right common factor matrices from multiple matrices, and Tucker2 decomposition [13,9] can be used to extract first (left) and second (right) modes common factor matrices from above type of dataset. We introduce a more flexible decomposition model called the linked Tucker2 decomposition (LT2D) in Figure 1. The LT2D does not extract only their common factor matrices but also their individual factor matrices at the same time. Thus, the LT2D decomposes set of 3D-tensors into first and second mode common and individual factor matrices, and N of core tensors.

In order to implement the LT2D, we imposed orthogonality constraints and applied the Alternating Least Squares (ALS) algorithm to optimize its criterion. Furthermore, we conducted several experiments to demonstrate their advantages.

The rest of this paper is organized as follows. In Section 2, a basic idea and related works on common and individual component analysis are introduced. In Section 3, a novel linked tensor decomposition and its ALS based algorithm are proposed. In Section 4, we present experiments using our new method and discuss the results of the experiments. Finally, we give our conclusions in Section 5.

2   Common and Individual Component Analysis for Multi-block Matrix Data

First, we consider a group of data as multi-block matrices $X^{(n)} \in \mathbb{R}^{I \times J_n}$ for $n = 1, ..., N$. In each matrix, the number of columns are the same but the number of rows are different. For example, each column is a pattern vector and there are J_n samples for each n-th matrix. Many techniques have been proposed to analyze such a dataset. If we want to extract some important components for each matrix, which are called individual components, principal component analysis (PCA) and related methods [6,16,7] can be applied to each $\boldsymbol{X}^{(n)}$, individually. Moreover, if we want to extract some 'common' important components for all matrices, partial least squares (PLS) regression [8,5] and canonical correlation analysis (CCA) can be useful for the case of $N = 2$, and the PCA can be also applied to concatenated matrix $[\boldsymbol{X}^{(1)}, \boldsymbol{X}^{(2)}, .., \boldsymbol{X}^{(N)}] \in \mathbb{R}^{I \times \sum_n J_n}$, simply. The population value decomposition (PVD) [4] can be used for the case of $J_1 = J_2 = \cdots = J_N$. In order to extract common and individual components, simultaneously, joint and individual variation explained (JIVE) algorithm [10] has been proposed. Furthermore, common and individual feature extraction (CIFA) [15] has been proposed as an algorithm to extract common and individual components, separately and sequentially.

3   Common and Individual Component Analysis for Multi-block Tensor Data

In this paper, we consider more general problem as above one. Thus, we consider a group of data as multi-block tensors $\underline{\boldsymbol{Y}}^{(n)} \in \mathbb{R}^{I_1 \times I_2 \times I_3^{(n)}}$ for $n = 1, ..., N$. For example each matrix is an $(I_1 \times I_2)$-pattern-matrix and there are $I_3^{(n)}$ samples for each n-th tensor. In order to extract left and right common components, the PVD, Tucker2 decomposition [13,9], and CP decomposition [2] can be applied to its concatenated $(I_1 \times I_2 \times \sum_n I_3^{(n)})$-tensor, however, the numbers of left and right components must be the same in the CP decomposition. As a previous study, we proposed the linked CP tensor decomposition (LCPTD) [14] which can decompose a multi-block tensor data into multi-way common and individual components and N diagonal core tensors. The LCPTD model is useful for multi-way blind source separation, low-rank approximation, and feature extraction for multi-block tensor data; however, the flexibility of the model is low due to the fact that numbers of left and right components must be the same. In this study, we propose a more flexible decomposition model called the "linked Tucker2 decomposition" (LT2D) illustrated in Figure 1.

In formula, the LT2D is given by

$$\underline{\boldsymbol{Y}}^{(n)} \cong \underline{\boldsymbol{G}}^{(n)} \times_1 \boldsymbol{A}^{(n)} \times_2 \boldsymbol{B}^{(n)} \tag{1}$$

for $n = 1, 2, ..., N$, where $\underline{\boldsymbol{Y}}^{(n)} \in \mathbb{R}^{I_1 \times I_2 \times I_3^{(n)}}$ is an observed 3D-tensor data, $\underline{\boldsymbol{G}}^{(n)} \in \mathbb{R}^{M_1 \times M_2 \times I_3^{(n)}}$ is a core tensor, $\boldsymbol{A}^{(n)} \in \mathbb{R}^{I_1 \times M_1}$ and $\boldsymbol{B}^{(n)} \in \mathbb{R}^{I_2 \times M_2}$ are the first and the second modes component matrices. Each $\boldsymbol{A}^{(n)}$ and $\boldsymbol{B}^{(n)}$ are composed as $[\boldsymbol{A}_C, \boldsymbol{A}_I^{(n)}]$ and $[\boldsymbol{B}_C, \boldsymbol{B}_I^{(n)}]$, respectively, where $\boldsymbol{A}_C \in \mathbb{R}^{I_1 \times L_C}$ and $\boldsymbol{B}_C \in \mathbb{R}^{I_2 \times R_C}$ are common components (CCs) corresponding for all tensors (i.e. all n), $\boldsymbol{A}_I^{(n)} \in \mathbb{R}^{I_1 \times L_I}$ and $\boldsymbol{B}_I^{(n)} \in \mathbb{R}^{I_2 \times R_I}$ are individual components (ICs), which characterize the difference or independence of individual tensors (i.e., $M_1 = L_C + L_I$ and $M_2 = R_C + R_I$). This model can separate common and individual multi-way sources from group of tensors, and also reduce the multi-way dimensionalities of original tensors via the extracted core-tensors.

3.1 Orthogonal LT2D Algorithm

In this paper, we consider to impose the orthogonality constraint into each mode component matrix, and formulate the following optimization problem

$$\underset{\Theta}{\text{minimize}} \quad C(\Theta) := \sum_{n=1}^{N} ||\underline{\boldsymbol{Y}}^{(n)} - \underline{\boldsymbol{G}}^{(n)} \times_1 \boldsymbol{A}^{(n)} \times_2 \boldsymbol{B}^{(n)}||_F^2,$$

$$\text{subject to} \quad \boldsymbol{A}^{(n)T}\boldsymbol{A}^{(n)} = \boldsymbol{I}_{M_1}, \ \boldsymbol{B}^{(n)T}\boldsymbol{B}^{(n)} = \boldsymbol{I}_{M_2}, \tag{2}$$

for all n, where $C(\Theta)$ is the cost function and Θ represents a set of parameters (i.e., $\{\boldsymbol{A}^{(n)}, \boldsymbol{B}^{(n)}, \underline{\boldsymbol{G}}^{(n)}\}_{n=1}^{N}$). Substituting one of the KKT conditions $\underline{\boldsymbol{G}}^{(n)} = \underline{\boldsymbol{Y}}^{(n)} \times_1 \boldsymbol{A}^{(n)T} \times_2 \boldsymbol{B}^{(n)T}$ into (2) and transforming this, then we can obtain a simplified optimization problem with respect to each-mode basis matrix. The optimization problem for the first-mode (left) components is given by

$$\underset{\{\boldsymbol{A}^{(n)}\}_{n=1}^{N}}{\text{minimize}} \quad \sum_{n=1}^{N} ||\boldsymbol{X}_A^{(n)} - \boldsymbol{A}^{(n)} \boldsymbol{A}^{(n)T} \boldsymbol{X}_A^{(n)}||_F^2,$$

$$\text{subject to} \quad \boldsymbol{A}^{(n)T}\boldsymbol{A}^{(n)} = \boldsymbol{I}_{M_1} \text{ for all } n, \tag{3}$$

where $\boldsymbol{X}_A^{(n)} := [\underline{\boldsymbol{Y}}^{(n)} \times_2 \boldsymbol{B}^{(n)T}]_{(1)} \in \mathbb{R}^{I_1 \times M_2 I_3^{(n)}}$. Thus we consider to factorize $\boldsymbol{X}_A^{(n)}$ by the common and the individual left factor matrices in a similar way to JIVE algorithm [10]. To update left common components (CCs) $\boldsymbol{A}_C$, we run the truncated SVD (tSVD) as

$$[\boldsymbol{X}_A^{(1)} - \boldsymbol{A}_I^{(1)} \boldsymbol{A}_I^{(1)T} \boldsymbol{X}_A^{(1)}, \boldsymbol{X}_A^{(2)} - \boldsymbol{A}_I^{(2)} \boldsymbol{A}_I^{(2)T} \boldsymbol{X}_A^{(2)},$$
$$\dots, \boldsymbol{X}_A^{(N)} - \boldsymbol{A}_I^{(N)} \boldsymbol{A}_I^{(N)T} \boldsymbol{X}_A^{(N)}] \cong \boldsymbol{A}_C \boldsymbol{D}_C \boldsymbol{V}_C^T. \tag{4}$$

To update left individual components (ICs) $\boldsymbol{A}_I^{(n)}$, we run the tSVD one by one, individually, which is given by

$$[\boldsymbol{X}_A^{(n)} - \boldsymbol{A}_C \boldsymbol{A}_C^T \boldsymbol{X}_A^{(n)}] \cong \boldsymbol{A}_I^{(n)} \boldsymbol{D}_I^{(n)} \boldsymbol{V}_I^{(n)T}, (n = 1, 2, \dots, N). \tag{5}$$

In order to obtain optimal parameters, we iterate the common and the individual steps alternately until convergence.

Similarly, the optimization problem for the second-mode (right) components is given by

$$\underset{\{\boldsymbol{B}^{(n)}\}_{n=1}^{N}}{\text{minimize}} \quad \sum_{n=1}^{N} ||\boldsymbol{X}_B^{(n)} - \boldsymbol{B}^{(n)} \boldsymbol{B}^{(n)T} \boldsymbol{X}_B^{(n)}||_F^2,$$

$$\text{subject to} \quad \boldsymbol{B}^{(n)T}\boldsymbol{B}^{(n)} = \boldsymbol{I}_{M_2} \text{ for all } n, \tag{6}$$

where $\boldsymbol{X}_B^{(n)} := [\underline{\boldsymbol{Y}}^{(n)} \times_1 \boldsymbol{A}^{(n)T}]_{(2)} \in \mathbb{R}^{I_2 \times M_1 I_3^{(n)}}$. Then we can update matrices $\boldsymbol{B}_C$ and $\boldsymbol{B}_I^{(n)}$ in similar way. Finally, the algorithm for the LT2D decomposition can be summarized as Algorithm 1.

Algorithm 1. Orthogonal LT2D algorithm

---

1. **Input:** $\{\underline{Y}^{(n)} \in \mathbb{R}^{I_1 \times I_2 \times I_3^{(n)}}\}_{n=1}^N$, L_C, L_I, R_C, and R_I;
2. **Initialize:** random orthonormal matrices A_C, B_C, and $\{A_I^{(n)}, B_I^{(n)}\}_{n=1}^N$, where $A_C \perp A_I^{(n)}$ and $B_C \perp B_I^{(n)}$ for all n;
3. $A^{(n)} \leftarrow [A_C, A_I^{(n)}]$ for all n;
4. $B^{(n)} \leftarrow [B_C, B_I^{(n)}]$ for all n;
5. **repeat**
6. $\quad X_A^{(n)} \leftarrow [\underline{Y}^{(n)} \times_2 B^{(n)T}]_{(1)}$ for all n;
7. $\quad$ **repeat**
8. $\qquad Z_A \leftarrow [X_A^{(1)} - A_I^{(1)} A_I^{(1)T} X_A^{(1)}, X_A^{(2)} - A_I^{(2)} A_I^{(2)T} X_A^{(2)}, \dots, X_A^{(N)} - A_I^{(N)} A_I^{(N)T} X_A^{(N)}]$;
9. $\qquad A_C \leftarrow$ left-singular matrix of tSVD(Z_A, L_C);
10. $\qquad A_I^{(n)} \leftarrow$ left-singular matrix of tSVD($[X_A^{(n)} - A_C A_C^T X_A^{(n)}]$, L_I) for all n;
11. $\qquad A^{(n)} \leftarrow [A_C, A_I^{(n)}]$ for all n;
12. $\quad$ **until** $\sum_{n=1}^N \|X_A^{(n)} - A^{(n)} A^{(n)T} X_A^{(n)}\|_F^2$ converge
13. $\quad X_B^{(n)} \leftarrow [\underline{Y}^{(n)} \times_1 A^{(n)T}]_{(2)}$ for all n;
14. $\quad$ **repeat**
15. $\qquad Z_B \leftarrow [X_B^{(1)} - B_I^{(1)} B_I^{(1)T} X_B^{(1)}, X_B^{(2)} - B_I^{(2)} B_I^{(2)T} X_B^{(2)}, \dots, X_B^{(N)} - B_I^{(N)} B_I^{(N)T} X_B^{(N)}]$;
16. $\qquad B_C \leftarrow$ left-singular matrix of tSVD(Z_B, R_C);
17. $\qquad B_I^{(n)} \leftarrow$ left-singular matrix of tSVD($[X_B^{(n)} - B_C B_C^T X_B^{(n)}]$, R_I) for all n;
18. $\qquad B^{(n)} \leftarrow [B_C, B_I^{(n)}]$ for all n;
19. $\quad$ **until** $\sum_{n=1}^N \|X_B^{(n)} - B^{(n)} B^{(n)T} X_B^{(n)}\|_F^2$ converge
20. **until** $\sum_{n=1}^N \|\underline{Y}^{(n)} - \underline{Y}^{(n)} \times_1 A^{(n)} A^{(n)T} \times_2 B^{(n)} B^{(n)T}\|_F^2$ converge
21. $\underline{G}^{(n)} = \underline{Y}^{(n)} \times_1 A^{(n)T} \times_2 B^{(n)T}$ for all n;
22. **Output:** $\{\underline{G}^{(n)}, A^{(n)}, B^{(n)}\}_{n=1}^N$

---

4   Simulations

First, we show an experiment for linked multiway blind source separation by using synthetic data. Original CCs and ICs were given as sine/cosine time series or square waves and sparse core tensors were generated randomly. Two tensors were generated by two CCs and one ICs for each tensor (i.e., $L_C = R_C = 2$ and $L_I = R_I = 1$). Then, the size of each tensor is $(100 \times 100 \times 10)$. The LT2D was applied to extract CCs and ICs from the generated tensors. The T2D and the PVD were also applied to extract only CCs from the same data. Figure 2 shows the performance of extracted CCs and ICs by individual all methods. Comparing the results, the LT2D algorithm could extract correctly sources, however the T2D and the PVD failed.

Next, we show the convergence properties of objective function for 1000 random initialization in the LT2D algorithm by using the ORL faces dataset [12]. The computer, used in this simulation, has 3.9 Gbytes of memory and Intel Core 2 Duo CPU E8200 2.6 GHz×2. Ubuntu 12.10 with 32 bit was installed, and the simulation was conducted by the MATLAB software. We selected 10 faces from

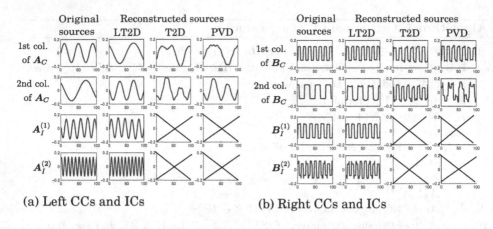

(a) Left CCs and ICs   (b) Right CCs and ICs

Fig. 2. Results of Linked Multiway Blind Source Separation

the dataset and the size of multi-block tensor data is given as $(112{\times}92{\times}1){\times}10$. Parameters were set as $L_C = L_I = R_C = R_I = 3$. Fig. 3 shows the functional boxplot of all simulations. A central solid line is the median curve for all 1000 curves, upper and lower dashed lines with circles show 25 and 75 percentile curves, and two outside dotted lines with cross marks show 5 and 95 percentile curves for all simulations. Stopping criterion is considered that the absolute value of changes of objective function is smaller than 1e-8, and the maximum iteration of internal loop and external loop are set as 20 and 500, respectively. Average and standard deviation of convergence time were 36.39 and 13.89 seconds, respectively. Average and standard deviation of number of iterations for convergence were approximately 121 and 46 times, respectively. From the result, we can confirm that the algorithm was stable and converged to desired solution.

Figure 4 (a) shows the variation dispersion of the solution of the LT2D algorithm for each 20 Monte-Carlo-runs. We generated two synthetic tensors randomly by orthonormal CCs, orthonormal ICs, and sparse core tensors changing parameters $L_C, L_I, R_C, R_I \in \{1, 2, ..., 5\}$, and estimated CCs and ICs by using the LT2D algorithm. Sum of number of components were fixed as $L_C + L_I = R_C + R_I = 6$. Amari distance [1] was used for evaluating the distance between true components and estimated components. When the number of common components is small, we can estimate the common sources accurately, however, the accuracy of individual components decreases, and vice versa. This result implies that the LT2D algorithm is useful when we extract only a few common or individual components in an accurate way. Furthermore, there is a possibility of improvements by an iterative procedure of the deflation and the LT2D with a few common and many individual components.

Figure 4 (b) shows the changes of Amari distance between original CCs and estimated CCs of the PVD, the Tucker2 decomposition (T2D), and the LT2D. We generated ten tensors randomly by four orthonormal CCs ($L_C = R_C = 4$), several orthonormal ICs changed in $L_I, R_I \in \{1, ..., 7\}$, and sparse core tensors, and extracted CCs by using above three methods. Length of each bar shows

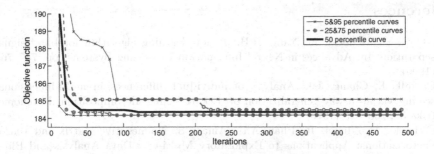

Fig. 3. Convergence properties of proposed algorithm

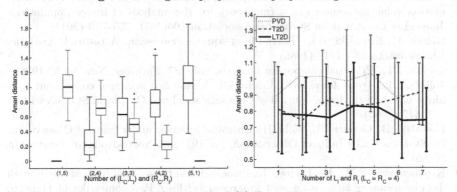

(a) Errors of common and individual (b) Error comparison of common compo-
components in LT2D   nents in PVD, T2D, and LT2D

Fig. 4. Estimation errors for each rank decompositions

average and standard deviation for 20 Monte-Carlo-runs. From this result, the
LT2D outperformed the other algorithms. Moreover, the LT2D algorithm extracts not only common components but also individual components.

5 Conclusion

In this paper, we proposed a new multi-block tensor decomposition model called
the LT2D, and implemented it by imposing the orthogonality constraints into
left and right components. The LT2D is a promising decomposition model for
multi-block tensor data analysis because of its flexibility and simplicity. There
are several possibilities that the LT2D can be applied to the feature extraction,
classification, clustering, tensor data completion, data denoising and compression. Especially, it could be applied as the robust common and individual component analysis into the multi-block tensor data such as EEG, MRI, and fMRI.
However, it needs some more improvements for uniqueness and computational
cost, and there are possibilities of imposing alternative constraints such as sparsity, nonnegativity, and statistical independency. These are topics of our future
works.

References

1. Amari, S., Cichocki, A., Yang, H.H.: A new learning algorithm for blind signal separation. In: Advances in Neural Information Processing Systems, pp. 757–763 (1996)
2. Carroll, J., Chang, J.J.: Analysis of individual differences in multidimensional scaling via an n-way generalization of 'Eckart-Young' decomposition. Psychometrika 35, 283–319 (1970)
3. Cichocki, A., Zdunek, R., Phan, A.H., Amari, S.: Nonnegative Matrix and Tensor Factorizations: Applications to Exploratory Multi-way Data Analysis and Blind Source Separation. Wiley Publishing (2009)
4. Crainiceanu, C.M., Caffo, B.S., Sheng, L., Zipunnikov, V.M., Punjabi, N.M.: Population value decomposition, a framework for the analysis of image populations. Journal of the American Statistical Association 106(495), 775–790 (2011)
5. Geladi, P., Kowalski, B.R.: Partial least-squares regression: A tutorial. Analytica Chimica Acta 185, 1–17 (1986)
6. Jolliffe, I.T.: Principal component analysis, vol. 487. Springer, New York (1986)
7. Jolliffe, I.T., Trendafilov, N.T., Uddin, M.: A modified principal component technique based on the lasso. Journal of Computational and Graphical Statistics 12(3), 531–547 (2003)
8. Kowalski, B., Gerlach, R., Wold, H.: Chemical Systems under Indirect Observation. In: Systems Under Indirect Observation, pp. 191–209. North-Holland, Amsterdam (1986)
9. Kroonenberg, P.M., De Leeuw, J.: Principal component analysis of three-mode data by means of alternating least squares algorithms. Psychometrika 45(1), 69–97 (1980)
10. Lock, E.F., Hoadley, K.A., Marron, J., Nobel, A.B.: Joint and individual variation explained (JIVE) for integrated analysis of multiple data types. The Annals of Applied Statistics 7(1), 523 (2013)
11. Phan, A., Cichocki, A.: Tensor decompositions for feature extraction and classification of high dimensional datasets. Nonlinear Theory and Its Applications, IEICE 1(1), 37–68 (2010)
12. Samaria, F., Harter, A.: Parameterisation of a stochastic model for human face identification. In: Proceedings of 2nd IEEE Workshop on Applications of Computer Vision (1994)
13. Tucker, L.R.: Some mathematical notes on three-mode factor analysis. Psychometrika 31(3), 279–311 (1966)
14. Yokota, T., Cichocki, A., Yamashita, Y.: Linked PARAFAC/CP tensor decomposition and its fast implementation for multi-block tensor analysis. In: Huang, T., Zeng, Z., Li, C., Leung, C.S. (eds.) ICONIP 2012, Part III. LNCS, vol. 7665, pp. 84–91. Springer, Heidelberg (2012)
15. Zhou, G., Cichocki, A., Xie, S.: Common and individual features analysis: beyond canonical correlation analysis. arXiv preprint arXiv:1212.3913 (2012)
16. Zou, H., Hastie, T., Tibshirani, R.: Sparse principal component analysis. Journal of Computational and Graphical Statistics 15(2), 265–286 (2006)

Celebrity Face Image Retrieval Using Multiple Features

Jie Jin and Liqing Zhang

Key Laboratory of Shanghai Education Commission for Intelligent Interaction and
Cognitive Engineering
Dept. of Computer Science and Engineering, Shanghai Jiao Tong University, China
jinjie@sjtu.edu.cn, zhang-lq@cs.sjtu.edu.cn

Abstract. Large scale face image retrieval is a hot topic in the field of internet
retrieval. There exist a number of interesting applications of face image
processing, such as hair-style design. In this paper, we propose a content-based
face image retrieval system aiming at finding similar photos of celebrities to a
user input image using a novel fusion of features and evaluation of results. After
image preprocessing such as cropping facial parts and feature extraction with
some exoteric methods, an algorithm we propose remolds the traditional features
based on their statistic characteristics. The remolded features are then fused to a
novel image representation to retrieve face images more effectively. A large
number of experiments based on the dataset collected on the Internet demonstrate the good performance of our method in Mean Average Precision (MAP).

Keywords: face image retrieval, feature remolding, feature fusion, similarity
definition.

1 Introduction

Due to the expansion of digital images and the need of recognizing them, content
based image retrieval (CBIR) has attracted many researchers who work on diversi-
form aspect of it [13, 14, 15]. In this paper, we focus on developing the face image
retrieval (FIR) method for the problem of finding the photos of celebrities that looks
most similar to the query image.

While the field of image retrieval has grown tremendously in the past decades [1],
human face image retrieval is still a challenging problem since the geometrical configuration of facial structures are quite the same [2] and the variation within faces of the
same person could be large due to different hair styles, poses or facial expressions[11,
12]. On the other hand, finding celebrities whom one looks like does not really mean
to find the particular super star. That is, any similarity in pose, hair or expression that
manifests celebrity temperament is considered "alike" to some extent in this work,
thus the second difficulty mentioned above is eliminated from our consideration.

In this paper, we propose a method of face image retrieval with high precision
utilizing hybrid image features of the face, including HSV (hue, saturation and value),
HOG (Histograms of Oriented Gradients) and LBP (Local Binary Patterns) histogram.
As shown in Fig. 1, the retrieval system mainly contains three parts: face region detection, feature extraction, and hybrid feature matching.

C.K. Loo et al. (Eds.): ICONIP 2014, Part III, LNCS 8836, pp. 119–126, 2014.
© Springer International Publishing Switzerland 2014

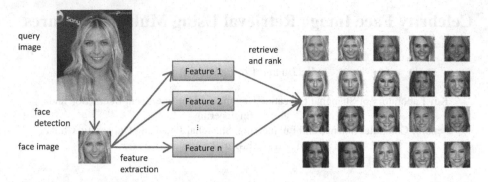

Fig. 1. Chief procedure of the proposed retrieval system

For the sake of our selection of features and the evaluation criterion, we define a face similarity measure in which the three factors of hair, gesture and expression should all be considered. In order to meet these demands, multiple features accomplished in color, shape and texture are extracted to characterize the outline of cheek/hairline, the color of hair/skin/eye and the subtle variety of facial features.

The contribution of our work consists of following parts: propose a novel algorithm to improve retrieval efficiency by remolding features of HSV and HOG according to the information gain of each dimension; increase face description accuracy by weighing hybrid contribution of global feature and local features in retrieval stage; define a characteristic evaluation on retrieving results for celebrity faces.

2 Face Representation with Multiple Features

In this section, we introduce the face representation module of our system, which mainly includes face detection and multiple feature extraction.

2.1 Detection of Face Region

Given an image with face, the first procedure is to find the facial region. In this part, we choose to apply the robust real-time face detection [3] method, whose accuracy and rapidness is universally acknowledged. Then we resize the face to 64 * 64 pixels.

2.2 Multiple Description of Face

Since gesture, expression and color are three components in defining the face similarity, we need to define multiple image features reflecting these factors. In this work, HSV, HOG and LBP features are chosen as color, shape and texture descriptors respectively to represent the face extracted from the original image, and each of them are remolded to satisfy the demand of retrieving similar faces.

Compressed HSV Color Density. A. R. Smith proposed a human perceptual color space of HSV in 1978 [5].

In our work, the distribution of HSV serves as a global feature. The original 256 bins of each layer are combined to a 768-dimension feature vector. To reduce the dimension and to improve the representation of faces, we provide a method to select the scope of each bin, as depicted in Fig. 2.

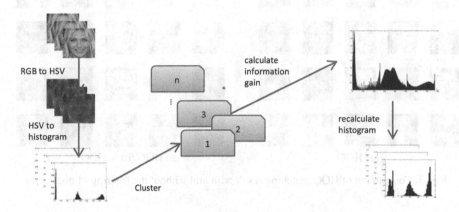

Fig. 2. Steps of extracting and compressing HSV features

First of all, we get RGB face images from our dataset, gray images presented as duplicated gray scale value in each layer to get only V value in HSV space. Secondly, the intensity is counted after conversion to HSV, 256 bins in each layer. Then, all faces represented in a 768 dimension feature are clustered to get a temporary class label thus faces with similar color distribution flock together. After that, information gain of each dimension is calculated to testify their ability of distinguishing faces of different color distribution. Finally, less discriminative bins are merged together.

Histogram of HOG. The HOG descriptor was first proposed in 2005 for human detection, whose feature extraction chain is described in [6].

In this work, we choose the cell width of 8 pixels and block width of 2 cells on our 64-pixel-wide face image, splitting gradients into 9 unsigned orientations, resulting in 1764 dimensions of feature. As gradient information is calculated in each cell, which concludes the hair regions, facial parts and the boundaries between them, HOG descriptor has high capability in describing the hair style, pose and facial expression based on contour.

Although the result is satisfactory, the high dimension of HOG feature would cause low retrieval efficiency. Consequently, we provide a compromised solution: before matching the HOG descriptor as a local feature, we calculate its global distribution. For each orientation, we quantify the descriptors' density into 100 bins. Then the process in

Fig. 2 is applied again to the histogram of HOG reckoning the most appropriate ranging edges.

Followed by regular matching of HOG descriptors, the filtering step on histogram accelerates retrieving while the effectiveness preserves, as demonstrated in Fig. 3.

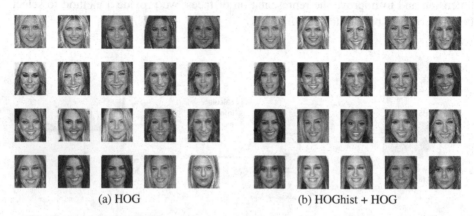

(a) HOG   (b) HOGhist + HOG

Fig. 3. Comparison of HOG matching result with and without the filtering of histogram

Patched LBP histogram. LBP is employed in this work to describe the details of the face which [7] performed well in texture classification at prime tense, and was later applied to representing face recognition [8, 9]. The LBP operator labels every pixel in the image by thresholding its 3*3 neighbors with the center pixel value producing a coded image, followed by computing the histogram with predefined bins according to rotation invariant and uniform patterns [7]. In our work, the histogram is then calculated on 121 8*8-pixel overlapping patches of the face image, whose sequence forms the LBP representation of the query and database faces.

3   Face Image Similarity on Hybrid Features

Considering time complexity and referring to the experiment demonstrated in [4], we choose L1 similarity measurement, which is highly efficient and performs relatively well:

$$D(X,Y) = \sum_i |X_i - Y_i| \tag{1}$$

The retrieving phase can be divided mainly into two steps: step 1, filtering with global features forsaking most candidates; step 2, re-ranking with local features. In the first step, we match through all images in the dataset with compressed HSV and HOG histogram, leaving only thousands of faces that are most likely to be similar to the query face. This approach promises to be within one second and could be adapted to parallel as the dataset grows. In the second step, detailed matching of HOG and LBP is employed, which used to be time consuming.

At the moment of fusing the measurements of different features, we follow the discipline hereinafter:

$$D(X,Y) = \prod_j \left(1 + \alpha_j D_j(X,Y)\right) \qquad (2)$$

$$\text{where } \sum_j \alpha_j = 1$$

in which α_j adjusts the significance of feature j.

On deciding the exact values of α, we build a validation set with 17 subjects in our dataset. For each subject, we select 50 face images of similar pose, expression and hair style, which sum up a labeled set of 850 faces. On this set, we test the retrieval system with different parameters, and then plot the precision-recall curve to get the best settings, as illustrated in Fig. 4.

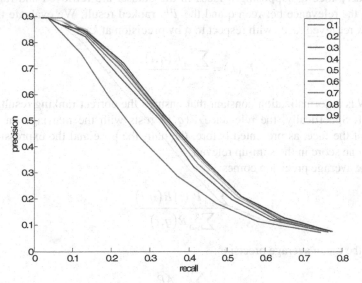

Fig. 4. The precision-recall trade-off with α of HOG varying from 0.1 to 0.9, based on which we decide the value of 0.7 to be adopted

Subsequently, the weights for HOG, LBP, and HSV are set to 0.63, 0.27, and 0.1 respectively, indicating that shape feature do plays a more important role in face description while texture and color offer supplementary information.

4   Experiments

In this section, we first introduce the dataset used in the experiments, and then compare the retrieval performance of the proposed method with existing methods.

4.1   Dataset

The whole dataset is constructed by online images of the world celebrities. We browse through the Celebrity 100 List[1] of year 2011 and 2012, and then collect hundreds of web images for each of them by Google Image[2]. After face detection, we get 286903 face images in total.

In the following experiments, 49 celebrities with one picture for each are selected arbitrarily as a testing set to examine the performance of different combinations of previously introduced methods with the evaluation criterion presented below.

4.2   Evaluation Criterion

Drawing on the experience of [10], we follow a rank-based criterion for evaluation. Given a query face q, supposing n faces in the dataset are retrieved and ranked, let $R(q, i)$ be the relevance between q and the ith ranked result. We evaluate the ranking of top k retrieved faces with respect to q by precision at k:

$$P_q(k) = \frac{\sum_{i=1}^{k} R(q,i)}{kN} \tag{3}$$

in which N is a normalization constant that ensures the correct ranking results in precision of 1. Specifically, the relevance $R(q, i)$ rests with the marriage on multiple attributes of the face, as presented before, the hair, the pose and the expression, each voting for one score in the sum-up relevance.

Then the average precision comes:

$$AP_q = \frac{\sum_{i=1}^{n} P_q(i) R(q,i)}{\sum_{i=1}^{n} R(q,i)} \tag{4}$$

Finally the mean average precision:

$$MAP = \frac{\sum_{q=1}^{Q} AP_q}{Q} \tag{5}$$

where Q stands for the number of query faces.

4.3   Result

By experimenting on the single-features of HOG, LBP and SIFT, we found that HOG performs best, so we perform experiments with different combinations of features and methods including: HOG-only, HOG with histogram filtering, unweighted multiple features, and weighted multiple features. The evaluation is based on the criterion of precision at k, averaged by the 49-picture testing set, as plotted in Fig. 5(a).

---

[1] http://www.forbes.com/celebrities/
[2] http://images.google.com/

The second and third line of HOG-only and HOG&hist almost twining together testifies that the acceleration by histogram filtering does not reduce much precision. The unweighted line stands for combination of HOG, LBP and HSV features with histogram filtering and the same parameter α, while the weighted line stands for the same combination except that the values of α are tuned with the validation set. Compared with HOG methods, the advantage of weighted fusion and the disadvantage of unweighted fusion are both visible, thus the multiple-feature method outperforms if and only if the parameters are finely tuned.

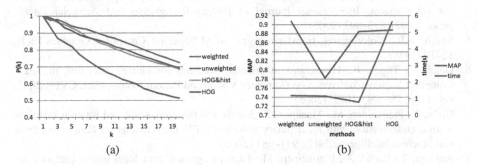

(a)   (b)

Fig. 5. (a) The precision plot of different combinations of features and methods. (b) The red line represents MAP in major coordinate, while the blue one represents time consumption in minor coordinate.

To illustrate the balance between time and precision, Fig. 5(b) displays the MAP and time consumption of different methods, in which the time consumption includes feature extraction and distance calculation. Since α is calculated once, this part of timing is not recorded here. As plotted, the histogram filtering on HOG saves much time than original HOG distance comparison with acceptable precision loss. Weighted feature fusion method ranks first in precision with MAP of 0.907 and costs a little more time than single HOG feature filtered with its histogram.

5   Conclusions

In this work, we provide a celebrity face image retrieval system by using novel remold and fusion of multiple features, which improves efficiency and accuracy. The evaluation criterions of precision at k and MAP for similar faces are defined. The final MAP exceeds 0.9 on testing set which is sharply quicker than using the original single features and evidently more accurate than their average combination.

Acknowledgement. The work was supported by the national natural science foundation of China (GrantNo. 91120305, 61272251).

References

1. Datta, R., Joshi, D., Li, J., Wang, J.Z.: Image retrieval: Ideas, influences, and trends of the new age. ACM Computing Surveys (CSUR) 40(2), 5 (2008)
2. Lee, H., Chung, Y., Kim, J., Park, D.: Face image retrieval using sparse representation classifier with gabor-LBP histogram. In: Chung, Y., Yung, M. (eds.) WISA 2010. LNCS, vol. 6513, pp. 273–280. Springer, Heidelberg (2011)
3. Viola, P., Jones, M.J.: Robust real-time face detection. International Journal of Computer Vision 57(2), 137–154 (2004)
4. Shih, P., Liu, C.: Comparative assessment of content-based face image retrieval in different color spaces. International Journal of Pattern Recognition and Artificial Intelligence 19(07), 873–893 (2005)
5. Smith, A.R.: Color gamut transform pairs. ACM Siggraph Computer Graphics 12(3), 12–19 (1978)
6. Dalal, N., Triggs, B.: Histograms of oriented gradients for human detection. In: IEEE Computer Society Conference on Computer Vision and Pattern Recognition, CVPR 2005, vol. 1, pp. 886–893. IEEE (2005)
7. Ojala, T., Pietikainen, M., Maenpaa, T.: Multiresolution gray-scale and rotation invariant texture classification with local binary patterns. IEEE Transactions on Pattern Analysis and Machine Intelligence 24(7), 971–987 (2002)
8. Ahonen, T., Hadid, A., Pietikäinen, M.: Face recognition with local binary patterns. In: Pajdla, T., Matas, J(G.) (eds.) ECCV 2004. LNCS, vol. 3021, pp. 469–481. Springer, Heidelberg (2004)
9. Ahonen, T., Hadid, A., Pietikainen, M.: Face description with local binary patterns: Application to face recognition. IEEE Transactions on Pattern Analysis and Machine Intelligence 28(12), 2037–2041 (2006)
10. Liu, S., Song, Z., Liu, G., Xu, C., Lu, H., Yan, S.: Street-to-shop: Cross-scenario clothing retrieval via parts alignment and auxiliary set. In: IEEE Conference on Computer Vision and Pattern Recognition (CVPR), pp. 3330–3337. IEEE (2012)
11. Wu, Z., Ke, Q., Sun, J., Shum, H.-Y.: Scalable face image retrieval with identity-based quantization and multireference reranking. IEEE Transactions on Pattern Analysis and Machine Intelligence 33(10), 1991–2001 (2011)
12. Chen, B., Chen, Y., Kuo, Y., Hsu, W.: Scalable face image retrieval using attribute-enhanced sparse codewords, 1 (2012)
13. Cao, Y., Wang, C., Zhang, L., Zhang, L.: Edgel index for large-scale sketch-based image search. In: IEEE Conference on Computer Vision and Pattern Recognition (CVPR), pp. 761–768. IEEE (2011)
14. Smith, B.M., Zhu, S., Zhang, L.: Face image retrieval by shape manipulation. In: IEEE Conference on Computer Vision and Pattern Recognition (CVPR), pp. 769–776. IEEE (2011)
15. Zhou, R., Chen, L., Zhang, L.: Sketch-based image retrieval on a large scale database. In: Proceedings of the 20th ACM International Conference on Multimedia, pp. 973–976. ACM (2012)

A Novel Adaptive Shrinkage Threshold
on Shearlet Transform for Image Denoising

Sheikh Md. Rabiul Islam, Xu Huang, and Kim Le

Faculty of Education, Science, Technology and Mathematics
University of Canberra, Australia
{Sheikh.Islam,Xu.Huang,Kim.Le}@canberra.edu.au

Abstract. Shearlet is a new multidimensional and multiscale transform which is optimally efficient in representing image containing edges. In this paper an adaptive shrinkage threshold for image de-noising in shearlet domain is proposed. Experimental results show that images de-noised with the proposed approach had higher qualities than those produced with some of the other de-noising methods like wavelet-based, bandlet-based, shearlet-based and curvelet-based.

Keywords: UIQI, Denoising, Kurtosis, Skewness, Shearlet transform, Adaptive threshold.

1 Introduction

The wavelet transform, one of the computational harmonic analysis methods, has been successfully used in image denoising field. There is a vast literature on image denoising using the wavelet threshold or shrinkage that was first introduced by Donoho and Johnstone [3]. The most well-known thresholds are those of Donoho and Johnstone and Sure [4]. However, when wavelet is used for image denoising, oscillation occurs along edges, and therefore wavelet fails to capture the geometric regularity along the singularities of surfaces. In order to overcome this limitation of the traditional wavelet, several image representations have been proposed to capture the geometric regularity of a given image, including curvelet [5], contourlet [6], bandlet [7] and shearlet transform[8]. An efficient denoising algorithm, which was adapted to the scales and orientations of an image based on shearlet transform, was proposed in [9]; the authors introduced a shrinkage thresholding function with three factors K, $\varepsilon_{j,l}$ and σ, where K is a constant, $\varepsilon_{j,l}$ denotes the average energy distribution of white noise in the shearlet coefficient, and σ is the standard deviation of white noise. Other authors [10] proposed a joint multi-scale algorithm based on the auto-adaptive Monte Carlo threshold for curvelet; they also considered a shrinkage thresholding function with three factors K, $\sigma_{j,l}$ and σ, where the second factor, $\sigma_{j,l}$, is different from [9] and is the curvelet coefficient of white noise.

In this paper, we propose an efficient de-noising method using an adaptive shrinkage threshold for the shearlet transform based on our previous work [11]. Since edges are usually the most prominent features in natural and scientific images, the

C.K. Loo et al. (Eds.): ICONIP 2014, Part III, LNCS 8836, pp. 127–134, 2014.
© Springer International Publishing Switzerland 2014

localization of edges is a fundamental low level task for higher level applications such as shape recognition, 3D reconstruction, data enhancement and restoration. In most common edge detector schemes, to watch out for the interference of noise, an image is first smoothed out or mollified. For example, in the classical Canny edge detection algorithm [12], an image is first convolved with a scalable Gaussian filter. Our proposed de-noising approach can be applied in association with any edge detection application.

To evaluate the effectiveness of the proposed de-noising method, we made an experiment on a set of images (about 30 images) selected from open sources as follows: For each selected image, a) Introduce a fixed level of noise to the image; b) Apply some de-noising methods, including our proposed approach, to the noisy image; c) Assess the qualities of the de-noised images using an image quality assessment scheme [13]. Experimental results show that the images which were de-noised by our proposed approach had higher qualities than those produced with some other de-noising methods like wavelet-based [1], bandlet-based [7], shearlet-based [2] and curvelet-based [5].

The paper is organised as follows. Sections 2 & 3 give a review of the shearlet transform and the Donoho universal threshold using for de-noising. Section 4 describes our proposed de-noising approach with an adaptive shrinkage threshold on the shearlet transform. Experimental results are reported in Section 5. The paper ends with a brief conclusion.

2   Shearlet Transform

The theory of composite wavelets provides an effective approach for combining geometry and multi-scale analysis by taking advantage of the classical theory of affine systems. In the dimension of two, i.e., $n = 2$, the affine systems with composite dilations are the collections of the form $\{\Psi_{AB}(\psi)\}$, where $\psi \in L^2(\mathbb{R}^2)$, and A, B are 2×2 invertible matrices with $|\det B = 1|$, and

$$\Psi_{AB}(\psi) = \left\{\psi_{j,k,l} = |\det A|^{j/2}\psi(B^l A^j x - k) : j, l \in \mathbb{Z}, k \in \mathbb{Z}^2\right\}$$

The elements of this system are called composite wavelets if $\Psi_{AB}(\psi)$ forms a Parseval frame for $L^2(\mathbb{R}^2)$.

The shearlet is a special example of composite wavelets in $L^2(\mathbb{R}^2)$. There are collections of the form $\{\Psi_{AB}(\psi)\}$ where $A = A_0 = \begin{pmatrix} 4 & 0 \\ 0 & 2 \end{pmatrix}$, the anisotropic dilation matrix, and $B = B_0 = \begin{pmatrix} 1 & 1 \\ 0 & 1 \end{pmatrix}$, the shear matrix. For any $\xi = (\xi_1, \xi_2) \in \mathbb{R}^2, \xi_1 \neq 0$, let $\psi^{(0)}$ be given by

$$\hat{\psi}^{(0)}(\xi) = \hat{\psi}^{(0)}(\xi_1, \xi_2) = \hat{\psi}_1(\xi_1)\hat{\psi}_2\left(\frac{\xi_2}{\xi_1}\right) \tag{1}$$

Where $\hat{\psi}_1\hat{\psi}_2 \in C^\infty(\mathbb{R})$, $supp\ \hat{\psi}_1 \subset [-1/2, -1/16] \cup [1/16, 1/2]$ and $\hat{\psi}_2 \subset [-1,1]$. This implies that $\hat{\psi}^{(0)}$ is C^∞ and compactly supported with $\hat{\psi}^{(0)} \subset [-1/2, -1/2]^2$. In addition, we assume that

$$\sum_{j\geq0}\left|\hat{\psi}_1(2^{-2j}\omega)\right|^2 = 1 \text{ for } |\omega| \geq \frac{1}{8} \tag{2}$$

and, for each $j \geq 0$

$$\sum_{l=-2^j}^{2^j-1}\left|\hat{\psi}_2(2^j\omega - l)\right|^2 = 1 \text{ for } |\omega| \leq 1 \tag{3}$$

The equations (2) and (3) imply that

$$\sum_{j\geq0}\left|\hat{\psi}^{(0)}(\xi A_0^{-j}B_0^{-l})\right|^2 = \sum_{j\geq0}\sum_{l=-2^j}^{2^j-1}\left|\hat{\psi}_1(2^{-2j}\xi_1)\hat{\psi}_2\left(2^j\frac{\xi_2}{\xi_1} - l\right)\right|^2 = 1$$

Because of the fact that $\hat{\psi}^{(0)}$ is supported inside $\left[-\frac{1}{2}, -\frac{1}{2}\right]^2$ the collection:

$\left\{\psi_{j,k,l}^{(0)} = 2^{3j/2}\hat{\psi}^{(0)}(B_0^l A_0^j x - k): j \geq 0, -2^j \leq l \leq 2^j, k \in \mathbb{Z}^2\right\}$ is a Parseval frame for $L^2(\mathcal{D}_0)^V = \{L^2(\mathbb{R}^2): supp\ \hat{f} \subset \mathcal{D}_0\}$. From the conditions on the support of $\hat{\psi}_1$ and $\hat{\psi}_2$, the functions $\psi_{j,k,l}$ have frequency support.

Similarly we can construct a Parseval frame for $L^2(\mathcal{D}_1)^V$, where $\mathcal{D}_1 = \left\{(\xi_1, \xi_2) \in \mathbb{R}^2: |\xi_1| \geq \frac{1}{8}, \left|\frac{\xi_2}{\xi_1}\right| \leq 1\right\}$ is the vertical cone.

Let $A_1 = \begin{pmatrix} 2 & 0 \\ 0 & 4 \end{pmatrix}$, $B_1 = \begin{pmatrix} 1 & 0 \\ 1 & 1 \end{pmatrix}$ and $\hat{\psi}^{(1)}$ be given by $\hat{\psi}^{(1)}(\xi) = \hat{\psi}^{(1)}(\xi_1, \xi_2) = \hat{\psi}_1(\xi_1)\hat{\psi}_2\left(\frac{\xi_2}{\xi_1}\right)$, where $\hat{\psi}_1$ and $\hat{\psi}_2$ are defined above. Then the collection

$\left\{\psi_{j,k,l}^{(1)} = 2^{3j/2}\hat{\psi}^{(1)}(B_1^l A_1^j x - k): j \geq 0, -2^j \leq l \leq 2^j, k \in \mathbb{Z}^2\right\}$ is a Parseval frame for $L^2(\mathcal{D}_1)^V$. Finally, let $\varphi \in L^2(\mathbb{R}^2)$ satisfy $|\hat{\varphi}(\xi)|^2 + \sum_{j\geq0}\sum_{l=-2^j}^{2^j-1}|\hat{\psi}^{(0)}(\xi A_0^{-j}B_0^{-l})|^2 + \sum_{j\geq0}\sum_{l=-2^j}^{2^j-1}|\hat{\psi}^{(1)}(\xi A_1^{-j}B_1^{-l})|^2 = 1$ for $\xi \in \mathbb{R}^2$. This implies that $\hat{\varphi} \subset \left[-\frac{1}{2}, -\frac{1}{2}\right]^2$, with $|\hat{\varphi}(\xi)| = 1$ for $\xi \in \left[-\frac{1}{16}, \frac{1}{16}\right]^2$ and the set $\{\varphi(x - k): k \in \mathbb{Z}^2\}$ is a Parseval frame for $L^2\left(\left[-\frac{1}{16}, \frac{1}{16}\right]^2\right)^V$. For more details refer to [2] [8].

3 Donoho Universal 'VisuShrink' Threshold

Finding an optimal threshold is a challenge for any efficient and effective de-noising algorithm. Donoho proposed a threshold with a parameter $\delta_{j,l}$ which is called the universal 'VisuShrink' threshold [14]:

$$\delta_{j,l} = \sigma\sqrt{2log(N)} \tag{4}$$

where N is the number of pixels of the image and σ is the standard deviation of the noise level. Donoho and Johnstone proposed a simple wavelet-based de-noising scheme called *VisuShrink* using Equation (4). Many researchers have shown that the original Donoho threshold given in Equation (4) has some weakness. In [2], Donoho proposed an improvement on the threshold as follows:

$$\delta_{j,l} = \sigma \sqrt{2log(N)} \times 2^{\frac{(j-l)}{2}} \tag{5}$$

For shearlet transform de-noising, the Donoho universal threshold does not produce images with smooth visual appearances. We propose a modified threshold which is presented in the next section.

4   Image De-noising with Adaptive Shrinkage Threshold on Shearlet Transform

First, we modify the Donoho universal threshold $\delta_{j,l}$ as follows

$$\delta_{j,l} = tanh\left(\sigma\sqrt{2log(N)} \times 2^{\frac{(j-l)}{2}}\right) = \frac{2}{\left(1+e^{-2\sigma\sqrt{2log(N)}\times2^{\frac{(j-l)}{2}}}\right)} - 1 \tag{6}$$

Equation (6) is called hyperbolic tangent sigmoid function [15] or active function. The noise standard deviation in each scale and direction of a normalized noise image is inferred using Monte Carlo techniques. If we take $N \times N$ random samples based on g, the Monte Carlo simulation [16] can be used to estimate the noise standard deviation $\sigma_{\gamma_{j,l}}$ as follows

$$\sigma_{\gamma_{j,l}} = \frac{1}{N}\sqrt{\sum_{j=1}^{N}\sum_{l=1}^{N} f_{\gamma}^{Shearlet}.f_{\gamma}^{Shearlet^*}} \tag{7}$$

where $f_{\gamma}^{Shearlet}$ denotes shearlet coefficients of noisy image and $f_{\gamma}^{Shearlet^*}$ is the complex conjugate of $f_{\gamma}^{Shearlet}$.

Suppose that a noisy image can be expressed as $f = I + D$ where I is the original image and D is a combination of the popular three different noises, namely white Gaussian noise, Poison noise and impulse noise. With hard thresholding, a de-noised shearlet coefficient $\hat{f}(j,l)$, where j and l are the scale index and the direction index respectively, is calculated as follows:

$$\hat{f}(j,l) = \begin{cases} f(j,l) & |f(j,l)| \geq T(j,l) \\ 0 & |f(j,l)| < T(j,l) \end{cases} \tag{8}$$

In (8), $T(j,l)$ is a threshold and $f(j,l)$ is a shearlet coefficient of the noisy image. In the traditional shearlet hard thresholding, small shearlet coefficients are simply set to zero; this usually brings non-smooth shearlet-like artefacts. To overcome this drawback, we propose an adaptive shrinkage threshold function for smooth regions and edge preservation into complexity of image based on the multi-scale and multi-directional characteristics of shearlet transform. The proposed threshold function is defined as

$$T(j,l) = K\sigma^l\left(\delta_{j,l}\right)^m\left(\sigma_{\gamma_{j,l}}\right)^n \tag{9}$$

Where, σ is the standard deviation of the noisy image, $\delta_{j,l}$ is the Donoho threshold parameter as described in Equation (6), and $\sigma_{\gamma_{j,l}}$ is the noise standard deviation as given in Equation (7). The constant K often has a value in the range [2, 3] for different scales. The values of l, m and n are about 1. However, they can be tuned to get optimal quality indexes (QI). The de-noising method with our proposed adaptive shrinkage threshold is summarised as follows:

De-noising Algorithm
For each noisy image
  1) Perform the discrete shearlet transform to decompose the noisy image and obtain shearlet coefficients $I_k(j, l)$ where j is the scale index and l is the direction index.
  2) Using Equation (6) to calculate $\delta_{j,l}$ of the noise shearlet coefficients.
  3) Using Equation (7) to calculate the noise standard deviation variance $\sigma_{\gamma_{j,l}}$ from the noise shearlet coefficients after the shearlet transform using the Monte Carlo simulation.
  4) Using Equation (9) to calculate the threshold $T(j, l)$ at each scale and each direction, and then apply this adaptive hard threshold to the de-noised shearlet coefficients $\hat{f}(j, l)$ in Equation (8).
  5) Perform the inverse discrete shearlet transform to get the de-noised image.

  To evaluate the efficiency of the de-noising approach using the proposed adaptive shrinkage threshold on shearlet transform, we made the following experiment on a set of images selected from open sources.

Experiment procedure:
For each experimental image
  1) Introduce noise to the image. The noise is a combination of three popular noises including fixed Poisson noise at $\lambda = 0.9686$, 100% impulse noise density, and white Gaussian noise with a fixed deviation level $\sigma \in [10, 20, 30, 40, 50]$
  2) Perform some de-noising approaches, including our proposed algorithm
  3) Use an index, e.g. UIQI, to evaluate the quality of the de-noised images.
  4) Compare the efficiency of the proposed de-noising approach with those of other methods based on their quality indexes.

5   Experimental Results

This section presents experimental results performed with our proposed de-noising algorithm on images selected from open source databases [17]. The images are of 8-bit 512×512 pixels. We used different image quality indexes: PNSR, UIQI, Q (Skewness), Q (Kurtosis) [13] to evaluate the qualities of the de-noised images. For comparison, besides our proposed method, other de-noising techniques were also used, including Shearlet transform with hard thresholding, Bandlet transform with hard thresholding, Curvelet transform with curvelet thresholding, and wavelet thresholding techniques. Figure 1 shows the quality indexes of de-noised images obtained with different de-noising methods performed on the Lena image. For our proposed method, we initially chose $l = m = n = 1$ and $K = 2.8$. The experimental results clearly show that the proposed method out-performed all other four methods for all values of the noise deviation in the range [10,50].

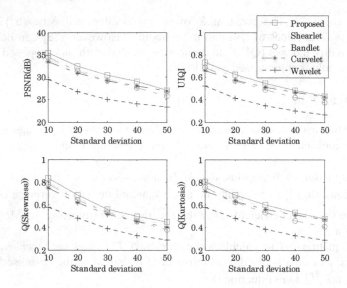

Fig. 1. Plots of four image quality indexes PSNR, UIQI, Q(Skewness), Q(Kurtosis) for Lena image versus noise standard deviations in the range [10,50] for five different de-noising methods including the proposed approach

We also made experiments to obtain optimal values for l, m and n in Equation (9) using a similar approach for tuning the three gains G_P, G_I and G_D of a PID (Proportional, Integral and Differential) control. Recall that the control action of a PID control is given by

$$C = G_P E(t) + G_I \int E(t)d(t) + G_D \frac{dE(t)}{dt}$$

where $E(t)$ is the error from a set-point. By taking the logarithm of Equation (9), $\log T(j, l) = \log K + l \log \sigma + m \log \delta_{j,l} + n \log \sigma_{\gamma_{j,l}}$, we derive a form similar to that of PID control. In this form, the powers l, m and n can be considered as the gains.

Figure 2 shows the visual comparison of de-noised images while tuning for optimal values of l, m and n. The comparison is clearer with data plotted in Figure 3, which shows the relationship between the image quality index UIQI and the values of l, n and m for different standard deviations performed on an X-ray image.

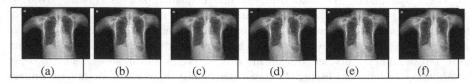

| (a) | (b) | (c) | (d) | (e) | (f) |

Fig. 2. Visual comparison of proposed de-nosing method while tuning for optimal parameters: a) Input image, b) Noisy image with noise deviation $\sigma = 10$, c) De-noised image with $m = 1$ and $l = n = 0$, d) De-noised image with $m = 1$ and $l = n = 1.6$, e) De-noised image with $m = 0$ and $l = n = 1$, f) De-noised image with $m = 1$ and $l = n = 1$

In Figure 3, with $K = 2.8$, $\sigma = 10$, when $m = 1$, the optimal point is $(UIQI, l/n) = (0.56, 1.4)$ and when $l = n = 1$, the optimal point is $(UIQI, m) = (0.56, 1)$. These two optimal points are plotted (points 21) in Figures 4a & 4b, which also plot optimal points for 29 other images. Figures 4a & 4b show that for different images, the optimal quality indexes UIQI were obtained when the values of l, m and n are about unity. A reasonable choice is $(l = n = 1; m = 0.9)$.

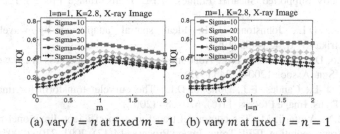

(a) vary $l = n$ at fixed $m = 1$ (b) vary m at fixed $l = n = 1$

Fig. 3. The relationship between the image quality (UIQI) and tuned optimal values for different standard deviation for an X-ray images with $K = 2.8$

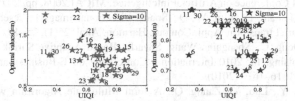

(a) Optimal points $(UIQI, l/n)$ at $m = 1$ (b) Optimal points (UIQI, m) at $l = n = 1$

Fig. 4. Optimal points for 30 experimental images 1) Peppers(grey), 2) Lena (grey), 3) Barbara (grey), 4) Boat (grey), 5) Aerial (grey), 6) Airplane-1 (grey), 7) Couple (grey), 8) Elaine (grey), 9) Tank (grey), 10) Truck (grey), 11) CT-1 cancer (grey)., 12) Baboon (grey), 13) Cat (grey), 14) Man (grey), 15) Monkey (grey), 16) Airplane-2 (grey), 17) Heritage (grey), 18) Fish (grey), 19) House (grey), 20) Infrared (grey), 21) X-ray cancer (grey), 22) MR cancer (grey), 23) Lena (colour), 24) Peppers (colour), 25) Barbara (colour), 26) Airplane (colour), 27) Boat(colour), 28) House(colour), 29) Baboon(colour), 30) CT-2 cancer (grey).

6   Conclusion

In this paper, an effective and efficient de-noising approach is proposed based on the shearlet transform. The experimental results show that the proposed method has significantly improved noise removal for the three popular types of noises: Gaussian, Poisson and impulse noises; it gives better results with smoothness and edge preservation at the shearlet coefficients. The results also demonstrate that the shearlet transform is very competitive for de-noising tasks, and noise removal significantly depends on noise levels of images. Image compression frameworks and edge detection schemes may need to consider in determination of threshold to further improve de-noising performance.

References

1. Pizurica, A., Philips, W.: Estimating the probability of the presence of a signal of interest in multiresolution single- and multiband image denoising. IEEE Trans. Image Process. 15(3), 654–665 (2006)
2. Lim, W.-Q.: The Discrete Shearlet Transform: A New Directional Transform and Compactly Supported Shearlet Frames. IEEE Trans. Image Process. 19(5), 1166–1180 (2010)
3. Donoho, D.L., Johnstone, J.M.: Ideal spatial adaptation by wavelet shrinkage. Biometrika 81(3), 425–455 (1994)
4. Donoho, D.L., Johnstone, I.M.: Adapting to unknown smoothness via wavelet shrinkage. J. Am. Stat. Assoc., 1200–1224 (1995)
5. Starck, J.-L., Candes, E.J., Donoho, D.L.: The curvelet transform for image denoising. IEEE Trans. Image Process. 11(6), 670–684 (2002)
6. Do, M.N., Vetterli, M.: The contourlet transform: an efficient directional multiresolution image representation. IEEE Trans. Image Process. 14(12), 2091–2106 (2005)
7. Villegas, O.O.V., De Jesus Ochoa Dominguez, H., Sanche, V.G.C.: A Comparison of the Bandelet, Wavelet and Contourlet Transforms for Image Denoising. In: Seventh Mexican International Conference on Artificial Intelligence, MICAI 2008, pp. 207–212 (2008)
8. Easley, G., Labate, D., Lim, W.-Q.: Sparse directional image representations using the discrete shearlet transform. Appl. Comput. Harmon. Anal. 25(1), 25–46 (2008)
9. Chen, X., Deng, C., Shengqian-Wang: Shearlet-Based Adaptive Shrinkage Threshold for Image Denoising. In: 2010 International Conference on E-Business and E-Government (ICEE), pp. 1616–1619 (2010)
10. He, J., Sun, Y., Luo, Y., Zhang, Q.: A Joint Multiscale Algorithm with Auto-adapted Threshold for Image Denoising. In: Fifth International Conference on Information Assurance and Security, IAS 2009, vol. 2, pp. 505–508 (2009)
11. Islam, S.M.R., Huang, X., Liao, M., Srinath, N.K.: Image Denoising Based on Wavelet for IR Images Corrupted by Gaussian, Poisson & Impulse Noises. IJCSNS Int. J. Comput. Sci. Netw. Secur. 13(6), 59–70 (2013)
12. Canny, J.: A Computational Approach to Edge Detection. IEEE Trans. Pattern Anal. Mach. Intell. PAMI-8(6), 679–698 (1986)
13. Islam, S.M. R., Huang, X., Le, K.: A Novel Image Quality Index for Image Quality Assessment. In: Lee, M., Hirose, A., Hou, Z.-G., Kil, R.M. (eds.) ICONIP 2013, Part III. LNCS, vol. 8228, pp. 549–556. Springer, Heidelberg (2013)
14. Donoho, D.L.: De-noising by soft-thresholding. IEEE Trans. Inf. Theory 41(3), 613–627 (1995)
15. Yonaba, H., Anctil, F., Fortin, V.: Comparing Sigmoid Transfer Functions for Neural Network Multistep Ahead Stream flow Forecasting. URNAL Hydrol. Eng. © ASCE, 275–283 (April 2010)
16. Bhadauria, H.S., Dewal, M.L.: Medical image denoising using adaptive fusion of curvelet transform and total variation. Comput. Electr. Eng. 39(5), 1451–1460 (2013)
17. SIPI Image Database - Misc

Perception of Symmetry in Natural Images

A Cortical Representation of Shape

Ko Sakai, Ken Kurematsu, and Shouhei Matsuoka

Department of Computer Science, University of Tsukuba, Japan
{sakai,kurematsu,matsuoka}@cvs.cs.tsukuba.ac.jp

Abstract. Symmetry has long been considered as an influential Gestalt factor for grouping and figure-ground segregation. As natural contours are not precisely symmetric in terms of geometry, we proposed a quantification of the degree of symmetry (DoS) that is applicable for arbitrary contours in natural images. DoS showed an agreement with the perception of symmetry in judgment of symmetry axis. Multi-dimensional scaling, together with similarity tests among natural contours, showed that DoS is a quantitative perceptual measure that accounts for the shape of contour. These results indicate that DoS reflects the perception of symmetry in natural contours, and further suggest that DoS is a plausible candidate for representing shape in the cortex.

Keywords: vision, perception, cognitive science, cortical representation, natural image, visual psychophysics.

1 Introduction

Gestalt factors, such as convexity, closure, parallel and symmetry, have been known as cues for grouping and figure-ground segregation that are crucial bases for the perception of shape and object. Symmetry has long been considered as an influential Gestalt factor because symmetry is frequently observed among living creatures and artificial products [e.g., 1, 2]. However, natural contours are not precisely symmetric in terms of geometry, thus no quantitative analysis on symmetry has been studied with natural images. Quantification of *the degree of symmetry* needs to be proposed for investigating the perception in natural images. Focusing on symmetry in natural contours, we established a computational index that describes the degree of symmetry (DoS) inherent in arbitrary contours. DoS was computed based on the degree of the overlap of contours between two sub-images divided by the optimal symmetry axis that was searched thoroughly. Our psychophysical experiment showed that the proposed DoS agreed with the perception of symmetry in the judgment of symmetry axis. To assure that DoS is a quantitative perceptual measure, we performed similarity tests between a variety of natural contour patches, and analyzed whether DoS accounts for the similarity. Multi-dimensional scaling (MDS) analyses showed that DoS, together with convexity and closure, is indeed a perceptual measure. These results indicate that DoS reflects the perception of symmetry in natural contours. Together with the recent evidence on adaptation [3], our result also supports the cortex representation of

C.K. Loo et al. (Eds.): ICONIP 2014, Part III, LNCS 8836, pp. 135–141, 2014.
© Springer International Publishing Switzerland 2014

symmetry as a basis for shape perception. The proposed DoS will greatly help studying symmetry in the perception of natural images.

2   Quantification of Symmetry

We propose the *degree of symmetry* as a quantitative measure to describe how much a local contour is close to the axial symmetry. Because natural contours are barely symmetric in terms of geometry, quantification of the degree of symmetry needs to be proposed for investigating the perception in natural images. We established a computational index that describes the degree of symmetry inherent in arbitrary contours.

2.1   Definition of the Degree of Symmetry

We consider the degree of symmetry for local contour patches. The degree of symmetry is computed based on the degree of the overlap of contours between the two sides divided by the optimal symmetry axis, as illustrated in Fig. 1. We thoroughly search the optimal symmetry axis, by rotating and translating the axis and computing the overlap of contours between the two sides. The axis is represented by the rotation, θ, and the translation in x. Note that the axis could be placed anywhere in the patch. The overlap of contours between the two sides (a & b) is given by:

$$dos_{\theta,x} = \frac{\sum_{i=1}^{(N-1)/2}(\sum_{j=1}^{(N-1)/2}(a_{ij}b_{((N-1)/2)+1-i,j}))}{length}$$

Eq. 1

where i and j correspond to x and y directions of the rotated/translated patch, respectively, with the origin at the left-top corner. Note that the original patch is rotated and translated so that the symmetry axis is vertical and located at the center. N is the spatial extent of the patch in pixel ($N=69$ throughout this article). The degree of overlap is normalized by the length of contour in the patch (*length*). The optimal symmetry axis of a patch, which is described by θ and x, is given by: 1

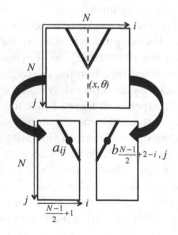

Fig. 1. An illustration of the computation of DoS. A contour patch (top) was divided by an axis (dotted line). DoS was given by the degree of the overlap of the contours between the two sides (bottom panels). If a contour is perfect symmetry, DoS is one.

$$OSA = \operatorname{argmax}(dos_{\theta,x}) \qquad\qquad \text{Eq. } 2$$

We normalize *dos* by the largest *dos* (max(*dos*)) among 1302 patches taken from Berkeley Segmentation Dataset (BSD) [4] (see section 3.1), and define it as the degree of symmetry of the patch (k) :

$$DOS_k = \frac{dos_{OSA,k}}{\max\limits_{k}(dos_{OSA,k})} \qquad\qquad \text{Eq. } 3$$

We confirmed that the patch with DoS=1 showed perfect symmetry.

2.2   Degree of Symmetry in Natural Contours

We computed DoS for the patches from BSD, as a few examples of the optimal symmetry axis and DoS shown in Fig. 2. From visual inspection, the optimal axes and DoS appear to naturally represent symmetry. The distribution of DoS for all patches ranged between about 0.4 and 1.0, as shown in Fig. 2. In the following sections, we examine quantitatively whether the optimal axis and DoS agree with the perception of symmetry. If DoS is the perceptual measure of symmetry, it will suggest the cortical representation of symmetry in a form similar to DoS.

3   Perception of Symmetry

We defined DoS as a computational measure of axial symmetry applicable for arbitrary contours in natural images. In this section, we examine whether DoS agrees with the perception of symmetry. Specifically, we performed psychophysical experiments to test whether the optimal symmetry axis derived by DoS matches with the human judgment of symmetry axis.

3.1   Methods

We performed psychophysical experiments to obtain the perceptual axis of symmetry in natural contours. We presented a series of contour patches taken from natural

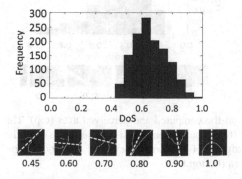

Fig. 2. Six examples of contour patch, DoS and the optimal symmetry axis (dotted lines) determined from DoS (bottom). A contour with higher DoS appears more symmetric. The histogram of DoS (top).

Fig. 3. An illustration of the experimental procedure. Following a mask (left), a contour patch together with a probe bar (red solid line) was presented (center). Participants rotated and translated the probe (right) to indicate the line that is most likely the symmetry axis.

images [4], and asked participants to determine a symmetry axis. The experimental procedure is illustrated in Fig. 3. We chosen systematically 1302 patches from BSD so that their curvature and closeness varied for a wide range. A single patch of 4x4 degrees in visual angel was presented on a liquid crystal display. Participants placed a bar as if it constitutes the best symmetry axis. Nine participants with normal or corrected-to-normal vision in their age of twenties repeated the task twice, therefore 18 axes were obtained for each patch. The experiment was approved by the research ethical committee of the institute.

3.2   Perception and the Degree of Symmetry

The perceptual axes of symmetry were obtained for the 1302 patches of natural contours, and compared with the optimal axes determined by DoS, as a few examples shown in Fig. 4(Left). Perceptual axis often varied among participants and trials, in such a case, multiple axes would represent the perceptual symmetry of a patch. To evaluate the consistency of the axis among trials and participants, we defined the *consistency* that is given by the degree of overlap among the axes. The consistency is one if all 18 axes are identical. A few examples and the distribution of the consistency are shown in Fig. 4(Right).

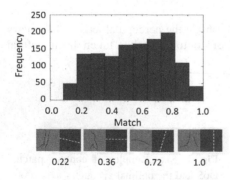
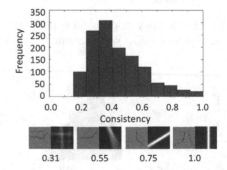

Fig. 4. Left: The histogram of the match between the computed and perceived axes (top). The bottom panels show four examples of contour (left), the computed axis (dotted lines) and the perceived axis (red solid lines). Right: The histogram of the consistency (top), and four examples of contour and superimposed perceptual axes (bottom).

We analyzed how often the computed axis matches with the perceptual axis. The histogram of the match (Fig. 4(Left)) shows a wide distribution, which would indicate that DoS does not match with perception in many cases. However, note that the patches consisted of a wide range of contours including those do not appear symmetry. Because the participants were forced to place an axis for all patches, the analysis should disregard those patches without symmetrical contours. It is difficult to rank the degree of symmetry without DoS. We come to utilize the consistency of axis as a measure to disregard non-symmetric contours. Because the consistency should be low if participants barely perceive symmetry in a patch, and high if they clearly perceive symmetry. We examined whether the match increases as the consistency increases. If this is the case, DoS can be considered as it correctly reflects the perception. We plotted the match as a function of consistency, as shown in Fig. 5. We observe a clear tendency that match increases with the consistency. Most of patches with the consistency > 0.6 show the match > 0.8. These results show that the proposed DoS agrees with the perception of symmetry.

4 Perceptual Representation of Symmetry --- Similarity Judgment of Contours ---

To assure that DoS is a quantitative perceptual measure, we determined psychophysically the multi-dimensional configuration that represents the perceptual shape of local contour, and analyzed whether DoS could be an axis of the configuration. Specifically, we performed similarity tests between a variety of natural contour patches, and analyzed whether DoS accounts for the similarity by multi-dimensional scaling (MDS) analysis.

4.1 Methods

We performed psychophysical experiments to obtain perceptual similarity of local, natural contours. We presented a pair of the contour patches, following a mask with a fixation aid, as shown in Fig. 6. We chose 54 patches from those used in the previous

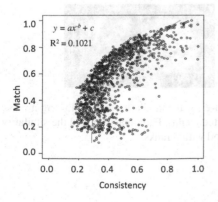

Fig. 5. The match between the computed and perceived axes as a function of the consistency. As the consistency increases, the match increase. The coefficient of determination for the nonlinear regression is shown in the inset.

experiment (section 3.1). The patches were chosen by visual inspection so as to assure a wide variety in contour shape. All pairs of the patches (1431) were presented in a random order. Six participants judged the similarity of pairs by the subjective scaling method with 5 ranks. The other experimental conditions were similar to those in 3.1. We calculated the perceptual similarity between all pairs of patches for each participant.

We applied MDS [5] to the perceptual similarity between the pairs of patches, and obtained the spatial configuration of the perception in 1 and 2 dimensions. This configuration provides coordinates for all patches, in which similar patches are close to each other, and dissimilar patches are distant. Thus, the spatial configuration represents the perception of natural contours.

4.2   Multi-Dimensinal Scaling Analysis

We tested whether the perceptual configuration (PC) of the patches agrees with the configuration of DoS (DC). For instance, the two-dimensional (2D) PC has two axes, meaning that contour shape can be described by two factors. We examined whether one of the factor could be DoS. We computed the overall pair-wise Euclidian distance of patches between the PC and DC, following the minimization of the distance by the Procrustes rotation method. We performed statistical tests to examine whether the distance between the PC and DC was significantly smaller than the distance between the PC and the random configuration. We carried out the test for each participant. For 1D configuration, 3 out of 6 participants showed a significant difference between PC-DC and PC-random distances. The result indicates that DoS accounts for the similarity in natural contours in half of participants with 1D configuration.

As we discussed in the previous section, the patches consisted of a wide range of contours including those do not appear symmetry. For this reason, we introduced another axis (factor) that would account for the similarity [6, 7]. We used either convexity or closeness in addition to DoS, and performed 2D analysis. The statistical tests showed that all participants showed a significant difference between PC-DC and PC-random distances in both cases (convex-DoS & closeness-DoS), as summarized in Table 1. These results show that DoS accounts for the similarity in natural contours, indicating that DoS reflects the perception of symmetry in natural contours.

 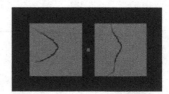

Fig. 6. An illustration of the experimental procedure for the similarity test. Following a mask with a fixation aid (left), a pair of patch was presented (right). Participants judged the similarity between the patches by the subjective scaling method with 5 ranks.

Table 1. Summary of the statistical tests for 1D and 2D configurations in MDS analysis

1 dimension	2 dimensions
Symmetry 3/6 participants were significant	Symmetry & Closure 6/6 were significant
	Symmetry & Convexity 6/6 were significant

5   Conclusions and Discussions

We proposed a computational index, DoS, that describes the degree of symmetry inherent in arbitrary contours. Our psychophysical experiment showed that DoS agrees with the perception of symmetry in the judgment of symmetry axis. To assure that DoS is a quantitative perceptual measure, we performed similarity tests between a variety of natural contour patches, and analyzed whether DoS accounts for the similarity. MDS analyses showed that DoS, together with convexity and closure, is indeed a perceptual measure. These results show that DoS accounts for the similarity in natural contours, indicating that DoS reflects the perception of symmetry in natural contours. Together with the recent evidence on adaptation [3], our result supports the cortex representation of symmetry as a basis for shape perception.

Acknowledgements. This work was supported by grant-in-aids from JSPS (KAKENHI 26280047), and grant-in-aids for Scientific Research on Innovative Areas, "Shitsukan" (No. 25135704) from MEXT, Japan.

References

[1] Hung, C.C., Carlson, E.T., Connor, C.E.: Medial axis shape cording in Macaque inferotemporal cortex. Neuron 74, 1099–1113 (2012)

[2] Hatori, Y., Sakai, K.: Early representation of shape by onset synchronization of border-ownership-selective cells in the V1-V2 network. J. Opt. Soc. Am., A 31, 716–729

[3] Gheorghiu, E., Bell, J., Kingdom, F.A.A.: Visual adaptation to symmetry. VSS 2014 23, 227 (2014)

[4] Martin, D., Fowlkes, C., Tal, D., Malik, J.: A database of human segmented natural images and its application to evaluating segmentation algorithms and measureing ecological statistics. Proc. ICCV 2, 416–423 (2001)

[5] Kruscal, J.B.: Multidimensional scaling by optimizing goodness of fit to a nonmetric hypothesis. Psychometrika 29, 1–27 (1964)

[6] Fowlkes, C.C., Martin, D.R., Malik, J.: Local figure-ground cues are valid for natural images. J. Vision 7(8), 2 (2007), doi:10.1167/7.8.2

[7] Sakai, K., Nishimura, H., Shimizu, R., Kondo, K.: Consistent and robust determination of border ownership based on asymmetric surrounding contrast. Neural Networks 33, 257–274 (2012)

Image Denoising with Rectified Linear Units

Yangwei Wu, Haohua Zhao, and Liqing Zhang*

Key Laboratory of Shanghai Education Commission for Intelligent Interaction and
Cognitive Engineering,
Dep. of Computer Science & Engineering, Shanghai Jiao Tong Univ., Shanghai, China
{wuyangwei,haoh.zhao}@sjtu.edu.cn, zhang-lq@cs.sjtu.edu.cn

Abstract. Deep neural networks have shown their power in the image
denoising problem by learning similar patterns in natural images. However, the traditional sigmoid function has shown its limitations. In this
paper, we adopt the rectified linear (ReL) function instead of the sigmoid
function as the activation function of hidden layers to further enhance
the ability of neural network on solving image denoising problem. Our
experiment shows that by better capturing patterns in natural images,
our model can achieve better performance and less time consumption
than those using sigmoid units. A large number of experiments show
that our approach can achieve the state-of-the-art performance.

Keywords: Rectified Linear units, Deep Learning, Neural Networks,
Image Denoising.

1 Introduction

Image denoising is a basic problem in image processing which takes a noisy image
as input and a noise reduced image as output. Although it is an old problem, in
recent years many good approaches have been proposed.

Up to now, image denoising approaches are mainly classified to three categories by the information source they use. The first category of approaches
solves the problem from the input image locally such as median filter, mean
filter, Gaussian filter. Advanced approaches of this type make use of edges, textures and other local information in natural images. The second category contains approaches that solve the problem from the entire input image such as the
non-local model[1] and BM3D [2] which is considered to be the state-of-the-art
approach for image denoising. The key idea is that there are generally many
similar patterns in a natural image. By grouping similar patches together and
denoising them collaboratively, we can expect to get a good result.

The third category contains the approaches that use a set of images for training models or bases. One way of these is to train a set of overcomplete sparse
bases such as K-SVD[3] and OTSC[4,5]. Another way is to train a neural network. In Burger's work[6], large plain neural networks are trained using large
amount of training data, which achieve better performance than BM3D.

---

* To whom all correspondence should be addressed.

C.K. Loo et al. (Eds.): ICONIP 2014, Part III, LNCS 8836, pp. 142–149, 2014.
© Springer International Publishing Switzerland 2014

Recently, due to the fast development of deep neural network[7], many new types of neural networks have been applied to the image denoising problem such as stacked sparse auto-encoder[8], convolutional networks[9], which both have shown good performance.

So far, standard neural networks use sigmoid functions such as the hyperbolic tangent (tanh) function or the logistic function as their activation function. Before the idea of deep neural networks, neural networks with such activation functions cannot be well trained when the number of layers is high due to the vanishing gradient problem[10]. The idea of layer-wised training solves this problem by training each layer in turn and combining them together by a final tuning step[7]. Recently, another way to solve the training problem of deep networks has been proposed by using the rectified linear (ReL) function: $max(0, x)$[11], which is shown in Fig.1, instead of sigmoid functions for hidden layers. Deep networks with rectifier nonlinearities have been shown to perform well in speech recognition[12,13] and image recognition[11]. However, to the best of our knowledge, no work on image denoising has used neural networks with ReL units. In

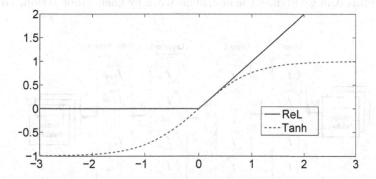

Fig. 1. Nonlinearity functions used in neural network hidden layers. The hyperbolic tangent (tanh) function is the typical choice while some recent works prefer the rectified linear (ReL) function.

this work, we evaluate neural networks with rectified linear function as their activation function for the image denoising problem. We will demonstrate the reason why we prefer the rectified linear function to sigmoid functions in image denoising problem, which includes better sparsity and faster convergence speed. A large number of simulations will be given to show the effectiveness of this modification.

We will first review image denoising with neural networks and introduce our model in section 2. Then we will show the simulations comparing ReL networks with sigmoid networks and other methods separately in image denoising problem in section 3. Finally, the conclusion will be given in section 4.

2   Model Description

In this section, we will first review the process how we use neural networks in image denoising. Then we will introduce the reason why we prefer to use the rectified linear function instead of sigmoid functions as the activation function.

2.1   Neural Networks in Image Denoising

The idea of image denoising using neural network is to learn a neural network that maps noisy image patches onto clean image patches where the noise is reduced or even removed. The parameters of the neural network can be estimated by training on pairs of noisy and clean image patches using stochastic gradient descent.

More precisely, we randomly pick a clean patch y from an image dataset and generate a corresponding noisy patch x by using the corruption procedure, such as adding additive white Gaussian noise on it. Then we use the noisy patch x as the input vector of the neural network and the clean patch y as the output vector. And we update the neural network by back-propogation. To make

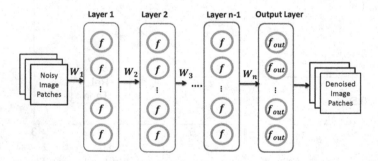

Fig. 2. The framework of neural network denoising. f is the activation function of hidden layers and f_{out} is the activation function of the output layer. We propose to use the ReL function instead of the hyperbolic tangent function in the hidden layers. The activation function of the output layer is the hyperbolic tangent function in our model.

back-propogation more efficient, some initializations and tricks in training are needed[14]:

1. Data normalization: Ideally, the input of neural networks should be transformed to that have approximate mean zero and variance one. However, given the noisy patch, it is hard to predict the mean and variance of the noise free patch especially when noise is strong. So we uniformly map the range of pixel values from $[0, 1]$ to $[-0.8, 0.8]$ to get an approximate mean zero over the dataset and use hyperbolic tangent function as the activation function of the output layer.

2. Weight initialization: The weights of layer i are sampled randomly from a uniform distribution in $[-4 * \sqrt{n_{i-1} + n_i}, -4 * \sqrt{n_{i-1} + n_i}]$ where n_i denotes the number of units in layer i. By this way, we can adapt its initial distribution to the number of units of the corresponding layers.

3. Other configurations: We use mini-batch training where the mini-batch size is set to 100. We use experience value as initial learning rate and after each epoch of mini-batch training we will multiply the learning rate by 0.99. The momentum coefficient is set to 0.5.

The given noisy image is decomposed into overlapping patches and each patch is denoised by the neural network separately. The denoised image is obtained by placing the denoised patches at the locations of their noisy counterparts, then averaging on the overlapping regions. To reduce time consumption on prediction, we use the sliding-window method in which the stride size is 3. In general, we can denoise an image of size 512*512 within half a minute without much optimization, which is much faster than BM3D[2]. The reason is that we directly use existing trained models while BM3D doesn't.

2.2 Rectified Linear Units (ReLU)

Instead of using sigmoid function as the activation function of hidden layers in neural network, we propose to use the rectified linear function: $rectifier(x) = max(0, x)$. This function is more similar to the function of the common neural activation function motivated by biological data which keeps zero when input current is below some threshold and gradually increases when input current is beyond the threshold[11]. This change also brings several advantages which make it more suitable for our problem.

First, deep neural networks with ReLU can be trained with back-propogation directly without "pretraining". Pretraining was proposed to solve the vanishing gradient problem[10] when using sigmoid units[7]. However for ReLU the gradient of $max(0, x)$ is a very simple step function. In our experience even with random initialization deep networks can be trained successfully.

Second this type of units has some inherent sparsity. Sparsity has become a concept of interest in image processing. For neural networks, the activation values in the hidden layer can be viewed as a representation of the input. For many reasons, we prefer to get a sparse representation of the input and many types of sparse constraints on weights and activation values have been proposed to enhance the sparsity of the activation values. For the rectified linear function, if the input x is less than zero, the result is zero, which is very helpful to generate a sparse representation.

Third, networks with ReLU need less training time than those with sigmoid functions. For a single unit, ReLU can be computed much faster as they do not require exponentiation and division operations. To quantify this, we randomly generate a vector x with a hundred billion elements between $[-1, 1]$ and compute $max(0, x)$ and $tanh(x)$ in MATLAB. The former calculation costs 0.1637 seconds while the latter one costs 0.8055 seconds. Given the same number of iterations for training networks, using ReLU can get an overall speed up of 25% in our experiment. Another benefit is, given the same training data, the same network size, the same network config as described previously, networks using ReLU need less iterations to converge, which will be shown in the next section.

3   Simulations

In this section, we will first compare two type of units in image denoising with networks of a small size. Then we will compare our method with some previous image denoising methods. We use the same image dataset with [5] which contains 200 architecture images. We use 150 of them as the source of training patches and the rest 50 of them as test images. The patch size and number of patches we grab from each image depend on the size of our neural network. We use the standard additive Gaussian white noise with different noise levels for evaluation.

3.1   Comparison of Two Units in Denoising

To compare the two types of activation functions, we use a simple network structure which has a single hidden layer and the patch size is set to just 8*8. No constraint is added as we just want to compare the performance of the two units. The network size is set to $[64, 640, 64]^1$ using a MATLAB implementation. The standard variance of the noise images is set to 75 (assuming pixel value is in [0,255]). From each image in our training image set, we extract 2000 patches and we get in total 300000 patches for training. The number of training epochs is set to 500 which is enough for the simple structure to converge well. Then we train two models separately and analysis their performance.

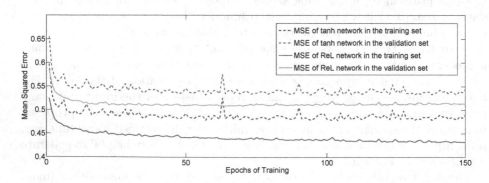

Fig. 3. Mean square error in both the training set and the validation set during the training procedure of two types of networks

---

[1] The network has 64 units in input layer, 640 units in the first hidden layer and 64 units in output layer.

We record the training curves of two models in Fig.3. We can see the network using the ReL function can better fit the training set than the one using tanh function. First, it can achieve a very low mean square error (MSE) which the tanh network may not be able to achieve. Second, it can reach the low MSE in a small number of epochs, which shows its learning speed.

We illustrate the filters learned by the two models after 500 epochs of training in Fig.4. we can clearly see that filters learned by ReL networks have much more Gabor-like filters than those learned by tanh networks. As no sparse constraint is added, the number of Gabor-like filters can be a measurement for the inherit sparsity of the network.

Then we compare the two networks using average Peak Signal-to-Noise Ratio (RSNR) on the test set. The network using tanh function has a PSNR of 23.779 ± 1.34 while the network using ReL function has a PSNR of 24.469 ± 1.56, which means the latter has much better denoising performance than the former.

Since no constraint that may influence their performance is added in this experiment, the result shows that networks with ReL units provide a better baseline for later optimization. Another reason is that we need much less time to train a ReL network well than to train a sigmoid network of the same size, which means we can train more complex models with the same resources.

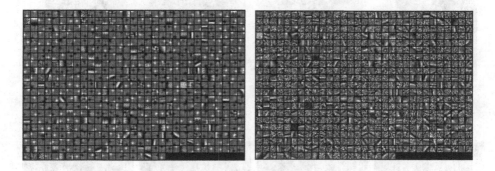

Fig. 4. Filters learned by neural networks with hyperbolic tangent functions (left) and rectified linear functions (right) as the activation function of hidden layers

3.2   Comparison with Other Models

We then compare our method with some previous models by training a large network using ReLU with a size of $[144, 720, 720, 720, 144]$ with an GPU implementation using toolbox *convnet*[2]. Denoising results are displayed in Table.1. We will mainly compare our method with BM3D below.

We first compare two models with average Peak Signal-to-Noise Ratio on the test set. We can see that when noise is strong, our method outperforms BM3D. When the standard variance of noise is 25, performance of our model is not as

---

[2] https://code.google.com/p/cuda-convnet/.

Table 1. Comparison of the denoising performance. Performance is measured by Peak Signal to Noise Ratio (PSNR). Results are averaged over the test set.

Method	Standard deviation σ/PSNR of noise images		
	25/PSNR=20.61	50/PSNR=15.11	75/PSNR=12.21
KSVD	29.19 ± 1.68	24.62 ± 1.56	21.35 ± 1.44
BM3D	**30.40±2.03**	26.07 ± 1.86	23.17 ± 1.66
Ours	29.93 ± 2.02	**26.81±2.02**	**24.89±1.78**

good as BM3D. In fact, we need a very large model and a very large training data set to outperform BM3D like what Burger has done using a large sigmoid network[6].

From Fig.5 we can see different preferences of the two methods. The results of BM3D are more smoother than our method while ours reserve more image details. For instance, in the result of the first figure, the sky in the BM3D result is more smoother while the outline of the bricks in our result is reserved better.

Fig. 5. Denoising performance on two images in the test set. We note that denoising result generated by BM3D are more smoother than our method, while our method reserves more details than BM3D.

4   Conclusion

In this paper, we propose deep neural networks with rectified linear units as hidden units in the image denoising problem. We find that by using rectified linear units we can achieve better performance and faster convergence than using sigmoid units. The comparison between our method and previous models indicates our model can achieve better denoising performance when additive noise is high.

Acknowledgement. The work was supported by the national natural science foundation of China (Grant Nos. 91120305, 61272251).

Reference

1. Buades, A., Coll, B., Morel, J.M.: A non-local algorithm for image denoising. In: IEEE Computer Society Conference on Computer Vision and Pattern Recognition(CVPR), vol. 2, pp. 60–65. IEEE (2005)
2. Dabov, K., Foi, A., Katkovnik, V., Egiazarian, K.: Image denoising by sparse 3-d transform-domain collaborative filtering. IEEE Transactions on Image Processing 16(8), 2080–2095 (2007)
3. Aharon, M., Elad, M., Bruckstein, A.: k-svd: An algorithm for designing overcomplete dictionaries for sparse representation. IEEE Transactions on Signal Processing 54(11), 4311–4322 (2006)
4. Ma, L., Zhang, L.: Overcomplete topographic independent component analysis. Neurocomputing 71(10), 2217–2223 (2008)
5. Zhao, H., Luo, J., Huang, Z., Nagumo, T., Murayama, J., Zhang, L.: Image denoising based on overcomplete topographic sparse coding. In: Lee, M., Hirose, A., Hou, Z.-G., Kil, R.M. (eds.) ICONIP 2013, Part III. LNCS, vol. 8228, pp. 266–273. Springer, Heidelberg (2013)
6. Burger, H., Schuler, C., Harmeling, S.: Image denoising: Can plain neural networks compete with bm3d? In: 2012 IEEE Conference on Computer Vision and Pattern Recognition (CVPR), pp. 2392–2399 (June 2012)
7. Hinton, G.E., Salakhutdinov, R.R.: Reducing the dimensionality of data with neural networks. Science 313(5786), 504–507 (2006)
8. Xie, J., Xu, L., Chen, E.: Image denoising and inpainting with deep neural networks. In: NIPS, pp. 350–358 (2012)
9. Jain, V., Seung, H.S.: Natural image denoising with convolutional networks. In: NIPS, vol. 8, pp. 769–776 (2008)
10. Bengio, Y., Simard, P., Frasconi, P.: Learning long-term dependencies with gradient descent is difficult. IEEE Transactions on Neural Networks 5(2), 157–166 (1994)
11. Glorot, X., Bordes, A., Bengio, Y.: Deep sparse rectifier networks. In: Proceedings of the 14th International Conference on Artificial Intelligence and Statistics, JMLR W&CP, vol. 15, pp. 315–323 (2011)
12. Zeiler, M.D., Ranzato, M., Monga, R., Mao, M., Yang, K., Le, Q., Nguyen, P., Senior, A., Vanhoucke, V., Dean, J., et al.: On rectified linear units for speech processing. In: IEEE International Conference on Acoustics, Speech and Signal Processing (ICASSP), pp. 3517–3521. IEEE (2013)
13. Maas, A.L., Hannun, A.Y., Ng, A.Y.: Rectifier nonlinearities improve neural network acoustic models. In: Proceedings of the ICML (2013)
14. Lecun, Y., Bottou, L., Orr, G.B., Müller, K.R.: Efficient backprop (1998)

Shape Preserving RGB-D Depth Map Restoration

Wei Liu[1], Haoyang Xue[1], Yun Gu[1], Jie Yang[1], Qiang Wu[2], and Zhenhong Jia[3]

[1] The Key Laboratory of Ministry of Education
for System Control and Information Processing
Shanghai Jiao Tong University, Shanghai, China
[2] School of Computing and Communications, University of Technology,
Sydney, Australia
[3] School of Information Science and Engineering, Xinjiang University, Urumqi, China
{liuwei.1989,xuehaoyangde,geron762,jieyang}@sjtu.edu.cn,
Qiang.Wu@uts.edu.au, jzhh@xju.edu.cn,

Abstract. The RGB-D cameras have enjoined a great popularity these years. However, the quality of the depth maps obtained by such cameras is far from perfect. In this paper, we propose a framework for shape preserving depth map restoration for RGB-D cameras. The quality of the depth map is improved from three aspects: 1) the proposed *region adaptive bilateral filter* (RA-BF) smooths the depth noise across the depth map adaptively, 2) by associating the color information with the depth information, incorrect depth values are adjusted properly, 3) a *selective joint bilateral filter* (SJBF) is proposed to successfully fill in the holes caused by low quality depth sensing. Encouraging performance is obtained through our experiments.

Keywords: depth map restoration, joint bilateral filter, diffusion, Kinect.

1 Introduction

Recent years, growing attention has been paid to the RGB-D images. In particular, the great success of low cost structured-light camera such as Kinect [1] has brought lots of RGB-D based applications like gaming [2], and new research area such as object recognition [9].

However, due to the simple depth measuring mechanism, the quality of the obtained depth map is far from perfect and mainly suffers from three problems: 1) *Invalid pixels* which do exist but are not sensed by the depth sensor, i.e. zero depth values or close to zero. In our work, a pixel is classified as an invalid pixel once its depth value is zero. The rest of the pixels on the depth map are *valid pixels*. Invalid pixels always form "holes" (black regions) on the depth map. 2) *Region various noise* on the original depth map-the noise on the area close to depth edges is much heavier than that away from the depth edges. We use *depth edges* to denote the edges on the depth map. Though the noise normally follows the quadratic law both in theory and experiments [11] [5], in our work,

C.K. Loo et al. (Eds.): ICONIP 2014, Part III, LNCS 8836, pp. 150–158, 2014.

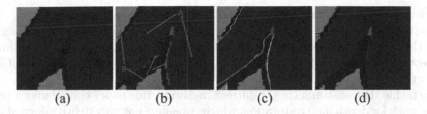

Fig. 1. (a) original depth map, (b) and (c) incorrect pixels (in pink) illustration, (d) incorrect pixels corrected by the proposed method

it is shown that region various noise is more obvious and this is also confirmed by [14]. 3) *Incorrect pixels* which have incorrect depth values (different from noise). Incorrect pixels exist along some regions (but not all the regions) of depth edges on the original depth map. Figure 1(b) illustrates incorrect pixels: regions in black labeled with "A" are foreground, regions in grey labeled with "B" are background, and regions in pink labeled with "C" are fake foreground, i.e. incorrect pixels.

The goal of the restoration is to restore a noise free depth map while the holes are properly filled in without altering the shape of the objects on the depth map. Depth map restoration has two categories: 1) restoration based on the depth information only and 2) restoration based on both the depth and color information. For each category, the operation can be carried out based on the information of current frame or based on the information of multiple frames. Restoration in [10] utilized a cascade of two modified median filters based on the depth information of current frame, while the method in [8] restored the depth map with normalized convolution and the guided filter taking the depth information of multiple frames into account. However, both methods in [10] [8] produced results of limited quality because only depth information was considered. Methods in [14] [15] [3] [13] took both the depth information and color information of current frame into account for restoration and promising results were shown in their papers. Especially, the method in [14] introduced the concept of depth layer and produced results of state of art performance. Methods in [5] [6] [4] took the depth and color information of multiple frames into account for the restoration. However, these methods were mainly designed for scenes with static background and dynamic foreground. Though the approach in [12] based on the motion analysis and the non-causal spatial-temporal median filter could handle dynamic scene, it is time consuming and cannot produce accurate restoration. The hole issue appearing on depth map is mainly caused by invalid pixels. There are two kinds of holes. The first kind is the *small holes* which can be properly filled in by considering the depth information in the neighboring area and the corresponding color information provided by the RGB-D camera. The second kind is the *large holes* which normally appear along the regions of depth edges on the original depth map. Applying the methods in [15] [3] [13] for filling in such holes may cause jagged or blurring depth edges or extra incorrect pixels on the restored depth map.

Previous work more focused on smoothing the noise and properly filling the holes [10] [8] [15] [3] [13]. However, all these methods above could not well maintain the object shape during the process of noise smoothing and hole filling. Moreover, these methods did not have a solution to correct depth values of incorrect pixels.

To tackle the constraints in the current methods, this paper contributes a new framework for depth map restoration, which considers not only depth information plus color information but also the depth discontinuity information in order to preserve the object shape. First, we propose a *region adaptive bilateral filter* (RA-BF) to smooth the noise. Then we correct the error depth values of the incorrect pixels with the help of both the depth and color information. Finally a novel *selective joint bilateral filter* (SJBF) is proposed to properly fill in the holes.

2   The Proposed Method

Our method consists of three steps: region various noise smoothing followed by incorrect pixels correction plus depth discontinuity map refinement, and finally holes filling.

In the following sections, we use *depth discontinuities* to denote the positions in the real world where distance between the objects and the depth sensor changes. The map of depth discontinuities is different from the map of depth edges defined in Section 1 mainly because of the holes and the incorrect pixels. The depth edge map of a completely accurate depth map is the same with the depth discontinuity map. In fact, we use depth discontinuities to refer the edges on the depth discontinuity map in the following sections. Uppercase letter with a subscript denotes either the pixel or the value of the pixel at the position indexed by the subscript. The uppercase letter with a hat and a subscript denotes the evaluated value of the pixel at position indexed by the subscript.

2.1   Region Various Noise Smoothing with Region Adaptive Bilateral Filter (RA-BF)

In our work, noise smoothing is carried out only on the regions of valid pixels on the original depth map. This operation is formulated as Equation (1):

$$\hat{D}_i = \frac{1}{Z_i} \sum_{D_j \in \mathcal{N}_i} D_j \cdot e^{-\left(a \cdot \frac{|D_i - D_j|^2}{f(\theta)} + b \cdot |i-j|^2\right)} \tag{1}$$

where $\mathcal{N}_i$ $(D_j \in \mathcal{N}_i)$ is the valid pixels set in the $w \times w$ patch on the original depth map , in which the center is D_i, and Z_i is a normalization constant which is the sum of the coefficient of D_j in Equation (1), and a, b are also constant values, the region adaptive term $f(\theta)$ is designed to consider the region various noise where θ is the perpendicular distance of D_i to its nearest depth edge on the original depth map. $f(\theta)$ is defined as Equation (2):

$$f(\theta) = \begin{cases} C_1, & \theta \leq r; \\ C_2, & otherwise \end{cases} \tag{2}$$

where constants $C_1 > C_2$. This means the smooth strength for D_i is larger if D_i is closer to the depth edges. Otherwise, the smooth strength is smaller. This can be implemented by firstly dilating the edge map obtained from the original depth map with a *disk* basic element of radius r. Then the pixels within the dilated edges are smoothed with $f(\theta) = C_1$. Otherwise, we set $f(\theta) = C_2$.

2.2 Incorrect Pixels Correction and Depth Discontinuity Map Refinement

Depth discontinuities can be described using edge information on the depth map. However, edges on the original depth map cannot well describe the depth discontinuities mainly due to the holes and incorrect pixels on the depth map. Figure 2(b) shows the edge map obtained from the original depth map. According to [15] [3] [7], depth discontinuities often simultaneously appear at the same locations on a depth map and the corresponding color image. We initialize the depth discontinuity map as follows: the edge map obtained on the depth map is processed by dilate operation, and the output is combined with the edge map of the corresponding color image through AND binary operation. The result of AND operation is regarded as the initial depth discontinuity map. Figure 2(c) shows an initial depth discontinuity map. It is shown that some edge points on the initial depth discontinuity map are not correct. We call these incorrect edge points *introduced texture edges*. It is observed that introduced texture edges exist inside flat depth areas. We design Equation (3) to remove these fake depth discontinuities.

$$g(E_i) = sgn\left(\max_{j \in \Omega_i} \left|D_j^\Delta\right| - T_\Delta\right), i \in \Lambda \tag{3}$$

where Λ is the set of coordinates on the initial depth discontinuity map where the values of the binary pixels equal one. E_i represents the pixel at position i on the initial depth discontinuity map. $sgn(\cdot)$ is a sign function where $sgn(x) = 1$ for $x \geq 0$ and $sgn(x) = 0$ for $x < 0$. T_Δ is a given threshold. D_j^Δ is the Laplacian of the smoothed depth map which is obtained from Section 2.1. Ω_i is the neighboring area of position i. In fact, Ω_i is a straight line that is perpendicular to the edge on the initial depth discontinuity map and across the position i. The length of Ω_i is determined by $[-R, R]$. $-R$ and R mean we consider D_j^Δ in Ω_i on two sides of the edge.

We compute Equation (3) for all E_i ($i \in \Lambda$) on the initial depth discontinuity map pixel by pixel. If $g(E_i) = 0$, E_i belongs to the introduced texture edge and is removed. If $g(E_i) = 1$, E_i belongs to the depth discontinuity and is kept. And then we get the *refined depth discontinuity map*. The refined depth discontinuity map is regarded as an approximation of the ideal depth discontinuity map. Figure 2(d) shows the refined depth discontinuity map. It is shown that most introduced texture edges have been removed.

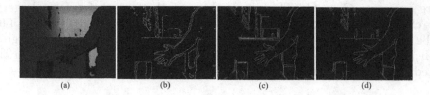

Fig. 2. (a) original depth map, (b) depth map discontinuity map, (c) initial depth discontinuity map, (d) refined depth discontinuity map

For all $g(E_i) = 1$, if there exists incorrect pixels around the position i (incorrect pixels only exist along *some* regions of depth edges), we further correct the incorrect pixels on the smoothed depth map obtained in Section 2.1. Figure 1(c) shows incorrect pixels (regions in pink) when we draw the refined depth discontinuity map on the smoothed depth map. If we have $k = \arg\max_j \max_{j \in \Omega_i} |D_j^{\Delta}|$, then the correction will be performed as $D_s \leftarrow D_k$ for all $s \in \Omega_i$, and s is between i and k. D_s is the depth value of the pixel on the smoothed depth map at the same position where D_s^{Δ} lies. In this way, we refine the initial depth discontinuity map and obtain a depth map with the incorrect pixels corrected at the same time. The obtained depth map is denoted as *refined depth map*. Figure 1(d) illustrates the refined depth map of Figure 1(a).

2.3 Holes Filling with Selective Joint Bilateral Filter (SJBF)

We fill the holes with the help of color information as well as the refined depth discontinuity map obtained in Section 2.2. Each time we only fill the invalid pixels with at least one valid pixel in the 8-neighborhood. This is to gradually diffuse the valid pixels into the holes . We start the diffusion at all holes simultaneously. A diffusion process at a hole area is terminated once it meets either valid pixels or the edge defined in the refined depth discontinuity map. The whole diffusion process will be terminated till all the diffusion processes meet valid pixels or the edges. The filling is implemented as Equation (4):

$$\hat{D}_i = \frac{1}{Z_i'} \sum_{D_j \in \mathcal{SN}_i} D_j \cdot e^{-\left(c \cdot \sum_{z \in C} \left| I_i^z - I_j^z \right|^2 + d \cdot |i-j|^2 \right)} \tag{4}$$

where Z_i' is a normalization constant which is the sum of the coefficient of D_j in Equation (4), C is the index set of color channels in RGB color space and I_i^z represents the value of the z channel of pixel i, the *selected neighbors* $\mathcal{SN}_i$ of invalid pixel D_i is the set of valid pixels in the $w \times w$ patch on the depth map, in which the center is D_i. $\mathcal{SN}_i$ is defined as Equation (5):

$$\mathcal{SN}_i = \left\{ D_j \left| e^{-(\tilde{D}_i - D_j)^2} \geq T \right. \right\} \tag{5}$$

where $\tilde{D}_i$ is the average value of the valid depth pixels in the 8-neighborhood of D_i, $T \in [0, 1]$ is a given threshold that represents how similar are the valid pixels in $\mathcal{SN}_i$ to D_i .

Unlike joint bilateral filter (JBF) which utilizes all the valid pixels in the $w \times w$ patch centered at D_i for filling, SJBF uses Equation (5) to *select* the neighbors in the patch first, i.e. the main difference between SJBF and JBF is the final neighbors used for hole filling. For the small holes which mainly lie on small flat regions on the depth map, SJBF is very similar to JBF because the neighbor selection based on Equation (5) seldom eliminates any valid pixels in the $w \times w$ patch because of similar depth on the small flat area. While for the large holes, SJBF can outperform JBF especially when the depth discontinuity passes through the patch and the corresponding colors of the valid pixels on two sides of the depth discontinuity are similar. Without neighbor pixels selection as in SJBF, JBF treats valid pixels on two sides of the depth discontinuity equally and causes blurring or jagged edge or even incorrect pixels on the restored depth map. However, the neighbor selection based on Equation (5) in SJBF can select proper neighbors: most valid pixels on the same side of the depth discontinuity with D_i are kept because of their similar depth values with $\tilde{D}_i$, while most of the valid pixels on the other side of the depth discontinuity are eliminated due to the obvious difference between their depth values and $\tilde{D}_i$. Thus the properly selected neighbors can successfully avoid the problems of JBF as mentioned above when filling in large holes.

3 Experiments

We compare our proposed method with [15] [3]. The data was produced by Kinect [1]. All the three methods restore the depth map with the help of the depth and color information of current frame. [15] only considered the noise and the invalid pixels while it did not take the incorrect pixels problem into account. [3] proposed a region-adaptive JBF taking the advantages of the edge information in the corresponding color image as structure guidance for adaptive support region selection. Figure 3 shows the experiment results. Figures in the first column are original depth maps. Their corresponding color images are shown in the second column. Figures in the third, fourth and fifth column show the restored depth maps by [3], by [15] and by our proposed method respectively. The first two rows in Figure 3 show results of two testing cases. The results by [3] can well preserve the shape of objects mainly due to taking advantages of the edge information in the corresponding color image. However, the restored depth map clearly 'copies' edges from the color image. In Figure 3(c1), clear depth edge exists between the ceiling and the wall on the left while there is no depth change in this region.

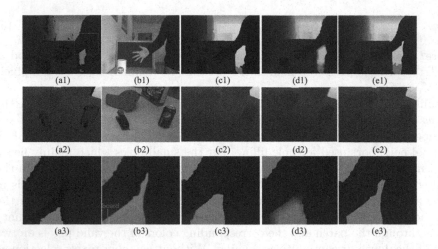

Fig. 3. original depth maps (column 1), corresponding color images (column 2), results by [3] (column 3), results by [15] (column 4), and results by our proposed method (column 5)

The situation is the same with the door. Obviously, blurring edges exist in Figure 3 (d1) and (d2), while there are also jagged depth edges on the top of the board behind the hand in Figure 3 (d1) and around the hat in Figure 3 (d2). Both results by [3] and [15] introduce incorrect pixels in the region between the elbow and the chest and the region on the top of the beverage can.

To further explain the experiment results, we zoom in the region inside the red box shown in Figure 3(a1) and show them in the third row in Figure 3. The variants of JBF in [15] [3] only consider color information together with spatial correlation and do not select the neighbors when filling in the holes. As labeled in Figure 3(b3), pixels at two sides of the depth discontinuity of the wall and the board have similar colors. Thus the result by [3] introduces incorrect pixels (labeled with "A"), while the result by [15] has blurring edges on the restored depth map. Our SJBF can well handle this case because it takes the advantage of Equation (5) to select the neighbors first. Thus our result has no incorrect pixels like [3] and blurring edges like [15]. Results by [15] [3] also have extra incorrect pixels in the region between the elbow and the chest because they do not correct the incorrect pixels before the holes filling while our method does (illustrated in Figure 1(d)). It is shown that our result has much fewer incorrect pixels than the other two results. Additionally, the region various noise is also well smoothed by the proposed RA-BF according to our experiment results.

4   Conclusion

In this paper, we analyze the problems on the depth map obtained by the RGB-D camera: the holes formed by invalid pixels, the region various noise and the incorrect pixels. RA-BF is proposed to smooth the noise. Then incorrect pixels are corrected and a refined depth discontinuity map is obtained at the same time . Finally the holes are properly filled in using SJBF with the help of the refined depth discontinuity map. The experiment demonstrates that the proposed method can greatly improve the quality of the original depth maps.

Acknowledgment. This research is partly supported by NSFC, China (No: 6127325831100672, 61375048,), Ph.D. Programs Foundation of Ministry of Education of China (No.20120073110018).

References

1. Microsoft corporation. kinect for xbox 360
2. Bleiweiss, A., et al.: Enhanced interactive gaming by blending full-body tracking and gesture animation. In: ACM SIGGRAPH ASIA 2010 Sketches, p. 34. ACM (2010)
3. Chongyu, C., et al.: A color-guided, region-adaptive and depth-selective unified framework for kinect depth recovery. In: 2013 IEEE 15th International Workshop on Multimedia Signal Processing (MMSP), pp. 7–12. IEEE (2013)
4. Camplani, M., Mantecon, T., Salgado, L.: Accurate depth-color scene modeling for 3d contents generation with low cost depth cameras. In: 2012 19th IEEE International Conference on Image Processing (ICIP), pp. 1741–1744. IEEE (2012)
5. Camplani, M., Mantecon, T., Salgado, L.: Depth-color fusion strategy for 3-d scene modeling with kinect. IEEE Transactions on Cybernetics 43(6), 1560–1571 (2013)
6. Camplani, M., Salgado, L.: Efficient spatio-temporal hole filling strategy for kinect depth maps. In: IS&T/SPIE Electronic Imaging, p. 82900E. International Society for Optics and Photonics (2012)
7. Diebel, J., Thrun, S.: An application of markov random fields to range sensing. In: Advances in Neural Information Processing Systems (NIPS), pp. 291–298 (2005)
8. Jakob, W., Sebastian, B., Joachim, H.: Real-time preprocessing for dense 3-d range imaging on the gpu: defect interpolation, bilateral temporal averaging and guided filtering. In: 2011 IEEE International Conference on Computer Vision Workshops (ICCV Workshops), pp. 1221–1227. IEEE (2011)
9. Kevin, L., et al.: A large-scale hierarchical multi-view rgb-d object dataset. In: 2011 IEEE International Conference on Robotics and Automation (ICRA), pp. 1817–1824. IEEE (2011)
10. Andrew, M., et al.: Enhanced personal autostereoscopic telepresence system using commodity depth cameras. Computers & Graphics 36(7), 791–807 (2012)
11. Fabio, M., et al.: Geometric investigation of a gaming active device. In: SPIE Optical Metrology, p. 80850G. International Society for Optics and Photonics (2011)
12. Sergey, M., et al.: Temporal filtering for depth maps generated by kinect depth camera. In: 2011 3DTV Conference: The True Vision-Capture, Transmission and Display of 3D Video (3DTV-CON), pp. 1–4. IEEE (2011)

13. Fei, Q., et al.: Structure guided fusion for depth map inpainting. Pattern Recognition Letters (2012)
14. Shen, J., Cheung, S.C.S.: Layer depth denoising and completion for structured-light rgb-d cameras. In: 2013 IEEE Conference on Computer Vision and Pattern Recognition (CVPR), pp. 1187–1194. IEEE (2013)
15. Yang, J., Ye, X., Li, K., Hou, C.: Depth recovery using an adaptive color-guided auto-regressive model. In: Fitzgibbon, A., Lazebnik, S., Perona, P., Sato, Y., Schmid, C. (eds.) ECCV 2012, Part V. LNCS, vol. 7576, pp. 158–171. Springer, Heidelberg (2012)

Online Detection of Concept Drift in Visual Tracking

Yichen Liu and Yue Zhou

Institute of Image Processing and Pattern Recognition,
Shanghai Jiao Tong University, Shanghai 200240
lycsjtu@163.com, zhouyue@sjtu.edu.cn

Abstract. In the field of data mining, detecting concept drift in a data stream is an important research area with many applications. However the effective methods for concept drift detection are seldom used in visual tracking in which drifting problems appear frequently. In this paper, we present a novel framework combining concept drift detection with an online semi-supervised boosting method to build a robust visual tracker. The main idea is converting updated templates to a data stream by similarity learning and detecting concept drift. The proposed tracker is both robust against drifting and adaptive to appearance changes. Numerous experiments on various challenging videos demonstrate that our technique achieves high accuracy in real-world scenarios.

Keywords: Concept Drift, Visual Tracking, Similarity Learning, Semi-Supervised Boosting.

1   Introduction

The distribution of a data stream is often not stable but changes with time, often these changes lead to performance degeneration in the old model. Thus detection of drifting concepts has extensive applications in data mining, such as spam filtering [1, 2]. To our knowledge, the first exploration of combining concept drift detection with visual tracking was introduced in [3]. They proposed a simple Bayesian approach to detect drift points. However, their method is applicable in limited situations when abrupt drift happens, such as light mutation. In this paper, we present an online learning method combined with concept drift detection to finish the tracking task in real-world scenarios.

There exists one key problem in online learning method for tracking: drifting. Slight inaccuracies in the tracker can lead to incorrectly labeled training examples. Each update to the tracker may introduce an error which may accumulate over time resulting tracking failure. To tackle the drifting problem, extensive techniques have been proposed. Grabner et al. [4] proposed a semi-supervised online boosting method which is based on the idea of [5]. In their approach, a fixed prior classifier which is trained from some labeled examples is used for supervising the update process. This method tackles the drifting problem by restricting the update in a certain range. But the tracker may fail when the target has a significant appearance change. In this case, one can employ concept drift detection techniques to revalidate the tracker.

C.K. Loo et al. (Eds.): ICONIP 2014, Part III, LNCS 8836, pp. 159–166, 2014.

In order to detect concept drift in visual tracking, the updated target templates need to be converted into a data stream. In this paper, we treat the similarity between templates and concept (i.e. prior classifier) as the data to be mined. Learning similarity functions is an area which has received considerable attentions in machine learning. Our learning approach is inspired by the work of Leistner et al. [6]. They proposed a similarity learning method based on semi-supervised boosting. Their technique enables us to measure the distance between newly labeled samples and the concept in feature space. As shown in Fig.1, concept drift manifests an apparent trend from the view of similarity.

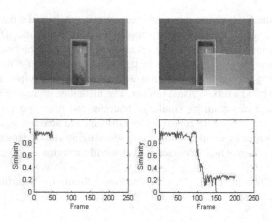

Fig. 1. An example of concept drift (occlusion) in visual tracking

Fig.1 demonstrates a special scenario (sudden occlusion) where only abrupt drift is considered. However in visual tracking, drifting types can be varied, thus a detection method which can accommodate for different situations has to be utilized. As data is generated constantly with the ongoing tracking process, the underlying distribution of data stream may change over time. In this paper, our change-detection algorithm is based on a two-window paradigm. Successive data points are maintained in two fixed-size windows: current window and reference window. We employ a statistical approach called L_1-distance-test [7] to verify whether the distributions of data points in the two windows are close or not. This test makes no assumption about the structure of the distributions and performs well in the application in visual tracking.

The remainder of this paper is organized as follow. After an introduction to the semi-supervised boosting method for similarity learning in Section 2, we introduce the drift detection algorithm L_1-distance-test and its application in visual tracking in Section 3. Section 4 presents the entire tracking framework of our concept drift detection based method. Section 5 demonstrates some experiments and results. Finally, our work concludes with Section 6.

2   Semi-supervised Boosting for Similarity Learning

Similarity learning is a key step in our concept drift detection. It measures the similarity between samples and outputs a similarity score which is added to a data stream. The change of distribution in the data stream indicates the occurrence of concept drift in the tracking process. Usually in visual tracking, the target to track is manually selected in the first frame. Our prior classifier $H^P(x) \in [-1, 1]$ is trained from the original labeled data using a boosting method. The prior classifier measures the similarity between updated templates and the target, generating a data stream which consists of similarity scores. As a confidence score is obtained from the prior classifier, according to [6], a distance measure is defined as

$$d(x_i, x_j) = |H^P(x_i) - H^P(x_j)| \tag{1}$$

This means samples that are close in distance share similar confidence scores. Then the distance measure is converted to a similarity measure by

$$S(x_i, x_j) = e^{\left(-\frac{d(x_i, x_j)^2}{\delta^2}\right)}, \tag{2}$$

Where δ is the scale parameter.

3   Online Detection of Concept Drift in a Data Stream

3.1   Testing Closeness of Distributions

If two unknown distributions over an n elements set are given, how to test whether they are statistically close is an interesting question. In this paper, we use the L_1-distance-test proposed in [7] for online distribution closeness testing. This method makes no assumption about the distributions and runs in time linear in the sample size. Our experiments show that the L_1-distance-test achieves high accuracy in the concept drift detection in visual tracking.

In the original L_1-distance-test algorithm, some elements appearing less than certain times are discarded before the test is performed. However, we omit this step in our test because the element set we use is small. Thus we give our simplified version of L_1-distance-test algorithm.

Table 1. The L_1-Distance-Test Algorithm

```
Algorithm 1 L₁-distance-test⟨p, q, ε, δ⟩
 1: Sample p⃗ and q⃗ for
 2:     M=O(max(ε⁻², 4)n^(2/3) log n) times
 3: Let Sₚ and S_q be the sample sets
```

4: Let n_i^p and n_i^q be the times element i appears in S_p and S_q
5: for m=1,2,..,M do
6:   update n_i^p by checking m-th element in S_p
7:   update n_i^q by checking m-th element in S_q
8: end for
9: Output 1 if $\sum_i |n_i^p - n_i^q| > \varepsilon M/8$
10: Otherwise output 0

In Algorithm 1, the parameters p and q are elements sets in two distributions, and parameters ε and δ can be tuned for adjusting the testing accuracy. Parameter n is the number of all possible elements. The presented algorithm runs in time complexity of O(M) if hashing technique is utilized. It has been proved in [7] that the L_1-distance-test generates a correct output with probability at least 1-δ.

3.2 Detecting Concept Drift in a Data Stream

Our change-detection algorithm is based on a two-window paradigm. Successive data points are maintained in two fixed-size windows: current window and reference window. The reference window focuses on the original data points that share high similarity with the target, however the current window focuses on the most recent data points, and slides forward whenever a new data point is added. Also the reference window is updated with each detected change. The L_1-distance-test is used to verify whether the distributions of data points in the two windows are close or not.

Table 2. The Concet Drift Detection Algorithm

Algorithm 2 Online Detection of Concept Drift
1: $c_0 \leftarrow 0$
2: for i=1…k do
3:   $Window_{1,i} \leftarrow$ first $m_{1,i}$ points from time c_0
4:   $Window_{2,i} \leftarrow$ next $m_{2,i}$ points in stream
5: end for
6: while new data is added to the stream do
7:   Slide $Window_{2,i}$ by 1 point
8:   if L_1-distance-test$\langle Window_{1,i}, Window_{2,i}, \varepsilon, \delta\rangle$=1
9:     $c_0 \leftarrow$current time
10:     report change at time c_0
11:     clear all windows and GOTO step 2
12:   end if
13: end while

Note that our detection algorithm processes the data stream in a discrete manner, so before adding the similarity score to the data stream, it needs to be discretized by mapping the similarity score to a corresponding integer in the element set.

4   Robust Tracking under the Framework of Concept Drift

4.1   Online Semi-supervised Boosting for Tracking

To finish the task of online semi-supervised boosting, Grabner et al. [8] introduced "selectors". Each selector $h^{sel}(x)$ contains a set of weak classifiers. At every training iteration t, a weak classifier $H_t(x)$ with lowest training error is picked. Thus in each selector, we can set the label and weight for unlabeled example by

$$y_t = sign(\tilde{z}_t(x)) \ and \ \lambda_t = |\tilde{z}_t(x)| \tag{3}$$

where y_t is the pseudo-label and λ_t is the corresponding weight. And $\tilde{z}_t(x)$ is the the pseudo-soft-label which is defined by

$$\tilde{z}_t(x) = \frac{\sinh(H^P(x) - H_{t-1})}{\cosh(H^P(x))} = \tanh(H^P(x)) - \tanh(H_{t-1}(x)) \tag{4}$$

In the semi-supervised boosting based tracking approach [4], the tracking problem is formulated as binary classification between the target and background. Often in visual tracking, the target to track is manually selected in the first frame. A prior classifier is initialized by taking positive training samples and negative training samples from the target and background respectively. At every iteration, the classifier is evaluated in the local neighborhood to generate a confidence map which will be analyzed to find the target position. Note that the update process is restricted by the prior classifier.

However, the tracking approach described above has a major drawback. When the target has a significant appearance change, the tracker may fail. In this case, one can employ concept drift detection techniques to revalidate the tracker.

4.2   Tracking with Concept Drift Detection

The structure of our proposed tracker is depicted in Fig.2.

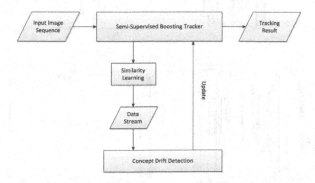

Fig. 2. The structure of our proposed tracker

In the semi-supervised boosting tracking, a template is evaluated by the prior classifier before updating. The updated templates restricted by the prior classifier are similar to the initial appearance, thus the tracker can't adapt to significant appearance changes. Using the drift detection algorithm discussed in Section 3, one can detect a major concept drift when significant appearance change happens. Our proposed tracker is both robust against drifting and adaptive to appearance changes.

5   Experiments

We compare our tracker with three other trackers, including SemiT[4], CT[9], MIL[10]. We include SemiT into our experiment because our tracker is implemented based on it. By adding drift detection module to SemiT, we can see a significant improvement in the tracking accuracy.

5.1   Quantitative Comparison

The four test sequences (*Dudek, football, girl, shaking*) for quantitative analysis exhibit extensive challenging properties: illumination changes, pose variations, occlusions, background clutters and so on. We use the center location error which is defined as the Euclidean distance of the center positions between the tracked target and the ground truth. The tracking results are shown in Fig.3.

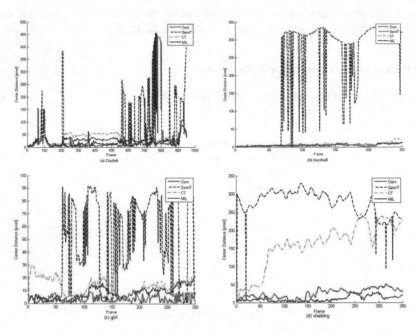

Fig. 3. The quantitative tracking results on the benchmark sequences

As we can see from the graphs, the online drift detection technique greatly improves the tracking accuracy of SemiT. The average center location errors of the tested trackers are shown in Table 1. The best and second best results are shown in red and blue respectively.

Table 1. The average center location errors of the four trackers

	Dudek	football	girl	shaking
SemiT[4]	124.3	232.4	63.5	275.1
CT[9]	38.2	4.1	13.9	147.8
MIL[10]	22.7	3.6	9.8	13.3
Ours	13.6	4.7	6.5	20.3

5.2   Qualitative Comparison

In this section, we perform qualitative evaluation on five testing sequences (*Dudek, girl, football, shaking, David*). Fig.4 shows the tracking results. Note that we won't draw the bounding box if the tracker loses the target.

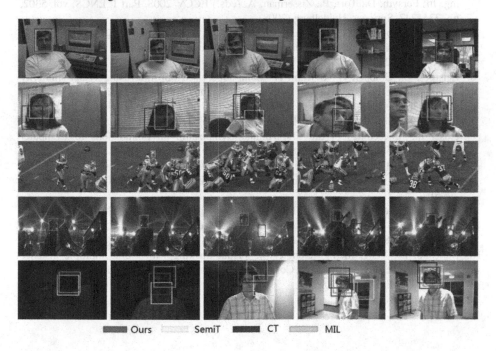

Ours   SemiT   CT   MIL

Fig. 4. The tracking results on *Dudek, girl, football, shaking, David* respectively

In general, our approach and MIL yield the best tracking results among the four. SemiT performs worst because it can't handle appearance changes properly and loses the target frequently. Both CT and MIL get a drift problem after the target is occluded in some frames. Besides, CT seems to have difficulty handling significant illumination changes.

6   Conclusion

In this paper, we propose a framework that combines concept drift detection with visual tracking to improve tracking performance. Experiments show that our tracker is of high accuracy in real-world scenarios. Further research could be done on exploring more powerful concept drift techniques to get a better tracking result.

References

1. Delany, S.J., Cunningham, P., Tsymbal, A., Coyle, L.: A case-based technique for tracking concept drift in spam filtering. Knowledge-Based Systems 18(4), 187–195 (2005)
2. Cunningham, P., Nowlan, N., Delany, S.J., Haahr, M.: A case-based approach to spam filtering that can track concept drift. In: The ICCBR, pp. 3–16 (2003)
3. Chen, L., Zhou, Y., Yang, J.: Object tracking within the framework of concept drift. In: Lee, K.M., Matsushita, Y., Rehg, J.M., Hu, Z. (eds.) ACCV 2012, Part III. LNCS, vol. 7726, pp. 152–162. Springer, Heidelberg (2013)
4. Grabner, H., Leistner, C., Bischof, H.: Semi-supervised on-line boosting for robust tracking. In: Forsyth, D., Torr, P., Zisserman, A. (eds.) ECCV 2008, Part I. LNCS, vol. 5302, pp. 234–247. Springer, Heidelberg (2008)
5. Kumar Mallapragada, P., Jin, R., Jain, A.K., Liu, Y.: Semiboost: Boosting for semi-supervised learning. IEEE Transactions on Pattern Analysis and Machine Intelligence 31(11), 2000–2014 (2009)
6. Leistner, C., Grabner, H., Bischof, H.: Semi-supervised boosting using visual similarity learning. In: IEEE Conference on Computer Vision and Pattern Recognition, CVPR 2008, pp. 1–8. IEEE (2008)
7. Batu, T., Fortnow, L., Rubinfeld, R., Smith, W.D., White, P.: Testing that distributions are close. In: Proceedings of the 41st Annual Symposium on Foundations of Computer Science, pp. 259–269. IEEE (2000)
8. Grabner, H., Bischof, H.: On-line boosting and vision. In: 2006 IEEE Computer Society Conference on Computer Vision and Pattern Recognition, vol. 1, pp. 260–267. IEEE (June 2006)
9. Zhang, K., Zhang, L., Yang, M.-H.: Real-time compressive tracking. In: Fitzgibbon, A., Lazebnik, S., Perona, P., Sato, Y., Schmid, C. (eds.) ECCV 2012, Part III. LNCS, vol. 7574, pp. 864–877. Springer, Heidelberg (2012)
10. Babenko, B., Yang, M.H., Belongie, S.: Visual tracking with online multiple instance learning. In: IEEE Conference on Computer Vision and Pattern Recognition, CVPR 2009, pp. 983–990. IEEE (June 2009)

Temporally Regularized Filters for Common Spatial Patterns by Preserving Locally Linear Structure of EEG Trials

Minmin Cheng[1], Haixian Wang[1,*], Zuhong Lu[1], and Deji Lu[2]

[1] Key Lab. of Child Development and Learning Science of Ministry of Education,
Research Center for Learning Science,
Southeast University, Nanjing, Jiangsu 210096, China
[2] Medical Electronics Lab. , Southeast University, Nanjing, Jiangsu 210096, China
{mmcheng,hxwang,zhlu}@seu.edu.cn,deji.lu@gmail.com

Abstract. Common spatial patterns (CSP) is a commonly used method of feature extraction for motor imagery–based brain computer interfaces (BCI). However, its performance is limited when subjects have small training samples or signals are very noisy. In this paper, we propose a new regularized CSP: temporally regularized common spatial patterns (TRCSP), which is an extension of the conventional CSP by preserving locally linear structure. The proposed method and CSP are tested on data sets from BCI competitions. Experimental results show that the TRCSP achieves higher average accuracy for most of the subjects and some of them are up to 10%. Furthermore, the results also show that the TRCSP is particularly effective in the small–sample data sets.

Keywords: brain–computer interfaces (BCI), common spatial patterns (CSP), locally linear structure, regularization.

1 Introduction

Brain computer interfaces (BCI) have emerged as a promising way of non-muscular communication with external world for severely paralyzed persons [1]. Electroencephalogram (EEG)–based BCI transfers intents of an individual, reflected in distinguishable EEG signals directly, into control commands of an assistive device. The successful decoding of the mental tasks heavily relies on a robust classification of the EEG signals. Among the plenty of decoding methods [2], common spatial patterns (CSP) is a widely used feature extraction method that can learn spatial filters maximizing the discriminability of two classes. Its effectiveness has been demonstrated by the BCI competitions [3], [4].

Despite its popularity and efficiency, CSP is also known to be highly sensitive to noise and outliers [5]. Mathematically, CSP is formulated as the simultaneous diagonalization of two covariance matrices. There is an inherent drawback for the estimation of covariance matrices in using the conventional strategy. Specifically,

---

* Corresponding author.

C.K. Loo et al. (Eds.): ICONIP 2014, Part III, LNCS 8836, pp. 167–174, 2014.

CSP does not take the temporal structure information of EEG time courses into account in the estimation of covariance matrices. In other words, CSP is a time-independent global method, and the temporal information is completely ignored.

In this paper, we propose a temporally regularized CSP (TRCSP), which incorporates the temporal structure information into the CSP learning process under the umbrella of regularization [6]. The temporal structure of EEG trials is characterized by using local linear embedding (LLE) [7]. Considering the advantage of LLE in successful discovery of manifold structure in machine learning, we aim to capture the locally linear structure of EEG trials with the LLE–based regularization. It is expected that such a prior information would help finding discriminative spatial filters, even with noisy EEG signals or small number of training samples, since the locally linear structure explicitly considers the temporal manifold behind the generation of EEG signals.

The framework of this paper is arranged as follows. Section 2 describes the conventional CSP algorithm and the proposed TRCSP algorithm. Section 3 gives details about the EEG data sets used for evaluation. Then the comparison results of the two methods are presented in Section 4. And finally Section 5 concludes the paper.

2   Methods

2.1   Common Spatial Patterns

Common spatial patterns(CSP) uses a linear transform to project multi–channel EEG data points into a low–dimensional spatial subspace with a projection matrix, of which each row consists of weights for channels. This transformation is to maximize the variance of band–pass filtered EEG signals of one class while minimizing the variance of EEG signals of the other class. Let $X^i = \left\{ x_l^i \in R^d | l = 1, 2, \cdots, s \right\} (i = 1, 2, \cdots, n_x)$ be the EEG trials of one class, and $Y^j = \left\{ y_l^j \in R^d | l = 1, 2, \cdots, s \right\} (j = 1, 2, \cdots, n_y)$ another class, where d denotes the number of channels, s is the number of samples within a trial, and n_x and n_y are the numbers of trials corresponding to the two classes. The trial segments are assumed to be already band–pass filtered, centered and scaled. The spatial covariance matrices of the two classes are calculated as

$$\overline{C_x} = \frac{1}{n_x} \sum_{i=1}^{n_x} \frac{X^i X^{iT}}{tr\left(X^i X^{iT} \right)} \qquad \overline{C_y} = \frac{1}{n_y} \sum_{j=1}^{n_y} \frac{Y^j Y^{jT}}{tr\left(Y^j Y^{jT} \right)} \tag{1}$$

where T represents the transpose operator, tr is the trace operator that sums up the diagonal entries of a matrix. The CSP approach aims to find a spatial filter $\omega \in R^d$ to extract discriminative features. Mathematically, the spatial filter of CSP is formulated by maximizing (or minimizing) the criterion[8], [9]

$$J(\omega) = \frac{\omega^T \overline{C_x} \omega}{\omega^T \overline{C_y} \omega} \tag{2}$$

The spatial filter is solved by the generalized eigenvalue equation

$$\overline{C_x}\boldsymbol{\omega} = \lambda\overline{C_y}\boldsymbol{\omega} \tag{3}$$

The few eigenvectors associated with eigenvalues from two ends of the eigenvalue spectrum are employed as spatial filters. The variances (possibly after a log–transformation) of the spatially filtered EEG data points are used as features for the purpose of classification.

2.2   Temporally Regularized Common Spatial Patterns

In this subsection we formulate the proposed TRCSP algorithm, which seeks to include temporal structure information into the learning process of the CSP. The EEG samples within a trial are actually a time course of signals. The temporally close samples usually correlated when recording a task–cued brain activity. It is beneficial to make use of the intrinsically temporal correlation to provide supplementary information and then regularize the computation of spatial filters. In other words, we try to keep the intrinsically temporal structure of EEG trials during the CSP filtering.

The temporal structure of EEG trials is captured by using LLE, which is well developed in the field of machine learning and has shown effective in manifold modeling. The basic idea is that we utilize LLE to consider temporally local relationship of EEG samples within the time course of EEG epochs. The relationship is expressed in terms of locally linear representation. Mathematically, LLE models each sample as a linear combination of its k nearest neighbors, and try to preserve this locally linear relationship in a transferred low–dimensional space. Different from the conventional LLE, in which the k nearest neighbors are identified with respect to Euclidean distance, we choose the k nearest neighbor EEG samples in terms of time points since we are interested in the temporal structure information of EEG time course. The reconstruction error is then measured by the cost function

$$\varepsilon\left(\boldsymbol{S}\right) = \sum_{l=1}^{s}\left\|\boldsymbol{x}_l - \sum_{m=1}^{s}S_{lm}\boldsymbol{x}_m\right\|^2 \tag{4}$$

where $\boldsymbol{S}$ is a matrix with real entries denoting representational weights. The weights S_{lm} summarize the contribution of the mth sample to the reconstruction of the lth sample in terms of linear representation. To compute the weights S_{lm}, we minimize the cost function subject to two constraints: (a) Each sample $\boldsymbol{x}_l$ is reconstructed only from its k nearest neighbors, resulting in $S_{lm} = 0$ if $\boldsymbol{x}_m$ does not belong to this set; (b) The row entries of the weight matrix sum to one, i.e., $\sum_{m=1}^{s}S_{lm} = 1$ for the purpose of transitional invariance. The matrix of weights $\boldsymbol{S}$ reflects the temporal structure information. Once $\boldsymbol{S}$ is obtained, LLE seeks a low–dimensional filtered space that preserves the temporal structure information of EEG trials as faithfully as possible. Let $\boldsymbol{z}_l$ $(1 \leq l \leq s)$ be the filtered signal of $\boldsymbol{x}_l$ $(1 \leq l \leq s)$ via the linear transformation $\boldsymbol{z}_l = \boldsymbol{\omega}^T\boldsymbol{x}_l$. One wishes to minimize the cost function

$$\varPhi\left(\boldsymbol{Z}\right) = \sum_{l=1}^{s}\left\|\boldsymbol{z}_l - \sum_{m=1}^{s}S_{lm}\boldsymbol{z}_m\right\|^2 \tag{5}$$

where $Z = [z_1, z_2, \cdots, z_s]$. Note that the weights matrix S is fixed here and the transformation matrix Z is to be optimized. By substituting $z_l = \omega^T x_l$ into (5), it follows that

$$\Phi(\omega) = \sum_{l=1}^{s} \left\| \omega^T x_l - \sum_{m=1}^{s} S_{lm} \omega^T x_m \right\|^2 \tag{6}$$

With some matrix operations, (6) can be rewritten as

$$\Phi(\omega) = \omega^T X L X^T \omega \tag{7}$$

where $X = [x_1, x_2, \cdots, x_s]$, $L = (I_s - S^T)(I_s - S)$, and I_s is an $s \times s$ identity matrix.

We now incorporate $\Phi(\omega)$ into the objective function of the classical CSP in order to penalize solutions such that the temporal structure information is preserved. Formally, the objective function of our TRCSP is given by

$$J(\omega) = \frac{\omega^T \overline{C_x} \omega}{\omega^T \overline{C_y} \omega + \alpha(\omega^T X L X^T \omega)} = \frac{\omega^T \overline{C_x} \omega}{\omega^T (\overline{C_y} + \alpha X L X^T) \omega} \tag{8}$$

Maximizing $J(\omega)$, would leads to the minimization of $\Phi(\omega)$, thus modifying spatial filters so as to satisfy the prior information. The parameter α is a user-defined positive constant which adjust the influence of the regularization term $\Phi(\omega)$. The higher the value of α is, the more favor the regularization term is given. The corresponding eigenvalue equation of (8) boils down to

$$\overline{C_x} \omega = \lambda(\overline{C_y} + \alpha X L X^T) \omega \tag{9}$$

Thus, the filters ω maximizing $J(\omega)$ are the leading eigenvectors corresponding to the largest eigenvalues. In the other hand, we need to accordingly maximize the dual objective function

$$J(\omega) = \frac{\omega^T \overline{C_y} \omega}{\omega^T \overline{C_x} \omega + \alpha(\omega^T X L X^T \omega)} = \frac{\omega^T \overline{C_y} \omega}{\omega^T (\overline{C_x} + \alpha X L X^T) \omega} \tag{10}$$

Eventually, the spatial filters used are the leading eigenvectors corresponding to the eigenvalue problems of (8) and (10).

It is noted that in the above formulation of TRCSP, X denotes a general EEG trial. In implementation, we exploit the temporally local information of all the training trials. Specifically, we sum up all the locally linear structure expression as the final regularization term. Besides, TRCSP has two parameters: k which defines the number of the nearest neighbor samples, and α which defines the level of regularization. In the following experiments, the two parameters are specified with ten-fold cross validation on the training data. And we adopt linear discriminant analysis (LDA) as the classifier.

3   Materials for Evaluation

Three EEG data sets from public BCI competitions, recorded from totally 17 subjects, are used to assess the proposed TRCSP, Its performance is compared to the classic CSP algorithm.

3.1 EEG Data Sets

Data set IVa of BCI competition III is of two–class motor imagery (MI) paradigm by recording 5 subjects. Imagination of right hand and foot movements was performed after a visual cue per trial. The EEG measurements were recorded using 118 electrodes and sampled with 100 Hz. For each subject, there are totally 280 trials for two classes, 140 per class. Among them, 168, 224, 84, 56 and 28 training trials are respectively for subject 1 through 5.

Data set IIIa of BCI competition III contains EEG signals from 3 subjects, who performed 4 classes cued motor imagery, i.e., left hand, right hand, foot, and tongue MI. The EEG measurements were recorded using 60 sensors by a 64–channel EEG amplifier from Neuroscan. The EEG was sampled with 250 Hz and filtered between 1 and 50 Hz with Notchfilter on. In our study, only EEG data corresponding to right and left hands MI are used. In both of the training and testing sets, 45 trials per class are used for subject B1, and 30 trials per class for subject B2 and B3.

Data set IIa of BCI competition IV was constructed by recording 9 subjects, who carried out left hand, right hand, both feet and tongue MI tasks. 22 EEG channels were recorded. Signals were sampled with 250 Hz and bandpass filtered between 0.5 and 100 Hz with Notchfilter on. Only EEG signals of left and right hands MI are used for the present study. Each subject participated a training and a testing session, both sessions containing 72 trials for each class.

Table 1. Classification performances of CSP and TRCSP. The best percentage accuracy is displayed for each subject in the two Data sets of BCI competition III.

| | BCI competition III | | | | | | | | Overall | |
| | Data set IVa | | | | | Data set IIIa | | | | |
Subject	A1	A2	A3	A4	A5	B1	B2	B3	Mean	std
CSP	66.07	91.07	53.6	71.88	52.78	96.67	61.67	96.67	73.8	17.3
TRCSP	68.75	100	62.2	82.14	85.71	96.67	68.33	96.67	82.56	13.8

Table 2. Classification performances of CSP and TRCSP. The best percentage accuracy is displayed for each subject in Data set IIa of BCI competition IV.

| | Data set IIa, BCI competition IV | | | | | | | | | Overall | |
Subject	C1	C2	C3	C4	C5	C6	C7	C8	C9	Mean	std
CSP	86.11	57.64	96.5	70.1	60.42	70.14	82.64	93.0	93.75	78.92	13.9
TRCSP	87.5	63.89	97.9	70.1	65.97	68.75	81.94	95.83	92.36	80.47	12.7

3.2 Preprocessing

The EEG signals are band–pass filtered with cutoff frequencies 8 Hz and 30 Hz by using a fifth order Butterworth filter as recommended in [10]. Following the winner of BCI competition IV and [11], we use the time interval from 0.5 s to 2.5 s after the visual cue that indicates the start of imaginary as samples on all of the three data sets.

4   Results and Discussion

We use CSP and TRCSP to extract features on the data sets. Compared with CSP, there are two parameters in TRCSP which need to be configured. The parameters are selected by using ten–fold cross–validation method on the training sets. For each subject, the spatial filters are learnt on the training set available. As suggested in [7], three pairs of spatial filters for feature extraction are calculated in CSP and TRCSP. Then the log–variances of the spatially filtered EEG signals are used as input features for LDA. The results of classification accuracies and mean accuracies, as well as the corresponding standard deviations, are reported in Tables 1 and 2.

All the classification accuracies performed by TRCSP are larger than 60%. On average, TRCSP achieves better classification accuracies (mean: 81.45 ± 13.3) than CSP (mean: 76.51 ± 15.8). Whereas, it seems that TRCSP does not give high increase in classification accuracy for subjects who already have good performances (except for A2). With a closer look, results show that, for some subjects, using TRCSP leads to dramatic increase in performance as high as 10%, even higher than 30% for the subject whose performance is close to random by CSP (A5). It is interesting that the classification accuracy for A2 is always kept in 100% when using TRCSP in a wild range of parameters. Especially for the data set IVa of BCI competition III, the performance of TRCSP is much better than CSP. For the subjects A3, A4 and A5, TRCSP significantly enlarges the classification accuracies compared with CSP. It is probably because of the very small training set for these three subjects. It implies that adding a prior information, here a locally linear preserving penalty can help to find spatial filters despite the limited amount of training data, as agreed with [12].

Surprisingly, TRCSP leads to poorer performance than CSP on a few subjects, focusing on data set IIa of BCI competition IV. This might be due to the instability of EEG signal itself and the playing condition of subjects. Besides,

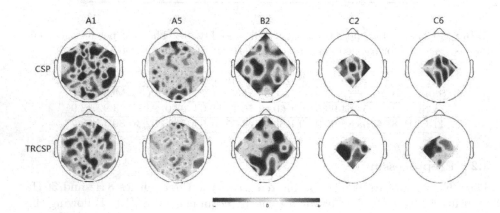

Fig. 1. Mappings of spatial filters obtained with CSP and TRCSP, for some subjects: A1, A5 (118 electrodes), B2 (60 electrodes), and C2, C9 (22 electrodes).

there is a very important point we can not ignore. That is, the best parameter may not be exactly found by the cross–validation strategy. It means that the real classification capacity TRCSP could possibly achieve better performances.

Some mappings of spatial filters obtained with both of CSP and TRCSP are presented in Fig. 1. The deeper color represents the greater weights. That is, they are more important for classification. In general, these pictures show that the CSP filters with large weights distribute in the whole brain, roughly and irregularly. Relatively, the TRCSP filters are generally smoother and more in line with the physiological characteristics. As expected from cerebral physiological theory, the weights are stronger over the motor cortex area. This suggests that the TPCSP algorithm lead to filters with more neurophysiological reality.

5 Conclusion

In this paper, we propose a new approach, called TRCSP, for optimizing spatial filers by incorporating temporal structure information to the conventional CSP. We add a locally linear regularization term to the CSP objective function. The experimental results confirm that TRCSP has the ability to obtain improved accuracies. In the future, much work is still needed to tune the appropriate parameters of TRCSP.

Acknowledgments. This work was supported in part by the National Basic Research Program of China under Grant 2015CB351704, the National Natural Science Foundation of China under Grant 61375118, the Natural Science Foundation of Jiangsu Province under Grant BK2011595, and the Program for New Century Excellent Talents in University of China under Grant NCET-12-0115. The authors thank Dr. F. Lotte for providing the code of filter mapping.

References

1. Wolpaw, J.R., Birbaumer, N., McFarland, D.J., Pfurtscheller, G., Vaughan, T.M.: Brain-computer interfaces for communication and control. Clinical Neurophysiology 113(6), 767–791 (2002)
2. Hwang, H.J., Kim, S., Choi, S., Im, C.H.: EEG-based brain-computer interfaces: A thorough literature survey. International Journal of Human-Computer Interaction 29(12), 814–826 (2013)
3. Blankertz, B., Muller, K.R., Curio, R., et al.: The BCI competition2003: Progress and perspectives in detection and discrimination of EEG single trials. IEEE Transactions on Bio-medical Engineering 51(6), 1044–1051 (2004)
4. Blankertz, B., Muller, K.R., Krusienski, D.J., et al.: The BCI competition III: Validating alternative approaches to actual BCI problems. IEEE Transactions on Neural Systems and Rehabilitation Engineering 14(2), 153–159 (2006)
5. Grosse-Wentrup, M., Liefhold, C., Gramann, K., Buss, M.: Beamforming in non-invasive brain computer interfaces. IEEE Transactions on Biomedical Engineering 56(4), 1209–1219 (2009)

6. Blankertz, B., Kawanabe, M., Tomioka, R., Hohlefeld, F., Nikulin, V., Muller, K.R.: Invariant common spatial patterns: Alleviating nonstationarities in brain-computer interfacing. Advances in Neural Information Processing Systems 20, 113–120 (2008)
7. Roweis, S.T., Saul, L.K.: Nonlinear dimensionality reduction by locally linear embedding. Science 290, 2323–2326 (2000)
8. Blankertz, B., Tomioka, R., Lemm, S., Kawanabe, M., Muller, K.R.: Optimizing spatial filters for robust EEG single-trial analysis. IEEE Signal Processing Magazine 25(1), 41–56 (2008)
9. Parra, L.C., Spence, C.D., Gerson, A.D., Sajda, P.: Recipes for linear analysis of EEG. NeuroImage 28(2), 326–341 (2005)
10. Ramoser, H., Muller-Gerking, J., Pfurtscheller, G.: Optimal spatial filtering of single trial EEG during imagined hand movement. IEEE Transactions on Rehabilitation Engineering 8(4), 441–446 (2000)
11. Lotte, F., Guan, C.: Regularizing common spatial patterns to improve BCI designs: Unified theory and new algorithms. IEEE Transactions on Biomedical Engineering 58(2), 355–362 (2011)
12. Lotte, F., Guan, C.: Spatially regularized common spatial patterns for EEG classification. In: Proceedings of the 20th International Conference on Pattern Recognition (ICPR), Istanbul, Turkey (2010)

Interactive Color Correction of Display by Dichromatic User

Hiroki Takagi[1], Hiroaki Kudo[1,*], Tetsuya Matsumoto[1],
Yoshinori Takeuchi[2], and Noboru Ohnishi[1]

[1] Graduate School of Information Science, Nagoya University,
Furo-cho, Chikusa-ku, Nagoya 464-8603, Japan
{takagi,kudo,matumoto,ohnishi}@ohnishi.m.is.nagoya-u.ac.jp
[2] Department of Information Systems, School of Informatics, Daido University
10-3, Takiharu-cho, Minami-ku, Nagoya 457-8530, Japan
ytake@daido-it.ac.jp

Abstract. Applications supporting dichromats based on confusion loci are proposed. We propose an interactive method to correct display color by measuring confusion color pairs for using such an application. The method measures 11 confusion color pairs on a display that shows an unknown color gamut. It estimates the most similar pattern from the confusion loci database, which is composed of scaling up/down one of R, G and B. It corrects display color to the sRGB gamut. We showed a tendency of confusion loci pattern for scale change and measured results for a dichromat. It improved the color difference in six of eight color settings.

Keywords: dichromatism, confusion loci, color correction, color gamut.

1 Introduction

Recently, lots of applications to assist dichromats have been released. Some are intended for dichromats to use by themselves. They use a color appearance model for dichromats [1] to present the regions that are confusion colors in a captured image or a web page. Then, it converts them to easily discriminative colors (e.g. [2]). The confusion colors of the model are aligned on the line in the chromaticity diagram of color space called confusion lines or loci. The algorithms based on the confusion lines may not always work well for devices in different settings, e.g. a tablet outdoors, a laptop PC for presentation in a dim room, a desktop monitor in an office. The color characteristics or settings for devices or illuminants are quite different. Brightness and chromaticity do not correspond on each device, although the application outputs the same RGB signals. Therefore, it is necessary to calibrate the color of a device display whenever a dichromat utilizes such an application. Interactive correction of color profile has been proposed (e.g. [3]), but it was not designed for this use.

Here, we focused on the strength of the R, G, B light sources of the device.

---

* Corresponding author.

C.K. Loo et al. (Eds.): ICONIP 2014, Part III, LNCS 8836, pp. 175–182, 2014.

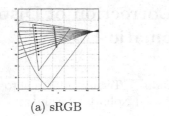   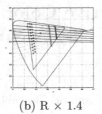

(a) sRGB   (b) R × 1.4

Fig. 1. Shift of chromaticity and confusion loci. (□ : sRGB, ∗ : scaled state).

2   Proposed Method

2.1   Confusion Loci

Each confusion locus represents the position of confusion colors for dichromats in the chromaticity diagram of some color space. In the $u'v'$ chromatic diagram, a locus does not distribute on a curve, but does on a line. If some confusion color pairs are given, the lines are almost converged at one point. The convergence points are different according to the type of dichromatism. If the display is set in sRGB correctly, the locus pattern is obtained as in Fig. 1(a). This figure shows the pattern of protan type, and coordinates of its conversion point are $(u', v') = (0.656, 0.502)$.

We propose an interactive method of color calibration of the display based on measurements of confusion color pairs by a dichromatic user. We utilize the change of the geometrical pattern of the confusion loci according to the strengths of R, G, B signals.

We model the status of the display, which is not set to sRGB, as the state is scaling up/down with one of three signals for the standard color. For example, R×1.4 means that the R signal is 1.4 times sRGB, and G and B are raw values. Supposing we use such a display, R signal is high-intensity scaling of 1.4, and we measure the confusion loci for the same R, G, B signals of sRGB settings; then we obtain the locus pattern as in Fig. 1 (b).

2.2   Database of Confusion Loci Pattern Changing Scaling of One Primal Color

As the confusion color pairs, we selected the colors on two parallel lines to the line that passes through the coordinates of green and blue primary colors. They correspond to the left vertex and bottom vertex of a triangle in Fig. 1(a). The positions of selected colors are shown by square marks in Fig. 1(a). The center of the triangle corresponds to the white point. We defined the position of the left line as the ratio 3:7 of the edge connected between green and blue to the white point. Also, we defined the right line at the position of three times its distance from the white point. Eleven color pairs are selected on the right line by equal spaces. Thus, we defined 11 confusion color pairs, and we set the brightness to $Y = 20.0$. The selected confusion color pairs are shown in Table 1.

We constructed the database of loci patterns as follows. First, the scaling coefficient is defined. We set the scaling (w) from 0.5 to 2.0 every 0.1 in sRGB signals.

Table 1. Confusion color pairs

u'	v'	R	G	B	u'	v'	R	G	B
0.176	0.301	86	110	246	0.333	0.366	226	0	188
0.173	0.327	83	115	226	0.331	0.384	221	40	175
0.170	0.352	80	120	207	0.328	0.401	216	54	163
0.166	0.378	78	124	189	0.326	0.418	212	64	150
0.163	0.404	75	127	171	0.324	0.436	208	71	137
0.160	0.430	73	129	153	0.322	0.453	205	78	124
0.157	0.455	71	132	135	0.320	0.470	201	83	110
0.153	0.481	70	134	116	0.318	0.488	198	88	95
0.150	0.507	68	136	94	0.315	0.505	195	92	78
0.147	0.533	66	137	68	0.313	0.523	192	95	56
0.144	0.558	65	139	19	0.311	0.540	189	99	0

(Y=20.0)

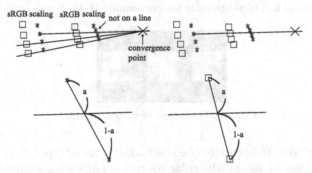

Fig. 2. Calculation of the shift of a confusion locus

It can transform the scaling for the linear RGB signals with the calculation of $w^{2.2}$. Here, we supposed the gamma of the display is set to 2.2. Then, the RGB signal is presented by the form of $(w_r R, w_g G, w_b B)$. One of three coefficients is scaling up/down. The others are set to 1.0.

If we displayed one pair of confusion colors on the display setting at sRGB correctly, the positions of original colors in the $u'v'$ chromaticity diagram are located on the line as in the top left figure in Fig. 2. Under the condition of scaling up/down of one of R, G, B, the colors that are located at asterisk marks are displayed; then they and the convergence point are not aligned on a line. The point of the right asterisk mark is not on a line.

Here, we consider that the color of the left side asterisk is fixed. Also, we'll find the confusion color pair on the interpolate line of the right side groups of colors with asterisk marks in the top right figure. It is found in the line that passes through the left side asterisk and convergence point. Next, we calculate the interpolate point between the positions of asterisk colors as in the bottom left figures. This applies the transformation of the original color, which is shown by the right side square with the interpolate ratio. Thus, we calculate the line (a confusion locus) that passes through the position of the left square mark and the interpolated point for right side squares at the bottom right figure. We stored the slope of the line as locus patterns in the database. For 11 confusion

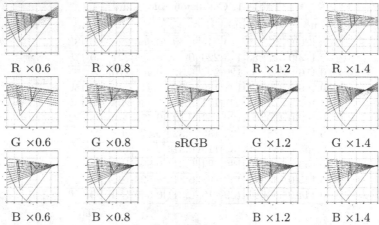

Fig. 3. Confusion loci of protanopia for one among R, G, B was scaled up/down

Fig. 4. Illustration of visual stimuli

color pairs, we tried to calculate the slope and, if an interpolation of right side asterisks could not be found, the color confusion pairs were excluded. A part of the database is shown in Fig. 3.

2.3   Color Matching by Dichromatic User

A dichromatic user performs a color-matching task with a display that has to be calibrated for color property.

The visual stimuli on the display are shown in Fig. 4. Three squares are presented. The center one is painted a reference color that corresponds to the left side confusion color. Its color is greenish or bluish. Then, it does not change until completion of a trial. Both end squares are painted the test color to match. Their colors are assigned in the neighborhoods of its confusion color pair. Their colors are reddish. A dichromatic user judges which of the test patterns is more similar to the reference color in the center square. After judgment, the non-selected test pattern is changed to a color that is the averaged color of the test patterns. The sides of the test patterns are changed randomly, and presented again; the user again judges the colors of the test patterns. The bisection method was adopted. When the user judges that the colors of test patterns are not discriminative, the trial is completed. If both initial test patterns are not perceived as similar to the reference color, its trial is skipped and recorded as such. Eleven confusion color pairs are measured. We calculate the slope of a line that passes through the reference color and confusion color, from which are obtained the color matching.

2.4   Evaluating Similarity between Measured Confusion Loci Patterns and Ones in Database

To estimate the color status of the display, we evaluate the similarity between the measured confusion loci patterns and confusion loci patterns in the database, which are composed by 11 slopes of lines of confusion color pairs.

The line of confusion colors is described as follows. To measure confusion lines, it holds that

$$A_i u' + B_i v' + C_i = 0. \tag{1}$$

The index of i represents the i_{th} confusion color pair. Similarly, for the database,

$$D_{s,i} u' + E_{s,i} v' + F_{s,i} = 0. \tag{2}$$

The index of s represents the status of the display. That is, the status of scaling up/down for one of three signals (R, G, B) (e.g. R ×1.4). Here, we consider the perpendicular vectors for each of the lines, $\mathbf{x}_i = (B_i, -A_i)^T$, $\mathbf{y}_{s,i} = (E_{s,i}, -D_{s,i})^T$. With these, we calculate the following evaluation function.

$$e_s = \alpha_s \sum_i \left(1 - \frac{|\mathbf{x}_i \cdot \mathbf{y}_{s,i}|}{\|\mathbf{x}_i\| \|\mathbf{y}_{s,i}\|} \right) \tag{3}$$

We estimate the settings s that minimized e_s as the color status of the display. Here, α_s is calculated by the ratio of 11 to the number of complete measured trials. If no trials are completed for some setting, then it is excluded from the estimation candidates.

2.5   Color Correction for Display

We constructed the database changing the scaling up/down for one of three primary color signals in the sRGB setting. The scaling factor is transformed to the scaling factor under the linear RGB space by $(w^{2.2})$ with γ-value. We supposed the relation of RGB and CIE XYZ as follows.

$$\mathbf{A} \begin{pmatrix} R \\ G \\ B \end{pmatrix} = \begin{pmatrix} X \\ Y \\ Z \end{pmatrix} \tag{4}$$

For the given scale and RGB signals, CIE XYZ is calculated by the following equation.

$$\mathbf{P} \mathbf{A}^T = \mathbf{Q}, \tag{5}$$

where $\mathbf{P} = \begin{pmatrix} R_1 & G_1 & B_1 \\ R_2 & G_2 & B_2 \\ \vdots \\ R_n & G_n & B_n \end{pmatrix}$, $\mathbf{Q} = \begin{pmatrix} X_1 & Y_1 & Z_1 \\ X_2 & Y_2 & Z_2 \\ \vdots \\ X_n & Y_n & Z_n \end{pmatrix}$. We estimated the matrix of $\mathbf{A}^T$

by the least square method. Then, we obtained equations $\mathbf{A}^T = \left(\mathbf{P}^T \mathbf{P} \right)^{-1} \mathbf{P}^T \mathbf{Q}$, $\hat{\mathbf{A}} = \mathbf{Q}^T \mathbf{P} \left(\mathbf{P}^T \mathbf{P} \right)^{-1}$. Thus, the following equation is derived.

Fig. 5. Confusion loci under the raw color status of the monitor

$$\hat{A}^{-1}\begin{pmatrix} X \\ Y \\ Z \end{pmatrix} = \begin{pmatrix} R \\ G \\ B \end{pmatrix}. \tag{6}$$

We define $\hat{A}^{-1}$ for each status of scaling. We can estimate the color $(X, Y, Z)^T$ should be displayed with $\hat{A}^{-1}$ and $(\hat{R}, \hat{G}, \hat{B})^T$. With this transformation, we can calibrate the color status of the display.

3 Experiment

3.1 Experimental Setup

The proposed method is implemented as the program components with C++ (VisualStudio) and G++ (Cygwin) on a personal computer (OS: Windows 7). We used an LCD Monitor (Sharp LL-T2015H) with the following specifications: Screen size, 408[mm] × 306[mm]; Resolution, 1,600[pixels] × 1,200[pixels]; Color Scale, 256[steps] for each color. We measured the chromaticity for raw primary color (R, G, B) signals by luminance color meter (Konica Minolta, CS-200). We calculated the confusion loci from the raw color status (standard) of this monitor. It is shown in Fig. 5. There are shifts from the loci pattern of sRGB (Fig. 1(a)).

The measurement was performed in a darkroom. We set the monitor at a distance of 0.45[m] from the user. Visual angle of a square's side is 11 [arc deg].

We set the six kinds of contrast of the monitor's settings. We denote them as R+, R++, G+, G++, B+, B++. Even if we index R, G or B, it does not mean the color shift caused by only one signal. Because shifts from sRGB have already occurred, the multi colors affect them. We also measured for the two preset settings as 'cool color' and 'warm color'.

3.2 Result

A dichromat user who had consented to participate in a psychological experiment served as the subject. The confusion loci measured by luminance color meter (left), user's response (center) and the estimated pattern from the database (right) for each setting are shown in Fig. 6. The results for the preset color setting are shown in Fig. 7. To show quantitative estimation, we calculate the color difference in the $u'v'$ chromaticity diagram for the test colors. Test colors and color differences in test colors are shown in Tables 2 and 3.

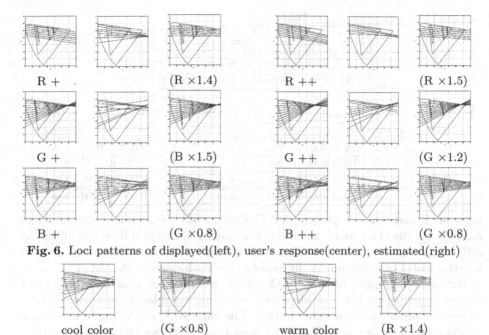

R +   (R ×1.4)   R ++   (R ×1.5)

G +   (B ×1.5)   G ++   (G ×1.2)

B +   (G ×0.8)   B ++   (G ×0.8)

Fig. 6. Loci patterns of displayed(left), user's response(center), estimated(right)

cool color   (G ×0.8)   warm color   (R ×1.4)

Fig. 7. Loci patterns of user's response(left) and estimated(right)

First, we will discuss Fig. 6 and Table 3. If we obtain the locus patterns for the left and right columns, figures are quite similar for each setting, the estimation is correct, and the color differences are reduced from the state before correction.

For the results of changing R, we can see that the convergence point is located outside of the figure. As we see in Fig. 3, for a larger scale factor of R, or smaller one of G, the convergence point is located more at the right side because the u' axis shows reddish and greenish components, which are in a complementary color relationship to each other. Therefore, we can conjecture that it is reasonable that the shift of the convergence point along the axis was observed. It holds that this is similar to the converse shift (left side) for a small factor of R or a large one of G. The user's loci disperse little, and estimated patterns are similar to the figures of the displayed patterns. Certainly, color differences are reduced, and the strength of scale is reflected. The scale of R++ is larger than that of R+.

For the results of changing G, we can see the large fluctuation of the user's response confusion pairs in the figure of G+. The estimated loci pattern is B ×1.5. It is not a good estimation. The color difference was expanded as in Table 3. Although the fluctuation also can be in G++, we can reduce the color difference in the estimated pattern G ×1.2.

For the results of changing B, we can see the abrupt changes of slope of lower confusion loci. It is caused by the saturation by reaching the maximum output of the light source B. Therefore, it is a larger factor, i.e. B++ has more effect. If we exclude the user's loci that correspond to saturation, the arrangement of the remaining loci is similar in both cases. Both settings estimate the scale of G ×0.8. This monitor color is biased to reddish, as in Fig. 5. As the relationship

Table 2. Test colors

u'	v'	R	G	B	u'	v'	R	G	B
0.15	0.50	66	135	101	0.18	0.53	106	130	67
0.17	0.32	76	115	232	0.23	0.33	158	89	221
0.34	0.45	214	68	125	0.33	0.51	201	88	70

(Y=20.0)

Table 3. Estimation for each setting and color difference of before/after correction

setting	estimation	before	after	success	setting	estimation	before	after	success
R+	R ×1.4	0.059	0.034	Yes	R++	R ×1.5	0.086	0.045	Yes
G+	B ×1.5	0.053	0.088	No	G++	G ×1.2	0.083	0.041	Yes
B+	G ×0.8	0.025	0.024	Yes	B++	G ×0.8	0.051	0.029	Yes
cool	G ×0.8	0.012	0.032	No	warm	R ×1.4	0.050	0.029	Yes

between G and R is complementary, G ×0.8 means the monitor color shifts to reddish. On the other hand, as in Fig. 3, changing of factor B does not affect the shift of the convergence point. Finally, it is reasonable that the scale of G ×0.8 is estimated. Color differences are reduced for both cases, certainly.

For preset settings, the scale of G ×0.8 is estimated in cool color, and color difference reduction was not obtained. We think the blue component is more or less strong in the 'cool color' setting. The saturation in the user's loci graph is observed. Together with B+, B++ and this result, we think parameters in equation (3) do not work well for such a factor (i.e. saturation). Therefore, we think it is of value to consider parameters in future work. For a warm color setting, we obtained good results such that the scale of R ×1.4 is reddish and color difference was reduced.

4   Conclusions

We proposed an interactive method to correct display color by a dichromatic user measuring confusion color pairs. The method measures 11 confusion color pairs on a display that has an unknown color gamut. The system outputs the most similar pattern from the confusion loci database, which is composed of scaling up/down one of the primary color signals (R, G, B). We showed a tendency of confusion loci pattern for scale change and measured results for a dichromatic user. We found that the color differences in six of eight color settings were reduced.

References

1. Brettel, H., Viénot, F., Mollon, J.D.: Computerized Simulation of Color Appearance for Dichromats. J. Opt. Soc. Am. A 14(10), 2647–2655 (1997)
2. Tanaka, G., Suetake, N., Uchino, E.: Yellow-Blue Component Modification of Color Image for Protanopia or Deuteranopia. IEICE Trans. E94-A(2), 884–888 (2011)
3. Farup, I., Hardeberg, J.Y., Bakke, A.M., Kopperud, S., Rindal, A.: Visualization and Interactive Manipulation of Color Gamuts. In: 10th IS&T/SID Color Imaging Conference, pp. 250–255 (2002)

A Neural Ensemble Approach for Segmentation and Classification of Road Images

Tejy Kinattukara and Brijesh Verma

Central Queensland University, Brisbane, Australia
{kjtejy,b.verma}@cqu.edu.au

Abstract. This paper presents a novel neural ensemble approach for classification of roadside images and compares its performance with three recently published approaches. In the proposed approach, an ensemble neural network is created by using a layered k-means clustering and fusion by majority voting. This approach is designed to improve the classification accuracy of roadside images into different objects like road, sky and signs. A set of images obtained from Transport and Main Roads Queensland is used to evaluate the proposed approach. The results obtained from experiments using proposed approach indicate that the new approach is better than the existing approaches for segmentation and classification of roadside images.

Keywords: Artificial Neural Networks, Support Vector Machines, Hierarchical Segmentation, Classification, Clustered Ensemble.

1 Introduction

Recently, both automatic road objects detection and recognition have been the subject of many studies. Roadside objects recognition is important for detecting the various risk factors on the roads and improving the overall safety of the road. Many factors need to be taken into account in an automatic traffic image recognition system. The objects appearance in an image depends on several aspects, such as outdoor lighting conditions, camera settings and camera itself. Also the images taken from moving vehicles can produce blurred images because of the vehicle motion.

Many roadside objects do not have specific shapes and the color also varies which create problems during segmentation and classification. The problems considerably affect the segmentation step, which is the initial stage in detection and recognition systems. In this paper, the aim of segmentation is to extract the road objects from the images, as this is crucial in achieving good classification results. Many segmentation methods have been presented in the literature using various image processing techniques.

A quantitative comparison of several segmentation methods used in traffic sign recognition is presented in [1]. The methods presented are colour space thresholding, edge detection, and chromatic decomposition. A simple and effective method that accurately segments road regions with a weak supervision provided by road vector data is presented in [2]. A factorisation based segmentation algorithm is applied to achieve this. An algorithm for real time detection and recognition of signs using

C.K. Loo et al. (Eds.): ICONIP 2014, Part III, LNCS 8836, pp. 183–193, 2014.
© Springer International Publishing Switzerland 2014

geometric moments is presented in [3]. A video segmentation algorithm for ariel surveillance is shown in [4]. It uses a mixture of experts for obtaining the segmentation results. A method for detection, measurement, and classification of painted road objects is presented in [5]. The features are extracted using dark light transition detection on horizontal line regions and robust method. An active vision system for real time traffic sign recognition is presented in [6]. The recognition algorithm is designed by intensively using built-in functions of an off-the-shelf image processing board for easy implementation and fast processing. A method to develop a computer vision system capable of identifying and locating road signs is explained in [7].

A fast and robust framework for incrementally detecting text in road signs is presented in [8]. The framework applies a divide and conquers strategy to decompose original task to sub tasks. It presents a novel method to separate text from video. A new adaptive and robust method for colour road segmentation is presented in [9]. A fitting and predicting approach is used to extend the features to the whole image. A system to detect and interpret traffic signs in colour image sequences is presented in [10]. The colour segmentation of the incoming images is performed by high order neural network. Various methods to evaluate segmentation methods are shown in [11].

A method for developing a computer vision system capable of identifying and locating road signs using colour segmentation strategy is shown in [12]. A method that combines the decisions of weak classifiers is shown in [13]. It presents a road sign identification method based on ensemble learning approach. The methods mentioned above are mainly meant for images with high quality and the objects to be separated are road signs.

In this paper, we discuss a new approach based on neural ensemble to segment and classify roadside image into different class of objects. The ensemble learning process combines the decisions of multiple classifiers created by clustering. The images used in the experiments were provided by Queensland Transport and Main Roads and were obtained from Australian countryside highways.

The remainder of this paper is organised as follows. Section 2 explains the proposed neural ensemble approach. Section 3 presents the methodology adopted using proposed neural ensemble for extraction and classification of road objects. Section 4 briefly describes the previous approaches used for road image segmentation. Section 5 describes the data collection part and experiments. Section 6 details the comparison of experimental results between various approaches. Section 7 details the conclusions from experimental analysis and directions for future research.

2   Proposed Ensemble Approach

The proposed approach for generating ensemble of classifiers is based on the concept of clustering and fusion. The initial task is to cluster the road images into multiple segments and use a set of base classifiers to learn the decision boundaries among the patterns in each cluster. This process of clustering partitions a dataset into segments that contains highly correlated data points. These correlated data points always tend to stay close together geometrically. Also these data points are difficult to classify when patterns from multiple classes overlap within a cluster. When clustering is applied on datasets associated with a class, two types of segments are produced atomic and non-atomic. An atomic cluster contains patterns that belong to the same class whereas a non- atomic cluster is composed of patterns from multiple classes.

Following the clustering process, classifiers are trained based on the patterns of non-atomic clusters and class labels are assigned for the atomic clusters. The class of a test pattern is predicted by finding the suitable cluster based on its distance from the cluster centres and using corresponding class label for atomic cluster and then by using suitable classifier for non-atomic cluster. So clustering helps in identifying difficult to classify patterns. Once clustering operation has been performed and clusters are identified a neural network classifier is trained for each cluster grouping.

In k-means clustering the assignment of pattern to a cluster can be different based on seeding mechanism (initial state of cluster centres) where the number of k-means clusters is different to the actual number of clusters in the data. If multiple clustering can be performed with different seeding points then pattern might go to different cluster each time. When a new clustering operation is performed with different initial seeding it is called layering and these clusters form a layer. This cluster alignment will be different from one layer to the next. So a classifier can be trained on these non-atomic clusters for each layer and the result of the classifiers fused together by the majority vote algorithm to create an ensemble. Thus the layers provide a means of introducing diversity in ensemble and making it easier to classify non-atomic patterns.

3   Research Methodology

The proposed research methodology adopted is shown in Fig. 1 followed by explanation of each step done in the process. In this paper we focus on the classification of road objects like road, sky as well as road signs. It is hard to classify roadside images as there are multiple objects in a single image and also many objects scattered and mingled with one another.

3.1   Segmentation

During segmentation, we take into account the characteristic features related to change in the colour components. The first step of segmentation is to measure the colour features. At first the road images are segmented into two colour channels: white and non-white. In this approach k-means is used with k=2 for road image segmentation. Segmentation produces white segments for lanes, sky, dry vegetation and road signs. The non white segment contains road, coloured road signs and green vegetation. In this stage the potential road objects are located by their position in the image. The road extraction process is done by block based feature extraction method on bottom part of the image. The sky region is separated by confining the search to top section of the image.

3.2   Feature Extraction

The segmented image is then subjected to block based feature extraction. At first a block size of 64*64 is defined and using the defined block size the image is divided

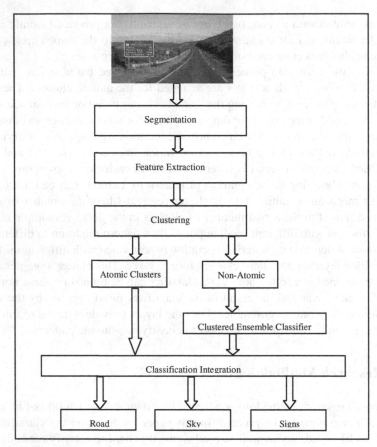

Fig. 1. Research methodology using proposed ensemble approach

into equal no of blocks. For road segment extraction the image is divided at the bottom part of image. Then each block is labelled into different classes as road, non–road, and background. Features are then extracted from each block.

For sky segment extraction the image is divided at top of the image. Then each block is labelled as sky, non-sky and vegetation. Finally feature extraction is performed. Following types of road signs were extracted from the image: green signs, light blue signs, yellow signs, and speed signs. Colour ranges are used to extract regions from the images. Then the boundaries of the regions are extracted. The regions are then further filtered by comparing the signature of each blob with those obtained from the reference shapes.

3.3   Clustering and Classification

The dataset for the road images is clustered by using k-means clustering algorithm. This produces both atomic clusters where only one class membership exists and non-atomic clusters where more than one class is present. A neural network is then trained on the non-atomic clusters and this produces a layer (a trained classifier based on a particular dataset created through training). This process is repeated for a number of clustering operations where the cluster initialization point (seeding) is different. So each classifier layer can be trained to recognize different decision boundary for the non-atomic clusters.

After the training operation is completed the network is tested. The ensemble classifier during testing evaluates as to which class a test pattern belongs in two steps. In the first step the cluster membership is determined by examining the said patterns distance to the cluster centroid. If the pattern belongs to an atomic cluster then the class label for that cluster is returned. If the pattern belongs to a non-atomic cluster then the class label from trained network is obtained. Finally a majority vote is used to return the decision from the ensemble classifier.

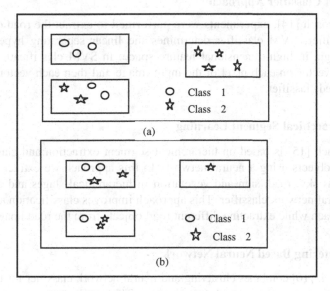

Fig. 2. Same dataset clustered at two layers

Fig. 2 shows the method of layering that occurs by varying the seeds of the cluster centers. In Fig. 2(a) three clusters have been defined and the class membership indicates that two atomic clusters and one non-atomic cluster have been formed. The non-atomic cluster contains multiple classes so a neural network classifier needs to be trained on this cluster. The atomic clusters are easier to classify and their class labels signify the classifying test patterns. In the second part of diagram Fig. 2(b) the cluster seeding are changed and the layer produces three clusters and cluster memberships are very much different to the previous layer. This difference in patterns creates diversity during neural network training which helps improving performance of ensemble classifier.

3.4   Classification Integration

The outputs obtained from different neural networks trained on each clustered layer are integrated by majority vote as this mechanism improves the overall output when compared to the individual accuracy of each classifier. The majority vote selects the output with highest number of votes.

4   Existing Classification Approaches

This section describes the previous approaches used in the experiments to extract the different road objects and compare the results with proposed ensemble approach. The different approaches for road objects extraction are detailed in the following sections.

4.1   SVM Classifier Approach

In this approach [14], experiments were performed to extract the road objects using SVM classifier. SVM classifier determines the linear separating hyper-plane with largest margin in high-dimensional feature space. In SVM classification we extract the feature vector on each pixel of the input image and then each vector is classified by the trained classifier.

4.2   Hierarchical Segment Learning

This approach [15] is based on hierarchical segment extraction and classification of segmented objects using a neural network. In this approach we extract different objects such as sky, road, sign and vegetation in hierarchical stages and classify them using a neural network classifier. This approach improves classification accuracy over SVM approach while extracting different road objects from the road images.

4.3   Clustering Based Neural Network

This approach [16] combines clustering and neural network classifier for the classification of road images into road and sky segments. This approach first creates clusters for each available class and then uses these clusters to form subclasses for each extracted road image segment. The integration of clusters in the classification process is designed to increase the learning abilities and improve the accuracy of the classification system.

5   Data Collection and Experiments

The road images were extracted from TMR videos using Matlab version 7 on Windows platform. The original videos are in avi format with MJPEG codec. The videos were converted into avi format with MPEG codec using Prism Software. The image frames obtained were saved as JPEG files and subjected to different processing methods. All the images used in our experiments have a resolution of 960×1280. We need to identify the best segmentation method. The best segmentation method is considered to be the one which gives the best recognition results. For good recognition results we have defined criteria as high recognition rate, low number of lost signs, high speed, and low number of false alarms. To test the performance more than 400 images were obtained from different sequences while driving at nominal speed. A database was constructed from the images which were taken under different settings and lightning conditions. A sample set of images used is shown in Fig. 3.

Fig. 3. Sample set of roadside images

In the hierarchical approach [15], a neural network is used for segmentation and classification of road objects. A block size of 64×64 and $(960 \times 1280)/(64 \times 64)$ segments per image was used. The images are then subjected to clustering and feature extraction process. The extracted features are used to train classifiers to classify the different road objects. Matlab neural network toolbox was used for neural network training by varying the number of hidden units. In the experiments using SVM classifier [14] we replaced neural network by SVM classifier and performance was evaluated.

In clustering based neural network approach [16], clustering is used to create clusters or sub classes within existing classes and integrates these clustering based new classes with the training process. The classifier used in this approach is a multilayered perceptron classifier with a single hidden layer. The classifier is trained on each cluster and results are integrated. Training of the weights was achieved using the backpropagation learning algorithm. The following parameter setting was used during the training process for all datasets: 1) Learning rate = 0.01; 2) Momentum = 0.2; 3) Epochs, i.e., no. of iterations = 55; and 4) RMS goal = 0.01. The best parameter settings on datasets were found by trial and error.

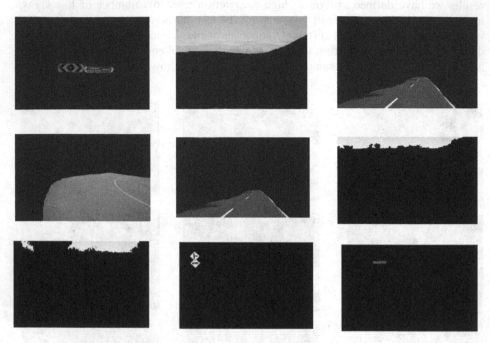

Fig. 4. Sample set of extracted road objects using proposed ensemble

6   Results and Analysis

Many images from videos on different routes and under different lightning conditions were captured. Each sequence includes hundreds of images. To analyse the different situations and the problems encountered, we extracted several sets, including some frames that were difficult to segment.

Each set represented different problems in segmentation like low illumination, rainy conditions and similar background colour. For results in this paper, a total of 400 images selected from thousands of 960×1280 pixel images were analysed. These sets are representative of the segmentation problems that arises during segmentation. Various methods to measure segmentation performance are shown in [11] but they do not specify a standard. They also cause excessive execution time.

In this paper we propose an evaluation process based on the performance of the whole classification system using four measures /criteria. The criteria or measures for evaluation are described below.

1) Rate of correct classification: The sum of all correctly classified objects divided by the total number of objects. The value of 100 indicates that all possible objects were correctly classified.

2) Number of lost objects: This relates to number of objects that were not classified in any way.

3) Number of maximum scores: This indicates the number of times a method achieved maximum score.

4) False recognition rate: This represents percentage of objects wrongly classified by a method with respect to number of total objects classified.

The experiments were conducted using the above measures and results were compared for proposed and exiting approaches. As shown in Table 1, the classification rate is higher in the ensemble learning approach which is around 91.2%. It shows considerable improvement when compared to other approaches. The number of objects lost and falsely detected is also lower in the case of ensemble approach. Fig. 4 shows the sample set of extracted road objects. Fig. 5 indicates the comparison between the approaches for the various measures.

Table 1. Performance measures of various approaches

Measures	SVM	Hierarchical	Clustering	Ensemble
Classification (%)	80.2	81.5	88.4	91.2
Lost	4	5	3	2
Max	6	9	4	12
False (%)	2.34	3.4	2.1	0.00

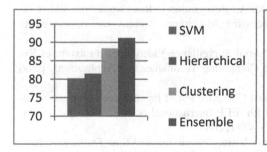

(a) Rate of Correct Classification (%)

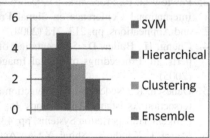

(b) Number of Lost Objects

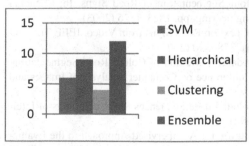

(c) Number of Maximum Scores

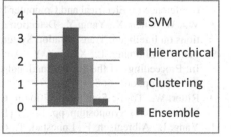

(d) False Recognition Rate

Fig. 5. Comparative analysis of results using various approaches

7   Conclusions

In this paper we have presented an ensemble technique and evaluated it on the roadside images provided by TMR. The tests prove the significant improvement in classification rate over existing approaches for roadside image segmentation and classification. The proposed approach uses different seeding points to partition data into various layers and generates clusters with different patterns at each layer. A classifier is trained on each cluster for each layer. The outputs are fused using majority vote to create an ensemble clustered network. The experimental results show that the proposed approach using clustered ensemble can correctly segment and classify different road images and extract the objects with classification rate of 91.2% which is higher than the classification rates obtained by existing approaches for roadside object extraction. In our current work, we have conducted evaluation for road, sky and sign objects so further experiments and evaluation with more roadside objects at different conditions (cloudy, rainy and night) will be conducted in future research.

References

1. Gomez-Moreno, H., Maldonado-Bascon, S., Gil-Jamanez, P., Lafeuente-Arroyo, S.: Goal Evaluation of Segmentation Algorithms for Traffic Sign Recognition. IEEE Transactions on Intelligent Transportation Systems 11, 917–930 (2010)
2. Yuan, J., Cheriyadat, A.M.: Road Segments in Ariel Images by exploiting Road Vector Data. In: IEEE Conference on Geospatial Research and Application, pp. 16–23 (2013)
3. Pawlowski, P., Proszynski, D., Dabrowski, A.: Recognition of Road signs from Video. In: International Conference on Signal Processing Algorithms, Architecture, Arrangements and Applications, pp. 213–218 (2008)
4. Cheng, H., Butler, D.: Segmentation of Arial Surveillance Video using a mixture of Experts. In: Proceedings of Digital Image Computing Techniques and Applications, p. 66 (2005)
5. Danescu, R., Nedevschi, S.: Detection and Classification of Painted Road Objects for Intersection Assistance Applications. In: 13th IEEE International Annual Conference on Intelligent Transportation Systems, pp. 433–438 (2008)
6. Miura, J., Kanda, T., Shirai, Y.: An Active Vision System for Real-time Traffic Sign Recognition. IEEE Transactions on Intelligent Transportation Systems, 52–57 (2000)
7. Benallal, M., Meunier, J.: Real-time Colour Segmentation of Road Signs. In: Canadian Conference on Electrical and Computer Engineering, pp. 1823–1826 (2003)
8. Wu, W., Chen, X., Yang, Y.: Detection of Text on Road Signs from Video. IEEE Transactions on Intelligent Transportation Systems, 378–390 (2005)
9. Fan, C., Zhuming, L., Xiuqing, Y.: An Adaptive Method of Colour Road Segmentation. In: Proceedings of the 7th International Conference on Computer Analysis of Images and Patterns, pp. 661–668 (1997)
10. Ritter, W.: Traffic Sign Recognition in Colour Image Sequences. In: Proceedings of Intelligent vehicles Symposium, pp. 12–17 (1992)
11. Yang, L., Albregtsen, F., Lnnestad, T., Grttum, P.: A Supervised Approach to the Evaluation of Segmentation Image Segmentation Methods. In: Hlaváč, V., Šára, R. (eds.) CAIP 1995. LNCS, vol. 970, pp. 759–765. Springer, Heidelberg (1995)

12. Benallal, M., Meunier, J.: Real-time Colour Segmentation of Road Signs. In: IEEE Canadian Conference on Electrical and Computer Engineering, vol. 3, pp. 1823–1826 (2003)
13. Kouzani, A.Z.: Road Sign Identification using Ensemble Learning. In: Proceedings of IEEE Intelligent Vehicles Symposium, pp. 438–443 (2007)
14. Rahman, A., Verma, B., Stockwell, D.: An Hierarchical Approach towards Road Image Segmentation. In: International Joint Conference on Neural Networks, pp. 1–8 (2012)
15. Kinattukara, T., Verma, B.: Hierarchical Segment Learning method for Road Objects Extraction and Classification. In: 13th IEEE International Conference on Computer and Information Technology, pp. 432–438 (2013)
16. Kinattukara, T., Verma, B.: Clustering based Neural Network Approach for Classification of Road Images. In: 5th International Conference of Soft Computing and Pattern Recognition, p. 178 (2013)

Using Biologically-Inspired Visual Features To Model The Restorative Potential Of Scenes

James Mountstephens

School of Engineering and Information Technology, Universiti Malaysia Sabah
james@ums.edu.my

Abstract. This paper describes novel, interdisciplinary work towards learning the properties of visual scenes that restore our directed attention from fatigue. A groundtruth dataset of images rated for restorative potential was constructed and validated using human subjects, and biologically-inspired image features were used to train a number of regression models for this rating. The trained models were used to predict the restorative potential of unseen images and the predictions were tested using human subjects, with promising results.

Keywords: Biologically-inspired Vision, Attention, Restoration, Interdisciplinary.

1 Introduction

Computer Vision, whether biologically-inspired or based on mathematical or engineering principles, has made considerable progress in the automatic extraction of objective information from digital images but has so far paid little attention to the effect that certain images can have on a human viewer. Emprical work in Environmental Psychology has demonstrated that viewing specific visual scenes (even as photographs) can be beneficial, both subjectively and objectively. In comparison to most built scenes, viewing certain natural environments has been shown to improve cognitive performance, mood, the ability to plan, and sensitivity to interpersonal cues, as well as physiological levels such as stress and arousal [1,2,3,4,5,6,7]. One important way to understand these effects is as a form of restoration, the renewal of "physical, psychological and social capacities that have become depleted in meeting ordinary adaptational demands" [1]. The occurrence of restoration is measurable [1,2,3,4,5,6,7] and may be thus distinguished from simply liking a scene or feeling good whilst viewing it. In [3] for example, subjects were given a sustained cognitive task until fatigue, shown a slideshow of either nature or built scenes, and then asked to perform the task again. Only the subjects who viewed scenes of nature performed well on retest, demonstrating restored capabilities. In general, viewing natural scenes is more restorative than built scenes but restorative potential is not so clear cut: not all natural scenes restore and certain manmade scenes have also been found to be restorative [5].

If such benefits are available it would be worthwhile to model the properties of restorative scenes to better understand the phenomenon and to use the model practically

C.K. Loo et al. (Eds.): ICONIP 2014, Part III, LNCS 8836, pp. 194–201, 2014.

to evaluate arbitrary scenes. Software systems able to automatically evaluate the restorative potential of a given image would literally add a new dimension to image analysis and selection. If we can reliably identify them, we might actively use restorative images in our endeavours, be they website design or home decoration. Mobile phone apps presenting images predicted to restore might help us meet the challenges of day-to-day life. In architecture and urban planning, competing designs of living spaces might be selected based on restorative potential or the rating might even be incorporated into the design process itself. But what properties of a scene make it restorative? Its colour, shape, texture? Its layout, organisation and viewing distance? The presence of curves rather than straight lines? The objects present? The number of available visual features and statistics to explore is overwhelming. However, since human responses to stimuli are a product of both the properties of the stimulus itself and of the characteristics of the human perceptual system, it makes sense to first explore features inspired by biological visual systems. For example, Riesenhuber and Poggio's HMAX [11] is a physiologically-plausible computational model of object recognition in cortex, intended to explain cognitive phenomena in terms of simple and well-understood computational processes. HMAX is a purely feedforward model and has been shown capable of capturing the invariance properties and shape tuning of neurons in macaque inferotemporal cortex. In this paper, HMAX is explored for its potential to provide features able to characterise a scene's restorative potential.

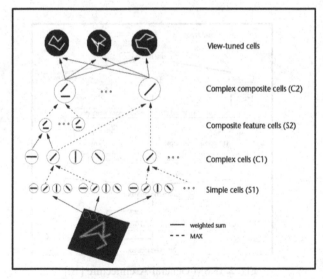

Fig. 1. HMAX General Architecture [11]

Another line of guidance towards identifying restorative features comes from a key theory in Environmental Psychology itself. Attention Restoration Theory (ART) explains the cognitive benefits of viewing nature in terms of their effect on the crucial cognitive resource of attention [2]. ART is predicated on a distinction between two main modes of attention, which is also recognised by cognitive science and computational modelling. Involuntary (or exogenous) attention is driven largely bottom-up by sensory stimuli. It is responsible for effortless orientation to salient stimuli and is thought to be mediated subcortically in the superior colliculi. Voluntary (or endogenous, directed) attention involves top-down inhibition of involuntary attention and the excitation of task-relevant locations. It is mediated by a number of cortical areas forming a "dorsoparietal network" and is crucial for intentional action and concentrating on tasks. However, voluntary attention requires effort to sustain and long-term demands deplete this resource, leading to what ART calls directed attention fatigue (DAF). DAF leaves us unhappy, unable to plan, insensitive to interpersonal cues and increases our likelihood of errors in performance [2,3]. ART claims that natural scenes contain stimuli that facilitate a move into involuntary mode, allowing directed attention to recover. By this account, natural scenes are more fascinating, containing patterns and objects that attract attention effortlessly but are not so stimulating as to require effortful focus and decision-making. Although this explanation of the benefits of nature still requires further substantiation, the "black box" occurrence of attention restoration is well-supported using methods of assessment devised within ART for that purpose [3]. The suggestion from ART that attention and bottom-up features are important may be useful guidance. Therefore, in addition to HMAX, Itti and Koch's saliency-based model of visual attention (abbreviated here as IKSM) [10] will also be explored here as a source of visual features. Being bottom-up and biologically-inspired makes IKSM appropriate for the task at hand.

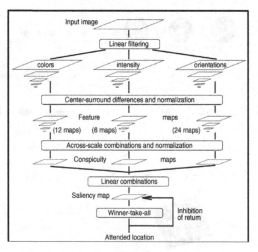

Fig. 2. IKSM General Architecture [10]

This paper is a pilot study in a novel area. The approach used here will be to i) build a dataset of scenes rated for restorative potential by human subjects using a scale developed within ART, ii) validate the ratings by testing whether they actually do produce restoration, iii) extract features from the labelled scenes using HMAX and IKSM and train regression models using them, iv) predict the restorative potential of unseen scenes, and v) test whether the predictions do produce restoration.Groundtruth Restorative Images

To learn restorative features of scenes, a dataset of images rated for restorative potential must be acquired and validated. The basic source of images was the 2688 image '8 Scene Categories' set constructed by Torralba and Oliva in their work on the 'gist' of a scene [12]. It was chosen here because it covers a wide range of both natural and built scene types. A subset of 72 scenes was manually selected from the master dataset in a way intended to provide coverage of natural vs built and the potential for restorative vs nonrestorative scenes. These 72 images were supplemented with a further 8 scenes of industrial views, taken from the web, which the source dataset was considered to lack. 11 Malaysian undergraduates (5 male, 6 female, aged 23-25) were given the Perceived Restorativeness Scale (PRS) [8] and asked to rate each of the 80 scenes using an online system custom developed for this purpose. The results of the rating are shown in Figure 3 below. Ratings for images in each row of the figure increase from left to right and each row is a continuation of the previous one. Inspection demonstrates that the subjects did not simply distinguish between natural and built scenes. Some natural scenes have low ratings and some built scenes are rated as moderate-to-highly restorative. This is consistent with the ART literature and suggests that finding features to capture the distinction between scenes with high and low restorative potential may be challenging.

Fig. 3. 80 groundtruth scenes arranged row-wise in order of Perceived Restorativeness

To ensure that the ratings reflect genuine restorative potential, validation was carried out following the 'SART-slideshow-SART' protocol of Berto [3]. SART (sustained attention response test) is a challenging test designed to fatigue directed attention. Subjects were required to do one session of SART, watch a slideshow of scenes rated either high or low for perceived restorativeness, and then do another session of SART. Four measures characterising performance were calculated: sensitivity to target detection d', reaction time (in ms), number of correct instances and number of incorrect instances. The 25 top and 25 bottom rated images from the 80 were selected for the slideshow. Paired t-tests on each of the four performance measures across sessions 1 and 2 were conducted. Subjects viewing images rated high on the PRS demonstrated a significant ($p < 0.05$) increase in d' and number of correct responses, and significant decreases were found in the same measures after viewing the low-rated scenes. This is consistent with [3] and, small sample size notwithstanding, we will consider the groundtruth validated for this pilot study.

2   Learning Restorative Potential

The groundtruth dataset was used to learn a regression model intended to be capable of predicting the restorative potential of a given image. Matlab provided the basic software infrastructure used in these experiments. Functions from the image processing, optimisation and statistics toolboxes were utilised at pre-and and post-processing stages. The freely-available Saliency Toolbox (STB), a Matlab implementation of IKSM [13] was used to calculate and extract conspicuity maps (CM) for colour, intensity and orientation, which were combined into a single vector of 768 components for each image. A basic implementation of HMAX, hmin [14], was used in the same manner to calculate feature vectors of 8150 components. The pretrained dictionary of S2 features included with hmin was used in the calculations but by default no colour information was used. Both types of feature vectors were fed into the freely-available SVM library libSVM [15], training in nu-SVR regression mode. Linear, polynomial, RBF and sigmoid kernels were trained. After training, the full 2688 images (minus the 72 groundtruth images) of the '8 Scene Categories' dataset were processed and the trained models were used to predict the level of restorative potential for each scene. The following three feature/kernel combinations were manually chosen for study based on inspection of their output: i) HMAX with a linear kernel, ii) CM with a polynomial kernel and iii) HMAX with a polynomial kernel. The results for the top and bottom 25 ratings for each combination are shown in Figure 4 (overleaf).

Inspection shows that each feature/kernel combination has captured a different aspect of the groundtruth. HMAX with a linear kernel makes a clear distinction between built and natural scenes. It appears to have associated high levels of activity in neurons trained to respond to to bar-like stimuli with low levels of restorativeness and is therefore biased towards scenes of skyscrapers. However, it has not captured the low restorative natural scenes or the high restorative built scenes. CM with a polynomial kernel is mixed although there is a bias towards open nature in the high-rated scenes. Interestingly, although no sunsets were in the groundtruth, it predicts them as highly restorative. This is not currently understood. HMAX with a polynomial kernel focusses on a distinction between spaciousness and detailed natural texture. Whether the emptiness of the low-rated scenes and the busy texture of the high-rated scenes is appropriate remains to be seen.

Fig. 4. Scenes predicted to have low (top half) and high (bottom half) restorative potential by (columns) i) HMAX with linear SVM kernel, ii) Conspicuity Maps with polynomial kernel and iii) HMAX with polynomial kernel

Manual inspection may yield insight but the only way to verify how successful these predictions are is to test them for restorativeness. The predictions made by each setup were validated using the same SART-slideshow-SART protocol used in constructing the groundtruth. HMAX with a linear SVM kernel showed significant (p < 0.05) improvement between sessions on correctness and reaction time, suggesting that it may have captured some aspects of restorativeness. The other combinations showed no significant improvement.

3   Conclusion

This paper has decribed the early stages of work in a novel area and results are suggestive rather than definitive. As a pilot study it may be considered successful but more work is required and is currently underway; in particular, a more detailed analysis of the properties of the features and classifier that were most successful here.

This analysis is expected to be challenging. Although biologically-inspired features can have the advantage of greater plausibility when modelling human phenomena, high-dimensional feature vectors consisting of individual cell responses are arguably harder to interpret than higher-level image processing descriptors such as colour, shape and texture, or mathematical properties such as entropy and complexity. There is, of course, no guarantee that reality should be easily understood but these more intuitive approaches to modelling restorative scenes are being investigated concurrently. We also intend to incorporate higher-level processes of scene and object recognition into the characterisation if necessary, including the popular 'gist' scene descriptor due to Oliva and Torralba [12].

It is expected and hoped that there will be considerable congruence between the various levels of description. We believe that attention restoration reflects a human response to a deep property of nature. It is conjectured that restorative scenes will be found to possess an optimal level of complexity – a balance between order and disorder; not so disordered as to require effort to perceive, and not so orderly as to be exhausted with a few glances. Characterising restorative scenes in this more abstract way may also help to explain why manmade scenes can also restore.

Lastly, work is underway to extend Itti and Koch's system to directly model the fatigue of directed attention, a feature missing from all known computational models of attention. It is hoped that this will allow a grounded extension of ART which may be of value to both computational modelling and Environmental Psychology.

References

1. Hartig, T., Staats, H.: Linking preference for environments with their restorative quality. In: Tress, B., Tress, G., Fry, G., Opdam, P. (eds.) From Landscape Research to Landscape Planning. Aspects of Integration, Education and Application, pp. 279–292. Springer, Dordrecht (2005)
2. Kaplan, S.: The Restorative Benefits of Nature: Toward an Integrative Framework. Journal of Environmental Psychology 15, 169–182 (1995)
3. Berto, R.: Exposure to restorative environments helps restore attentional capacity. Journal of Environmental Psychology 25, 249–259 (2005)
4. Hartig, T., Evans, G.W., Jamner, L.J., Davis, D.S., Gärling, T.: Tracking restoration in natural and urban field settings. Journal of Environmental Psychology 23, 109–123 (2003)
5. Korpela, K., Hartig, T.: Restorative qualities of favorite places. Journal of Environmental Psychology 16, 221–233 (1996)
6. Kaplan, S., Kaplan, R.: Cognition and the environment. Functioning in an uncertain world. Ann Arbor. Ulrich, MI (1981)
7. Kaplan, S., Talbot, J.F.: Psychological Benefits of a Wilderness Experience. In: Altman, I., Wohlwill, J.F. (eds.) Behaviour and the Natural Environment, vol. 6, pp. 163–203. Plenum Press, New York (1983)
8. Hartig, T., Korpela, K., Evans, G.W., Gärling, T.: A measure of restorative quality in environments. Scandinavian Housing and Planning Research 14, 175–194 (1997)
9. Robertson, I.H., Manly, T., Andrade, J., Baddeley, B.T., Yiend, J.: 'Oops!': Performance correlates of everyday attentional failures in traumatic brain injured andnormal subjects. Neuropsychologia 35, 747–758 (1997)

10. Itti, L., Koch, C., Niebur, E.: A Model of Saliency-Based Visual Attention for Rapid Scene Analysis. IEEE Transactions on Pattern Analysis and Machine Intelligence 20(11), 1254–1259 (1998)
11. Riesenhuber, M., Poggio, T.: Hierarchical Models of Object Recognition in Cortex. Nature Neuroscience 2, 1019–1025 (1999)
12. Oliva, A., Torralba, A.: Modeling the shape of the scene: a holistic representation of the spatial envelope. Int. Journal of Computer Vision 42(3), 145–175 (2001)
13. Walther, D., Koch, C.: Modeling attention to salient proto-objects. Neural Networks 19, 1395–1407 (2006)
14. http://cbcl.mit.edu/jmutch/hmin/ (accessed July 30, 2014)
15. http://www.csie.ntu.edu.tw/~cjlin/libsvm/ (accessed July 30, 2014)

Classification of Stroke Patients' Motor Imagery EEG with Autoencoders in BCI-FES Rehabilitation Training System

Mushangshu Chen, Ye Liu, and Liqing Zhang*

Key Laboratory of Shanghai Education Commission for Intelligent Interaction and
Cognitive Engineering, Department of Computer Science and Engineering, Shanghai
Jiao Tong University, China
zhang-lq@cs.sjtu.edu.cn

Abstract. Motor imagery based Brain Computer Interface (BCI) system is a promising strategy for the rehabilitation of stroke patients. Common Spatial Pattern (CSP) is frequently used in feature extraction of motor imagery EEG signals and its performance depends heavily on the choice of frequency component. Moreover, EEG of stroke patients, which is full of noise, makes it hard for traditional CSP to extract discriminative patterns for classification. In order to deal with the subject-specific band selection, in this paper, we adopt denoising autoencoders and contractive autoencoders to extract and compose robust features from CSP features filtered in multiple frequency bands. We compare our method with traditional methods on data collected from two months clinical rehabilitation. The results not only demonstrate its superior recognition performance but also evidence the effectiveness of our BCI-FES rehabilitation training system.

1 Introduction

Brain Computer Interface (BCI), as an alternative communication channel between human brain and external devices, is a good way that combines Electroencephalography (EEG) signals with motor control [1]. Recently some studies have demonstrated that motor imagery based BCI is a very promising method in rehabilitation training of strokes [2]. One of the most effective algorithms for motor imagery based BCI is Common Spatial Pattern (CSP) [3]. The spatial filters generated by CSP reflect the specific activation of cortical areas. However, the performance of CSP heavily depends on the proper selection of frequency bands and channels [4][5].

It is generally considered that motor imagery of normal people attenuates EEG μ and β rhythm over sensorimotor cortices [6][7]. However, for special populations suffering from neurophysiological diseases (e.g., stroke), some studies recently found that the μ and β rhythm in motor imagery EEG of stroke patients have been modulated [8]. In our analysis, a similar regularity is observed

---

* Corresponding author.

C.K. Loo et al. (Eds.): ICONIP 2014, Part III, LNCS 8836, pp. 202–209, 2014.

that the most informative frequency bands for classifying left and right motor imagery have deviated from normal α and β rhythms in different stages of the clinical experiments [9]. Furthermore, our analysis also illustrates that irregular discriminative patterns from impaired cortex is different from those of the normal subjects, and there frequently exists messy imagination contents in the motor imagery EEG of stroke patients [10]. In consequence, conventional methods suitable for normal subjects usually achieves a relatively low level classification accuracy [11].

To make full use of latent spectral information in EEG of stroke patients and extract robust features from noisy data, in this paper, we adopt denoising autoencoders [12] and contractive autoencoders [13] to obtain a encoding of the raw features of stroke patients' signals.

The rest part of this paper is organized as follows: a detailed introduction about our rehabilitation training system, experiment setup and data collection are given in Section 2. Section 3 describes the deep learning method for extracting EEG features. Section 4 demonstrates a comparative result when applying our method, frequency boosting and CSP-SVM on data collected from clinical experiments. Apart from the classification performance, we describe a phenomenon of contribution of band changing during rehabilitation, which may reveal some mechanisms of stroke patients' recovery in frequency aspect. Finally we give a brief conclusion in Section 5.

2 Experiment Paradigm

2.1 BCI-FES Rehabilitation Training System

Our BCI-FES rehabilitation system is consists of 5 modules: real-time data acquisition module, data storage and analysis module, visualization module, multimodal feedback module and human effect training module [14].

In general, the system aims at restoring motor functions of paralyzed limbs for post-stroke patients by active motor imagery directed by training tasks. EEG signal is collected by data acquisition module during subject's imagery and label of each segment of subject's motor imagery is recognized online after feature extraction and classification in data storage and analysis module. Multi-modal feedback module gives a corresponding feedback including visual, auditory and tactile response given the classification result and visualization module gives a real-time observation concurrently.

In order to improve training effect, we adopt a new rehabilitation training paradigm which attempts to reconstruct the motor sensory feedback loop [15]. During experiments the subject is required to reconfirm the label and can correct the label when necessary.

2.2 Experiment Setup and Data Collection

We conduct our clinical rehabilitation on seven participated subjects in hospital training with our BCI-FES rehabilitation system. Another three patients only receiving regular clinical treatments are considered as control group to assess the effectiveness of our system and rehabilitation paradigm.

In general, each subject is required to participate in 3 days' training per week. Each day's training consists of 8 sessions which contains 15 trials of motor imagery tasks. Each trial lasts for 4 seconds and is cut into 25 1s sliding windows with step length 0.125s for online classification. At the end of experiment cycle, a post-training section consists of 2 sessions will be conducted to evaluate rehabilitation efficacy. Raw EEG data is recorded by a 16-channel(FC3, FCZ, FC4, C1-C6, CZ, CP3, CPZ, CP4, P3, PZ and P4) g.USBamp amplifier under a sample rate of 256 Hz. After removing artifacts, we filter the EEG into α(8-12 Hz), β(12-30 Hz), γ(30-45 Hz) band for feature extraction.

3   Method

3.1   Common Spatial Pattern

The goal of CSP [3] is to design spatial filters that lead to optimal variances for the discrimination of two populations of EEG related to left and right motor imagery.

We denote raw EEG data as a $ch \times time$ matrix E, where ch is the number of channels and $time$ is the number of samples per channel. The filtered signal matrix S is $S = WE$ or $S(t) = We(t)$, where $W \in \mathbb{R}^{d \times ch}$ is spatial filter matrix.

First the sum spatial covariance can be eigen decomposed as $\Sigma_1 + \Sigma_2 = \frac{1}{n_L} \sum_{i=1}^{n_L} \frac{E_{L_i} E'_{L_i}}{trace(E_{L_i} E'_{L_i})} + \frac{1}{n_R} \sum_{i=1}^{n_R} \frac{E_{R_i} E'_{R_i}}{trace(E_{R_i} E'_{R_i})} = UDU^T$.

Then the whitened covariance by $P = \sqrt{D^{-1}}U^T$ can be decomposed as $\hat{\Sigma}_1 + \hat{\Sigma}_2 = P(\Sigma_1 + \Sigma_2)P^T = V(\Lambda_L + \Lambda_R)V^T$.

Finally, by selecting first and last m eigenvectors in V, the CSP filter is obtained as $W = P^T V \in \mathbb{R}^{2m \times ch}$. The filtered signal matrix is given by $s(t) = We(t) = (s_1(t) \ldots s_d(t))^T, d = 2m$. And feature vector $x = (x_1, x_2, \ldots, x_d)^T$ is calculated by $x_i = \log(\frac{var[s_i(t)]}{\sum_{j=1}^{d} var[s_j(t)]})$.

3.2   Autoencoder

A simplest autoencoder (AE) [16] is composed of two parts, an encoder and a decoder. The encoder is a function f that maps an input $x \in \mathbb{R}^d$ to a hidden representation $h \in \mathbb{R}^{d'}$ through a deterministic mapping $h = f_{\theta_1}(x) = s(Wx+b)$, parameterized by $\theta_1 = \{W, b\}$. The resulting latent representation h is mapped back to a reconstruction y, where $y = g_{\theta_2}(h) = s(W'h + b')$ with $\theta_2 = \{W', b'\}$.

Autoencoder training consists in optimizing parameters $\theta = \{W, b, W', b'\}$ to minimize the average reconstruction error on a training set D_n: $\mathcal{J}_{AE}(\theta) = \sum_{x \in D_n} L(x, g(f(x))) + \beta \sum_{j=1}^{d'} KL(\rho || \hat{\rho}_j)$, where L is the reconstruction error, ρ is sparsity parameter, $\hat{\rho}_j$ is the average activation of hidden unit j and β controls the weight of the sparsity penalty term $KL(\rho || \hat{\rho}_j)$. In case of s being the sigmoid, L is cross-entropy loss:$L(x, y) = - \sum_{i=1}^{d} x_i \log(y_i) + (1 - x_i) \log(1 - y_i)$.

Denoising Autoencoder(DAE). To enforce robustness to noisy inputs, denoising autoencoder [12] first corrupts input x, then train the autoencoder to reconstruct the clean version. The objective function is

$$\mathcal{J}_{DAE}(\theta) = \sum_{x \in D_n} \mathrm{E}_{\tilde{x} \sim q(\tilde{x}|x)}[L(x, g(f(\tilde{x})))] \tag{1}$$

where the expectation is over corrupted $\tilde{x}$ obtained from a corruption process $q(\tilde{x}|x)$.

Contractive Autoencoder(CAE). CAE [13] is obtained with the regularization term $||J_f(x)||_F^2 = \sum_{ij}(\frac{\partial f_j(x)}{\partial x_i})$, giving objective function

$$\mathcal{J}_{CAE}(\theta) = \sum_{x \in D_n} (L(x, g(f(x))) + \lambda||J_f(x)||_F^2) \tag{2}$$

The basic autoencoders can be stacked into deep networks. Greedy layer-wise training is a good way to initialize a stacked autoencoder. The features from the stacked autoencoder can be used for classification by feeding the last layer's output to a softmax classifier.

In our analysis, EEG signals are filtered into α, β, γ band and form a *channel* $\times$ *time* $\times$ *window* $\times$ *band* format data. Then CSP is applied to each band's data *channel* $\times$ *time* $\times$ *window* to extract frequency specific features. These features are normalized to $[0, 1]$ using $NormalizedFeature = \frac{Feature - \min(Feature)}{\max(Feature) - \min(Feature)}$ and then fed into a stacked autoencoder.

We adopt two kinds of autoencoder described above to pretrain each layer in turn with first week's data and finetune the whole model using the same data with labels. Then subsequent data are split into 2 parts: The first 7 sessions' data in each day is used to finetune the network in order to adapt the model to the pattern changing during rehabilitation and the last session's data is for testing. Figure 1 shows the structure of our model.

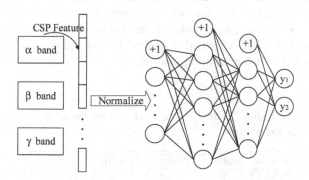

Fig. 1. The structure of our model. EEG data filtered by each band are transformed into a 4-dimensional feature by CSP, and fed into a neural network with $12 \times 25 \times 16 \times 2$ units after being normalized.

4   Result

Considering the training parameters of DAE for all the subjects, we set learning rate 0.5, scaling factor 0.99, momentum 0.5 and a binary masking noise with fraction 0.3 as the input. The sparsity constraint and penalty factor is subject specific.

In terms of CAE, we set λ 0.05, learning rate 0.5, scaling factor 0.99 and momentum 0.5 for all the subjects.

In order to evaluate the performance of our method, we also apply traditional CSP-SVM method and frequency boosting method on dataset of 4 patients. We calculate classification accuracies of stroke patients EEG in the chosen 6 weeks out of 2 months and finally the mean test session accuracy of each week is calculated. Table 1 shows test accuracies of 4 patients pretrained using DAE.

Table 1. Test accuracies of sliding windows using DAE. Note that the data collected in first week are used to pretrain each layer's autoencoder.

Subject	Age	2nd Week	3rd Week	4th Week	5th Week	6th Week
1	65	0.57	0.63	0.61	0.73	0.79
2	71	0.49	0.51	0.51	0.56	0.76
3	50	0.56	0.55	0.51	0.60	0.69
4	65	0.52	0.58	0.53	0.62	0.72

Compared with other methods, our method achieves a better accuracy (Fig. 2, Table 2). We consider the hidden layer's output as the hidden encoding of the original features. The hidden encoding of autoencoders provides a new representation of the original data in the subspace defined by the weights. Our experiment result shows that such representations are more robust to noisy data of stroke patients and give us much useful discriminative information for classification, which attributes to sparse and smooth representation.

Figure 2 also shows a rising tendency in terms of test accuracy. It's worth mentioning that the whole result also shows the feasibility and effectiveness of our BCI-FES rehabilitation training system.

Table 2. Best Test Session Accuracy. The best session for all subjects all appeals in the last week of our experiment.

Method	Subject 1	Subject 2	Subject 3	Subject 4
CSP-SVM	0.64	0.818	0.626	0.671
F-Boost	0.843	0.843	0.669	0.737
CSP-DAE	**0.941**	**0.913**	**0.779**	0.72
CSP-CAE	0.923	0.779	0.711	**0.749**

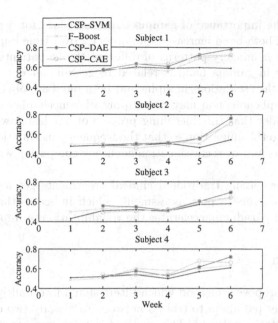

Fig. 2. Mean test accuracy of CSP-SVM, Frequency Boosting, CSP-DAE and CSP-CAE over time. Note that: (1) For autoencoders, data of first week is used for pretraining. (2) CSP-DAE achieves a higher accuracy in most cases. (3) A rising tendency can be observed over time.

To analyze the band contribution to classification, we calculate the relative importance of each feature using connection weights [17]. Figure 3 shows the gradual changes of importance during experiment of four subjects.

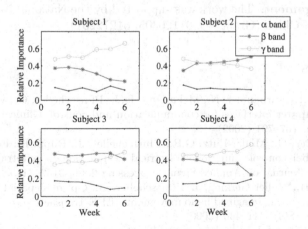

Fig. 3. The contribution changes of each band over rehabilitation process

The fact that the importance of gamma band increased for 3 patients while the contribution of beta band increased for subject 2 over time implies the gradual changes of motor imagery patterns of different stroke patients. Noting that oscillatory activity in gamma band is related to gestalt perception and cognitive functions and the oscillations in alpha and beta band are obvious indicators of movement, this phenomenon may reveal potential mechanisms about stroke recovery. We consider that the finetuning process of our last 5 weeks' training adapts the model to this migration so that the frequency modulations can be detected by CSP, thus improves classification accuracy comparing with traditional method.

Three patients receiving traditional clinical treatments get a lower clinical rehabilitation parameters in post assessment, which indicates that our system and active training paradigm accelerates the rehabilitation of impaired cortex.

5   Conclusion

In this paper, we propose a method which filters signal into multiple bands and use the autoencoder paradigm to train a network to classify two classes' motor imagery EEG of stroke patients. This method detects important structure in the raw common spatial patterns by using a local unsupervised criterion to pretrain each layer in the network and captures discriminative pattern changes during rehabilitation process by keeping finetuning the model. Compared with traditional CSP-SVM classifier, our method achieves a better result on both accuracy over time and optimal session accuracy. The analysis of band changing during rehabilitation provides a prior knowledge about motor imagery pattern of stroke patients. Furthermore, the comparison of experimental group with control group demonstrates the effectiveness of our multi-modal BCI-FES rehabilitation training system and active training paradigm.

Acknowledgement. The work was supported by the National Natural Science Foundation of China (Grant No. 91120305, 61272251).

References

1. Wolpaw, J.R., Birbaumer, N., McFarland, D.J., Pfurtscheller, G., Vaughan, T.M.: Brain–computer interfaces for communication and control. Clinical Neurophysiology 113(6), 767–791 (2002)
2. Pfurtscheller, G., Müller-Putz, G.R., Pfurtscheller, J., Rupp, R.: Eeg-based asynchronous bci controls functional electrical stimulation in a tetraplegic patient. EURASIP Journal on Applied Signal Processing 2005, 3152–3155 (2005)
3. Ramoser, H., Muller-Gerking, J., Pfurtscheller, G.: Optimal spatial filtering of single trial eeg during imagined hand movement. IEEE Transactions on Rehabilitation Engineering 8(4), 441–446 (2000)
4. Zhang, H., Zhang, L.: Spatial-spectral boosting analysis for stroke patients' motor imagery eeg in rehabilitation training. CoRR, Vol. abs/1310.6288 (2013)

5. Li, J., Zhang, L.: Regularized tensor discriminant analysis for single trial eeg classification in bci. Pattern Recognition Letters 31(7), 619–628 (2010)
6. Pfurtscheller, G., Neuper, C.: Motor imagery and direct brain-computer communication. Proceedings of the IEEE 89(7), 1123–1134 (2001)
7. Song, L., Gordon, E., Gysels, E.: Phase synchrony rate for the recognition of motor imagery in brain-computer interface. Advances in Neural Information Processing Systems 18, 1265 (2006)
8. Shahid, S., Sinha, R.K., Prasad, G.: Mu and beta rhythm modulations in motor imagery related post-stroke eeg: a study under bci framework for post-stroke rehabilitation. BMC Neuroscience 11(Suppl. 1), P127 (2010)
9. Liang, J., Zhang, H., Liu, Y., Wang, H., Li, J., Zhang, L.: A frequency boosting method for motor imagery EEG classification in BCI-FES rehabilitation training system. In: Guo, C., Hou, Z.-G., Zeng, Z. (eds.) ISNN 2013, Part II. LNCS, vol. 7952, pp. 284–291. Springer, Heidelberg (2013)
10. Wang, H., Liu, Y., Zhang, H., Li, J., Zhang, L.: Causal neurofeedback based BCI-FES rehabilitation for post-stroke patients. In: Lee, M., Hirose, A., Hou, Z.-G., Kil, R.M. (eds.) ICONIP 2013. LNCS, vol. 8226, pp. 419–426. Springer, Heidelberg (2013)
11. Liu, Y., Li, M., Zhang, H., Li, J., Jia, J., Wu, Y., Cao, J., Zhang, L.: Single-trial discrimination of eeg signals for stroke patients: A general multi-way analysis. In: EMBC 2013, pp. 2204–2207. IEEE (2013)
12. Vincent, P., et al.: Stacked denoising autoencoders: Learning useful representations in a deep network with a local denoising criterion. The Journal of Machine Learning Research 9999, 3371–3408 (2010)
13. Rifai, S., Vincent, P., Muller, X., Glorot, X., Bengio, Y.: Contractive auto-encoders: Explicit invariance during feature extraction. In: Proceedings of the 28th International Conference on Machine Learning (ICML 2011), pp. 833–840 (2011)
14. Liu, Y., Zhang, H., Wang, H., Li, J., Zhang, L.: Bci-fes rehabilitation training platform integrated with active training mechanism. In: IJCAI 2013 Workshop on Intelligence Science (2013)
15. Li, J., Zhang, L.: Active training paradigm for motor imagery bci. Experimental brain research 219(2), 245–254 (2012)
16. Vincent, P., Larochelle, H., Bengio, Y., Manzagol, P.-A.: Extracting and composing robust features with denoising autoencoders. In: Proceedings of the 25th International Conference on Machine learning, pp. 1096–1103. ACM (2008)
17. Garson, G.D.: Interpreting neural-network connection weights. AI expert 6(4), 46–51 (1991)

Real-Time Patch-Based Tracking with Occlusion Handling

Jian Tian and Yue Zhou

Institute of Image Processing and Pattern Recognition
Shanghai Jiao Tong University
Shanghai, China

Abstract. A new method for real-time occlusion-robust tracking is proposed. By analyzing the process of occlusion occurrence, we present a fast and effective occlusion detection algorithm based on the spatio-temporal context information. As a result, we can always obtain correct target location using adaptive template matching with patch-based structure description, regardless of the occlusion situation. Our extensive experiments on many sequences verify the good performance of our algorithm. In addition, based on the framework of our algorithm and properties we find, more effective occlusion-robust tracking algorithms can be developed.

Keywords: Real-time tracking, occlusion detection, patch based model, structure description, context.

1 Introduction

In computer vision, visual object tracking is an important subject for a wide range of applications. Abrupt object motion, changing appearance patterns of both the object and the scene, non-rigid object structures, occlusion and camera motion challenge the tracking algorithms [1]. Occlusion is a common problem in tracking. It is different from other interference such as deformation and varying illumination. The tracker should update the appearance model to accommodate the object appearance, when the object orientation or illumination changes. However, the occluder should be regarded as another object. Therefore, occlusion in tracking is a particular interference and needs more attention to solve. Based on patch-based structure description and context information, we present a simple and fast tracking and a novel occlusion detection algorithm.

Our algorithm is a patch-based structure description method. It describes the spatial structure of tracking object and has good robustness for geometric variation. Previous work has presented many similar algorithms [2, 3, 4]. Junseok Kwon et al. [3] presented a non-rigid object tracking algorithm based on patch-based model and adaptive basin hopping monte carlo sampling (BHMC). It is robust for deformation effectively. However, the patch-based model in [3] can not resist scale variance and occlusion well.

Many tracking algorithms rely on learning to resist occlusion [5,6]. Kaihua Zhang et al. [6] presented a spatio-temporal context learning algorithm (STC). Spatio-temporal context makes tracking stable and accurate, context learning contributes to

C.K. Loo et al. (Eds.): ICONIP 2014, Part III, LNCS 8836, pp. 210–217, 2014.

occlusion-robust. However, learning algorithms detect occlusion based on similarity between image and object model, it is difficult to resist occlusion and other interference at the same time. Yi Deng et al. [7] proposed a patch-based occlusion detection algorithm. It focuses on patches obtained by segment at the boundary of target. Occlusion at the patches means target will be sheltered. Similar to it, our algorithm focuses on the boundary of target, however, we detect occlusion by analyzing the process of occlusion occurrence. It considers temporal context which is an essential information.

Our main contributions are: (a) Based on affine invariant patch-based model, we present a simple and fast tracking algorithm. (b) From a new perspective, we find a property to detect occlusion in tracking. (c) Based on the property, we present a patch-based occlusion detection algorithm, it is easy to compute and transplant.

2   Patch-Based Model and Real-Time Tracking

We propose to represent the object by many patches which can be moved, deleted or newly added. Each patch helps to locate the possible positions and scales of the object in the current frame. The patches are not based on an object model. To make such a representation more robust, a patch should be deleted if it belongs to the background, and a structure description is introduced to find it. We use template matching to move the patches because of its relative simplicity and low computational cost.

2.1   Patch Initialization

In order to get effective patches, first we use FAST [8] to find key points. It is an efficient key point detection algorithm. A patch is more likely to contain effective information if its position is close to target center. We set a key point as one of the patch's vertices which is farthest to the target center (Fig.1 (a)). Then we compare each patch with screen around target based on template matching. If a patch is similar to screen, it should be abandoned (Fig.1 (b)).

2.2   Examining the Patches by Structure Description

To update the patches, we use normalized correlation template matching. Other matching algorithms are available, such as LK matching [9], MSER [10]. Tracking procedure mainly includes two parts: (a) Rough localization. We use the whole target as template to match. It provides a stable target center to calculate proper search region and structure description similarity which will be introduced later. (b) Accurate adjustment. Based on patch similarity and structure description similarity, we remove inaccurate patches.

We use the ratio of each patch distance to average distance as structure description (Fig.1 (c)):

$$D_i = R_i / (\sum_{j=1}^{n} R_j / n), i = 1, 2, ..., n \tag{1}$$

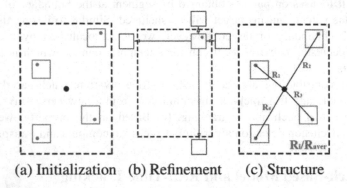

(a) Initialization   (b) Refinement   (c) Structure

Fig. 1. Patch-based model. (Black point is target center, red points are FAST points, red boxes are patches.).

where R_i is the distance between the center of patch i and target center. In Patch initialization, it is impossible to avoid generating new patches which do not belong to the object. Likewise, a patch may be updated to the background. These situations will reduce the accuracy of the appearance of the model. Wrong patches cannot be avoided, but it will exhibit motion characteristics different to other patches when object moves. We can use the center calculated by all patches except the current patch i to get R_i. Based on (1), if D_i is larger than it is in the last frame, we can consider it is a wrong patch and remove it.

2.3   Updating the Appearance Model

After updating patches, we can find those patches which are correctly tracked and locate the target in new frame.

At every frame, we can find a smallest rectangle which can include all well tracked patches, we call that rectangle the bounding rectangle:

$$BoundRect = \text{argmin}_r \, r.\text{area}() \quad \forall i \, r \, \& \, pathc[i] == patch[i] \tag{2}$$

, where r means an arbitrary rectangle in the frame on the condition that all N patches is in the rectangle r.

Estimation of the target displacement is the same as the BoundRect displacement and the width and height of the target can changed in proportion as the BoundRect. So the shape of the object can be updated adaptively.

In addition, performance of structure description depends on the number of patches, we should generate more patches when they are not enough. Scale of target often changes, we can properly adjust patch size based on scale of target.

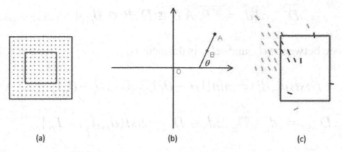

Fig. 2. Occlusion detection model

3 Occlusion Detection

As mentioned in Sec.1, most tracking algorithms merely rely on learning to resist occlusion. However, it is difficult to ensure the robustness for occlusion and other interference at the same time.

In this paper, we exploit the spatio-temporal context information to analyze the occlusion situation. Occlusion has two properties that the other interference do not have.

Property 1: Occlusion has the process that occluder relatively moves to object from screen. However, variation caused by other interference comes from object itself. Therefore, we can detect the process of occlusion from screen to tracking object.

Property 2: At the place occluder is moving from screen to target, local regions of screen and target move towards the target inner at the same time.Because property 2 considers the relation between frames and the relation between screen and target, it is spatio-temporal context information.

Based on the 2 properties, we present a simple model to detect occlusion in tracking. Fig.2 shows our model (it can be adjusted according to different tracking problems).

The tracker accepts a pair of images I_t, I_{t+1}, and a bounding box β_t, β_{t+1} (Sec. 2).Red box is the target region (Fig.2 (a)),Blue box has the same center as the red one, and its width and height is 2 times longer than the red one. A set of points is initialized on the blue rectangular grid. These points are then tracked by the affine lucas kanade feature tracker in [11] which generates a sparse motion flow between I_t and $I_t, \overrightarrow{AB}$ is an example of the motion flow of a point(Fig.2 (b)),Point O is the center of target which is moved by our tracker in Sec. 2. We select the motion flow D:

$$D = \{d_i = \overrightarrow{A_i B_i} \mid |\overrightarrow{OA_i} - \overrightarrow{OB_i}| > 0\} \tag{3}$$

And calculate the angle θ_i between $\overrightarrow{A_i B_i}$ and the horizontal axis. D is divided into two groups D_{in} and D_{out} (Fig.2 (c):

$$D_{in} = \{d_i = \overrightarrow{A_i B_i} \mid \overrightarrow{A_i B_i} \in D, B_i \in \beta_{t+1}\} \tag{4}$$

$$D_{out} = \{d_i = \overrightarrow{A_i B_i}, \overrightarrow{A_i B_i} \notin D, B_i \in \beta_{t+1}\} \tag{5}$$

The distance between d_i and d_j is defined as:

$$Dist(d_i, dj) = min\{| \theta_i - \theta_j |, 360 - | \theta_i - \theta_j |\} \tag{6}$$

$$D_{valid} = \{d_i \in D_{in}, \exists d_j \in D_{out}, Dist(d_i, d_j) < T_D\} \tag{7}$$

D_{valid} includes points that may be from the occluder. If the size of D_{valid} is larger than a threshold T_N , the target may be occluded and we should stop adding new patches to avoid adding patches belong to the occluder. If any patch intersects with an element in D_{valid}, we remove it. If the number of patches is less than a threshold T_C, the target is covered. Before the target is covered, we can still track the target through the remaining patches. Fig.3 shows the flowchart of our occlusion detection algorithm.

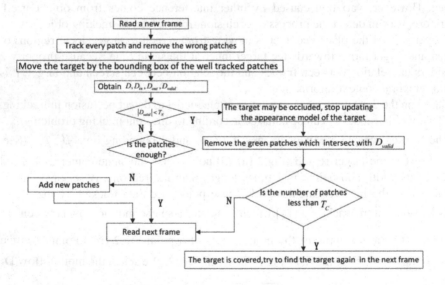

Fig. 3. The flowchart of our occlusion detection algorithm

4   Experimental Results

We conduct some experiments to verify the performance of our algorithm. Development environment of our algorithm is Visual Studio 2010 and Intel OpenCV library on Intel Core i3-2120 CPU and 4.00GB RAM. Experimental videos can be downloaded[1].

The main advantage of the proposed occlusion detection algorithm is that it can be applied to a range of trackers and is easy to implement. Kaihua Zhang et al. [6] presented a spatio-temporal context learning algorithm (STC). Spatio-temporal context makes tracking stable and accurate, context learning contributes to occlusion-robust.However, long-term occlusion will lead to tracking failure. With our occlusion detection, STC can promote the robustness of tracking under occlusions. We just stop learning Spatio-Temporal context model when the occlusion happens.

Fig.4 shows experimental results by STC, STC-OD and our patch-based tracker with occlusion detection (PBT-OD). The first row shows STC tracker can't handle heavy occlusion. The second row shows that with our occlusion detection algorithm STC-OD can recover the target .The third row shows the result by PBT-OD, which can also track the face well.

Fig. 4. Experimental results by STC, STC-OD and PBT-OD

We perform experiments on a wide range of video sequences downloaded from http://www.votchallenge.net/vot2013/evaluation_kit.html. This website provides many well-known short videos labeled with ground truth. We use Success plot in [12] to evaluate our algorithm. Given the tracked bounding box RT and the ground truth bounding box RA. The overlap score is defined as

$$S = \frac{RT \cap RA}{RT \cup RA} \qquad (8)$$

The numerator and denominator in (8) represent the number of pixels in the intersection and union of two regions. To measure the performance on a sequence of frames, we count the number of successful frames whose overlap S is larger than the given threshold. The success plot shows the ratios of successful frames at the thresholds varied

---

[1] Download link:
    ftp://tianjian2013:public@public.sjtu.edu.cn/iconip2014/

from 0 to 1. Using one success rate value at a specific threshold for tracker evaluation may not be fair or representative. Instead we use the area under curve (AUC) of each success plot to evaluate our algorithm. Fig.5 is the tracking sequences for evaluation and Fig. 6 shows the performance score of our tracker. The performance of the state-of-the-art trackers can be found in [12].

Fig. 5. Sample tracking results on four public challenging image sequences

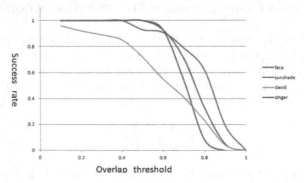

Fig. 6. Success rate plot

5   Conclusion

In this paper, we mainly focus on fast and occlusion-robust tracking problem. We present an affine invariant patch-based structure description and a fast tracking algorithm based on it. By analyzing the process of occlusion occurrence, we pre-sent two properties of occlusion detection and a portable algorithm based on patch-based context information. Comparative experiments verify the good per-formance of our algorithm. Our algorithm is simple to implement, and can run at real-time speed.

References

1. Yilmaz, A., Javed, O., Shah, M.: Object tracking: A survey. ACM Computing Surveys (CSUR) 38(4), 13 (2006)
2. Adam, A., Rivlin, E., Shimshoni, I.: Robust fragments-based tracking using the integral histogram. In: 2006 IEEE Computer Society Conference on Computer Vision and Pattern Recognition, vol. 1. IEEE (2006)
3. Kwon, J., Lee, K.M.: Tracking of a non-rigid object via patch-based dynamic appearance modeling and adaptive basin hopping monte carlo sampling. In: IEEE Conference on Computer Vision and Pattern Recognition, CVPR 2009. IEEE (2009)
4. Dihl, L., Jung, C.R., Bins, J.: Robust adaptive patch-based object tracking using weighted vector median filters. In: 2011 24th SIBGRAPI Conference on Graphics, Patterns and Images (Sibgrapi). IEEE (2011)
5. Kalal, Z., Mikolajczyk, K., Matas, J.: Tracking-learning-detection. IEEE Transactions on Pattern Analysis and Machine Intelligence 34(7), 1409–1422 (2012)
6. Zhang, K., et al.: Fast Tracking via Spatio-Temporal Context Learning. arXiv preprint arXiv:1311.1939 (2013)
7. Deng, Y., et al.: A symmetric patch-based correspondence model for occlusion handling. In: Tenth IEEE International Conference on Computer Vision, ICCV 2005, vol. 2. IEEE (2005)
8. Rosten, E., Drummond, T.W.: Machine learning for high-speed corner detection. In: Leonardis, A., Bischof, H., Pinz, A. (eds.) ECCV 2006, Part I. LNCS, vol. 3951, pp. 430–443. Springer, Heidelberg (2006)
9. Lucas, B.D., Kanade, T.: An iterative image registration technique with an application to stereo vision. IJCAI 81 (1981)
10. Donoser, M., Bischof, H.: Efficient maximally stable extremal region (MSER) tracking. In: 2006 IEEE Computer Society Conference on Computer Vision and Pattern Recognition, vol. 1. IEEE (2006)
11. Bouguet, J.-Y.: Pyramidal implementation of the affine lucas kanade feature tracker description of the algorithm. Intel Corporation 2, 3 (2001)
12. Wu, Y., Lim, J., Yang, M.-H.: Online object tracking: A benchmark. In: 2013 IEEE Conference on Computer Vision and Pattern Recognition (CVPR). IEEE (2013)

Blood Cell Image Retrieval System Using Color, Shape and Bag of Words

Mohammad Reza Zare[1] and Woo Chaw Seng[2]

[1]School of Information Technology, Monash University Malaysia
mohammad.reza@monash.edu
[2]Faculty of Computer Science and Information Technology, University of Malaya, Malaysia
cswoo@um.edu.my

Abstract. The ever increasing number of medical images in hospitals urges on the need for generic content based image retrieval systems. These systems are in an area of great importance to the healthcare providers. The first and foremost function in such system is feature extraction. In this paper, different feature extraction techniques have been utilized to represent medical blood cell images. They are categorized into two groups; low-level image representation such as color and shape analysis and local patch-based image representation such as Bog of Words (BoW). These features have been exploited for retrieving similar images. We have also used a generative model such as Probabilistic Latent Semantic Analysis (PLSA) on extracted BoW for retrieval task. Lastly, the retrieval results obtained from all the above features are integrated with one another to increase the retrieval performance. Experimental results using four different classes of 600 blood cell images showed 92.25% of retrieval accuracy.

Keywords: Image Retrieval. Feature Extraction. BoW. PLSA. CBIR.

1 Introduction

The growth of multimedia content production has been accelerated due to the wide availability of digital devices. Medical blood cell images are one of the most popular among the variety of multimedia content. The latest computer systems can store a very huge amount of medical images. Such medical image databases are the key component in diagnosis and preventive medicine. Therefore, there is an increased demand for a computerized system to manage these valuable resources. In addition, managing such data demands high accuracy since it deals with human life.

Traditional text based image retrieval is becoming infeasible due to the huge amount of blood cell images and it is also difficult to achieve a consistent diagnose during microscopic evaluation due to subjective impressions of observers. The solution to this problem is Content-Based Image Retrieval (CBIR); This would allow us to search based on the content rather than the keywords and meta-data descriptions. The CBIR performance strongly lays on the techniques used to represent images. There are two main approaches in CBIR: (i) those techniques that directly extract low level visual features from the images, and (ii) those techniques that represent images by

C.K. Loo et al. (Eds.): ICONIP 2014, Part III, LNCS 8836, pp. 218–225, 2014.

local descriptors. Low level visual features such as color, shape and texture have been used to classify and retrieve medical blood cell images in various studies [1-5].

Recently, more promising studies have been focused on local descriptors. SIFT feature is one of the most widely used local descriptor in object recognition tasks. With the advances of this local feature, researchers in the field of computer vision have attempted to resolve object classification problems by a new approach known as Bag of Words (BoW). In recent years, many studies have successfully exploited this feature in general scene and object recognition tasks [6-8] due to its simplicity, discrete representations and simple matching measures in preserving computational efficiency. The use of BoW model can also be found in medical image classification and retrieval tasks [9-12].The analysis on the results obtained from their studies proven that BoW performs better than other low level features. Andre *et al.* in [11] developed a system for endomicroscopy video retrieval. BoW has been employed in their study in order to produce visual and semantic outputs which are consistent with each other. In another medical image classification work, BoW was combined with other image representation techniques such as LBP, pixel value and Edge Histogram Descriptor with two different feature fusion schemes; Low Level and High Level [10]. The results obtained by this group clearly show that feature fusion methods outperform the results obtained by using a single feature in classification task. However, they have analyzed the results obtained by different feature extraction techniques and its proven that BoW features perform better than other feature representation used in that work.

BoW also works over probabilistic tools such as Latent Dirichlet Allocation (LDA) [18] and Probabilistic Latent Semantic Analysis (PLSA) [15], and has been successfully applied in image retrieval and annotation [13-14]. Although PLSA was originally proposed in the context of text document retrieval, it has also been applied to various computer vision problems such as classification and images retrieval where we have images as documents and the discovered topics are object categories (e.g. airplane, sky). Zare *et. al.* in [12] employed PLSA approach for classification of medical X-ray images in two different techniques; one to annotate medical X-ray images and the other one is to retrieve top five similar images to the query image. Classification accuracy rate obtained by this approach showed tremendous improvement compared to classification rate obtained by flat SVM classifier. Capturing meaningful aspects of images as well as generating low-dimensional and robust image representation can be considered as another ability of such tools which has been studied in various studies [14].

The purpose of this study is to improve the retrieval performance of the previous experiment [5] with a larger database. BoW has been employed as one of the feature extraction technique. PLSA based image retrieval has also been explored in this experiment. Then, intersections of the results obtained by this approach with the ones acquired by color and shape analysis are taken as final retrieval results.

The rest of the paper is organized as follows: Section 2 discusses the components of the proposed CBIR system for blood cell images. Experimental results and discussion are reported and analyzed in Section 3. Finally, the overall conclusion of this study is presented in Section 4.

2 Methodology

The proposed system is made up two modules; Feature Extraction and Image Retrieval, each has specific functionalities which will be described in detail below.

2.1 Feature Extraction Module

Feature extraction plays an important role in the performance of any image retrieval system because it can produce significant impact on the retrieval results. Different approaches for feature extraction have been employed in this experiment as explained below.

2.1.1 Color Analysis

Color is a fundamental characteristic of image content and it is a frequently used visual feature for CBIR. Color is a powerful descriptor that simplifies object recognition.

Histogram is a very commonly used color descriptor technique. Color histogram is obtained by quantizing the color space and counting the number of pixels that fall in each discrete color.

Color histogram is used to represent distribution of colors in an image in CBIR application. There are various techniques in measuring the dissimilarity between distributions of such features. In this research, Bhattacharya coefficient has been used to compare the color histograms between query image and images in the database. This is to indicate the relative closeness of the two sample images.

Calculating the Bhattacharya coefficient involves a rudimentary form of integration of the overlap of the two samples. The interval of the values of the two samples is split into a chosen number of partitions, and the number of members of each sample in each partition is used in the following formula:

$$Bhattacharya = \sum_{i=0}^{n} \sqrt{(\sum a_i \cdot \sum b_i)} \qquad (1)$$

where considering the samples a and b, n is the number of partitions, and a_i, b_i are the number of members of samples a and b respectively in the i'th partition. The Bhattacharya coefficient will range from 0 to 1 where 1 represents the completely similar image and 0 indicates that there is no similarity between two images. The concept of normalization will be used in Bhattacharya coefficient; normalization is a process that changes the range of pixel intensity values. The purpose is to bring the image with a different intensity values into a range that is more familiar and similar to the senses which in this case, the ranges is brought to values between 0 and 1 inclusive.

2.1.2 Shape Analysis

Another major image feature is the shape of an object. In shape-based image retrieval, the similarities of the shapes represented in images are measured. Generally, there are two categories of shape descriptors: boundary based and region based. In the boundary based shape descriptor, the focused is on the closed curve that surrounded the shape. There are various models describing this curve such as polygons, circular arcs, chain codes, etc. Region based shape descriptor such as Moment Invariants and Morphological Descriptor give emphasize to the entire shape region or the materials within the closed boundary. The dataset used in this research is blood cell images, where the aim is to determine the number of round objects. As such, one way to describe such images is to calculate the circularity ratio of the objects in an image. It represents how a shape is similar to a circle. The result of area and perimeter of an object inside each image will be used to form a simple metric indicating the roundness of an object using the following formula:

$$\text{Roundness Metric} = \frac{4\pi \times area}{perimeter^2} \tag{2}$$

This metric is equal to 1 only for a circle and it is less than 1 for any other shapes. The discrimination process can be controlled by setting an appropriate threshold. In this study, the threshold of 0.75 has been used since all the objects or bubbles in blood cell images are not completely round.

2.1.3   Bag of Words (BoW)

The process of BoW starts with detecting local interest point. Local interest point detectors have the task of extracting specific points and areas from images which are invariant to some geometric and photometric transformations. For the detection of local interest point, Difference of Gaussians (DoG) is used in this experiment. DoG detector proposed by Lowe [16] is invariant to translation, scale, rotation, and illumination changes and samples images at different locations and scales.

Next, distinctive feature that characterizes a set of keypoints for an image is extracted. Scale Invariant Feature Transform (SIFT) proposed by Lowe [17] is used to describe the grayscale image region around each keypoint in a scale and orientation invariant fashion. Each detected region is represented with the SIFT descriptor using the most common parameter configuration: 8 orientations and 4 × 4 blocks, resulting in a descriptor of 128 dimensions.

Next step in implementation of bag of visual words is the codebook construction where the 128-dimensional local image features have to be quantized into discrete visual words. This step uses k-means clustering method, and use cluster center as visual vocabulary term. Upon identification of cluster centers, each image is represented as histograms of these cluster centers by simply counting the frequency of the words appear in an image. To accomplish this task, each feature vector in an image is assigned to a cluster center using nearest neighbor with a Euclidean metric.

2.2   Retrieval Module

In this module, certain measurement techniques were used to determine the similarities between the feature extracted from query image and images in the database. Upon identifying the similarities, the respective images from the database will be then retrieved as similar images to the query image.

In color analysis, the color histogram of query image is compared with the color histogram of images in the database using Bhattacharya coefficient. The resulting value is a number ranging from 0 to 1 based on the normalization algorithm.

The threshold of 0.97 has been set; as such the top 10 images in the database where their similarity ratios are greater than the threshold value are selected as similar images to the query image.

Likewise, the same approach is used to retrieve the top 10 similar images to the query image based on shape descriptor of images. In this case, the ratio of circular objects existed in an image is compared with same ratio of all other images in the database. Then, those images having ratios close to the query image's ratio by ±3 are retrieved as similar images. In BoW based image retrieval approach, PLSA model is applied on extracted BoW to identify similar images to the query image as illustrated in Fig.3. These processes are explained in following:

Learning Phase:

1) Initially, PLSA model is trained on the set of images in the database with visual words (BoW) as an input to learn both $P(z_k|d_i)$ and $P(x_j|z_k)$.
2) **While** not converge **do**
   a) E-Step: Compute the posterior probabilities $P(z_k|d_i, x_j)$
   b) M-Step: Parameters $P(x_j|z_k)$ and $P(z_k|d_i)$ are updated from posterior probabilities computed in the E-Step.
   End While

Testing Phase:

1) The E-step and M-step are applied on the extracted BoW of the query image by keeping the probability of $P(x_j|z_k)$ learnt from the learning phase fixed.
2) Calculate the Euclidean distance between $P(z_k|d_i)$ and $(z_k|d_{query})$.
3) Those images with closest distance to $P(z_k|d_{query})$ will be retrieved as similar images.

The unobservable probability distribution $P(z_k|d_i)$ and $P(x_j|z_k)$ are learned from the data using the Expectation –Maximization (EM) algorithm. $P(z_k|d_i)$ denotes the probability of topic z_k given in document d_i. $P(x_j|z_k)$ denotes the probability of visual word x_j in topic z_k.

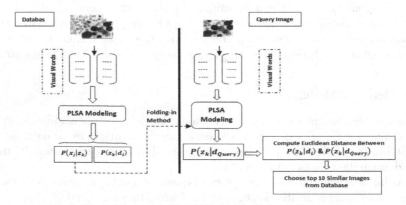

Fig. 1. Block Diagram of BoW-PLSA based Image Retrieval

Pseudo code of the retrieval process of 10 similar images to test images is demonstrated in following:

```
Start
   1.   Input [BoW]₆₀₀ₓ₅₀₀ to PLSA model in order to compute [XZ]₅₀₀ₓ₄ and [ZD]₄ₓ₆₀₀
   2.   Insert the visual feature extracted from the unseen query image ([BoW]₁ₓ₅₀₀ ) to PLSA
        model to compute matrix[ZD]₄ₓ₁, by keeping [XZ]₅₀₀ₓ₄ fixed from step 1.
   3.   The output is matrix[ZD]₄ₓ₁.
   4.   For i=1 to 600
                Compute Euclidean distance between [ZD]₄ₓᵢand [ZD]₄ₓ₁.
                The top ten (10) vectors in matrix [ZD]₄ₓᵢ with closet distance to vector [ZD]₄ₓ₁
        will be selected. Each vector represents one image.
                End For
End
```

3   Experimental Results and Discussion

Experiments were conducted to evaluate the retrieval performance obtained with respect to various image representation techniques. The database used in this research contains 600 blood cell images from four different classes; A1 Erythropoiesis, B1 Myeloid Cells Category, C1 Red Cell Disorder in the Neonate and Childhood , D1 Malarial Parasites Category. To evaluate the performance of the proposed retrieval system, we have randomly chosen 20 images from each class as query image. Precision is used to describe the accuracy of the proposed retrieval system.

Fig. 4 illustrates the average precision results obtained using the proposed retrieval approaches. The results showed that BoW-pLSA based retrieval approach outperformed the other two approaches.

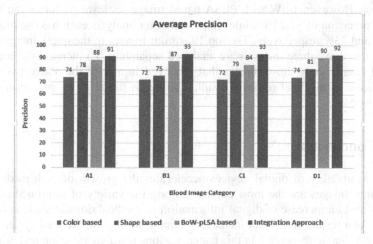

Fig. 2. Average Retrieval Precision

Analysis on the results shows that there are several irrelevant images retrieved using the color based approach which results in having lower retrieval precision. Further investigation on this results shows that most of these irrelevant retrieved images are having similar color histogram as the query image. This is due to the drawback of color histogram as it ignores the shape and texture of the images, i.e. two images from different category that happened to share color information can have similar color histograms. Fig. 5 illustrates this drawback; two images from Category B1 and one image from A1 category as well as their respective color histograms are presented.

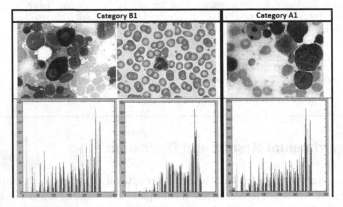

Fig. 3. Color Histogram of Sample Images from Two Categories

Comparing any retrieval systems is a difficult task especially in medical image retrieval domain. This is mostly due to the strong noises that exist in most of the medical images as well as intra-class variability and interclass similarities among them. In general, it is difficult to compare any two retrieval systems in the image retrieval domain. Due to the strong noise in most of the medical images as well as the existing similarities in the content of the images, it becomes imperative to use very precise descriptor. However, BoW with PLSA based image retrieval performs the best with average precision of 87.25% while shape and color analysis each has the precision of 78.25% and 73% respectively. The top 10 similar images to the query image retrieved from all the above three methods are analyzed separately. To increase the accuracy of retrieval results, only those images that retrieved in any of the above two methods were chosen as the final retrieval result. This leads the retrieval precision to be increased by 5%.

4   Conclusion

A wide availability of digital devices accelerates the growth of multimedia content production. Images are the most popular among the variety of multimedia contents. There is also an increase of digital information in medical domain such as blood cell images. As a result, there is an increased demand for a computerized system to manage these valuable resources. In this paper, various feature extraction techniques have been explored such as color and shape as well as bag of visual words. pLSA-BoW based image retrieval approach is also being exploited. A good performance has been obtained by combining the results obtained from the above feature extraction approaches.

References

1. Woo, C.S., Mirisaee, S.H.: Evaluation of a Content-Based Retrieval System for Blood Cell Images with Automated Methods. J. Med. Syst. 35, 571–578 (2011)
2. Hengen, H., Spoor, S., Pandit, M.: Analysis of Blood and Bone Marrow Smears using Digital Image Processing Techniques. In: SPIE Medical Imaging, San Diego, vol. 4684, pp. 624–635 (2002)
3. Cecilia, D.R., Andrew, D., Shahid, K.: Analysis of infected blood cell images using morphological operators. Image and Vision Computing 20, 133–146 (2002)
4. Pan, C., Yan, X., Zheng, C.: Recognition of blood and bone marrow cells using kernel-based image retrieval. International Journal of Computer Science and Network Security 6, 7 (2006)
5. Zare, M.R., Woo, C.S.: Integration of Color, Texture and Shape for Blood Cell Image Retrieval. Malaysian National Computer Confederation (2009),
6. http://www.mncc.com.my
7. Kesorn, K., Poslad, S.: An Enhanced Bag-of-Visual Word Vector Space Model to Represent Visual Content in Athletics Images. IEEE Trans. on Multimedia 14(1), 211–222 (2012)
8. Sui, L., Zhang, J., Zhuo, L., Yang, Y.C.: Research on pornographic images recognition method based on visual words in a compressed domain. IET Image Process 6(1), 87–93 (2012)
9. Zhou, W., Li, H., Lu, Y., Tian, Q.: Principal Visual Word Discovery for Automatic License Plate Detection. IEEE Trans. on Image Process. 21(9), 4269–4279 (2012)
10. Zare, M.R., Mueen, A., Woo, C.S.: Automatic Classification of Medical X-ray Images using Bag of Visual Words. IET Comp. Vision. 7(2), 105–114 (2013)
11. Dimitrovski, I., Kocev, D., Loskovska, S., Džeroski, S.: Hierarchical annotation of medical images. Pattern Recognition 44(10-11), 2436–2449 (2011)
12. Andre, B., Vercauteren, T., Buchner, A.M., Wallace, M.B., Ayache, N.: Learning Semantic and Visual Similarity for Endomicroscopy Video Retrieval. IEEE Trans. on Med. Imaging 31(6), 1276–1288 (2012)
13. Zare, M.R., Mueen, A., Woo, C.S.: Automatic Medical X-ray Image Classification using Annotation. J. Digit. Imag. (27), 77–89 (2014)
14. Yamaguchi, T., Maruyama, M.: Feature extraction for document image segmentation by PLSA model. In: DAS 2008 (2008)
15. Quelhas, P., Monay, F., Odobez, J.-M., Gatica-Perez, D., Tuytelaars, T., Gool, L.V.: Modeling Scenes with Local Descriptors and Latent Aspects. Paper presented at the Proceedings of the Tenth IEEE International Conference on Computer Vision (ICCV 2005), vol. 1 (2005)
16. Hofmann, T.: Unsupervised Learning by Probabilistic Latent Semantic Analysis. Mach. Learn. 42(1-2), 177–196 (2001)
17. Lowe, D.: Distinctive Image Features from Scale Invariant Key Points. International J. Comput.Vision 60(2), 91–110 (2004)
18. Lowe, D.G.: Object Recognition from Local Scale-Invariant Features. In: The Proceedings of the Seventh IEEE International Conference on Computer Vision (1999)
19. Blei, D.M., Ng, A.Y., Jordan, M.I.: Latent Dirichlet Allocation. J. Mach. Learn. Res. (3), 993–1022 (2003)

Analysis of OCT Images for Detection of Choroidal Neovascularization in Retinal Pigment Epithelial Layer

Sadaf Ayaz, Sadaf Sahar, Madeeha Zafar, Muhammad Usman Akram
, and Yasser Nadeem

College of Electrical and Mechanical Engineering,
National University of Sciences and Technology, Pakistan
{sadafayaz32,sadafsahar21,usmakram,yassernadeem}@gmail.com,
madeehazafar31eme@yahoo.com

Abstract. Choroidal Neovascularization (CNV) is an age related disease which deals with the Degeneration of Macular tissue. This degeneration causes acute drop in central vision as the age progresses. Therefore it is necessary to identify the changes caused by CNV for the Successful detection of this disease. In CNV the Retinal Pigment Epithelial (RPE) layer encounters changes in different attributes which can be identified with the help of Optical Coherence Tomography (OCT) Images. This paper focuses on analyzing the changes caused in the RPE layer due to CNV. The proposed system segments out RPE layer and observes the changes in RPE layer by calculating different features like Euler Number, Energy, Homogeneity and Correlation. The system is tested on locally gathered dateset of 50 images from different patients and has achieved an accuracy of 98%.

1 Introduction

Aging Macular Degeneration (AMD) is a common eye disease in people aging more than 50 years. This condition is caused because of the damage to the main visual field of the eye. This damage is caused because the macular tissues in the retina of the eye die out [1-2]. All over the world the population of the people suffering from age related diseases is around 25-30 million.

Digital fundus images are normally used for screening of retina to detect this disease but fundus images do not provide detailed imagery of retinal layers [2-4]. Optical Coherence Tomography (OCT) Imaging is a modern technique for taking cross sectional images with a very high resolution. OCT is just like ultrasound. The difference lies in method of imaging. Sound is used in ultra sound imaging whereas light is used when taking OCT images. OCT can provide images of tissues on a very small scale (a micron). OCT images has been very extensively used in the field of ophthalmology.[5]

Choroidal Neo Vascularization (CNV) is caused when the retinal tissues all deposit their waste normally in the form of fat particles in the choroid which

C.K. Loo et al. (Eds.): ICONIP 2014, Part III, LNCS 8836, pp. 226–233, 2014.

Fig. 1. (Left) Normal Vision, (Right) Loss of central Vision in patients with AMD

is the central visual area for the retina. These fat particles accumulate in the RPE layer resulting in growth of abnormal blood vessels and thickening of the capillaries in the retina as shown in Figure 2 [6].

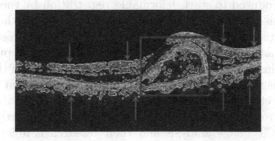

Fig. 2. OCT Image: RPE Layer (Red Arrows), Presence of CNV (Red Square)

In case of normal OCT the RPE layer is smooth with minimum distortions whereas in the OCT image of a patient suffering from CNV, the extra growth of capillaries can be seen in the RPE layer. Figure 3 shows the comparison of normal OCT image to the image of a patient suffering from CNV.

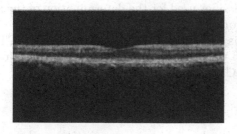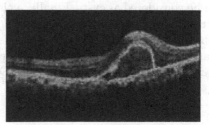

Fig. 3. (Left) OCT Image Of Normal Macula, (Right) OCT Image Of Patient With CNV

This paper consists of 5 sections. Some related work done is explained in section II. Section III explains the Methodology followed for the Segmentation and Features extraction purpose. This section also explains the classification of the complete Dataset of images on the basis of extracted attributes. Section IV discusses the results of proposed system. Finally conclusion is given in the last section.

2 Related Work

There are many methods and groups all over the world who have worked on computer aided diagnostic systems for retinal diseases using digital fundus images but limited work is done in this field for Analysis of OCT images for the characterization of different diseases [7,8]. But some researchers have done very useful addition in this field for example Chen et al. in [9] have proposed a new approach for deforming registration of OCT retinal images with higher accuracy in comparison of existing 3D methods during the segmentations of the layers. In addition to that a quite simple approach for performing scaling of OCT images has also been introduced to start deformable registration by these researchers.

Texture analysis contains the ability to provide a way for the diagnoses and differentiation of tissue and Ali A.pouyan et al. in [10] have presented a method which could have an important role for the enhancement of computer based OCT quantification technique in future. Like thickness and volume of the retinal layers can help in the diagnosis of disease. Also their results show that it is possible to separate out the retina layers without using systems with ultrahigh resolutions. Another group of researchers in [7,8] have also made very important contribution in the field of OCT image analysis and their research is able to analyse and separate out the RPE layer and quantifies the changes associated with CNV by calculating the values of area for the RPE layer and thickness etc. According to them CNV can be differentiated from the normal OCT image which is important in the use of many ophthalmological practices.

An approach for an automated methodology was presented by authors in [11] for cyst detection in OCT retinal images by using watershed algorithm for the detection of candidate regions in the images and after that the discard of all the possible regions to reduce eligible candidates, which due to some of the properties can be considered as cysts. Finally, a classifier used for the determination of their correspondence to cystic regions or not on the basis of texture features extracted from them. The research in this paper is also a contribution in the area of OCT images Analysis for the detection and classification of CNV on the basis of changes in RPE layer by Analysis of OCT images.

3 Proposed Methodology

The proposed system consists of three phases; image acquisition and pre-processing, Area of interest segmentation and attribute comparison and classification. Figure 4 shows all the phases included in the methodology of our proposed research. Pre-Processing phase is used for the initial processing in order to segment the area of interest from the original image. The CNV detection is done by the analysis of this area of interest.

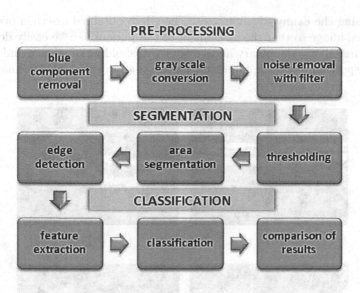

Fig. 4. Flow diagram of proposed system

3.1   Image Acquisition and Pre-processing

Fifty Persons were clinically examined by Ophthalmologist and OCT images are ob-tained for the observation of changes occurring in abnormal cases and then for the classification on the basis of changes in Normal and Abnormal cases. OCT images of size 240x480x3 of normal and patients suffering from CNV are selected for this study.

It is observed that the OCT image is a blend of primary colours red, green and blue as shown in Figure3. In the OCT images the blue component signifies the vitreous humour layer of the eye and does not provide any useful information hence the blue component is discarded from both normal and diseased image [9]. Only the images with green and red component are kept as further computations are based on these components of OCT images. The images are converted into gray scaled by averaging the red component and green component of the images.

3.2   Image Segmentation

A Threshold value is selected for the segmentation of the region of interest from the images. This threshold is calculated with reference to the Otsu's algorithm for calculating the gray level threshold for the images [13]. After the threshold is calculated the median filtered image is converted to a black and white image by assigning 1 to the values greater than threshold value and 0 to the values less than threshold value.

The Retinal Pigment Epithelial layer is segmented out as a result of thresholding. It is observed that the RPE layer is a thick layer and covers a minimum of 3000 pixels of the OCT image. This layer is segmented out and the boundary is created using the canny edge detector. The edges obtained are then overlaid on the original image so that the assessment of the patient can be easily done clinically. Figure6 shows the binary images after thresholding of normal and diseased images respectively and the highlighted normal and diseased OCT images.

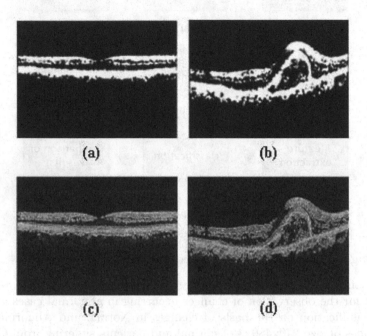

(a)   **(b)**

(c)   **(d)**

Fig. 5. (a)Binarized Image Of Normal OCT, (b)Binarized Image Of OCT With CNV, (c) Highlighted Edges Of Normal RPE Layer, (d) Highlighted Edges Of RPE With CNV

3.3   Attributes Extraction

For proper classification of OCT images as normal and abnormal, the proposed system extracts different features i.e. Euler Number, Energy, Homogeneity and Correlation.

1. Euler Number (f_1): The Euler Value of an Image gives a value proportional to the number of disruptions and holes in the segmented image.
2. Energy (f_2): Energy depicts how much uniform a certain image is. Energy of the normal OCT image is found out more than the one of the patient suffering from CNV [11].
3. Homogeneity (f_3): Homogeneity is a measure of how closely related are the pixels in an image. The lesser the difference of the values the more homogenous is an image.
4. Correlation (f_4): Correlation value returns a measure of how correlated a pixel is to its neighbor over the whole image. Correlation is 1 or -1 for a perfectly positively or negatively correlated image.

Table 1 shows the mean values of the extracted features both for set of normal images and diseased images. These feature values are used further for the classification purpose.

Table 1. Mean and standard deviation of feature values of normal and abnormal classes

Attributes	Mean ± Std Normal	Mean ± Std Diseased
Euler Number	940.14±200.05	1483.10±276.29
Energy	0.93±0.01	0.90±0.0155
Homogeneity	0.98±0.0044	0.96±0.0065
Correlation	0.5314±0.0551	0.4445±0.0597

The significance and contribution of each feature vector member is calculated using Box Plot analysis. Euler value and Homogeneity does not have any Outlier in Box Plot that is why these are considered as most significant features and are assigned more weights during classification

3.4   Classification

After features extraction of each OCT image collected for classification, Gaussian Mixture Models (GMM) has been used for the classification of infected images with CNV. This classifier uses four extracted features as the Feature Vector for classification. GMM makes use of Nave Bayes' theorem and follows probabilistic techniques for the classification of data into classes on the basis of their features. For application of GMM in classification, the data set have two categories i.e. normal and diseased (R_1 & R_2). Training and testing data are the subsets of both these classes and this data is randomly selected during training so that the biasness of the data can be eliminated.

Sample belongs to R_1 if $p(v|R_1)P(R_1) > p(v|R_2)P(R_2)$ else it belongs to class R_2. Where $p(v|R_n)$ is the probability density function(conditional) and is also called likelihood (v being the feature vector). $P(R_n)$ is the probability of occurrence of a class in the data set. GMM uses Gaussian functions for modeling the conditional probabilities of classes. These probabilities are represented as weighted sums of Gaussian functions and tell us the likelihood of the Gaussian Mixture Model:

$$p(v|R_i) = \sum_{j=1}^{k_i} N(v|\mu_j, \Sigma_j)w_j \tag{1}$$

Weights (w_j) are assigned on the basis of significance of different attributes used as feature vector [14]. Classification is also done by using some other classifiers for the comparison of results with GMM. This comparison is done using the Box Plot Thresholding, KNN, K-Means and Neural Networks.

4   Results and Discussion

The Complete Dataset of Images containing both Normal and Infected Eye images is classified into two classes named Normal and Diseased on the basis of extracted feature values using many different classifiers. The dataset is consisted of 50 total images having 41 Normal and 9 Diseased images. Feature Extraction is performed one by one on each image of Dataset and feature values of all images is collected as a result.

Out of total Images in the dataset, Half of Diseased and Same number of Normal images are used for the training of classifiers and the remaining others are used for the testing and Estimation of Accuracy of classifiers. The classification process is repeated 10 times and training images are selected randomly to avoid any kind of biasness which could affect the testing and accuracy of results. Table-2 shows the Sensitivity and Specificity values on the basis of mean value of total number of truly identified images. It also shows that the accuracy of GMM is much more as compared to others.

Table 2. Comparison of results between two classes using different classifier

Classifier	Sensitivity	Specificity	Accuracy
GMM	88.80%	100%	98%
Box Plot Thresholding	77.70%	95.10%	92%
KNN	50%	87.50%	92.50%
K-Means	33.30%	92.60%	82%
Neural Networks	0%	100%	90%

5   Conclusion

Analysis of Retinal Pigment Epithelium layer is necessary for the detection of Choroidal NeoVascularization causing vision loss with age. The work presented in this paper identifies changes in RPE layer due to CNV. These changes are identified on the basis of values of Euler Value, Energy, Homogeneity and Correlation of RPE layer. CNV has been differentiated from normal OCT image on the basis of these values. Higher Euler value suggests loss of continuity and presence of breaks in OCT image due to CNV. Also higher correlation value of normal image is due to fewer changes in RPE layer. Similarly lesser values of Energy and Homogeneity depict the loss of linearity in case of CNV. Classification of 50 images on the basis of these attribute values using various classifiers is done. The proposed system gives overall accuracy of 98% in the detection of CNV in OCT images.

References

1. Schmitt, J.M., Knuttel, A., Yadlowsky, M., Eckhaus, M.: Optical coherence tomography of a dense tissue - statistics of attenuation and backscattering. Phys. Med. Biol. 38 (1994)

2. Akram, M.U., Khalid, S., Tariq, A., Khan, S.A., Azam, F.: Detection and Classification of Retinal Lesions for Grading of Diabetic Retinopathy. Computers in Biology and Medicines 45(1), 161–171 (2014)
3. Tariq, M.U., Akram, A., Shaukat, S.A.: Khan, Automated Detection and Grading of Diabetic Maculopathy in Digital Retinal Images. Journal of Digital Imaging 26(4), 803–812 (2013)
4. Akram, M.U., Khalid, S., Khan, S.A.: Identification and Classification of Microaneurysms for Early Detection of Diabetic Retinopathy. Pattern Recognition 46(1), 107–116 (2013)
5. Fujimoto, J.G., Brezinski, M.E., Tearney, G.J., Boppart, S.A., Bouma, B., Hee, M.R., Southern, J.F., Swanson, E.A.: Optical biopsy and imaging using optical coherence tomography. Nat. Med. 1, 970–972 (1995) [PubMed]
6. Brezinski, M.E., Tearney, G.J., Bouma, B.E., Izatt, J.A., Hee, M.R., Swanson, E.A., Southern, J.F., Fu-jimoto, J.: Optical coherence tomography for optical biopsy: properties and demonstration of vascular pathology. Circulation 93, 1206–1213 (1996)
7. Prashanth, T.R., Paranjape, S.V., Ghosh, S., Dutta, P.K., Chatterjee, J.: Characterization of Changes in Retinal Pigment Epithelium Layer in Choroidal Neovasculari-zation through Analysis of Optical Coherence. In: 2010 IEEE Students' Technology Symposium (TechSym), pp. 39–43 (2010)
8. Shahidi, M., Wang, Z., Zelkha, R.: Quantitative thickness measurement of retinal layers imaged by optical coherence tomography. Am. J. Ophthalmol. 1056–1062 (2005)
9. Chen, M., Lang, A., Sotirchos, E., Ying, H.S., Calabresi, P.A., Prince, J.L., Carass, A.: Deformable registration of macular OCT using A-mode scan similarity. In: 2013 IEEE 10th International Symposium on Biomedical Imaging (ISBI), pp. 476–479 (2013)
10. Pouyan, A.A., Naseri, A., Kavian, N.: An Image Processing Technique to Detecting Retina Layers. In: 2010 International Conference on Signal and Image Processing (ICSIP), pp. 7–10 (2010)
11. González, B., Remeseiro, M., Ortega, M.G., Penedo, P.: Automatic Cyst Detection in OCT Retinal Images Combining Region Flooding and Texture Analysis. In: 2013 IEEE 26th International Symposium on Computer-Based Medical Systems (CBMS), pp. 397–400 (2013)
12. Arevalo, J.F., Garcia, R.A., Sanchez, J.G., Fernandez, C.F., Mendoza, A.J., Wu, L., Shields, C.L., Shields, J.A., Fujimoto, J.G., Materin, M.A.: Retinal Angiography and Optical Coherence Tomography. Springer, New York (2009)
13. Otsu, N.: A threshold selection method from gray-level histograms. IEEE Transactions on Systems, Man, and Cybernetics 9(1), 62–66 (1979)
14. Usman Akram, M., Tariq, A., Almas Anjum, M., Younus Javed, M.: Automated detection of exudates in colored retinal images for diagnosis of diabetic retinopathy. Applied. Optics. 51(20) (July 2012)

Online Object Tracking
Based on Depth Image with Sparse Coding

Shan-Chun Shen[1], Wei-Long Zheng[1], and Bao-Liang Lu[1,2,*]

[1] Center for Brain-Like Computing and Machine Intelligence,
Department of Computer Science and Engineering
Shanghai Jiao Tong Unviersity, Shanghai 200240 China
[2] Key Laboratory of Shanghai Education Commission for
Intelligent Interaction and Cognitive Engineering
Shanghai Jiao Tong University, Shanghai 200240 China
bllu@sjtu.edu.cn

Abstract. Online object tracking is a challenging problem because of changing environment including diverse illumination and occlusion conditions. The emergence of commercial real-time depth cameras like Kinect make online RGBD-based object tracking algorithm become a focus of research. In this paper, we propose a robust online depth image-based object tracking method with sparse coding. We introduce sigmoid normalization for local depth patch. In order to recovery from tracking failure in condition of heavily occlusion. we present a detection module based on PCA bases. Experiments show that our method exceeds original color image-based method in case of environment changes.

Keywords: object tracking, depth image, sparse coding, normalization.

1 Introduction

Object tracking is one of the key problems in computer vision and it has broad practical scenarios such as activity recognition, motion analysis and image compression. Although the performance of object tracking algorithm has been much improved recently, it's still a great challenge to develop a robust object tracking algorithm considering some problems caused by illumination varying and target object occlusions.

The object tracking algorithm generally consists of three basic modules [1]: 1) object shape representation; 2) image features that hold the characteristic of target object; 3) strategies for detection the objects in a scene. The availability of high quality and inexpensive video cameras has improved the development of a great amount of object tracking methods based on color image features. In this paper, we propose a robust object tracking method based on depth image. Hence, we only discuss key issues related to image type.

---

* Corresponding author.

C.K. Loo et al. (Eds.): ICONIP 2014, Part III, LNCS 8836, pp. 234–241, 2014.

Generally speaking, there are four types of common visual features extracted from color image including color, edges, optical flow and texture. Numerous algorithms based on these features performance well in some constrained situation. Paschos proposed a color based object tracking solution in RGB color space [2], but color features are easily influenced by illumination. Object boundaries often located where image intensities strongly change. The new variational framework for detecting and tracking multiple moving objects is a very popular edge detection approach [3], it uses a statistical framework based on a mixed model. It is robust to illumination change but when occlusions occur the edge based method would lose target.

Color camera can real-time collect color image stream at the cost of losing information by projection 3D to 2D. As a result, color image based features would easily crash with changes of illumination. A new device Kinect can real-time acquire both color and depth image stream. A face tracking method integrated color and depth image stream is implemented with ASM model and statistical methods [4].

An online depth image based face tracking method is proposed on the assumption that face shape is an ellipse in [5]. However when occlusion occurs, the tracking method will lose target.

In this paper, we propose a general object tracking method based on single depth image, which is robust to occlusion and illumination changes. Compared with color imaged methods, our algorithm is less influenced by illumination change. With sparse coding representation, we can keep tracking the target object until the occlusion area reaches 50% of the target object.

The rest of this paper is organized as follows. In Section 2, we review the theory of sparse coding in object tracking; in Section 3, we introduce our tracking method; in Section 4, we present qualitative and quantitative results of our tracker on a number of challenging image sequences. Finally we conclude the paper in Section 5.

2   Sparse Coding

Sparse coding is a popular solution to object tracking problems recently. We can simply classified it into three forms: 1) appearance modeling based on sparse coding (AMSC); 2) target searching based on sparse coding(TSSR); 3) combination of both. Jia and colleagues proposed a structural local sparse coding model [6]. Mei and Ling solved most challenges like occlusion through a set of positive and negative trivial templates [7]. By transferring object tracking problem to a sparse approximation problem, they proposed a robust algorithm. Wang and colleagues proposed a new method that views coefficients of trivial templates as a single factor of tracking performance [8]. Studies mentioned above have proved that sparse coding is a good solution to color image based object tracking. In this paper we apply sparse coding to depth image based tracking algorithm. As shown in Fig. 1, each target is represented by some bases and templates with sparse coding.

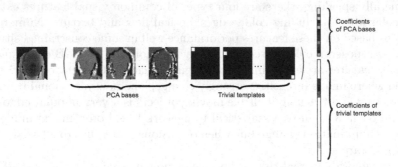

Fig. 1. sparse coding

In sparse coding framework, tracking problem is casted to finding the most likely patch among candidates by

$$y = Uz + e = [A \ I] \begin{bmatrix} z \\ e \end{bmatrix} = Bc \tag{1}$$

where y indicates the object vector, U denotes templates matrix, z represents coefficients of bases vectors and e is the coefficients of trivial templates. As is shown in Fig 1 , we assume that target object is sparsely represented by bases and trivial templates. We solve Eq. (1) via ℓ_1 minimization:

$$\min_{z,e} \frac{1}{2} \parallel y - Uz - e \parallel_2^2 + \lambda \parallel e \parallel_1 \tag{2}$$

where $\parallel \cdot \parallel_2^2$ and $\parallel \cdot \parallel_1$ are the ℓ_2 and ℓ_1 normal forms, respectively. Several works have been done on online subspace learning by learning and updating bases represented by A such as PCA and ICA. With an iteration algorithm, optimal z and e for each candidate are computed.

After getting the optimal z and e for each candidate, the object tracking problem is transferred to a statistical inference problem.

3   Tracking Algorithm

In this paper, we applied sparse coding to depth imaged object tracking. To some degree, depth image is the same as color image except for the meaning of each pixel. In color image, pixels represent the color of this point while in depth image they represent the distance from the point to camera. Aiming to reduce the influence of illumination change, we try to develop depth image based tracking methods, and in order to solve the occlusion problem we incorporate sparse coding. So we should design a new algorithm to adapt to depth image. The workflow of our algorithm is described in Fig. 2.

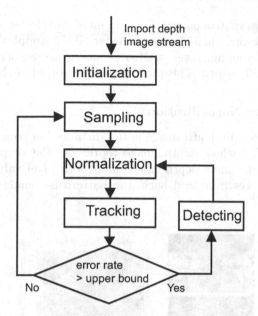

Fig. 2. Workflow of depth-image-based tracking algorithm with sparse coding

We adopt sparse coding method to tracking object. Firstly, we initialize the tracking by manually calibrating target position, computing PCA bases and setting other parameters such as patch size and bases number. Secondly, we sample in original depth image according to sampling parameters. The samples are size-adjustable to suit for demand of object front-back moving. To speed up the proposed tracking algorithm, we transfer all the samples to the same size patches. Then we consider the object tracking as a Bayesian task. By evaluating every patch, we find the patch with the highest posterior probability and return its location as the target. During the process, we compute the occlusion rate by coefficients of trivial templates. If the occlusion rate exceeds the upper bound, we discard the result and regard it as losing target. Then we startup the detection module. If not, we update the bases and go to next loop.

3.1 Alternative Box Sampling

Candidates are patches with size of 32*32, which is the result of trade-off between algorithm efficiency and accuracy rate. But it doesn't mean every patch is exactly a copy of a 32*32 patch in original depth image. According to the perspective relation, nearby object looks larger than distant one with the same size. So during the tracking process, the tracking sampling alternative boxes should be adjustable.

In detail, there are 5 parameters in tracking sampling stage: x and y denote transformation in plane, α and β are scale variation, and θ is angle rotation. Alternative boxes are uniformly distributed around the target. To adapt to the characteristics of the depth map, we set the α and β a little bigger. But too big α

and β mean more alternative boxes to be computed and slower processing speed. To speed up the tracking algorithm, we transfer all the samples to the same size by interpolation. In sampling stage, we don't concern the size of alternative boxes with regard to specific depth. This problem will be solved in the next section.

3.2 Depth Image Normalization

The meaning of pixels in depth image is the distance between camera and the point on object. The whole depth image represents the shape of the target. Transformation in the same depth can remain both pixel values and pattern. But once target moves front and back, the pattern is remained but the pixel values will shift .

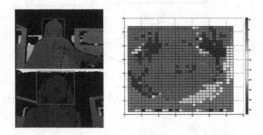

Fig. 3. Pixel value shifts of two frames. Left top is the face far from the camera. Left down is the nearer one. Right is the pixel value of two patches after interpolating.

As shown in Fig. 3, we should normalize the patch to eliminate the offset. The most common method of normalization is min-max normalization. If there are noises, they often deviate from average and become peaks of the image and finally their deviation results in extreme minimum or maximum. The existing of noise limits the performance of min-max normalization. So we adopt the sigmoid filter to normalize the patches [9]. The sigmoid function is a S-type function :

$$y = f(x) = \frac{1}{1 + e^{\frac{x-\beta}{\alpha}}} \tag{3}$$

In our method, α equals 1 and β is set to this median of each patch. β is set to the value because the median of a patch is not sensitive to noise. And a little peaks would not change the median much.

3.3 Restarting by Detecting

In this paper we solve the problem of occlusion by sparse coding. We estimate the occlusion by η which is the ratio of non-zero pixels and the number of occlusion map pixels. η is put forward to deal with partial model updating problem [8]. In our paper, we apply η to restart the tracking model when tracking failure occurs.

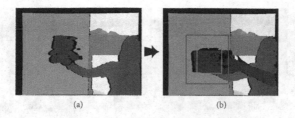

Fig. 4. Detecting sample method. (a) Image without occlusion and target is a bear; (b) Green box is the tracking sample range while the larger red one is for detecting sample range.

We set an upper bound and lower bound for η. Different values correspond to different tracking results. If η is larger than the upper bound, we view this situation as tracking failure, and we restart the tracking module. Since we are updating the bases of target, we don't adopt other detecting method. Instead, our detecting method is based on the recorded bases. Once η becomes larger than the upper bound, we startup the detecting module.

Our detecting method has similar idea as tracking method. Their difference lies on the sampling stage as shown in Fig. 4. On the assumption that when losing the target we still can find it in a wider scope centered on the original position, the sampling scope spreads to three times of the size of the original one. After sampling stage, the rest stages are the same as tracking method. By computing coefficients of bases and solving Bayesian task, we find the most likely patch among candidates. We compute error ratio η in detecting module. If η becomes lower than its upper bound we start the tracking module.

4   Experiment and Result

Our method is implemented in MATLAB on a Triple-Core Processor 2.10GHz with 6GB memory. The speed of our algorithm is related to sampling number. More sampling candidate boxes would slow down the processing speed. As a trade-off between computational efficiency and effectiveness, sampling number is set to 600. Our method is mainly compared with the original algorithm on the RGB image. We use 5 image sequences of a public dataset Princeton Tracking Benchmark [10] to test our algorithm as shown in Fig. 5. The challenges of each sequence and results are listed in Table 1. Our results are evaluated by average center error of pixels.

As shown in Table 1, the original color image based object tracking method and the depth image based object tracking method provide different performances in different cases.

- In cup sequences, challenge in this sequence is moving back and front. Their average center error of pixels are around 13. It means they both track the target closely.

Fig. 5. Image sequences of bear, cup, face, child and ball are listed from top to bottom (only RGB images listed)

Table 1. Results of experiments

test sequence	frame number	challenge	color image error	depth image error
cup	368	move back and front	13.93	**12.83**
face	330	occlusion	**15.30**	17.52
ball	117	illumination change	263.60	**14.49**
bear	281	heavily occlusion	192.99	**46.23**
child	164	no-rigid	**47.57**	135.34
average			106.67	**45.282**

- In face sequences, challenge is occlusion. A book may occlude most part of target. From Table 1 we can find that they provide good performance in face sequences. Because both methods are based on sparse coding.
- In ball sequence, we can find that the illumination changes when the ball rolls around. In color image sequence the method loses target in the fortieth frame as the ball rolls to another brighter room. While in depth image sequence, our method keeps tracking the ball through out the whole sequence.
- In bear sequences, heavily occlusion is the main challenge when a book occludes the target bear for a while. Heavily occlusion leads to losing target in color image and without restarting module in the rest images it fail to find the target again. In our methods we add the detecting module to detect the losing target and keep tracking again.
- Finally, in child sequences, no-rigid target tracking is the main challenge. From Table 1, we can find that both methods performs bad with large error pixels numbers. It illustrates that both of them lose target. It is because the target child is not a rigid object and his movements result in changing of target's shape.

5   Conclusions and Future Work

This paper proposes a robust tracking method based on the depth image with a sparse coding representation. We improve the performance of the color image based object tracking method with sparse coding representation by applying sigmoid normalization algorithm and by designing the detecting module. The two modifications are designed to acquire stable performance when illumination changes or occlusion occurs.

But we still leave the no-rigid object tracking problem unsolved, because our tracking patches are decomposed into PCA bases with different weights. The tracking method of no-rigid object is limited by the characteristics of the PCA bases. As a consequence, we plan to improve our method by adopting other type sparse representation in the future.

Acknowledgments. This work was partially supported by the National Natural Science Foundation of China (Grant No. 61272248), the National Basic Research Program of China (Grant No. 2013CB329401) and the Science and Technology Commission of Shanghai Municipality (Grant No. 13511500200).

References

1. Yilmaz, A., Javed, O., Shah, M.: Object tracking: A survey. Acm computing surveys (CSUR) 38(4), 13 (2006)
2. Paschos, G.: Perceptually uniform color spaces for color texture analysis: an empirical evaluation. IEEE Transactions on Image Processing 10(6), 932–937 (2001)
3. Paragios, N., Deriche, R.: Geodesic active contours and level sets for the detection and tracking of moving objects. IEEE Transactions on Pattern Analysis and Machine Intelligence 22(3), 266–280 (2000)
4. Cai, Q., Gallup, D., Zhang, C., Zhang, Z.: 3D deformable face tracking with a commodity depth camera. In: Daniilidis, K., Maragos, P., Paragios, N. (eds.) ECCV 2010, Part III. LNCS, vol. 6313, pp. 229–242. Springer, Heidelberg (2010)
5. Cao, Y., Lu, B.-L.: Real-time head detection with kinect for driving fatigue detection. In: Lee, M., Hirose, A., Hou, Z.-G., Kil, R.M. (eds.) ICONIP 2013, Part III. LNCS, vol. 8228, pp. 600–607. Springer, Heidelberg (2013)
6. Jia, X., Lu, H., Yang, M.-H.: Visual tracking via adaptive structural local sparse appearance model. In: 2012 IEEE Conference on Computer Vision and Pattern Recognition (CVPR), pp. 1822–1829. IEEE (2012)
7. Mei, X., Ling, H.: Robust visual tracking and vehicle classification via sparse representation. IEEE Transactions on Pattern Analysis and Machine Intelligence 33(11), 2259–2272 (2011)
8. Wang, D., Lu, H., Yang, M.-H.: Online object tracking with sparse prototypes. IEEE Transactions on Image Processing 22(1), 314–325 (2013)
9. Pei, S.-C., Lin, C.-N.: Image normalization for pattern recognition. Image and Vision Computing 13(10), 711–723 (1995)
10. Song, S., Xiao, J.: Tracking revisited using rgbd camera: Unified benchmark and baselines. In: ICCV (2013)

Real-Time Compressive Tracking
with a Particle Filter Framework

Xuan Yao and Yue Zhou

Institute of Image Processing and Pattern Recognition, Shanghai Jiao Tong University
yaoskd@sjtu.edu.cn

Abstract. Recently a real-time compressive tracking was proposed and achieved relative good results in terms of efficiency, accuracy and robustness. It belongs to the "tracking by detection" method. Slight inaccuracies in the tracker can lead to incorrectly labeled training examples in these algorithms, which degrade the classifier and usually cause drift. In this paper, we incorporate the motion model into the traditional compressive tracking where we utilize the particle filter. Therefore, our algorithm can handle drifting problem to some extent. Meanwhile, in order to improve the discriminative power of the classifier to relieve drifting problem radically, a modified naive Bayes classifier is proposed. The proposed algorithm performs favorably against state-of-the-art algorithms on some challenging video sequences.

Keywords: Compressive tracking, particle filter, naive Bayes classifier, tracking by detection.

1 Introduction

Object tracking is a well-studied problem in computer vision and has many applications. However, there is no single algorithm which will handle all circumstances, due to the complexity of the object and the environment, such as illumination change, pose and scale variation, occlusion.

As is stated in [1], a typical tracking system consists of three components:

- An appearance model, which can evaluate the likelihood that the object of interest is at some particular location.
- A motion model, which maintains the distribution of the locations of the object overtime.
- A search strategy for finding the most likely location in the current time.

Readers are recommended to [2] for a thorough overview of above three components.

Particle filter theory, also known as the bootstrap filter or sequential Monte Carlo filter, was first proposed from the field of signal processing[3], computer vision [4], and statistics[5], for solving the non-Gaussian and non-linear problems. It uses a set of weighted particles to approximate the location of the object. During tracking, particle filter maintains a distribution of the target's location and is characterized by the motion model.

C.K. Loo et al. (Eds.): ICONIP 2014, Part III, LNCS 8836, pp. 242–249, 2014.

Recently a large number of algorithms, known as the "tracking by detection", has thrived which focuses on the appearance model represented by the online learned classifier. Considering the successful application of Adaboost in the object detection [6], an on-line boosting algorithm [7][8][9] has been applied to the object tracking. In [7], a novel on-line Adaboost feature selection algorithm, known as OAB, is proposed for tracking. But it is sensitive to the noise and easy to drift. Later an online multiple instance learning (MIL) method [1] was proposed where samples are presented in sets, often called "bags" and labels are provided for the bags rather than the individual instances. The proposed method relieves the drifting problems and improves the accuracy of detection. The other relative method [10] uses a kernelized structured output support vector machine that learned online to provide adaptive tracking. Most of these above algorithms are suffered from heavy computational load that make it hard for real-time tracking and drift problems. Recently a novel tracking framework [11] was proposed that explicitly decomposes the tracking task into tracking, learning and detection. The PN learning constantly estimates the detector's error so that the algorithm can be applied to long-term tracking.

Compressive tracking [12], known as CT, adopts a very sparse measurement matrix to efficiently extract the features for the appearance model which makes it possible for real-time tracking and achieves relatively good results on some challenging video sequences. However, these algorithms often suffer from the drifting problem and the naive Bayes classifier used in the compressive tracking is quite simple. In this paper, to solve these problems, the key contribution of this work can be summarized as follows.

- We incorporate the particle filter into the traditional compressive tracking as the motion model to maintain the distribution of the location of the object. When drifting problem happens, the proposed algorithm will have a big chance to detect the object again.
- We calculate the correct rate of the classifier on each feature. The correct rate is then used as the weight of each feature's classifier. When the classifier classifies most of samples correctly, it is reasonable to believe the reliability of the classifier. Then the overall naive Bayes classifier is formulated as the weighted combination of each feature's classifier.

The paper is organized as follows: Section 2 briefly introduces the real-time compressive tracking proposed in [12]. Our method combining the above algorithms with particle filter is presented in Section 3. Then experimental results are shown in Section 4. Section 5 concludes the paper.

2   Compressive Tracking

A real-time compressive tracking method was proposed in [12]. In the paper, a very sparse measurement matrix was adopted to efficiently compress features from the foreground samples and background ones. The compression of feature is shown in Figure 1.

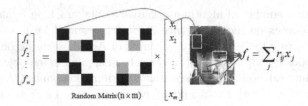

Fig. 1. Feature compression procedure, where **x** represents the original high-dimensional vector, matrix is a sparse random measurement matrix and **f** low-dimensional compressed feature

Then the tracking task was formulated as a binary classification problem with a simple naive Bayes classifier. The algorithm performs well in terms of accuracy, robustness, and speed.

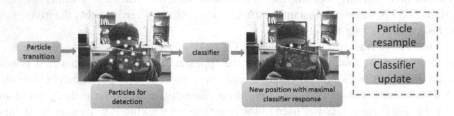

Fig. 2. Main components of our proposed algorithm at the t-th frame

3   Proposed Algorithm

3.1   Compressive Tracking Based on the Particle Filter Framework

In [12], the tracking problem is formulated as a detection task. To predict the object location in the next frame, samples are drawn from the neighborhood of the current target location. This may cause a problem that as long as drifting problems happens, the detection area will never cover the true neighborhood of the target which causes tracking failure. In our paper, to predict the object location, we draw samples from the particles. Thanks to the particle filter framework, after resampling step, there will always exists particles around the true location of target with relatively high weights, although maybe not the maximum weight. Once drifting happens, the target will be found with high probability in the subsequent frames. Meanwhile, it is worth considering that the examples for updating the classifier could be sampled from the particles. Furthermore, we use classifier response as the particle weight. Particle with the maximum classification score is determined as the object location in the next frame and particles with relative high weights are maintained while those with negligible weights discarded during resampling step. The main components of our algorithm are shown in Figure 2. Details are given in the following section.

3.1.1 Particle State Transition Model

We adopt a second-order autoregressive model

$$X_k - X_{k-1} = X_{k-1} - X_{k-2} + U_k \tag{1}$$

Where X_k represents the current states of particles and U_k the noise. It assumes that the motion between time k and time k-1 is the same as time k-1 and time k-2. In equation (1), we assume that the noise U_k are Gaussian.

$$U_k \sim N(0, \sigma_k) \tag{2}$$

where σ_k is the standard derivation. What should be mentioned is that when the target moves fast, σ_k should be increased. When the target moves slow, σ_k should be decreased. Furthermore, if σ_k is too large, more samples to predict will be included which will influence the determination of the classifier. If σ_k is too small, particles may not cover the whole area of the target.

3.1.2   Particle Weight

We use the classifier response as the weights of particles. However the classifier response often generates a negative value, which leads to a negative weight of a particle. Therefore, the next step is followed.

$$w^i = w^i - w_{\min} \tag{3}$$

From equation (3), all the weights will be positive. Then the normalization is performed and particles are resampled according to the weight so that more attention will be paid to the most likely area.

3.2   Improved Naive Bayes Classifier

The naive Bayes classifier in the compressive tracking is quite simple and its classification power is limited. We decide to take the correct rate of classifier on each feature into account.

We extract the expression that involve the specific feature from the overall Naive Bayesian classifier and the expression is defined as

$$p_i(i) = \log(\frac{p(v_i \mid y = 1)}{p(v_i \mid y = 0)}) \tag{4}$$

If $p_i(i) > 0$, the sample is considered as positive. If $p_i(i) \leq 0$, the sample is considered as negative. Here we define four parameters, n_c^+ (positive examples which are correct), n_f^+ (positive examples which are false), n_c^- (negative examples which are correct), n_f^- (negative examples which are false). Each feature's weight can be defined as

$$w_i = \frac{n_c^+ + n_c^-}{n_c^+ + n_c^- + n_f^+ + n_f^-} \tag{5}$$

The improved classifier

$$H(v) = \sum_{i=1}^{n} w_i \log\left(\frac{p(v_i \mid y=1)}{p(v_i \mid y=0)}\right) \tag{6}$$

3.3   The Algorithm

The pseudocode of our proposed algorithm is given below.

Algorithm. Compressive tracking based on the particle filter framework

Input: t-th video frame

1. Particles resampled in the (t-1)-th frame transit according to the second-order autoregressive model, and features with low-dimensionality are extracted from the particles.
2. Use classifier in (6) to each particle, the classifier response is assigned to the particle weight and the particle in the location l_t with the maximal classifier response is found.
3. Normalize particle weights and resample particles according to weights
4. Sample two sets of image patches $D^\alpha = \{z \mid \| \, l(z) - l_t \, \| < \alpha\}$ and $D^{\zeta,\beta} = \{z \mid \zeta < \| \, l(z) - l_t \, \| < \beta\}$ with $\alpha < \zeta < \beta$, where α, ζ, β are three parameters that we choose according to the experimental results, $l(z)$ is the center location of image patch used to update the classifier and D^α, $D^{\zeta,\beta}$ represent positive and negative samples respectively.
5. Extract the features with these two sets of samples and update the classifier.
6. Calculate the correct rate of each features' classifier on the above samples.

Output: Tracking location l_t and classifier parameters

---

4   Experiments

We evaluate our tracking algorithm with 3 state-of-the-art methods on some challenging video sequences. Four video sequences are presented in the Figure 3 to show advantages of our proposed algorithm over other methods. There are usually two evaluation methodologies which are the center location error (CLE) and bounding box overlap (BBO). We adopt the CLE for quantitative analysis which is showed in Table 1, for our algorithm is based on a fixed tracking scale and the ground truth usually a varied tracking scale so that the BBO methodology may not reflect the experimental results properly. Finally, our tracker is implemented in C++, which runs at 60 frames per second (FPS) on an Intel Core 3.20 GHz CPU with 4 GB RAM.

(a) Bolt

(b) Tiger 2

(c) Lemming

(d) David

Our algorithm ▪▪▪▪ CT   MIL Track ▪▪▪▪ OAB

Fig. 3. Screenshots of some sampled tracking results

Table 1. Center location error. Red fonts indicate the best performance while the blue fonts indicate the second best ones

Sequence	Our algorithm	CT	OAB	MIL Track
Bolt	12	82	370	107
Tiger 2	10	18	26	8
Lemming	24	139	115	77
David	10	9	46	27
Average CLE	14	62	139	55

The ability to handle the drifting problem. As is known to us, the "tracking by detection" methods suffer from drifting problems where incorrectly labeled examples may degrade the discriminative power of the classifier and cause drift. In our proposed algorithm, when the tracking box drifts, there will still be lots of particles with rather high weights gathering around the neighborhood of the target. In subsequent frames, those particles may get the maximum classifier response with a relative high probability over the other non-target regions. The target player, Bolt, as shown in Figure 3(a) is almost lost in the frame 237 in both our algorithm and the compressive tracking because of the drastic appearance change after the finishing line. But our algorithm actually maintains some particles around the true target. As a result, in the frame 239, the mistake is corrected by our tracker while the traditional compressive tracker loses the target and never finds it back. The same situation is shown in the Figure 3(b) between the frame 149 and 151.

The improved discriminative power of the classifier. We calculate the correct rate of the classifier on each feature after update and the overall naive Bayes classifier is formulated as the weighted combination of each feature's classifier, which means that more samples are classified correctly by a certain feature's classifier, more we can trust on it. As shown in Figure 3(c), in the frame 229, the target is not detected precisely by compressive tracker and later in the frame 1049, the situation happens again when the appearance of the target changes dramatically which causes tracking failures. However, in our algorithm adopting the improved classifier, the tracking result has improved significantly.

From Table 1, our proposed tracker has the least average center location error among some state-of-art algorithms including compressive tracking. In the video sequence, Bolt and Lemming, drift problem happens and causes a big center location error in the compressive tracking while our algorithms achieve a rather good result.

5   Concluding Remarks

In this paper, we incorporate the particle filter framework into the compressive tracking. When detecting the target in the next frame, instead of searching in a neighborhood region of the previous loacation, we search from the particles resampled in the previous frame and use the classifier response as the particle weight. Meanwhile, the simple naive Bayes classifier is also modified to improve the discriminative power.

Experiments show that   our proposed algorithm has the ability to handle the drifting problem and tracks object more robustly .

References

1. Babenko, B., Yang, M.H., Belongie, S.: Robust object tracking with online multiple instance learning. IEEE Transactions on Pattern Analysis and Machine Intelligence 33(8), 1619–1632 (2011)
2. Yilmaz, A., Javed, O., Shah, M.: Object tracking: A survey. Acm computing surveys (CSUR) 38(4), 13 (2006)
3. Gordon, N.J., Salmond, D.J., Smith, A.F.M.: Novel approach to nonlinear/non-Gaussian Bayesian state estimation. IEE Proceedings F (Radar and Signal Processing) 140(2), 107–113 (1993)
4. Isard, M., Blake, A.: Contour tracking by stochastic propagation of conditional density. In: Buxton, B.F., Cipolla, R. (eds.) ECCV 1996. LNCS, vol. 1064, pp. 343–356. Springer, Heidelberg (1996)
5. Liu, J.S., Chen, R.: Sequential Monte Carlo methods for dynamic systems. Journal of the American statistical association 93(443), 1032–1044 (1998)
6. Viola, P., Jones, M.: Rapid object detection using a boosted cascade of simple features. In: Proceedings of the 2001 IEEE Computer Society Conference on Computer Vision and Pattern Recognition, CVPR 2001, vol. 1, pp. I-511–I-518. IEEE (2001)
7. Grabner, H., Grabner, M., Bischof, H.: Real-Time Tracking via On-line Boosting. BMVC 1(5), 6 (2006)
8. Grabner, H., Leistner, C., Bischof, H.: Semi-supervised on-line boosting for robust tracking. In: Forsyth, D., Torr, P., Zisserman, A. (eds.) ECCV 2008, Part I. LNCS, vol. 5302, pp. 234–247. Springer, Heidelberg (2008)
9. Stalder, S., Grabner, H., Van Gool, L.: Beyond semi-supervised tracking: Tracking should be as simple as detection, but not simpler than recognition. In: 2009 IEEE 12th International Conference on Computer Vision Workshops (ICCV Workshops), pp. 1409–1416. IEEE (2009)
10. Hare, S., Saffari, A., Torr, P.H.S.: Struck: Structured output tracking with kernels. In: 2011 IEEE International Conference on Computer Vision (ICCV), pp. 263–270. IEEE (2011)
11. Kalal, Z., Mikolajczyk, K., Matas, J.: Tracking-learning-detection. IEEE Transactions on Pattern Analysis and Machine Intelligence 34(7), 1409–1422 (2012)
12. Zhang, K., Zhang, L., Yang, M.-H.: Real-time compressive tracking. In: Fitzgibbon, A., Lazebnik, S., Perona, P., Sato, Y., Schmid, C. (eds.) ECCV 2012, Part III. LNCS, vol. 7574, pp. 864–877. Springer, Heidelberg (2012)

Image Super-Resolution with Fast Approximate Convolutional Sparse Coding

Christian Osendorfer*, Hubert Soyer*, and Patrick van der Smagt

Technische Universität München, Fakultät für Informatik, Lehrstuhl für Robotik und Echtzeitsysteme, Boltzmannstraße 3, 85748 München

Abstract. We present a computationally efficient architecture for image super-resolution that achieves state-of-the-art results on images with large spatial extend. Apart from utilizing Convolutional Neural Networks, our approach leverages recent advances in fast approximate inference for sparse coding. We empirically show that upsampling methods work much better on latent representations than in the original spatial domain. Our experiments indicate that the proposed architecture can serve as a basis for additional future improvements in image super-resolution.

Keywords: Image Processing, Sparse Coding, Convolutional Neural Networks.

1 Introduction

The term super-resolution in computer vision generally denotes the process of increasing the resolution of a given image or a set of images. Sparse Coding, a powerful dictionary learning method [1,2,3,4], was recently applied to single-image super-resolution in a very successful way [5,6,7,8]. Hereby, the sparse code couples two different kinds of dictionaries: One dictionary contains low-resolution atoms and one dictionary contains high-resolution atoms. Super-resolving an image patch is then performed in a straight-forward manner: Given the low-resolution patch, determine its sparse code relative to the low-resolution dictionary, and then apply this sparse code in the high-resolution generative model. Couzinie-Devy et al. [5] apply this idea for deblurring and super-resolution by processing each input image patch by patch. Yang et al. [6] follow a similar idea but propose different additions for face and natural images and combine their method with a global reconstruction constraint over the whole image. [7] and [8] take sparse coding for super-resolution even further, working not only with two dimensional images but processing even depth information.

Note that these approaches resolve the super-resolution problem in a very elegant *implicit* way: Upsampling the spatial data, as is necessary in the standard super-resolution approaches like bicubic interpolation [9], is achieved indirectly with the high-resolution dictionary. However, this entails that applying these

---

* Corresponding author.

C.K. Loo et al. (Eds.): ICONIP 2014, Part III, LNCS 8836, pp. 250–257, 2014.

methods to images larger than the training patches is cumbersome and computationally inefficient. Furthermore, finding the sparse code for a given image patch is a costly optimization problem itself and thus applying the mentioned approaches to large images with many patches is extremely slow. Yang et al. specify the time to enlarge a 85×86 image to 255×258 with their sparse coding model from [6] and a reasonably chosen set of parameters as approximately 30 seconds on a Core duo@1.83 Ghz with 2GB Ram.

We tackle exactly this problem: Our proposed super-resolution architecture leverages recent insights into fast approximate sparse coding and utilizes the natural characteristic of the convolutional operator. In this way, we can *train our model on exemplary image patches and scale it to arbitrarily sized test images without any additional cost.* We present our approach and the necessary preliminary work in section 2. Experimental details and results are described in section 3. Section 4 concludes with a brief outlook on future work.

2 Approach

Recently, Convolutional Neural Networks (CNN) [10] have gained a lot of attention due to their success in several large scale computer vision tasks [11,12,13]. Due to the nature of the convolutional operator, CNNs can be applied to inputs of arbitrary size, i.e. they are apriori not tied to the dimensionality of the samples from the training set. This property is often overlooked (see [14,15] for some notable recent exceptions), yet is one of the main ingredients in order to allow the transfer of learned *patch-based super-resolution* to *full image super-resolution*. However, the standard approach of the previously mentioned sparse coding based super-resolution methods is now no longer applicable: The *upscaling* of the data is encoded in the dictionary elements of these methods – this is not possible in a straight forward manner with a convolutional based approach.

Where could upsampling of an image happen? The common approach [9] is to upsample in the image domain. The problem of super-resolution then simply reduces to *learning an optimal deconvolutional* operator. However, if one considers the latent representation of an image (i.e. the convolutional sparse codes in our case), another option occurs: Upsampling this latent representation. Similar to standard signal processing we hypothesize that upsampling should be performed on the *adaptively learned* latent representation of an image and not on its original spatial representation.

Specifically our method consists of three parts: (i) Fast convolutional sparse coding of an input, (ii) upsampling of the sparse codes and (iii) convolutional decoding of the upsampled sparse codes. If the upsampling method is chosen in the right way, this architecture can be applied to inputs (i.e. images) of arbitrary dimensions.

2.1 Fundamentals

In recent years, a wide variety of sparse coding algorithms were developed that learn good feature representations of natural images [1,2,3,4]. A big practical hindrance of the standard sparse coding algorithms is that inference of a sparse code requires running a computationally expensive optimization algorithm. Utilizing the powerful approximation capabilities of neural networks, [16,17] propose an algorithm that can simultaneously learn an overcomplete dictionary for sparse coding and an approximator that predicts the optimal sparse representation.

Taking this idea even further [18] shows that by introducing convolutional operators a richer, more diverse set of features is learned. The objective function for their architecture is as follows:

$$\mathcal{L}(x, z, \mathcal{D}^{(d)}, \mathcal{D}^{(e)}) = \frac{1}{2} \underbrace{\| x - \sum_{k=1}^{K} \mathcal{D}_k^{(d)} * z_k \|_2^2}_{\text{decoder}} + \tag{1}$$

$$\underbrace{\sum_{k=1}^{K} \| z_k{}^\star - f(\mathcal{D}_k^{(e)} * x) \|_2^2}_{\text{encoder}} + \underbrace{\lambda |z|_1}_{\text{sparsity}}$$

where $x \in \mathbb{R}^{m \times n}$ is an input image, $z \in \mathbb{R}^{K \times o \times p}$ is a set of K many (2d) sparse codes (of dimension $o \times p$ each) with $z_k{}^\star$ being a version of sparse code k that is optimal with respect to the decoder part of the loss function. $\mathcal{D}^{(e)}$ and $\mathcal{D}^{(d)}$ are sets of encoder/decoder filters, f is a non-linear function, $*$ denotes convolution and $|z|_1$ is the l_1 norm over all sparse codes z. The *decoder* part combined with $|z|_1$ is equivalent to the standard convolutional sparse coding formulation. The *encoder* tries to produce representations that are similar to the optimal convolutional sparse codes. Given an input image, the encoder produces its corresponding sparse code and can therefore be seen as a single step approximator of the iterative sparse code optimization method, outperforming it significantly in speed.

Learning (i.e. finding $\mathcal{D}^{(e)}$ and $\mathcal{D}^{(d)}$) happens in an alternating manner: (i) First, by keeping $\mathcal{D}^{(e)}$ and $\mathcal{D}^{(d)}$ constant, minimize eq. 1 with respect to z. Starting from the initial value provided by $f(\mathcal{D}_k^{(e)} * x)$ (for all k) this can be done with various kinds of optimization algorithms. In our experiments, we employed the Fast Iterative Shrinkage-Thresholding Algorithm (FISTA) [19]. (ii) Second, based on this optimal sparse code, update $\mathcal{D}^{(e)}$ and $\mathcal{D}^{(d)}$ by one step of gradient descent.

2.2 Upsampling

Convolution produces results similar in size to its input when applied to an image. Naïvely employing the architecture from eq. 1 for super-resolution can therefore only be managed with a trick: Given a low-resolution image patch the encoder approximates a sparse code. The decoder then uses this sparse code to infer a high resolution of the *patch center only*. Super-resolving an image is then achieved by applying this process repeatedly to different areas of the low-resolution image followed by stacking together the results.

However, in order to have enough information in the sparse representation for upsampling the patch center, the filter sizes in the *encoder* have to be choosen very large in relation to the low-resolution image, which usually leads to learning averaging filters only. As expected, this idea yields very poor results which resemble only a very blurred upscaled version of the low-resolution image center.

For proper image super-resolution the model from eq. 1 requires some modifications: As already mentioned before hand, we introduce an upscaling layer between the encoding and decoding stage of the model, working on the sparse representation of an input:

$$\mathcal{L}(x^{(lr)}, x^{(hr)}, z, \mathcal{D}^{(d)}, \mathcal{D}^{(e)}, W) = \tag{2}$$

$$\frac{1}{2} \underbrace{\|x^{(hr)} - \sum_{k=1}^{K} \mathcal{D}_k^{(d)} * \widehat{W \overline{z}_k}\|_2^2}_{\text{decoder}} +$$

$$\underbrace{\sum_{k=1}^{K} \|z_k{}^\star - f(\mathcal{D}_k^{(e)} * x^{(lr)})\|_2^2}_{\text{encoder}} + \underbrace{\lambda |z|_1}_{\text{sparsity}}$$

where $x^{(lr)}$ and $x^{(hr)}$ are the low-resolution and high-resolution versions of the input x respectively and $W \in \mathbb{R}^{o_{(hr)} \cdot p_{(hr)} \times o_{(lr)} \cdot p_{(lr)}}$ is a matrix that scales the *flattened* sparse codes $\overline{z}_k$ up to their high-resolution version. After applying W the result has to be reshaped to the correct high-resolution sparse code shape $o_{(hr)} \times p_{(hr)}$ as denoted by $\widehat{W \overline{z}_k}$. This formulation of the model allows to use any upsampling method that is based on a linear transformation by choosing matrix W accordingly. Note that learning $\mathcal{D}^{(e)}$ and $\mathcal{D}^{(d)}$ proceeds exactly as in the original architecture from eq. 1. Figure 1 shows a graphical interpretation of the model with input data at different stages of the pipeline.

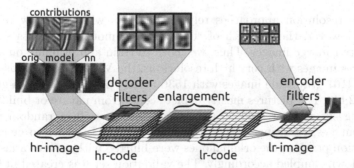

Fig. 1. On the right side, a low-resolution image is feed into the fast approximate convolutional sparse coding module, then upscaled and finally deconvoluted. See section 3.2 for more comments with respect to this Figure.

Albeit an arbitrarily structured matrix W would be the most flexible and powerful approach, the upsampling matrix W must be choosen as a convolutional operator itself. In our experiments, we considered four different kinds of upscaling operations for the sparse codes: bilinear interpolation, linear shifted interpolation, nearest neighbor interpolation and our own, non-standard, perforate interpolation. Figure 2 illustrates these methods graphically for the example of two-fold super-resolution: It shows the convolutional weights that are applied to a neighborhood of pixels in a low-resolution sparse code in order to generate a pixel in the high-resolution sparse code.

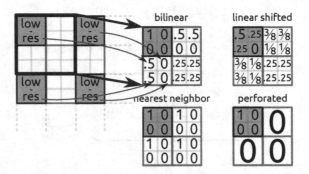

Fig. 2. Upsampling an image by a factor of 2: Every pixel in the low-resolution image is replaced by 4 pixels, indicated by the black square in the left-most image. How the values of these 4 pixels are actually computed depends on the specific upsampling scheme. Here we consider schemes that utilize 4 neighboring low-resolution pixels to compute one high-resolution pixel. To the right, we show the upsampling weights for the 4 methods mentioned in the text. Note that *perforated* upsampling induces additionally sparsification. It crudely approximates the *inverse of the widely-used max-pooling operator* from deep CNNs [12].

3   Experiments

Most super-resolution approaches rely on datasets with very low resolutions [6,20,21]. However, the strength of the presented model lies in its speed and applicability to large images. Thus, we chose to train and evaluate on a dataset that features images with very high resolutions, the Van Hateren dataset [22]. It comprises 4167 gray scale images with 1536×1024 pixels each and a gray scale depth of 12 bit. The pictures mostly depict scenes from nature or buildings. For the training set we extracted 20 patches of size 50×50 at random positions from 400 images, resulting in 8000 training samples. In order to generate the low-resolution patches, the original ones were blurred with an anti-aliasing filter and then down-sampled accordingly. The validation set was created in the same way but using a different set of 100 images. And finally the test set features 100 *unseen full-sized* images. Training and applying the other sparse coding based super-resolution algorithms cited earlier would not be tractable on (test) images of this size. We therefore compared our approach with standard super-resolution algorithms from the image processing domain [9]: bicubic spline based interpolation, bilinear interpolation and nearest neighbor interpolation. After training the model from eq. 2 is finished, we further *fine-tuned* the complete convolutional super-resolution model.

3.1 Training Details

To keep training time manageable the model was trained one sample (that is, a pair of low resolution and accompanying high resolution image patch) at a time, samples were chosen at random. The sparse codes were optimized with 5 iterations of FISTA at each training step. Less than 5 FISTA iterations decreased the final results noticeably while more iterations didn't have any influence on the results but increased training time significantly. The non-linearity f (see eq. 1) is set to a *soft threshold function* [17]. Filters were optimized with one step of gradient descent per model training iteration. All experiments were trained for 1 million epochs with an initial learning rate of $2 \cdot 10^{-4}$ that decayed as $\frac{2 \cdot 10^{-4}}{1.0 + (epoch/5000)}$. Finally, the model with the lowest objective function score on the validation set was further fine tuned with a learning rate of $1 \cdot 10^{-6}$ for another 16000 epochs.

3.2 Evaluation

There are a number of ways to evaluate the results of super-resolution: Some papers judge the quality of the results by their Mean Squared Error per pixel (MSE) to the ground truth [6], some use the related Peak-Signal-to-Noise Ratio (PSNR) [23] and others rely on Structured Similarity (SSIM) as a measure of error [20,21]. PSNR is logarithmically proportional to MSE and both can be argued to only inaccurately represent the human understanding of better or worse regarding the quality of an image-reconstruction. SSIM aims to tackle this shortcoming – we therefore report both PSNR and SSIM scores in our evaluation.

A qualitative impression of the learned architecture is shown in Figure 1: Typical filters for both the encoder as well as the decoder are shown, in this case for an architecture with 8 latent channels. On the left side, a super-resolved image patch (denoted by *model*) is shown, computed from a low-resolution image patch depicted at the right side. For comparison, the original (*orig*) high-resolution patch and the nearest neighbor interpolation (*nn*) is also shown.

Table 1 shows both the PSNR and SSIM scores for the various types of latent upsampling methods presented in section 2.2 (CNN-PF denotes *perforated interpolation*, CNN-BL denotes *bilinear interpolation*, CNN-NN denotes *nearest neighbor interpolation* and CNN-SH denotes *linear shifted interpolation*). BCI, BLI and NNI denote the classic bicubic, bilinear and nearest neighbor interpolation methods in the original spatial domain respectively. K, the number latent channels is set to 8 in all experiments: Smaller numbers (4, 6) resulted in

Table 1. Our perforated sparse code upsampling method performs best. For a full image from the test set the unoptimized version takes about 4 seconds, compared to approximately 2.5 seconds for bicubic interpolation. Larger numbers are better for both PSNR and SSIM.

	CNN-PF	CNN-BL	CNN-NN	CNN-SH	CNN-LD	BCI	BLI	NNI
PSNR	**32.55**	32.52	32.49	31.98	32.07	31.80	30.79	30.55
SSIM	**0.946**	0.945	0.942	0.941	0.944	0.935	0.922	0.913

inferior results, for larger numbers (12, 16, 32) training did not converge after 21 days and thus was stopped – FISTA proved to be the bottleneck for these larger models. All other hyperparameters were determined via the validation set. We also learned W (see eq. 2), which is resembled by the column CNN-LD.

Apart from the fact that CNN-PF outperforms all other approaches, in particular the widely used bicubic interpolation method, we point out the following two observations: (i) Both bilinear and nearest neighbor interpolation methods perform significantly better when applied to the latent representation, supporting our original hypothesis empirically. Hence, an obvious next step is to apply bicubic interpolation accordingly in the latent domain – however this can't no longer be written in the form of eq. 2 because now *non-linear features* need to be computed in the sparse domain. (ii) Fine-tuning did not improve the results. We assume that this is due to approximating FISTA with only one convolutional layer.

4 Summary and Outlook

We presented a single image super-resolution approach based on fast approximate sparse coding with convolutional neural networks. Our approach not only outperforms state-of-the-art super-resolution methods for large images but is also computationally efficient. As indicated by our experiments, unrolling the iterative convolutional FISTA algorithm in a way similar to [24] is a very promising future research direction. Extrapolating the observations from Table 1, latent bicubic upsampling, or even more general upsampling methods that can be realized through [25] should increase the performance of our framework considerably.

References

1. Olshausen, B.J., Field, D.J.: Sparse coding with an over complete basis set: a strategy employed by v1? Vision Research (1997)
2. Mairal, J., Bach, F., Ponce, J., Sapiro, G., Zisserman, A.: Discriminative learned dictionaries for local image analysis. In: CVPR (2008)
3. Lee, H., Battle, A., Raina, R., Ng, A.Y.: Efficient sparse coding algorithms. In: NIPS (2007)
4. Ranzato, M.A., Poultney, C., Chopra, S., Lecun, Y.: Efficient learning of sparse representations with an energy-based model. In: NIPS (2006)

5. Couzinie-Devy, F., Mairal, J., Bach, F., Ponce, J.: Dictionary learning for deblur-ring and digital zoom. arXiv preprint arXiv:1110.0957 (2011)
6. Yang, J., Wright, J., Huang, T.S., Ma, Y.: Image super-resolution via sparse representation. IEEE Transactions on Image Processing 19(11) (2010)
7. Hu, T., Nunez-Iglesias, J., Vitaladevuni, S., Scheffer, L., Xu, S., Bolorizadeh, M., Hess, H., Fetter, R., Chklovskii, D.: Super-resolution using sparse representations over learned dictionaries: Reconstruction of brain structure using electron microscopy. arXiv preprint arXiv:1210.0564 (2012)
8. Zhang, Y., Wu, G., Yap, P.T., Feng, Q., Lian, J., Chen, W., Shen, D.: Reconstruction of super-resolution lung 4d-ct using patch-based sparse representation. In: CVPR (2012)
9. Petrou, M., Petrou, C.: Image Processing: The Fundamentals. Wiley & Sons (2010)
10. Lecun, Y., Bottou, L., Bengio, Y., Haffner, P.: Gradient-Based Learning Applied to Document Recognition. Proceedings of the IEEE 86 (1998)
11. Krizhevsky, A., Sutskever, I., Hinton, G.E.: Imagenet classification with deep convolutional neural networks. In: NIPS (2012)
12. Ciresan, D.C., Meier, U., Schmidhuber, J.: Multi-column deep neural networks for image classification. In: CVPR (2012)
13. Girshick, R., Donahue, J., Darrell, T., Malik, J.: Rich feature hierarchies for accurate object detection and semantic segmentation. In: CVPR (2014)
14. Masci, J., Giusti, A., Ciresan, D.C., Fricout, G., Schmidhuber, J.: A fast learning algorithm for image segmentation with max-pooling convolutional networks. In: ICIP (2013)
15. Sermanet, P., Eigen, D., Zhang, X., Mathieu, M., Fergus, R., LeCun, Y.: Overfeat: Integrated recognition, localization and detection using convolutional networks. In: ICLR (2014)
16. Kavukcuoglu, K., Ranzato, M., LeCun, Y.: Fast inference in sparse coding algorithms with applications to object recognition. Technical report, Computational and Biological Learning Lab, Courant Institute, NYU (2008)
17. Kavukcuoglu, K., Ranzato, M., Fergus, R., Le-Cun, Y.: Learning invariant features through topographic filter maps. In: CVPR (2009)
18. Kavukcuoglu, K., Sermanet, P., Boureau, Y.L., Gregor, K., Mathieu, M., Cun, Y.L.: Learning convolutional feature hierarchies for visual recognition. In: NIPS (2010)
19. Beck, A., Teboulle, M.: A fast iterative shrinkage-thresholding algorithm for linear inverse problems. SIAM Journal on Imaging Sciences 2(1), 183–202 (2009)
20. He, L., Qi, H., Zaretzki, R.: Beta process joint dictionary learning for coupled feature spaces with application to single image super-resolution. In: CVPR (2013)
21. Lu, X., Yuan, H., Yan, P., Yuan, Y., Li, X.: Geometry constrained sparse coding for single image super-resolution. In: CVPR (2012)
22. van Hateren, J.H., van der Schaaf, A.: Independent component filters of natural images compared with simple cells in primary visual cortex. Proceedings: Biological Sciences 265(1394) (1998)
23. Zhang, K., Gao, X., Tao, D., Li, X.: Multi-scale dictionary for single image super-resolution. In: CVPR (2012)
24. Sprechmann, D., Bronstein, A.M., Sapiro, G.: Learning efficient sparse and low rank models. arXiv preprint arXiv:1212.3631 (2012)
25. Lin, M., Chen, Q., Yan, S.: Network in network. In: ICLR (2014)

Sparse Coding for Improved Signal-to-Noise Ratio in MRI

Fuleah A. Razzaq, Shady Mohamed, Asim Bhatti, and Saeid Nahavandi

Centre for Intelligent System Research,
Deakin University, Australia

Abstract. Magnetic Resonance images (MRI) do not only exhibit sparsity but their sparsity take a certain predictable shape which is common for all kinds of images. That region based localised sparsity can be used to de-noise MR images from random thermal noise. This paper present a simple framework to exploit sparsity of MR images for image de-noising. As, noise in MR images tends to change its shape based on contrast level and signal itself, the proposed method is independent of noise shape and type and it can be used in combination with other methods.

Keywords: Magnetic Resonance imaging(MRI), Sparse Coding, Signal-to Noise Ratio (SNR), Additive White Gaussian Noise (AWGN).

1 Introduction

MRI is an imaging technique employed in advanced medical facilities to study and generate images of internal structures of the human body. Most of the modern MRI machines use a super-conducting magnet to generate outer magnetic field B_0. Super-conducting magnets are not permanent magnets. Instead, these are electromagnets which means they work as magnets when electric current is passing through them. When RF pulse is applied, it creates a transverse Radio Frequency field. The Hydrogen atoms in human body absorb energy and go into excitation state. Later, RF coils are used to receive RF signals. MRI can only achieve limited Signal-to-Noise ratio (SNR) due to its physical and hardware limitations [1]. The SNR in MRI is dependent on image acquisition time and resolution or volume of object in spatial domain [2]. The magnetic signals are acquired using Radio Frequency (RF) sensors and the spatial domain is mapped into frequency data i.e. K-space. The data is collected in two channels real and imaginary. Due to hardware issues as well as thermal noise from patient [3], these channels get affected by Additive White Gaussian Noise (AWGN). Later, this frequency data is converted using Inverse Discrete Fourier transform and magnitude images are calculated using absolute values from real and imaginary data components. During this process, the noise distribution also gets effected and the Gaussian noise transforms into signal dependent Rician noise [4]. Managing and removing noise in MRI is a difficult because the noise is dependent on signal itself. Moreover, the noise in MRI varies spatially. The simple additive Gaussian noise in original signal tends to vary spatially in resultant magnitude

C.K. Loo et al. (Eds.): ICONIP 2014, Part III, LNCS 8836, pp. 258–265, 2014.

image. The noise in high intensity regions remains Gaussian while in low intensity image regions it acts as Rayleigh distribution [1].

The idea of sparsity deals with the amount of useful information within a signal. To construct MR image from sensors, frequency domain is used. There is only a small amount of coefficients which is actually significant and is used to represent the image. Whereas, others coefficients are of no use at all or they have small significance that the effect of discarding them is negligible. This idea leads to another domain which says that if the total useful information lies within few significant coefficients then an image can compressed to a very high level. Two types of sparsity are observed and hence used in image reconstruction techniques. Strongly Sparsified data set category is the one in which most of the coefficients are exactly zero and they are almost zero in case of weakly Sparsified data set. The Sparsity of MRI is previously exploited by researchers for Rapid MRI [5,6].

This paper presents a noise removal method based on sparsity of MR images. In MRI, sparsity distribution or curve can be predicted to some extend and a simple framework is defined to minimise the number of image coefficients. Wavelets are used as sparsifying transform domain. Noisy images were sparsified regionally and results are presented here. This method does not try to replace previous methods which are proposed in literature. It tries to improve and enhance previous methods and can be used in combination with other noise removal methods.

2 Noise in MRI

MRI machines reads signals from RF coils and captures data in frequency domain. These readouts have two components for each sample, real and imaginary.

$$Sig(j) = Sig_{real}(j) + \iota Sig_{Imaginary}(j) \tag{1}$$

Here Sig is the required signal at location j in K-space. While, Sig_{real} and $Sig_{Imaginary}$ are the real and imaginary components of the signal and $\iota = \sqrt{-1}$. Due to physical factors and patient's body temperature, thermal noise is introduced in the signal which is additive white Gaussian noise. This AWGN affect both real and imaginary component of the signal.

$$Sig(j) = (Sig_{real}(j) + Noise(j)) + \iota(Sig_{Imaginary}(j) + Noise(j)) \tag{2}$$

When data is in complex form Gaussian noise corrupts both real and imaginary components. The distribution of Gaussian noise for any random variable x mean μ and variance σ^2 can be described as

$$pdf(x) = 1/(\sigma\sqrt{2\pi})e^{-(x-\mu)^2/2\sigma^2} \ where \ x \in (-\infty, \infty) \tag{3}$$

This distribution shows a bell shaped distribution with a peak in center. This noise is easy to remove and handle.

However, this raw data is not available in most of case. MRI frequency data is converted into images using Discrete Fourier Transform (DFT). Fourier transform transfer the noise into image components without effecting its shape [1].

In next step, magnitude images are calculated and the complex data is discarded.

$$m(j) = |y(j)| \tag{4}$$

Now for each pixel j m(j) is combination of noise and real signal. This process changes the shape of noise distribution and makes it Rician Distribution which is signal dependent.

The signal dependent noise is hard to predict and remove but this is the final form of MR image data and in most cases only magnitude images are available. Noise removal is not only difficult in this form but also very crucial for most of MRI application. Furthermore, noise varies spatially in magnitude images. In high contrast or high magnitude images, it tends to take shape of Gaussian distribution for low contrast images Rician distribution tends to shape like Rayleigh Distribution because s becomes zero [1] .

3   De-noising MRI

The Signal to Noise Ratio of MR images is restricted by hardware and application limitations. Thus, noise removal methods are used to enhance imaging. It was suggested to use complex MRI data for noise removal rather than magnitude images. This makes noise removal easy as complex data only has additive Gaussian noise. However, in most real time cases complex MRI data is not readily available [7]. One major category of such methods is based on Gaussian filter and spatial pattern redundancy which is most often used in functional MRI(fMRI). However, it causes blur edges. Later on to avoid these issues, edge preserving filters were introduced into this method [8,9]. The edge preserving filters caused missing features for the low magnitude image areas.

3.1   Wavelets Based Noise Removal Methods

Another category of de-noising methods used wavelets to exploit its multi scale representation for de-noising. The basic procedure is to convert image into wavelet domain, using the transformed wavelet coefficients for noise removal and converting the de-noised wavelet data back into image. Wavelets were used in different range of methods from thresholds to complex filtering [1,10]. A wavelets based thresholding was applied in [11] . In another approach coefficients were squared which made noise independent of signal and thus easily removable [1]. In another method, the multi-scale representation of wavelets was used as correlation information for noise removal [10] Wiener filtering was also applied in wavelet domain for de-noising [12]. However, wavelets based processing generates artefacts which are dependent on the type of wavelets being used [13].

3.2   De-noising MR Images Using Local Sparsity Constraints

When images holds the sparsity condition but measured signals have noise, in that case MR signals can be represented as

$$Y(i) = \begin{Bmatrix} \gamma & i \notin \varrho \\ val + \gamma & i \in \varrho \end{Bmatrix} \tag{5}$$

Here γ is the noise level at any spatial location. Due to sparsity, Y has only $S = K/N$ significant coefficients while rest are zero or nearly zero. If i is a non-significant coefficient and its value can be discarded than from sparsity point of view it only holds noise. Whereas, if i is a significant coefficient it hold coefficient value with added noise. From this sparse condition it can be concluded that $\Gamma = N - S$ percent coefficients are just noise and can be discarded or replaced by zero. Also lesser the value of S means higher value of Γ as N is a constant size of any image. Γ with a larger value means more coefficient can be discarded and noise can be reduced further. Thus, replacing S with S_L as it is less than S_g.

$$\Gamma = N - S_L \tag{6}$$

Γ is the percent of coefficients which are pure noise and have no-significant value. The higher value of Γ means more coefficients can be discarded and less noise. Using, local energy level estimation images were sparsified better thus making S_L a lesser value and a more useful measure in terms of de-noising.

4 Methodology

This section proposes a novel method to de-noise MR images based on the fact that MR images exhibits sparsity. Sparsity is previously used in literature of MRI for under-sampled data [14]. In under-sampling we have missing information but when image is fully-sampled and is corrupted by noise, it is needed to somehow extract only information bits and discarding the rest. The proposed method works on transform sparsity of MR images. This method basically reduces the number of coefficients that are used to represent image based on image sparsity information. It does not change or modify any values. It will either select a coefficient value or will discard it completely. Thus, it can be used in combination with other noise removal methods which estimate noise and modifies the data. This will further enhance the quality of resultant image. The prerequisites of this method are (a) generate regional map and find suitable threshold levels using a reference image such that the resolution of reference image is same as images under experimentation (b) finding sparsifying transform (c) find sparsity ratios for each region.

Input

- Noisy image I.
- Threshold vector τ and respective Sparsity vector S.
- Transform operator α.

Algorithm

– Transform I into ω using transform operator α.
– Generate a region/sparsity map of input image I based on threshold vector τ such that each element of τ is used to generate a sub-region in transform domain ω (i.e. wavelet).
– Select S_i percent highest values from i^{th} region and discard rest of values.
– Regenerate I from ω.

Output

– Output image with reduced noise levels.

5   Experimental Results

Literature shows that Wavelets sparsify MR images very well [14,15]. As, images are fully sampled and sampling was done in Fourier domain. For these experiments the regional sparsity of MR images is analysed in Wavelets and the regions are also defined in Wavelet domain unlike [14,15] where the regions were defined in Fourier domain based on Energy distribution of Fourier.

5.1   De-noising Using Local Sparsity Constraints

All the experiments that are presented in previous section are used for MR de-noising. The experiments helped in understanding the sparsity of MR images in Wavelets and helped in developing some generic key features which can be used for image de-noising. The basic idea is to select limited number of coefficients and to preserve the over-all energy shape. As, energy distribution shows same kind of curve for all different kinds of MR images. In Fourier it shows a high energy peak in center and low energy regions on both ends while in Wavelets it shows high energy peak in start and low energy region afterwards. All MR images roughly maintain this shape. Thus, it can be used as a generic feature and can be used for image de-noising.

Different kinds of noisy images were used and experimental results of previous sections were used as reference point. For any image resolution, reference image should have same resolution. Experiments were done on two sets of images 448x448 and 512x512. Both Fourier and Wavelets were used as sparsifying domains. Firstly noisy image was sparsified using one global level. Later for Local regions, 3 sub-regions were used for both Wavelets and Fourier. To quantify the results SNR is used. As, we are dealing with noisy data SNR gives an estimation of de-nosing. However, incomplete data set effects the results but both results are presented for better understanding of the proposed method. AWGN with different levels of σ was used and added to K-space. That K-space was then converted into magnitude images and those images were used for experiments.

Table 1. PSNR for de-noised Image where noise ratio for AWGN is $\sigma = 10$

	PSNR
Noisy Image	23.1 22.5 17.9 24.3 17.9 20.0
Fourier	23.0 22.9 18.1 25.2 18.2 20.2
Localised Fourier	21.9 22.4 18.1 25.1 18.2 20.2
Wavelets	23.4 22.9 18.0 25.0 18.1 20.2
Localised Wavelets	24.0 25.4 19.1 27.0 18.7 21.3

Table 1 shows PSNR for reconstructed images. All the results were averaged out based on image type. The noise level for this set of experiments was $\sigma = 10$. All images showed improved quality when Wavelets are used as their

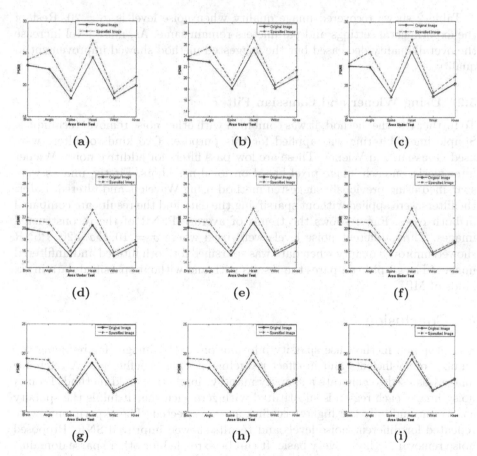

Fig. 1. PSNR for De-noising methods. First row is for noise level $\sigma = 10$ where (a) is quality index for sparsifying the noisy data (b) Sparsification with Gaussian Filter (c) Sparsification with Wiener Filter. Second row is for noise level $\sigma = 15$ and (d), (e), (f) are noisy data, Gaussian and Wiener filtered data respectively. Third row is for for noise level $\sigma = 20$.

sparse domain. Localise wavelets showed further improvement. Images like spine and wrist are the ones which were most effected by noise. Yet, all showed an improvement.

Table 2. PSNR for de-noised Image where noise ratio for AWGN is $\sigma = 20$

	PSNR
Noisy Image	17.9 17.2 13.3 18.4 13.4 15.2
Fourier	18.2 17.6 13.5 18.9 13.6 15.5
Localised Fourier	17.9 17.5 13.5 18.9 13.7 15.6
Wavelets	18.2 17.5 13.5 18.8 13.6 15.5
Localised Wavelets	19.1 18.9 13.7 20.0 13.8 15.7

Table 2 shows recovered image quality when noise level is $\sigma = 20$. Rest of the experimental settings and parameters remain same. As, noise level increase the overall quality decreased but the suggested method showed improvement in quality.

5.2   Using Wiener and Gaussian Filter

To further test the method, it was combined with other noise removal techniques. Simple linear filtering was applied for this purpose. Two kinds of filters were used Gaussian and Wiener. These are low pass filters for additive noise. Wiener filter works on each image pixel based on local neighbors. Firstly, images were sparsified using previously suggested method using Wavelets and filtered. Later, the filters were applied without sparsifying the data and the results are compared in both cases. Fig. 1 shows the trends of average PSNR of the reconstructed images. Three different noise levels were used where $\sigma = 10, 15 and 20$. PSNR showed improved quality when data was sparsified for both filtered and unfiltered images. The graphs compares the results with and without sparsified data in six kinds of MRI.

6   Conclusion

The proposed method use sparsity information of MR images for reducing the number of coefficients and in effect reducing the noise coefficients. A reference image was used to generate a sparsity map by simple threshold method. From a noisy image, each region is substituted with zero such that it fulfils the sparsity constraint and only the highest coefficients are selected. The experiments were repeated for different noise levels and results showed improved SNR. Proposed noise removal method is very basic. It can be extended for other sparse domains. Current work used 1-D division of image. For future work this technique can be extended with multi-dimensional sub-regions. The optimal way to define regions for noise removal is yet to be explored. Currently it was implemented alone and with Gaussian and Wiener Filtering. It can be extended and combined with other more complex noise removal methods.

References

1. Nowak, R.D.: Wavelet-based rician noise removal for magnetic resonance imaging. IEEE Transactions on Image Processing 8(10), 1408–1419 (1999)
2. Macovski, A.: Noise in mri. Magnetic Resonance in Medicine 36(3), 494–497 (1996)
3. Edelstein, W.A., Bottomley, P.A., Pfeifer, L.M.: A signal-to-noise calibration procedure for nmr imaging systems. Medical Physics 11(2), 180–185 (1984)
4. Drumheller, D.: General expressions for rician density and distribution functions. IEEE Transactions on Aerospace Electronic Systems 29, 580–588 (1993)
5. Parrish, T., Hu, X.: Continuous update with random encoding (cure): a new strategy for dynamic imaging. Magnetic resonance in medicine 33(3), 326–336 (1995)
6. Doyle, M., Walsh, E.G., Blackwell, G.G., Pohost, G.M.: Block regional interpolation scheme for k-space (brisk): A rapid cardiac imaging technique. Magnetic Resonance in Medicine 33(2), 163–170 (1995)
7. Alexander, M., Baumgartner, R., Summers, A., Windischberger, C., Klarhoefer, M., Moser, E., Somorjai, R.: A wavelet-based method for improving signal-to-noise ratio and contrast in mr images. Magnetic Resonance Imaging 18(2), 169–180 (2000)
8. Samsonov, A.A., Johnson, C.R.: Noise-adaptive nonlinear diffusion filtering of mr images with spatially varying noise levels. Magnetic Resonance in Medicine 52(4), 798–806 (2004)
9. Murase, K., Yamazaki, Y., Shinohara, M., Kawakami, K., Kikuchi, K., Miki, H., Mochizuki, T., Ikezoe, J.: An anisotropic diffusion method for denoising dynamic susceptibility contrast-enhanced magnetic resonance images. Physics in Medicine and Biology 46(10), 2713 (2001)
10. Pizurica, A., Philips, W., Lemahieu, I., Acheroy, M.: A versatile wavelet domain noise filtration technique for medical imaging. IEEE Transactions on Medical Imaging 22(3), 323–331 (2003)
11. Healy Jr, D.M., Weaver, J.B.: Two applications of wavelet transforms in magnetic resonance imaging. IEEE Transactions on Information Theory 38(2), 840–860 (1992)
12. Wirestam, R., Bibic, A., Lätt, J., Brockstedt, S., Ståhlberg, F.: Denoising of complex mri data by wavelet-domain filtering: Application to high-b-value diffusion-weighted imaging. Magnetic Resonance in Medicine 56(5), 1114–1120 (2006)
13. Yang, X., Fei, B.: A wavelet multiscale denoising algorithm for magnetic resonance (mr) images. Measurement Science and Technology 22(2), 025803 (2011)
14. Razzaq, F., Mohamed, S., Bhatti, A., Nahavandi, S.: Non-uniform sparsity in rapid compressive sensing mri. In: 2012 IEEE International Conference on Systems, Man, and Cybernetics (SMC), pp. 2253–2258 (October 2012)
15. Razzaq, F.A., Mohamed, S., Bhatti, A., Nahavandi, S.: Defining sub-regions in locally sparsified compressive sensing mri. In: The IASTED Int Conf. on Biomedical Engineering (BioMED), pp. 360–367 (2013)

Scalable Video Coding
Using Hybrid DCT/Wavelets Architectures

Tamer Shanableh

American University of Sharjah, College of Engineering, Sharjah, UAE
tshanableh@aus.edu

Abstract. This paper proposes the use of wavelet image transformation and polyphase downsampling in scalable video coding. A wavelet-based inter-frame
coding solution using the syntax and framework of both MPEG-4 H.264/AVC
and it is scalable extension, SVC. In the former codec, redundant slices are employed for coding the high frequency subbands of wavelet transformed imaged.
While in the latter codec, the wavelet subbands are arranged into separate
Coarse Grain Scalability (CGS) layers. Additionally, the paper proposes the use
of a modified polyphase downsampling in applications of scalability and error
resiliency. It is shown that the coding efficiency of the proposed solutions is
comparable to single layer coding.

Keywords: Digital Video Coding, Scalable Video Coding, MPEG.

1   Introduction

It is reported in [1] that the DCT block-based approach is suitable for coding wavelet
subbands. It was proposed to code the wavelet subbands in the base and enhancement
layer of MPEG-4 AVC/H.264 scalable video coding (SVC) [2]. The low frequency
band is coded in the base layer, the resultant quantization error and the high frequency
bands are arranged into one image and coded in the enhancement layer. Such an approach allowed for both SNR and dyadic spatial scalabilities. Both the base and enhancement layers are coded using the AVC intra-frame syntax. This paper extends the
reviewed work by proposing an inter-frame wavelet coding scheme in two different
coding arrangements using the framework of both AVC [3] and SVC.

For inter-frame wavelet coding, the high frequency subbands are time-variant because of the decimation process involved in the image wavelet decomposition. Thus,
translation motion in the pixel-domain image cannot be accurately estimated from the
wavelet coefficients. Complete-to-overcomplete Discrete Wavelet Transformation
(DWT) can be used to solve this problem. For instance, in [4] and [5] complete-to-
overcomplete DWT is applied to the locally decoded reference subbands. As a result,
each frequency subband ends up with 4 representations with different directions of unit
shifts. Motion estimation is then used to find a best match location in one of the four
reference representations. An extra syntactic field is needed to indicate the reference
subband representation to which the MV belongs. Clearly the complete-to-overcomplete

C.K. Loo et al. (Eds.): ICONIP 2014, Part III, LNCS 8836, pp. 266–275, 2014.
© Springer International Publishing Switzerland 2014

DWT of the reference subbands and the extra syntactical field violates the operations of the standardized codecs. Moreover, the results presented in [5] applies the above complete-to-overcomplete DWT in conjunction with pixel-accurate ME only. A traditional method for complete-to-overcomplete DWT was introduced in [6]. A time domain one dimensional signal is passed through a high pass and low pass filter followed by decimation to produce low and high frequency subbands. The original signal is also shifted by one unit and the decomposition procedure is repeated. In general at each decomposition level, the low frequencies are decomposed twice, with and without unit shifting. More advanced complete-to-overcomplete DWT methods are reported in [7] and [8].

In this paper, two solutions are proposed for interframe coding of wavelet coefficients. The first solution employs the redundant pictures of the AVC framework for the coding of wavelet subbands, while in the second solution, the wavelet subbands are coded in the enhancement layers of a SVC codec.

The paper is organized as follows. Section 2 introduces the proposed solution of using redundant pictures for video scalability. Section 3 introduces the proposed solution of using wavelet subbands with scalable video coding. Section 4 introduces the proposed polyphase downsampling approach to scalability. The experimental results are introduced in Section 5 and Section 6 concludes the paper.

2   Proposed Redundant Pictures Approach to Scalability

The AVC standard introduced the use of redundant pictures (or redundant slices) as an error resiliency tool. The idea is to allow the encoder to repeat the coding of a primary picture (or part of it) in a redundant picture syntax element. In case of transmission errors the decoder can choose to decode the redundant picture to conceal the error and alleviate picture drift. This paper proposes the use of the redundant pictures for the coding the high frequency wavelet subbands. The low frequency subbands on the other hand are coded using the primary picture syntax element. Note that the AVC standard indicates that a compliant decoder does not have to decode redundant pictures. Therefore, the proposed coding arrangement does not violate the standard in this regard.

In this arrangement, if a video server streams the primary pictures only then a low spatial resolution of the original video is received. This is fully compliant with any AVC decoder. On the other hand, if the server streams both the primary and redundant pictures then a scalable decoder will be able to reconstruct the video at a high spatial resolution.

The first stage of this solution is a pre-process in which the input images are transformed into the wavelet domain. High frequency subbands are then rearranged and coded as redundant pictures. This is illustrated in Figure 1 below. The rearrangement of high frequency subbands is necessary to guarantee that similar subbands are predicted from each other thus increasing the efficiency of motion estimation and compensation.

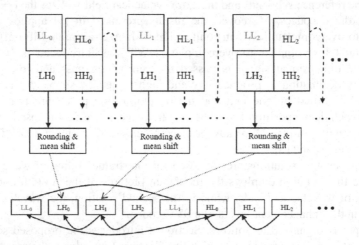

Fig. 1. Arrangement of wavelet subbands into primary and redundant pictures in the AVC framework

In the pre-processing it is also important to round and shift the mean of wavelet coefficients such that they can be represented with unsigned short data types. In this implementation and with one level of wavelet decomposition, the coefficients are represented by 10 bits only. The AVC implementation can be configured accordingly. Note that the rounding causes an imperfect reconstruction of the wavelet coefficients. Nevertheless it was noticed that loss in image quality is negligible. Empirically, the reconstructed rounded images have a PSNR of around 50 dB.

In the AVC coding stage, the standard specifies that primary pictures cannot be predicted from redundant ones. And a redundant picture cannot be predicted from its primary picture as well. Referring to Figure 1, clearly the prediction of say HL_0 (the subscript refers to the time index of the input image) from LL_0 is useless and the AVC coder will decide to perform an intra-frame coding instead. The rest of the high frequency subbands in this case i.e. HL_1 and HL_2 will be efficiently predicted from each other. Upon decoding, an extra post-process is required in which the decoded high frequency subbands and the decoded primary pictures are regrouped and inverse transformed into the higher spatial resolution.

3   Proposed Scalability Solution Based on Wavelet Subbands

In this proposed solution, the SVC scalable framework is used to encode both the low and high frequency subbands. The input images are DWT, rounded and mean shifted as in the aforementioned redundant pictures solution. The low frequency subband is coded as a base layer in this case. The high frequency subbands on the other hand are coded in separate SVC enhancement layers as illustrated in Figure 2 below.

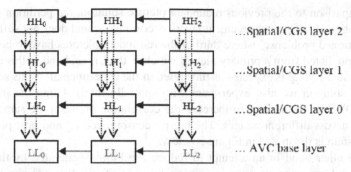

Fig. 2. Arrangements of wavelet subbands into 4 SVC layers

The SVC standard specifies that the spatial resolution of the enhancement layers can be greater than or equal to the spatial resolution of the base layer. In this case the upsamling filter of interlayer prediction is disabled and the deblocking of the base layer is omitted because the block boundaries between the layers are already aligned. Thus the arrangement of Figure 2 above is syntax friendly.

The prediction of high frequency subbands will naturally be intra-layer as opposed to inter-layer prediction. Nevertheless the vertical prediction lines in the figure indicate that other prediction modes can be applied. The SVC standard specifies a number of inter-layer predictions such as prediction of motion fields, prediction of MB partitioning, MB coding modes and so forth.

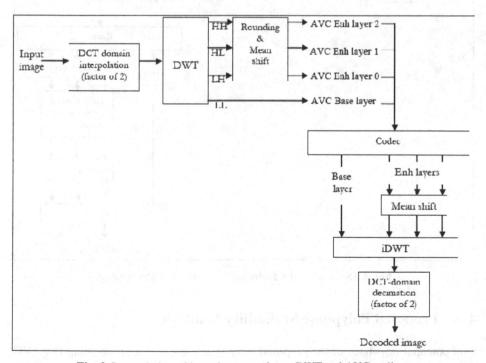

Fig. 3. Interpolation of input images prior to DWT and AVC coding

In comparison to the previous redundant picture solution, the perdition of high frequency subbands in the enhancement layers is continuous and does not suffer from the aforementioned problems   where third of the redundant pictures have to be either intra coded or predicted from a primary picture which is the LL subband in this case. Hence more efficient coding is expected as illustrated in the experimental results section.

In this solution we also experiment with spatially interpolating the input images prior to DWT in an attempt to increase the correlation between the same frequency subbands across different images. The pre-processing, coding and post processing of such a system is illustrated in Figure 3 below.

On the other hand in an attempt to reduce the bitrate generated by the high frequency band, an opposite solution can be thought of. Such bands can be spatially decimated prior to coding. However this arrangement will in some cases affect the efficiency of motion estimation and compensation. The pre-processing, coding and post processing of such a system is illustrated in Figure 4 below.

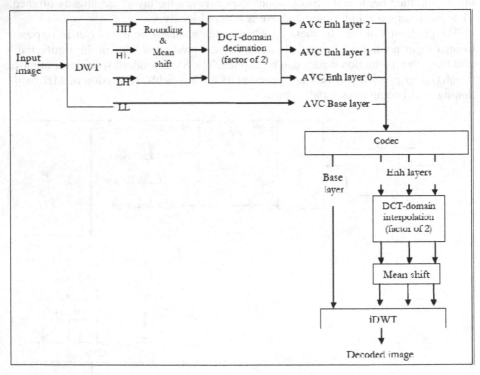

Fig. 4. Decimation of wavelet subbands prior to SVC coding

4   Proposed Polyphase Scalability Solution

One potential drawback of the proposed inter-frame wavelet solution suing the AVC redundant pictures is that it defeats the purpose of error resiliency. Thus one can think of an alternative solution in which redundant pictures can be used for both error resiliency and spatial scalability. The solution is based on polyphase downsampling which is usually used in Multiple Description (MD) coding [9,10,11].

In [12] a source video is polyphase down sampled and fed into separate AVC coders. The paper then focuses on transmission errors and proposes different concealment solutions and post processing to attenuate visual effects related to MD coding and transmission errors.

In this work we propose the use of polyphase downsampling as a scalability and error resiliency tool. One of the polyphase down sampled images (or descriptors) is used as a primary image within the AVC framework and the rest of the descriptors are used as redundant pictures. The redundant pictures can serve as an error resiliency tool because their visual content is very similar to the primary pictures. Likewise the redundant pictures can be used for enhancing the spatial scalability of the primary pictures.

For completeness, the concept of polyphase down sampling is illustrated in the Figure 5.

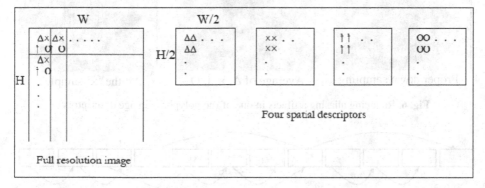

Fig. 5. Illustrating the concept of image polyphase downsampling.

A scalable solution based on such descriptors suffers from aliasing artifacts in the primary pictures (or base layer in this case) due to the lack of image filtering prior to down sampling. Hence this work proposes to replace the first descriptor (indicated by the 'Δ' samples) by the average of the four descriptors Δ, ×, 1 , O. This will provide a filtered and downsampled base layer which can be coded using the AVC primary pictures. Again, the rest of the descriptors are coded using redundant pictures. If all the descriptors are decoded then the original samples of the base layer descriptor can be recovered from the decoded average (in the primary pictures) and the '×', '1' and 'O' samples decoded from the redundant pictures.

For an alternative approach for filtering, an adaptive average can be used based on localized edge detection. In this case the 'Δ' samples are averaged with a predictor 'y' defined as:

$$y = \max(\times, 1) \quad \text{if } O \geq \max(\times, 1) \text{ or} \tag{1}$$
$$y = \min(\times, 1) \quad \text{if } O \leq \min(\times, 1) \text{ or}$$
$$y = \times + 1 - O \text{ (otherwise)}$$

Both methods of averaging and adaptive averaging generates similar results. However, it was noticed that the upsamling quality of the latter approach generated a higher PSNR (around 2 dB).

Figure 6 below visually shows the results of the proposed polyphase downsampling in comparison to the traditional approach. The aliasing artifacts on the background edges are evident in the descriptors generated from the 'x', 'l' and 'O' samples. However such aliasing affects are greatly attenuated in the averaged descriptor.

Figure 7 illustrates that similar to the inter-frame wavelet solution, the descriptors can be arranged into primary and redundant pictures in the AVC framework following the arrangement illustrated in Figure 1above. Notice that similar descriptors are grouped into one redundant picture group thus rendering the motion compensation process more efficient.

Proper down sampling   Average of Δ, ×, l, O.   All the 'x' samples

Fig. 6. Reducing aliasing artifacts in one of the polyphase image descriptors

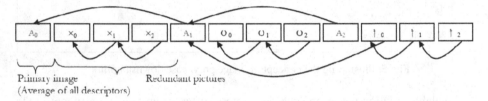

Primary image   Redundant pictures
(Average of all descriptors)

Fig. 7. Arrangement of image descriptors into primary and redundant pictures

5   Experimental Setup and Results

The experimental results used the following software; the JM reference software for (AVC) [13] and the JSVM reference software for SVC [14]. Both reference software are available on HHI institute, image and video coding website.

Figure 8 compares between the rate distortion curves of the proposed solutions against AVC single layer coding. Three test sequences are used; Crew and Harbour with a spatial resolution of 704x576 and IntoTree with a spatial resolutions of 1920x1080.

It is shown in the figure that in some cases the proposed inter-frame wavelet SVC solution outperforms single layer coding. In other cases, the proposed solution was slightly inferior to single layer coding. The figure also shows that the proposed SVC solutions slightly outperform the inter-frame wavelet coding based on redundant pictures (In the figure this is referred to as 'Proposed RP. AVC'). As mentioned previously, this is due to the fact that a redundant picture preceded by a primary picture will not be inter-frame coded. Again such pictures count for third of the redundant pictures.

Figure 9 on the other hand presents the results using the interpolation and decimation ideas of Figures 3 and 4 above. As for decimating the high frequency subbands prior to coding, the figure shows that a gain in PSNR was achieved for the Crew but not the Harbour sequence. This can be justified as follows. The Crew sequence is less spatially active than Harbour thus, the coarse representation of high frequencies by means of decimation means that more bits can be allocated to the low frequency band and therefore enhancing the overall image quality. In contrast, coarsely representing the high frequencies of the spatially active Harbour sequence has a counter effect on image quality.

Moreover, the figure shows that implementing the interpolation solution of Figure 3 above the opposite effect is observed. The Harbour sequence benefited from such a solution and the overall PSNR was higher than the proposed SVC solution. In conclusion it seems that the use of the interpolation and decimation techniques should be adaptive according to the spatial activity of the image content.

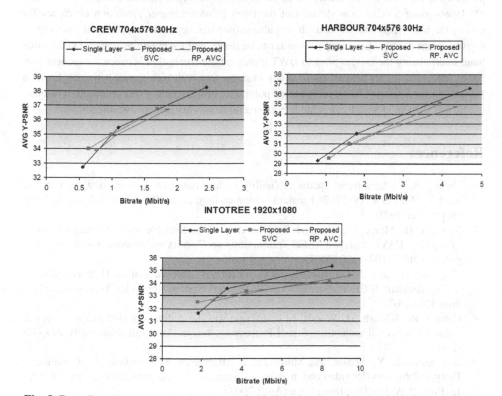

Fig. 8. Rate distortion curves for the proposed solutions in comparison to single layer coding

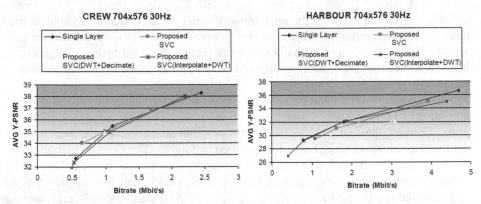

Fig. 9. Rate-distortion curves for the proposed interpolation and decimation solutions with inter-frame wavelet coding

6   Conclusion

This work proposed a number of novel arrangements for scalable video compression. It was proposed to high wavelet frequency subbands as either redundant pictures using AVC or scalable layers using SVC. It was shown that the latter provided higher prediction efficiency for coding the high frequency subbands. It was also shown that depending on the spatial activity of a given image the high frequency subbands can be decimated for bitrate reduction. On the other hand interpolating the images prior to DWT increased the correlation between subsequent subbands leading to higher prediction efficiency in sequences with high spatial activities. Lastly a framework for a solution based on modified polyphase downsampling was proposed. It is anticipated that such an approach can achieve both spatial scalability and error resiliency.

References

1. Hsiang, S.-T.: Intra-frame spatial scalability coding based on a subband/wavelet framework for MPEG-4 AVC/H.264 scalable video coding. In: ICIP 2007, San Antonio, Texas (September 2007)
2. Schwarz, H., Marpe, D., Wiegand, T.: Overview of the Scalable Video Coding Extension of the H.264/AVC Standard. IEEE Transactions on Circuits and Systems for Video Technology 17(9), 1103–1120 (2007)
3. Wiegand, T., Sullivan, G., Bjontegaard, G., Luthra, A.: Overview of the H.264/AVC video coding standard. IEEE Transactions on Circuits and Systems for Video Technology 13(7), 560–576 (2003)
4. Ohm, J.-R., Schaarb, M., Woods, J.: Interframe wavelet coding—motion picture representation for universal scalability. Signal Processing: Image Communication 19(9), 877–908 (2004)
5. Andreopoulos, Y., Schaar, M., Munteanu, A., Barbarien, J., Schelkens, P., Cornelis, J.: Fully scalable wavelet video coding using in-band motion-compensated temporal filtering. In: Proc. ICASSP 2003, Hong kong (April 2003)

6. Park, H.-W., Kim, H.-S.: Motion Estimation Using Low-Band-Shift Method for Wavelet-Based Moving Picture Coding. IEEE Trans. Image Processing 9(4), 577–587 (2000)
7. Andreopoulos, Y., Munteanu, A., Auwera, G., Schelkens, P., Cornelis, J.: A new method for complete-to-overcomplete discrete wavelet transforms. In: Proc. of 14th International Conference on Digital Signal Processing, vol. 2, pp. 501–504 (2002)
8. Li, X.: New results of phase shifting in the wavelet space. IEEE Signal Processing Letters 10(7), 193–195 (2003)
9. Caramma, M., Fumagalli, M., Lancini, R.: Polyphase down sampling Multiple Description Coding for IP Transmission. In: Proc. SPIE, vol. 4310, pp. 545–552 (December 2000)
10. Franchi, N., Fumagalli, M., Lancini, R., Tubaro, S.: A space domain approach for multiple description video coding. In: Proc. ICIP 2003, Barcelona, Spain (September 2003)
11. Gallant, M., Shiranit, S., Kossentiniq, F.: Standard-compliant multiple description video coding. In: ICIP 2001, Thessaloniki, Greece, vol. 1, pp. 946–949 (2001)
12. Bernardini, R., Durigon, M., Rinaldo, R., Celetto, L., Vitali, A.: Polyphase spatial sub-sampling multiple description coding of video streams with H264. In: Proc. ICIP 2004, Singapore, October, vol. 2, pp. 3213–3216 (2004)
13. JM reference software for H.264/MPEG-4 AVC is a video coding standard, HHI institute, http://iphome.hhi.de/suehring/tml/
14. JSVM reference software for the Scalable Video Coding (SVC), HHI institute, http://www.hhi.fraunhofer.de/fields-of-competence/image-processing/research-groups/image-video-coding/svc-extension-of-h264avc/jsvm-reference-software.html

Image Enhancement Using Geometric Mean Filter and Gamma Correction for WCE Images

Shipra Suman[1], Fawnizu Azmadi Hussin[1], Aamir Saeed Malik[1], Nicolas Walter[1], Khean Lee Goh[2], Ida Hilmi[2], and Shiaw hooi Ho[2]

[1] Electrical and Electronic Engineering, Universiti Teknologi PETRONAS, 31750 Tronoh, Perak, Malaysia
[2] Department of Medicine, university of Malaya, 50603, Kuala Lumpur, Malaysia
suman.shipra@ieee.org, {fawnizu,aamir_saeed}@petronas.com.my, {walter.nicolas.pro,klgoh56}@gmail.com, i_hilmi@um.edu.my, shooiho@yahoo.com

Abstract. The application of image enhancement technology to Wireless capsule Endoscopy (WCE) could extremely boost its diagnostic yield. WCE based detection inside gastrointestinal tract has been carried out over a great extent for the seek of the presence of any kind of etiology. However, the quality of acquired images during endoscopy degraded due to factors such as environmental darkness and noise. Hence, decrease in quality also resulted into poor sensitivity and specificity of ulcer and diagnosis. In this paper, a method based on color image enhancement through geometric mean filter and gamma correction is proposed. The developed method used geometric mean filtering to reduce Gaussian noise present in WCE images and achieved better quality images in contrast to arithmetic mean filtering, which has blurring effect after filtration. Moreover, Gamma correction has been applied to enhance small details, texture and contrast of the images. The results shown improved images quality in terms of SNR (Signal to Noise Ratio) and PSNR (Peak Signal to Noise Ratio) which is beneficial for automatic detection of diseases and aids clinicians to better visualize images and ease the diagnosis.

Keywords: Wireless capsule endoscopy (WCE), image enhancement, geometrical means filter, gamma correction.

1   Introduction

Wireless Capsule Endoscopy (WCE) is a recent technique (approved by Food and Drug Administration (FDA) in 2002)[1] that allows clinicians to inspect gastrointestinal (GI) tract. Earlier imaging modalities such as upper gastrointestinal endoscopy, colonoscopy, and push enteroscopy allowed examining the stomach, duodenum, colon and terminal ileum. However, these techniques are long procedures as they require preparation of the patient and are painful. Moreover, most of small intestinal parts could not be observed without performing surgery that is invasive [2, 3]. In 2000, a short paper published in nature [4] introduced an advanced form of endoscopy,

C.K. Loo et al. (Eds.): ICONIP 2014, Part III, LNCS 8836, pp. 276–283, 2014.

i.e. WCE, designed by the company Given Imaging® [5]. The apparatus employs re-modeling diagnosis process for GI tract and visualizes entire small intestine without sedation, pain or air insufflation. Hence, as the technology is not invasive, it has been promptly adopted by many practitioners and hospitals. Up to now, WCE has been used to detect many diseases [6-8] like small intestinal blooding, Crohn's diseases, ulcers, tumors, vascular lesions and colon cancers. It has also been reported by Given Imaging® that over 1,000,000 patients worldwide have already enjoyed the benefits of this device.

Fig. 1 describes WCE pill-shaped device including short-focal-length CMOS camera, light source, battery and radio transmitter. It is swallowed by the patient after about 12 hours fasting. This miniature device propelled by peristalsis of GI tract begins to work and record images at 2 frames per second while moving forward along the GI tract. At the same time, images are sent to a data recorder attached to the patient's waist wirelessly.

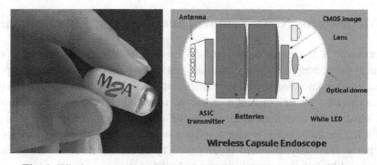

Fig. 1. Wireless capsule endoscopy capsule and its component [9]

The whole inspection process takes about 8 hours, before the image data can be processed. Finally, a physician performs analysis by watching the recorded data in the form of either video or images. However, the diagnosis process is time-consuming due to the huge amount of data (about 50,000 useful images per inspection). There-fore, the diagnosis is not a real-time process, making this situation a potential break-through for off-line post processing and computer aided detection.

Although clinical findings on WCE are encouraging, there still remains for a large gap for improving the automation [10]. For example, to reduce the image acquisition time, Olympus has been investigating a new generation of WCE such as self-propel capsule endoscopy [11]. One of the great challenges with the present WCE system is the image's quality. Qualities of WCE images are not ideal due to the following rea-sons. First, in order to reduce the communication bandwidth and save power, WCE images are not very clear due to high compression ratio [8]. Secondly, though CMOS image sensor has advantages of low power consumption and superior integration, the image quality it produced is not as good as that of CCD imagers [12]. Furthermore, the resolution of WCE image is only 256×256 due to volume limitation of encapsula-tion, especially power limitation, whereas traditional endoscopy has a superior performance on this aspect since no power limitation exists. Moreover, bad imaging conditions such as low illumination and complex circumstances in the GI tract will further deteriorate the quality of produced images. Finally, the short-focal-length camera pictures a low depth of focus, i.e. effects of depth will produce blurring.

The proposed method aims at enhancing the WCE images in order to improve the low illumination problem, sharpening the blurred parts and reduce the noise.. For the purpose of computer aided design (CAD) to ease the diagnosis of physicians [13], image filtering to reduce noise and gamma correction to enhance contrast of target images has been studied. Moreover, this method focused on local property of WCE images, leading to details enhancement. Results exhibit better performances of enhancement than conventional methods so as to assist diagnosis of physicians.

In the next Section is dedicated to the review of image enhancement techniques for noise removal and contrast enhancement. Methodology will be presented in Section 3 along with qualitative assessment. Section 4 provides experimental results and validation of the proposed methodology for real WCE sequences. Finally, Section 5 concludes the paper and presents future works.

2   Methods for Noise Removal and Contrast Enhancement

Noise removal is the process of removing noise from the image. Noise reduction techniques are conceptually very similar, regardless of the image being processed. However, a prior knowledge of the characteristics of the expected images gives better inference on the type of noise and eases the implementation of noise removal techniques. Mostly, the encountered noise in the acquired data exhibits a Gaussian-like distribution. Gaussian noise is characterized by his additive and zero-mean distribution property. Basically, the zero-mean property of the distribution allows such noise to be removed by locally averaging pixel values [14].

Contrast enhancement techniques improve the perception of objects in the scene by strengthen the brightness difference between objects and background. Contrast enhancements are typically performed as a contrast stretch followed by a tonal enhancement, although these could both be performed in one step. A contrast stretch improves the brightness differences uniformly across the dynamic range of the image, whereas tonal enhancements improve the brightness differences in the shadow (dark), midtone (grays), or highlight (bright) regions at the expense of the brightness differences in the other regions.

2.1   Geometric Mean Filtering

Additive white Gaussian noise is a standard model which is present in WCE images. It is an idealized form of white noise, which is caused by random fluctuations in the signal [15] in color cameras where more amplification is used in blue channel other than green and red channel. While facing Gaussian noise, each pixel of the image will be affected.

Noise is an unavoidable side effect. Fig. 2 describes the filtering process. It separates the red, green and blue channels. It is followed by introducing a gain to compensate the attenuation resulting from the filter. Each filtered channel is then combined to form resulting colored image.

Geometric mean filter replaces the colour value of each pixel with the geometric mean of colour pixel values from a larger region surrounding it, based on filter size (3x3 or 5x5) and yields a stronger filter effect. The geometric mean filter performed better than conventional methods such as arithmetic filters to remove Gaussian type noise and preserve edge features [16].

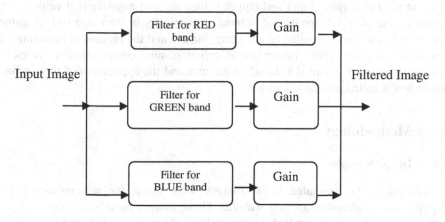

Fig. 2. Filteration of three band separately

Geometric filter is a simplest form of mean filter. Let's S_{xy} represents the set of coordinates in a sub-image window (neighborhood) of size m x n where m and n are equal, centered at point (x, y). The local image function f(x, y) is filtered image and g(s,t) is input image. In Geometric mean filter each restored pixel is given by the product of the pixels in the sub-image window, raised to the power 1/mxn as described in (1).

$$f(x, y)=[\pi(s,t)\epsilon \, Sxy \, g(s,t)]^{1/mn}$$ (1)

The purpose is to produce more objective images (ideally noiseless) for particular application than the original images and hence, increase the accuracy of further algorithms, making them more similar to the characteristics of human visual recognition system.

2.2 Gamma Correction

For color space transformation, the absolute separation between chrominance and luminance components is not achievable due to the cross talk of colour channels, i.e. colors are correlated. Compared to the above methods, gamma correction method has some advantages to overcome the effects of light distortion. However, it is often difficult to select suitable gamma values without a prior knowledge about illumination and the texture details are often lost because of over correction [17, 18]. Moreover, the varieties of images greatly challenge the performance of the traditional Gamma Correction Model (GCM) in applications.[19]

Gamma correction is a nonlinear operation used to encode the luminance in image systems. Gamma correction can be described for simple model, as follows:

$$V_{out} = A V_{in}^{\gamma} \qquad (2)$$

Where V_{in} is the input original image, V_{out} is the output corrected image and A is a constant used as a gain. Input and output values are non-negative real values; in the common case of $A = 1$, inputs and outputs are typically in the range 0–1. A gamma value $\gamma < 1$ is sometimes called an encoding gamma and the process of encoding with this nonlinear compressive power-law is called gamma compression; conversely a gamma value $\gamma > 1$ is called a decoding gamma and the application of the nonlinear power-law is called gamma expansion.

3   Methodology

3.1   Image Samples

In this paper, 11 annotated WCE images were taken for pre-processing from http://www.capsuleendoscopy.org website. These images have been labelled by experts to be used as gold standard during analysis. They labelled ulcerated and bleeding areas in each frame. We used these images along with more image samples for validation purpose in further experiment. Fig. 3 shows some samples of WCE images. Image 1 and image 4 have ulcerations highlighted by the blue ellipses whereas image 2 and 3 has bleeding underlined by the yellow ellipses.

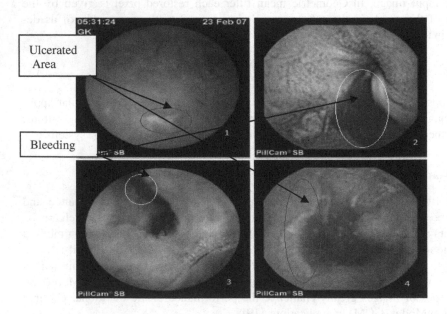

Fig. 3. Sample of WCE images

3.2   Image Enhancement

For image enhancement (see Fig. 4), filters such as geometric, harmonic mean and laplacian have been applied to improve the contrast. These WCE images have Gaussian noise, so mean filters are more suitable to remove this noise. Contrast stretching is performed by gamma correction of images after filtration, according to the method described in Section 2.

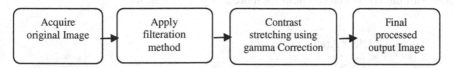

Fig. 4. Flow Chart of Image Enhancement process for WCE Images

3.3   Qualitative Analysis

To measure the quality degradation of an available distorted image with reference to the original image, a class of quality assessment metrics called full reference (FR) is considered. It can perform distortion measure having full access to the original image. The quality assessment metrics are estimated through computation of MSE (mean square error), RMSE (root mean square error), SNR (Signal to noise ratio) and PSNR (peak signal to noise ratio) using their standard formula for imaging.

4   Results and Discussions

Fig. 5 shows SNR and PSNR for the 11 reference images before and after noise removal. Here, SNR_1 illustrate output before filtration and SNR_2 after filtration. One can observe that SNR is increased, showing improvement in the quality of resulting images.

The size of the images used was 576×576 and the noise parameter was compared using SNR and PSNR. The 11 sample images are not taken from only one capsule endoscopy. Hence, the variations depicted in Fig. 5.

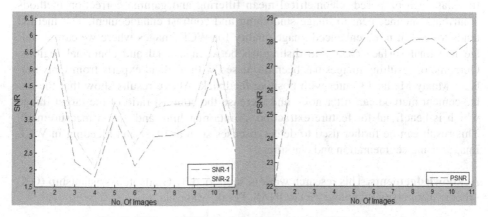

Fig. 5. SNR and PSNR, before (red) and After Filtering (green)

In Fig. 6, we can visualize effects of geometrical mean filtering and gamma correction at experimental level. The image on the left side is original WCE image with noise which is filtered by geometric mean filter to reduce noise and finally gamma correction is used to enhance contrast. Each pixel value has been changed as per suitable procedural filter, leading to more refined output images. If we visualize final images, we can see clearly more erythema patches which are not that clear in original image. Noise reduction also helps to make image quality better. It helps to visualize more villi pattern which can help for more accurate diagnosis.

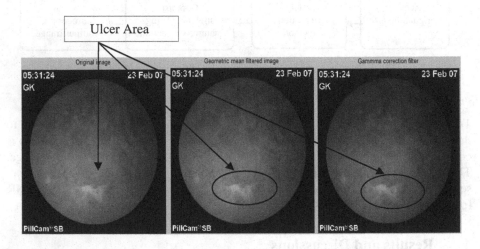

Fig. 6. Original Image (left), Geometric mean filtered image (centre) and gamma corrected image (right)

5   Conclusion

An image enhancement method based on geometric mean filtering and gamma correction has been proposed. Geometrical mean filtering and gamma correction methods contribute enhancement to image smoothing and contrast enhancement. The method tends to lead in more enhanced image quality for WCE images where we can visualize intestinal surface clearly to distinguish between normal and abnormal regions. Outcome of resulting images has been discussed with medical experts from University of Malay Medical Center with positive feedback. Above results show that the enhancement method can filter noise and increase the contrast ratio of the target image which is beneficial for feature extraction, points matching and vision measurement. This result can be further used to detect diseases such as ulcer and bleeding in WCE images using segmentation and classifiers.

Acknowledgements. This research work is supported by Graduate Assistantship (GA) scheme, Universiti Teknologi PETRONAS, Perak, Malaysia.

References

1. Chen, Y., Lee, J.: Ulcer detection in wireless capsule endoscopy video. In: Proceedings of the 20th ACM International Conference on Multimedia, pp. 1181–1184. ACM (2012)
2. Hwang, S., Celebi, M.E.: Polyp detection in wireless capsule endoscopy videos based on image segmentation and geometric feature. In: 2010 IEEE International Conference on Acoustics Speech and Signal Processing (ICASSP), pp. 678–681. IEEE (2010)
3. Penna, B., et al.: A technique for blood detection in wireless capsule endoscopy images. In: Proc of the 17th European Signal Processing Conference (EUSIPCO 2009), pp. 1864–1868. Citeseer, Glasgow (2009)
4. Iddan, G., et al.: Wireless capsule endoscopy. Nature 405, 417 (2000)
5. (2014), http://www.givenimaging.com/en-us/Innovative-Solutions/ Capsule-Endoscopy/Pillcam-SB/Pages/default.aspx (cited March 30, 2014)
6. Ge, Z.-Z., Hu, Y.-B., Xiao, S.-D.: Capsule endoscopy in diagnosis of small bowel Crohn's disease. World Journal of Gastroenterology 10(9), 1349–1352 (2004)
7. Lee, D., Poon, A., Chan, A.: Diagnosis of small bowel radiation enteritis by capsule endoscopy. Hong Kong Medical Journal= Xianggang Yi Xue Za Zhi/Hong Kong Academy of Medicine 10(6), 419–421 (2004)
8. Xiang, X., Li, G.-L., Wang, Z.-H.: Low-complexity and high-efficiency image compression algorithm for wireless endoscopy system. Journal of Electronic Imaging 15(2), 023017-1–023017-15 (2006)
9. Li, B., Meng, M.Q.-H.: Wireless capsule endoscopy images enhancement via adaptive contrast diffusion. Journal of Visual Communication and Image Representation 23(1), 222–228 (2012)
10. Meng, M.-H., et al.: Wireless robotic capsule endoscopy: state-of-the-art and challenges. In: Fifth World Congress on Intelligent Control and Automation, WCICA 2004, pp. 5561–5565. IEEE (2004)
11. Woo, S.H., et al.: Small intestinal model for electrically propelled capsule endoscopy. Biomed. Eng. Online 10, 108 (2011)
12. Holms, A., Quach, A.: Complementary Metal-Oxide Semiconductor Sensors (2010)
13. Feng, L., et al.: Automatic image enhancement based on multi-scale image decomposition. In: Fifth International Conference on Graphic and Image Processing. International Society for Optics and Photonics (2014)
14. Bini, A., Bhat, M.: A nonlinear level set model for image deblurring and denoising. The Visual Computer, 1–15 (2013)
15. McAndrew, A.: An introduction to digital image processing with matlab notes for scm2511 image processing. School of Computer Science and Mathematics, pp. 1–264. Victoria University of Technology (2004)
16. Hanumantharaju, M., et al.: Adaptive color image enhancement based geometric mean filter. In: Proceedings of the 2011 International Conference on Communication, Computing & Security, pp. 403–408. ACM (2011)
17. Guan, X., et al.: An image enhancement method based on gamma correction. In: Second International Symposium on Computational Intelligence and Design, ISCID 2009, pp. 60–63. IEEE (2009)
18. Cheng, Y., Wang, Y., Hu, Y.: Image enhancement algorithm based on Retinex for Small-bore steel tube butt weld's X-ray imaging. WSEAS Transactions on Mathematics 8(7), 279–288 (2009)
19. Liu, C., et al.: Gaussian fitting for carotid and radial artery pressure waveforms: comparison between normal subjects and heart failure patients. Bio-medical Materials and Engineering 24(1), 271–277 (2014)

Autoencoder-Based Collaborative Filtering

Yuanxin Ouyang[1,2], Wenqi Liu[1], Wenge Rong[1,2], and Zhang Xiong[1,2]

School of Computer Science and Engineering, Beihang University, China
Research Institute of Beihang University in Shenzhen, China
{oyyx@,wqliu@cse.,w.rong@,xiongz@}buaa.edu.cn

Abstract. Currently collaborative filtering is widely used in recommender systems. With the development of idea of deep learning, a lot of researches have been conducted to improve collaborative filtering by integrating deep learning techniques. In this research, we proposed an autoencoder based collaborative filtering method, in which pretraining and stacking mechanism is provided. The experimental study on commonly used MovieLens datasets have shown its potential and effectiveness in getting higher recall.

1 Introduction

With the explosion of information on the Internet, people relied on more and more recommender systems to solicit suggestions and/or make decisions, thereby solving the information overload problem. A lot of recommendation related techniques have been proposed and a notable one is collaborative filtering [1], which is widely lauded as a practical method for providing recommendations by utilising users' preference history to predict future preference. Generally, algorithms for collaborative filtering can be roughly divided into two general classes, i.e., memory-based and model-based approaches. Memory-based methods try to predict users' preference based on the ratings by other similar users, while model-based methods mainly rely on a prediction model by using Clusetering, Baysesian network and etc [3].

Currently, with the development of concept of deep learning, a new research area and has proven its success in speech and image recognition [4], researchers started to try to employ the inspiration of deep learning into collaborative filtering based recommender systems. For example, Salakhutdinov et al. proposed an approach employing Restricted Boltzmann Machines (RBM) [12] and Georgiev et al. further extended the original RBM-based model to a unified non-IID framework [5]. Truyen et al. explored joint modelling of users and items for collaborative filtering, but inside is an unrestricted version of Boltzmann Machines (BMs) [10]. Oord et al. used deep convolutional neural networks to provide music recommendation [9]. Gunawardana et al. described a tied Boltzmann Machine combining collaborative and content information [7].

Deep learning is also called feature learning due to its powerful ability to learn feature representations automatically. Besides, deep models can learn high-order features of input data which may be useful for recommendation as indicated in

C.K. Loo et al. (Eds.): ICONIP 2014, Part III, LNCS 8836, pp. 284–291, 2014.

[12]. Inspired by previous work, in this research we tried to employ another neural network model, autoencoder, into the collaborative filtering task. Experimental study on commonly used datasets is also conducted to present its potential and effectiveness.

The rest of the paper is organised as following. Section 2 will introduce the related work about collaborative filtering models and basic autoencoder. Then a modified autoencoder based collaborative filtering model will be illustrated in section 3. Section 4 will discuss the experimental study and Section 5 will conclude this paper and point out possible future work.

2 Related Work

Early approaches for collaborative filtering assume that similar users have similar interests, i.e., nearest neighbourhood based methods, which is normally called memory-based approaches. However, memory-based approaches do not scale well because they require access to the ratings of the entire set. Furthermore, there is another challenge that ratings are severe sparse making memory-based approaches perform unsatisfied. To overcome this challenges, model-based approaches such as singular value decomposition (SVD) have been proposed [6]. However, application of matrix factorization to sparse ratings matrices is still a non-trivial challenge. As such, Hoffman proposed a formal statistical model of user preferences using hidden variables over user-item-rating triplets [8].

Except for memory-based and model-based approaches, recently an alternative methods using idea of deep learning has been attached much importance. Among them autoencoder is a widely used deep learning model. Suppose we have only unlabelled training examples set $\{x^{(1)}, x^{(2)}, ...\}$, where $x^{(i)} \in \Re^n$. An autoencoder neural network is an unsupervised learning algorithm that applies backpropagation, setting the target values to be equal to the inputs. I.e., it uses $y^{(i)} = x^{(i)}$, as shown in Fig. 1.

The autoencoder tries to learn a function $h_{W,b}(x) \approx x$. In other words, it is trying to learn an approximation to the identity function, so as to output $\hat{x}$ which is similar to x. By placing constraints on the network, such as limiting the number of hidden units, interesting structure can be discovered about the data. For instance, if some of the input features are correlated, then this algorithm will be able to discover some of those correlations.

3 Autoencoders for Collaborative Filtering

3.1 Modeling User-Item Ratings

Suppose there are N users, M movies and integer ratings from 1 to K. An important problem in applying autoencoders to movie ratings is how to cope with the missing ratings efficiently. We cannot simply substitute missing values with 0 because the model will think that user give a rating of 0 and learn the negative preference, which is not the truth. In this paper we use a different autoencoder for each user, as shown in Fig. 2. Every autoencoder has the same number of hidden units, but an autoencoder only has input units for the movies rated by that user. As a result an autoencoder has few connections if that user rated few movies. Each hidden unit could then learn to model a significant dependency between the ratings of different movies. Each autoencoder only has a single training case, but all of the corresponding weights and biases are tied together. If two users have rated the same movie, their two autoencoders must use the same weights between the softmax input/output units for that movie and the hidden units. To simplify the notation, we will now concentrate on getting the gradients for the parameters of a single user-specific autoencoder. The full gradients with respect to the shared weight parameters can then be obtained by averaging over all N users.

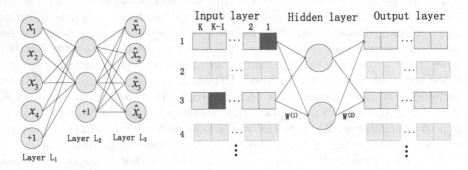

Fig. 1. Architecture of autoencoder

Fig. 2. An user-specific autoencoder for collaborative filtering

Learning. Suppose a user rated m movies. Let $a^{(1)}$ be a $K \times m$ observed binary indicator matrix with $a_i^{k(1)} = 1$ if the user rated movie i as k and 0 otherwise. We also let $a_j^{(2)}$, $j = 1, ..., F$, be the values of hidden variables. Here we choose activation function to be the sigmoid function. In feedforward step, the only difference is that because the output layer is of the same structure as the input layer, we compute the output unit $a_i^{k(3)}$ as:

$$a_i^{k(3)} = \frac{f(\sum_j w_{ij}^{k(2)} a_j^{(2)} + b_i^{k(2)})}{\sum_k f(\sum_j w_{ij}^{k(2)} a_j^{(2)} + b_i^{k(2)})}. \tag{1}$$

where $w_{ij}^{k(2)}$ denotes the weight associated with connection between $a_j^{(2)}$ and $a_i^{k(3)}$, $b_i^{k(2)}$ is the bias of $a_i^{k(3)}$. The denominator is the normalisation term which insures $\sum_k a_i^{k(3)} = 1$.

In backpropagation step, for a single training example (x, y), we define the cost function $J(w, b; x, y)$ to be squared-error function. Then given a training set of m examples, we define the overall cost function as:

$$J(w,b) = \frac{1}{m} \sum_i J(w,b;x^i,y^i) + \frac{\lambda}{2} \parallel w \parallel^2 . \qquad (2)$$

The first item in the definition of $J(w,b)$ is an average sum-of-squares error term. the second term is a regularization term (also called a weight decay term) which tends to decrease the magnitude of the weight and helps prevent overfitting problem. Our goal is to minimise the total cost function $J(w,b)$. Here we train our autoencoder using batch gradient descent.

Making Predictions. Given the training set of one user and a new query item q, we can initialise the input layer of the autoencoder with known ratings and carry out feedforward step. Specific units $a_q^{k(3)}$ in the output layer represents the probability which item q will be rated value k. As such the expected rating for item q is computed as:

$$r_q = \sum_k k \cdot a_q^{k(3)}. \qquad (3)$$

3.2   Initialisation of Parameters

As the optimisation problem of neural networks is nonconvex, the standard way to train autoencoders using backpropagation to reduce the reconstruction error is difficult to optimise the weights. Autoencoders with random small initial weights typically find poor local minima. A popular solution to this problem is greedy "pretraining" procedure. In this paper we use a two-layer network called Restricted Boltzmann Machine (RBM) to pretrain autoencoders. A RBM is a specific type of undirected bipartite graphical model consisting of two layers of binary variables: hidden and visible with no intra-layer connections. Training an autoencoder with RBM pretraining takes the following steps:

1) Train a RBM with analogous structure of autoencoder using input data.

2) Use the trained parameters of RBM to initialize corresponding weights and biases of autoencoder.

3) Fine-tune the weights using Backpropagationfor for optimal reconstruction of each user's ratings.

The key idea is that the greedy learning algorithm will perform a global search for a good, sensible region in the parameter space [11]. Therefore, with this pretraining, we will have a good data reconstruction model. Backpropagation is better at local fine-tuning of the model parameters than global search. So further training of the entire autoencoder using backpropagation will result in a good local optimum.

3.3   Deep Generative Models

A stacked autoencoder is a neural network consisting of multiple layers of autoencoders in which the outputs of each layer is wired to the inputs of the successive layer. A good way to obtain good parameters for a stacked autoencoder is to use greedy layer-wise training [2]. To do this, first we train the first layer on raw input to obtain parameters and transform the raw input into a vector consisting of activation of the hidden units. The second layer is then trained on this vector. Repeat for subsequent layers, using the output of each layer as input for the subsequent layer. While training each layer, we can also use RBM pretraining method to get better local optimum as mentioned in the previous subsection.

This method trains the parameters of each layer individually while freezing parameters for the remainder of the model. To produce better results, after training phase is complete, fine-tuning using backpropagation can be used by tuning the parameters of all layers are changed at the same time.

A stacked autoencoder enjoys all the benefits of any deep network of greater expressive power. Further, it often captures a useful hierarchial grouping or part-whole decomposition of the input. The first layer of a stacked autoencoder tends to learn first-order features of the raw input. Higher layers tend to learn even high-order features corresponding to patterns of previous-order features.

4   Experiments and Discussion

4.1   Datasets and Evaluation Metrics

We evaluated the above-described autoencoders on two MovieLens datasets, which are commonly used for evaluating collaborative filtering algorithms.

The first dataset (MovieLens 100k) consists of 100,000 ratings for 1,682 movies assigned by 943 users, while the second one (MovieLens 1M) contains one million ratings for 3,952 movies by 6,040 users. Each rating is an integer between 1 (worst) to 5 (best). For both datasets, we use 80% to make training set and others to be testing set.

To evaluate the proposed method, we use both mean absolute error (MAE) and root mean squared error (RMSE). MAE measures the deviation of the predicted values p_i from their true ratings r_i, which computes the absolute difference over all N pairs. Compared with MAE, RMSE gives more weights for prediction with bigger errors. The evaluation rules are in the following form:

$$MAE = \frac{\sum_{i=1}^{N} |r_i - p_i|}{N} \qquad RMSE = \sqrt{\frac{\sum_{i=1}^{N} (r_i - p_i)^2}{N}} \qquad (4)$$

4.2   Results and Discussion

Fig. 3 shows the dependency of MAE on the number of units in the hidden layer when the autoencoders are trained for 200 epochs. We can see that models get lower MAE values with the increasing number of hidden units, which is not obvious after hidden units are more than 120. It can be imagined that overfitting will become an issue with continuously increasing number of hidden units. Next, Fig. 4 shows the dependency of MAE on the number of epochs when the autoencoders are trained for 100 hidden units. The curve is similar to the previous one. After the number of training epochs is larger than 250, the MAE value stays relatively stable. Training epochs needed to acquire stable MAE can be different if learning rate is changed or with stochastic gradient descendent.

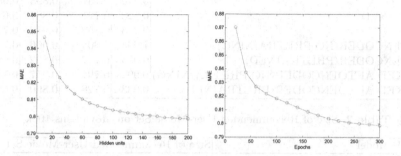

Fig. 3. Dependency of MAE on the Number of Hidden Units

Fig. 4. Dependency of MAE on the Number of Epochs

We further compared the prediction quality achieved by different methods using the MovieLens 100k and 1M datasets, respectively. Apart from the autoencoders described before, other collaborative filtering approaches based on nearest neighbours, SVD and RBM are included.

Table 1 shows the MAE and RMSE values of some basic models and autoencoder models on both datasets. From the results on MovieLens 100k, it can be seen that autoencoders without pretraining do not perform well enough, while RBM-pretrained autoencoders have similar performance with nearest neighbors and SVD models. Besides, the results of stacked autoencoders are slightly superior to autoencoders. But the gap is not obvious.

Evaluation measures on MovieLens 1M are shown on the right side. Apart from that models perform better than those on MovieLens 100k, the trend of results do not have big differences.

Finally, experiments are made to test the overlap degree between recommended user-movie sets from autoencoders and other CF models. We define recommended movie as that whose predict and real ratings are both higher than 4. Under this condition, the prediction is precise and users have positive references over these movies. Statistical data on MovieLens 100k are shown in Table 2.

USER-BASED & AUTOENCODER represents the intersection of recommended user-movie sets between user-based CF model and autoencoder model. Comparing the first, fourth and fifth row, we can find that there is still a large number of recommended movies beyond the intersection. Same phenomenon appears between autoencoders and other CF models.

After investigating the experimental result, some interesting findings can be revealed:

Table 1. Prediction Quality on MovieLens Dataset

CF Model	MovieLens 100k		MovieLens 1M	
	RMSE	MAE	RMSE	MAE
USER-BASED CF	0.937	0.736	0.915	0.709
ITEM-BASED CF	0.932	0.732	0.901	0.698
SVD	0.940	0.737	0.893	0.684
BIASED-SVD	0.926	0.721	0.887	0.681
RBM	0.953	0.752	0.918	0.710
AUTOENCODER(NO PRETRAINING)	1.004	0.804	0.966	0.754
AUTOENCODER(PRETRAINED)	0.939	0.737	0.892	0.688
STACKED AUTOENCODER(NO PRETRAINING)	0.992	0.791	0.957	0.747
STACKED AUTOENCODER(PRETRAINED)	0.933	0.728	0.890	0.684

Table 2. Size of Recommended User-Movie Set on MovieLens 100k

CF Model	Size of Recommended User-Movie Set
USER-BASED	3244
ITEM-BASED	3299
SVD	3316
AUTOENCODER	2622
USER-BASED & AUTOENCODER	1880
ITEM-BASED & AUTOENCODER	2138
SVD & AUTOENCODER	2056

1) Autoencoders are effective models for collaborative filtering as they have no worse performance than basic methods.

2) Pretraining with RBMs do make autoencoders get better local optimum as the results improve a lot.

3) Stacked autoencoders are superior, but not enough with small rating dataset alone which do not have enough high-order information.

4) Prediction quality of autoencoders remains consistent when the amount of training data increases by an order of magnitude, which is a good indication for potential practical applicability.

5) The results of autoencoders can be merged with other methods to get a higher recall without reducing precision. It is good news for that we usually combine different models in real circumstances but not use single model.

5   Conclusion and Future Work

In this paper we proposed a revised autoencoder models for collaborative filtering. To acquire better performance, we tried some improvements such as pretraining with RBM and stacking autoencoders together. Experimental study has been conducted on two commonly used datasets to prove that those models are effective and can be integrated with other CF models to get a higher recall.

There are several extensions to be considered. First our current models focus on modelling the correlation between item ratings. We can generate a similar model focusing on user ratings and then combine them together to get a better performance. Besides, we can introduce some content-based features into the model so that deep models may acquire more high-order information.

Acknowledgements. This work was partially supported by the National Natural Science Foundation of China (No. 61103095), the International S&T Cooperation Program of China (No. 2010DFB13350), and the Fundamental Research Funds for the Central Universities. We are grateful to Shenzhen Key Laboratory of Data Vitalization (Smart City) for supporting this research.

References

1. Adomavicius, G., Tuzhilin, A.: Toward the next generation of recommender systems: A survey of the state-of-the-art and possible extensions. IEEE Transactions on Knowledge and Data Engineering 17(6), 734–749 (2005)
2. Bengio, Y., Lamblin, P., Popovici, D., Larochelle, H., et al.: Greedy layer-wise training of deep networks. Advances in Neural Information Processing Systems 19, 153 (2007)
3. Breese, J.S., Heckerman, D., Kadie, C.M.: Empirical analysis of predictive algorithms for collaborative filtering. In: Proceedings of 14th International Conference on Uncertainty in Artificial Intelligence, pp. 43–52 (1998)
4. Deng, L.: An overview of deep-structured learning for information processing. In: Proceedings of Asian-Pacific Signal and Information Processing–Annual Summit and Conference (2011)
5. Georgiev, K., Nakov, P.: A non-iid framework for collaborative filtering with restricted boltzmann machines. In: Proceedings of the 30th International Conference on Machine Learning, pp. 1148–1156 (2013)
6. Goldberg, K., Roeder, T., Gupta, D., Perkins, C.: Eigentaste: A constant time collaborative filtering algorithm. Information Retrieval 4(2), 133–151 (2001)
7. Gunawardana, A., Meek, C.: Tied boltzmann machines for cold start recommendations. In: Proceedings of the 2008 ACM Conference on Recommender Systems, pp. 19–26. ACM (2008)
8. Hofmann, T.: Latent semantic models for collaborative filtering. ACM Transactions on Information Systems 22(1), 89–115 (2004)
9. van den Oord, A., Dieleman, S., Schrauwen, B.: Deep content-based music recommendation. In: Advances in Neural Information Processing Systems, pp. 2643–2651 (2013)
10. Phung, D.Q., Venkatesh, S., et al.: Ordinal boltzmann machines for collaborative filtering. In: Proceedings of the 25th Conference on Uncertainty in Artificial Intelligence, pp. 548–556. AUAI Press (2009)
11. Salakhutdinov, R., Hinton, G.: An efficient learning procedure for deep boltzmann machines. Neural Computation 24(8), 1967–2006 (2012)
12. Salakhutdinov, R., Mnih, A., Hinton, G.: Restricted boltzmann machines for collaborative filtering. In: Proceedings of the 24th International Conference on Machine Learning, pp. 791–798. ACM (2007)

Extended Laplacian Sparse Coding for Image Categorization

Mouna Dammak, Mahmoud Mejdoub, and Chokri Ben Amar

REGIM: REsearch Groups on Intelligent Machines
University of Sfax, National School of Engineers (ENIS)
Department of Electrical Engineering
Sfax, 3038, Tunisia
{mouna.damak,chokri.benamar}@ieee.org,
mah.mejdoub@gmail.com

Abstract. In image classification task, several recent works show that sparse representation plays a basic role in dictionary learning. However, this approach neglects the spatial relationships in the image space during dictionary learning. However, this approach neglects the neighboring relationship in dictionary learning. To alleviate the impact of this problem, we propose a novel dictionary learning based on Laplacian sparse coding method that profits from the neighboring relationship among the local features. For that purpose, we incorporate the matching between local regions in the Laplacian sparse coding formula. Moreover, we integrate statistical analysis of the distribution of the responses of each local feature to the dictionary basis in the final image representation. Our experimental results prove that our method performs existing background results based on sparse representation.

Keywords: Bag of visual words, Sparse coding, Image categorization, Image spatial information.

1 Introduction

Image classification framework consists in attributing one or more category labels to a given image. It is one of the most fundamental problems in computer vision and pattern recognition. Besides, it has a wide range of applications, such as image and video retrieval, video surveillance, biometrics, etc. In the recent literature, the Bag of Visual Words (BoW)[7] is the most popular approach in image classification task[9,4,3,2]. It has achieved the state-of-the-art performance in several databases. The original BoW [7] is based on K-means method to form the dictionary by quantifying the space of local features into a set of dictionary basis vectors. After that, each local feature is assigned to a single basis vector. We can note that the hard quantization is very strict and leads to error quantization especially if the features are located on the boundary proximity of divers basis vectors.

Sparse coding [15] aims to learn a dictionary and simultaneously find a sparse linear combination of basis vectors from this dictionary to represent the image

C.K. Loo et al. (Eds.): ICONIP 2014, Part III, LNCS 8836, pp. 292–299, 2014.
© Springer International Publishing Switzerland 2014

features. It has consistently enhanced the results on image classification problem by resolving efficiently the problem of hard quantization. Yang et al. [15]
proposed Sparse coding SPM (as referred ScSPM). They train the dictionary
and compute the sparse codes in the encoding step. In the pooling step, the max
pooled responses across different sub-regions are computed.

Sparse coding [15] treats local features independently, ensuing that the sparse
codes can vary greatly even for close features. To overcome this drawback, different extensions of sparse coding method [14,6,12] have been suggested recently
by adding some regularization or constraints in the sparse coding objective function. The Locality-constrained Linear Coding (LLC) [14] technique considers the
locality information in the feature coding process. Contrary to the sparse coding, LLC enforces locality instead of sparsity. It uses the k nearest neighbors
of features as the local basis vectors. This leads to smaller coefficient for the
basis vectors far away from each local feature. Laplacian Sparse Coding (LSC)
[6] learns an unsupervised dictionary, as well as the sparse representation that
preserves the conformity of close local descriptors in the data space. This method
has used histogram intersection similarity based on k-Nearest Neighbors (KNN)
method to construct a Laplacian matrix that tries to preserve the local con-
sistence in the feature space. Only the K-nearest local features are selected to
active the Laplacian matrix. This method obtains background results on several
object recognition.

After the encoding phase, the pooling step is applied in order to aggregate
the encoded features. Two major strategies are used: average pooling and max
pooling. The first strategy consists to take the average of the responses over the
region in a given visual word. It is applied generally after the BoW encoding
step. The second strategy considers the largest responses instead of its average and it is suitable to sparse encoded histograms. These two approaches have
two major drawbacks. Firstly, they ignore the spatial information when gathering the local features. As a solution, spatial pyramid representation is used in
[7,15,14,6] in order to incorporate the global spatial information into the pooling
step. Explicitly, each image is split progressively into finer cells. For every cell, a
histogram of basis vector is determined. These histograms are then mixed up using a weighting scheme depending on the level of the spatial pyramid. Secondly,
they consider a scalar result for each dictionary basis vector discarding the analysis of the distribution around each visual word. Avila et al. [1] enhance these
strategies by proposing Bag of Statistical Sampling Analysis (BoSSA) pooling.
It is applied to discretize the distance between K-means clusters and the local
features yielding a histogram of distances rather than a scalar. Each bin of this
histogram measures the average number of features assigned to a given visual
word, which discretized distance falls into this bin.

In this paper, the contributions can be summarized as follows:

1. In the encoding step, we propose a novel sparse coding method in order to
   enrich the image spatial information during the encoding phase. Compared
   to LSC that exploits the dependencies between local features only in the

feature space, we propose to exploit the dependencies among them in both feature and image spaces.

2. In the pooling step, inspired by the BoSSA [1] method that applies a statistical analysis on the distances between the local features and the k-means clusters, we develop a novel pooling method based on performing statistical analysis for the sparse codes.

2   Laplacian Sparse Coding Formula

Sparse coding method aims to reduce the problem of hard quantization. It finds a sparse linear combination of basis vectors for each image feature. Given the local feature space $X = [x_1, \ldots, x_N]$, $x_i \in \Re^{D \times 1}$, K basis vectors $U = [u_1, \ldots, u_K] \in \Re^{D \times K}$ generate the dictionary and the matrix of the sparse codes $V = [v_1, v_2, \ldots, v_N]$ where $v_i \in \Re^{k \times 1}$ and v_{ik} is the weight of the vector x_i in the basis vector u_k, the optimization problem of sparse coding can be rewritten as follows:

$$\min_{U,V} \|X - UV\|_F^2 + \lambda \sum_i \|v_i\|_1 \tag{1}$$

$$subject\ to\ \|u_j\| \leq 1; \forall j = 1, \ldots, K$$

The first term in Eq.1 is the reconstruction error, and the second term is used to control the sparsity of the sparse codes v_i. λ is the tradeoff parameter used to balance the sparsity and the reconstruction error. Sparse coding has proved its efficiency in feature quantization process. Yet, the major drawbacks of this coding method is that it neglects the consistency of the sparse codes for the close local descriptors.

Gao et al [6] proposed Laplacian sparse coding to incorporate the similarity among the local features in the feature space. They added a regularization term in the objective function of sparse coding. Given two local features x_i and x_j as well as their sparse codes v_i and v_j respectively, $W_{i,j}$ measures the similarity between these features, the function objective of LSC is described as follows:

$$\min_{U,V} \|X - UV\|_F^2 + \lambda \sum_i \|v_i\|_1 + \frac{\beta}{2} \sum_{i,j} \|v_i - v_j\|^2 W_{i,j} \tag{2}$$

Then, the formula 2 can be reformulated as:

$$\min_{U,V} \|X - UV\|_F^2 + \lambda \sum_i \|v_i\|_1 + \beta tr\left(VLV^T\right) \tag{3}$$

$$subject\ to: \ \|u_m\|^2 = 1$$

where β is the weight on the closeness restriction and L defines the Laplacian matrix.

3   Proposed Approach

In this section, we describe the details of the extended approach based on Laplacian sparse coding. First, the local features are extracted using dense SIFT [8] features. Then, the local regions are built around local features in order to incorporate the local spatial information during the sparse coding process. The local features are encoded, via our proposed approach, to sparse codes taking into account the consistency between the sparse codes and the local regions centred around their corresponding local features. Furthermore, we apply our proposed Sparse BoSSA Pooling (SBP) to give the final image representation. Finally, a multi-class non-linear SVM classifiers is trained for image category prediction. These steps are detailed in the next sections.

3.1   Feature Extraction

Several works [11] prove that sampling density is better than interest points. SIFT descriptor demonstrates its excellent results in image classification [13,5,6,15,14]. For that, we implement in our experiment dense SIFT features. Given a local region, SIFT descriptor is computed as 16 histograms of 8 gradient orientations. It gives a 128-dimensional vectors.

3.2   Proposed Extension of Laplacian Sparse Coding

Given the local feature space $X = [x_1, \ldots, x_N]$, $x_i \in \Re^{D \times 1}$ extracted as described in section 3.1. In order to take into account the local spatial information during the encoding phase, we form the local regions $R(x_i)$ centred around each local feature x_i. We consider the eight spatial neighbours $E(x_i)$ for each local feature x_i to form the local region $R(x_i) = \{x_i, E(x_i)\}$ as showed in Figure 1.

Original image   Local region image

Fig. 1. Illustration of the local spatial information extraction process

After that, we aim to learn the unsupervised dictionary and to compute the sparse code for each feature. In the classical Laplacian sparse coding, $W_{i,j}$ computes the similarity between local features x_i and x_j in order to realize the consistency between local features and sparse codes. In this paper, we propose to compute the similarity between x_i and x_j taking into account the similarity between their spatial neighborhood in the image. Explicitly, we fix $W_{i,j} = 1$

if the local region $R(x_i)$ is among the k-nearest neighbour of the local region $R(x_j)$, otherwise, we fix $W_{i,j} = 0$. To compute the similarity between $R(x_i)$ and $R(x_j)$, we define the similarity measure $S(R(x_i), R(x_j))$ as the summation of (1) the histogram intersection similarity between x_i and x_j and (2) the mean pairwise similarities between the matched local features in $E(x_i)$ and $E(x_j)$. For each local feature in $E(x_i)$, the matching is carried out by finding the closet local feature in $E(x_j)$ (in the sense of the histogram intersection similarity).

3.3   Proposed Sparse BoSSA Pooling Method

In the previous section, we have trained the unsupervised dictionary and we have coded each local feature by a sparse code. In this section, we will represent the final vector of a given image $I = \{x_i\}_1^M$ via these sparse codes. To measure the distribution of the responses of each local descriptor to the dictionary's vector basis, we adapt BoSSA [1] pooling strategy to our new sparse encoding scheme as referred sparse BoSSA pooling. For that purpose, we built a histogram $h_{k,b}$ of size B for each kth basis vector. Each bin of this histogram represents the occurrences of the absolute value of the sparse code weights that fall into this bin. The formula describes the computation of a given histogram h_k for an image I.

$$h_{k,b} = card\left(v_i \mid x_i \in I \text{ and } v_k^{min} + s \times b \leq |v_{ik}| \leq v_k^{max} + s \times (b+1)\right)$$

where

$$s = \frac{v_k^{max} - v_k^{min}}{B} \text{ and } b \in [0, \ldots, B-1]$$

B denotes the number of bins, v_k^{min} and v_k^{max} limit the range of activated sparse code weights $|v_{i,k}|$ over all descriptors x_i extracted from the images of the learning set and the step s corresponds to the length of the bin.

4   Experiments

4.1   Experimental Protocol

In our experiments, we extract densely SIFT features from 8×8 patches using a spatial stride equal to 4. After that, we form a local region for each local feature. Then, we learn the dictionary and we compute the sparse code for each local feature implementing our encoding method. Furthermore, we apply SPR in order to preserve the global spatial information and we apply SBP in each subregion. For fair comparison to [15,6], the splits of the SPR is $[(1 \times 1), (2 \times 2), (4 \times 4)]$. Also, the number of basis vectors is fixed to 1024 and the number of bins is fixed to $B = 3$. Two settings are included in our objective function λ: the sparsity of the sparse codes and β: the weight on the closeness restriction. λ and β are fixed by cross validation: we fix $\beta = 0.1$, $\lambda = 0.3$ for UIUC Sport and Caltech-256, and we fix $\beta = 0.2$, $\lambda = 0.4$ in Corel dataset. In the classification step, we train the histograms with the $chi - square$ non-linear SVM.

4.2 Datasets

We evaluate our approach for four datasets: UIUC-Sport, Corel-10 Dataset and caltech-256. For fair comparison, we keep the identical experimental properties as [15,6]. Table 1 summarizes the characteristics for all the datasets: the number of classes, the number of the images in the dataset, the number of training images per class and the number of test images.

Table 1. The general description of the datasets

	UIUC-sport	Corel	Caltech-256
# of classes	8	10	257
# of images	1792	1000	30607
# of training	70	50	15/30/45/60
# of test	remainder	50	remainder

4.3 Results

Impact of the Spatial Context (SC). Table 2 depicts the influence of the spatial context added in the regularization term of the objective function. We observe that the integration of the dependencies between local features both in feature space and image space is more important than the integration of only the dependencies between local features in the feature space.

Table 2. Impact of the spatial context on classification accuracy

Methods	UIUC-Sport	Corel	caltech-256
Without SC	$85.18 \pm .46$	$88.76 \pm .94$	$35.74 \pm .1$
With SC	$86.6 \pm .42$	$90.15 \pm .76$	$38.35 \pm .46$

Impact of SBP Pooling on Our New Encoding Method. In this experiment, we study the impact of our sparse BoSSA pooling method on classification accuracy. Table 3 shows that the proposed pooling method enhances the classification accuracy in all datasets. These results confirm the advantages introduced by SBP representation.

Table 3. Impact of sparse BoSSA pooling on image classification accuracy

Methods	UIUC-Sport	Corel	Caltech-256
Without SBP	$86.6 \pm .42$	$90.15 \pm .76$	$38.35 \pm .46$
With SBP	$87.85 \pm .46$	$91.33 \pm .94$	$39.64 \pm .53$

Table 4. Performance Comparison on Caltech-256 Dataset

Number of training images	15	30	45	60
Method	Average Classification rate(%)			
BoW [10]	23.5 ± 0.42	29.1 ± 0.38	32.17 ± 0.53	34.21 ± 0.24
ScSPM[15]	27.73 ± 0.51	34.02 ± 0.35	37.46 ± 0.55	40.14 ± 0.91
LLC [5]	27.74 ± 0.32	32.07 ± 0.24	35.09 ± 0.44	37.79 ± 0.42
LSC[6]	29.99 ± 0.15	35.74 ± 0.1	38.47 ± 0.51	40.32 ± 0.32
Our	33.72 ± 0.7	39.64 ± 0.53	42.16 ± 0.51	44.03 ± 0.63

Comparison with State-of-the-Art. We compare our approach to different image classification methods in the literature. The SPM baseline method and baseline methods based on sparse coding: ScSPM, LLC, LSC. Table 5 and 4 show that our method exceeds background performance on divers datasets. This demonstrates that the proposed method can improve the classical Laplacian sparse coding by taking into account the locality constraint among the local features in the encoding phase and the statistical distribution of the sparse code weights in the pooling step.

Table 5. Performance Comparison on UIUC-Sport and Corel datasets

Methods	UIUC-Sport	Corel
SPM [7]	79.98 ± 1.67	-
ScSPM [15]	82.74 ± 1.46	86.6 ± 1.01
LLC [14]	83.09 ± 1.3	87.93 ± 1.04
LSC [6]	85.18 ± 0.46	88.76 ± 1.04
Our	87.85 ± 0.46	91.33 ± 0.49

5   Conclusion

In this study, we aim to enhance the image classification task. For that, we propose a new sparse encoding method in order to improve the dictionary learning and the sparse coding process. Indeed, the incorporation of spatial locality among the features in the image space ensures the consistency between the sparse codes and the local regions centred around their corresponding local features. Furthermore, we propose a new pooling scheme that adapt BoSSA pooling on the novel sparse codes. This enables us to take into account the distribution of the sparse codes weights around each vector basis. Experimental results proves the efficiency of the proposed approach.

References

1. Avila, S., Thome, N., Cord, M., Valle, E., de Albuquerque Araújo, A.: Bossa: Extended bow formalism for image classification. In: 18th IEEE International Conference on Image Processing, pp. 2909–2912 (2011)
2. Ben Aoun, N., Mejdoub, M., Ben Amar, C.: Graph-based approach for human action recognition using spatio-temporal features. J. Visual Communication and Image Representation 25(2), 329–338 (2014)
3. Dammak, M., Mejdoub, M., Ben Amar, C.: A survey of extended methods to the bag of visual words for image categorization and retrieval. In: 9th International Conference on Computer Vision Theory and Application, pp. 676–683 (2014)
4. Dammak, M., Mejdoub, M., Zaied, M., Amar, C.B.: Feature vector approximation based on wavelet network. In: ICAART, vol. (1), pp. 394–399 (2012)
5. Gao, S., Tsang, I.W.H., Chia, L.T.: Sparse representation with kernels. IEEE Transactions on Image Processing 22(2), 423–434 (2013)
6. Gao, S., Tsang, I.W.H., Chia, L.T., Zhao, P.: Local features are not lonely: Laplacian sparse coding for image classification. In: 23th IEEE Conference on Computer Vision and Pattern Recognition, pp. 3555–3561 (2010)
7. Lazebnik, S., Schmid, C., Ponce, J.: Beyond bags of features: Spatial pyramid matching for recognizing natural scene categories. In: IEEE Conference on Computer Vision and Pattern Recognition, pp. 2169–2178. IEEE Computer Society (2006)
8. Lowe, D.G.: Distinctive image features from scale-invariant keypoints. Int. J. Comput. Vision 60(2) (2004)
9. Mejdoub, M., Ben Amar, C.: Classification improvement of local feature vectors over the knn algorithm. Multimedia Tools and Applications 64(1), 197–218 (2013)
10. Morioka, N., Satoh, S.: Learning directional local pairwise bases with sparse coding. In: Proceedings of the British Machine Vision Conference, pp. 1–11 (2010)
11. Nowak, E., Jurie, F., Triggs, B.: Sampling stra9th european conference on computer vision,tegies for bag of features features image classification. In: 9th European Conference on Computer Vision, pp. 490–503 (2006)
12. Ren, W., Huang, Y., Zhao, X., Huang, K., Tan, T.: Local hypersphere coding based on edges between visual words. In: Lee, K.M., Matsushita, Y., Rehg, J.M., Hu, Z. (eds.) ACCV 2012, Part I. LNCS, vol. 7724, pp. 190–203. Springer, Heidelberg (2013)
13. van de Sande, K.E.A., Gevers, T., Snoek, C.G.M.: A comparison of color features for visual concept classification. In: ACM International Conference on Image and Video Retrieval (2008)
14. Wang, J., Yang, J., Yu, K., Lv, F., Huang, T., Gong, Y.: Locality-constrained linear coding for image classification. In: 23th IEEE Conference on Computer Vision and Pattern Recognition, pp. 3360–3367 (2010)
15. Yang, J., Yu, K., Gong, Y., Huang, T.: Linear spatial pyramid matching using sparse coding for image classification. In: 22th IEEE Conference on Computer Vision and Pattern Recognition, pp. 1794–1801 (2009)

Stochastic Decision Making in Learning Classifier Systems through a Natural Policy Gradient Method

Gang Chen, Mengjie Zhang, Shaoning Pang, and Colin Douch

Victoria University of Wellington,
Unitec Institute of Technology, New Zealand
{aaron.chen,mengjie.zhang}@ecs.vuw.ac.nz,
ppang@unitec.ac.nz,
douchcoli@myvuw.ac.nz

Abstract. Learning classifier systems (LCSs) are rule-based machine learning technologies designed to learn optimal decision-making policies in the form of a compact set of maximally general and accurate rules. A study of the literature reveals that most of the existing LCSs focused primarily on learning *deterministic* policies. However a desirable policy may often be *stochastic*, in particular when the environment is partially observable. To fill this gap, based on XCS, which is one of the most successful accuracy-based LCSs, a new Michigan-style LCS called Natural XCS (i.e. NXCS) is proposed in this paper. NXCS enables direct learning of stochastic policies by utilizing a natural gradient learning technology under a policy gradient framework. Its effectiveness is experimentally compared with XCS and one of its variation known as XCS_μ in this paper. Our results show that NXCS can achieve competitive performance in both deterministic and stochastic multi-step problems.

1   Introduction

Originated from John Holland's seminal work on cognitive systems [5,6], learning classifier systems (LCSs) are rule-based machine learning technologies designed to learn optimal decision-making policies in the form of a compact set of maximally general and accurate rules (aka. classifiers) [13]. Among all LCSs developed to date, XCS, which was introduced by Wilson, is unarguably the most successful accuracy-based LCS [8]. In this paper, a new Michigan-style LCS with native support for stochastic decision making will be developed based on XCS.

LCSs have been successfully applied to solve *reinforcement learning* problems where a learning agent is situated in a multi-step environment often modeled as a Markov Decision Process (MDP) [11]. A study of the literature reveals that reinforcement learning is commonly conducted in LCSs by approximating the *state-action value function*, which is represented jointly by a group of classifiers.

Due to the value-function based approach, the aim of a LCS is to learn *deterministic* policies. However, learning *stochastic* policies is often shown to be more *reliable*, in particular when the environment is stochastic or *partially observable* [12]. To the best of our knowledge, few LCSs have ever attempted to

C.K. Loo et al. (Eds.): ICONIP 2014, Part III, LNCS 8836, pp. 300–307, 2014.

directly learn stochastic policies that explicitly associate with each action a suitable probability for it to be performed in every state.

In view of the gap in the literature, a new LCS called Natural XCS (i.e. NXCS) will be developed in this paper. NXCS enables direct learning of stochastic policies by utilizing a *natural gradient learning technology* under a *policy gradient framework* [1]. Inspired by several temporal-difference based natural learning algorithms [2,10], this paper presents the first study of natural gradient learning in LCSs.

The remainder of this paper is organized as follows. A short introduction to the XCS classifier system can be found in Section 2. Based on XCS, NXCS will be further developed in Section 3. The performance of NXCS is experimentally compared with XCS and its recent variation known as XCS$_\mu$ [9] in Section 4. Finally Section 5 concludes this paper.

2   XCS Classifier System

XCS is an effective reinforcement learning method in which generalization is obtained through evolving a *population* $[P]$ of classifiers. A detailed algorithmic description of XCS can be found in [4]. At any discrete time t, a learning agent receives sensory inputs from the current environment state s_t. It reacts by performing an action a chosen from A. The environment then transits to a new state at $t + 1$, i.e. s_{t+1}, and a reward r_{t+1} is provided as feedback to the agent. The goal of the agent is to maximize the amount of reward obtained in the long run. We briefly review the four key components of XCS in this section.

Classifier: In XCS, each classifier cl has a condition c_{cl}, an action a_{cl}, and several other parameters, including 1) the prediction p_{cl} that estimates the average payoff upon using the classifier; 2) the prediction error ϵ_{cl} ; and 3) the fitness F_{cl} that estimates the average relative accuracy of classifier cl.

Performance Component: Whenever a decision is to be made at any time t, XCS creates a *match set* $[M]_t$ containing all classifiers in the population that match the current sensory input from state s_t. For every action $a \in A$, the agent calculates the predicted value of performing a, i.e. $P_t(a)$, based on the prediction from every classifier belonging to $[M]_t$ [4]. After that, an action a will be selected and the corresponding group of classifiers recommending a will form the *action set* at time t, i.e. $[A]_t$. The selected action will then be performed and the reward r_{t+1} will be received subsequently. During learning, the ϵ-*greedy* selection method will be exploited to randomize action selection. During testing, however, an exploitation strategy will be employed such that the action a with the highest $P_t(a)$ will always be selected.

Reinforcement Component: Upon reaching a new state s_{t+1}, the parameters of those classifiers in $[A]_t$ will be updated according to [4]. In particular, the prediction p_{cl} of a classifier $cl \in [A]_t$ will be updated based on:

$$p_{cl}(t + 1) \leftarrow p_{cl}(t) + \beta \left(r_{t+1} + \gamma \max_{a \in A} P_{t+1}(a) - p_{cl}(t) \right) \qquad (1)$$

where β is a fixed learning rate.

GA component: On a regular basis, a genetic algorithm will be applied to those classifiers in $[A]_t$. In particular, proportional to their fitness, two classifiers from $[A]_t$ will be randomly selected to produce offspring classifiers, which are further modified through the crossover and mutation operations.

3   Natural XCS Classifier System

Aimed at learning stochastic policies, based on XCS, a new NXCS classifier system will be developed in this section. We organize our discussion into three subsections. Subsection 3.1 introduces the concept of stochastic policy. The reinforcement component of NXCS is further presented in Subsection 3.2. Finally, Subsection 3.3 develops a policy parameter learning component.

3.1   Stochastic Policy

In comparison with XCS, each classifier cl in NXCS includes an additional *policy parameter*, denoted as θ_{cl}. At any time t, using all classifiers in the match set $[M]_t$, the probability of taking any action $a \in A$ is determined according to (2) below.

$$\pi_t(s_t, a) = \frac{\prod\limits_{cl \in [M]_t^a} e^{\theta_{cl}}}{\sum\limits_{b \in A} \left(\prod\limits_{cl \in [M]_t^b} e^{\theta_{cl}} \right)} \tag{2}$$

where $\pi_t(s, a)$ refers to a *stochastic policy* that assigns a certain probability for performing any action a in state s_t at time t.

We construct a policy parameter vector $\boldsymbol{\theta}_t$ to include θ_{cl} of every classifier cl belonging to the match set $[M]_t$, assuming a pre-defined global order on these classifiers. The policy $\pi_t(s, a)$ is subsequently viewed as a function of $\boldsymbol{\theta}_t$.

3.2   Prediction Reinforcement

NXCS follows a similar learning procedure as XCS. During a single learning step at time t, based on the match set $[M]_t$, an action will be selected according to its probability defined in (2). The chosen action is then performed. As a result, the environment transits to a new state s_{t+1} and a scalar reward r_{t+1} is observed. r_{t+1} is then applied to update the prediction of each classifier $cl \in [A]$ using the updating rule below.

$$p_{cl}(t+1) \leftarrow p_{cl}(t) + \beta \cdot$$

$$\left(r_{t+1} + \gamma \sum_{a \in A} \pi_t(s_{t+1}, a) \cdot P_{t+1}(a) - p_{cl}(t) \right) \tag{3}$$

The prediction updating in NXCS is different from that of XCS as shown in (1). This is because every policy in NXCS is stochastic. Hence prediction cannot be updated by assuming that, at time $t + 1$, the action that gives the highest prediction will always be performed.

3.3 Learning Policy Parameters

In this subsection, a policy gradient framework is adopted to learn policy parameters. In particular, because the learning performance J (i.e. the discounted accumulated reward in the long run) can be treated as a function of $\boldsymbol{\theta}$, a straightforward approach is to learn $\boldsymbol{\theta}$ based on

$$\boldsymbol{\theta}_{t+1} \leftarrow \boldsymbol{\theta}_t + \lambda \cdot \nabla_{\boldsymbol{\theta}_t} J \tag{4}$$

where λ is a fixed learning rate. Practical application often shows that learning through (4) can be slow and unstable [10]. Instead of using $\nabla_{\boldsymbol{\theta}_t} J$, a *natural gradient* concept proposed by Amari can be very helpful [1]. Theoretically, stochastic policies learned through NXCS are equivalent to a family of statistical models situated in a Riemannian parameter vector space of $\boldsymbol{\theta}$. Each point in the space corresponds to a specific stochastic policy. In such a *Riemannian space*, learning should be performed through the *natural gradient* of J, i.e. $\tilde{\nabla}_{\boldsymbol{\theta}_t} J$. Particularly, we have

$$\boldsymbol{\theta}_{t+1} \leftarrow \boldsymbol{\theta}_t + \lambda \cdot \tilde{\nabla}_{\boldsymbol{\theta}_t} J \tag{5}$$

where

$$\tilde{\nabla}_{\boldsymbol{\theta}_t} J = G(\boldsymbol{\theta}_t)^{-1} \cdot \nabla_{\boldsymbol{\theta}_t} J \tag{6}$$

$G(\boldsymbol{\theta}_t)^{-1}$ is the inverse matrix of $G(\boldsymbol{\theta}_t)$. $G(\boldsymbol{\theta}_t)$ stands for the *Fisher information matrix* [1] of the stochastic policy represented by $\boldsymbol{\theta}_t$. In line with (5) and (6), it can be shown eventually that

$$\tilde{\nabla}_{\boldsymbol{\theta}_t} J \propto \cdot \delta_t \cdot \nabla_{\boldsymbol{\theta}_t} \log \pi_t(s, a) \tag{7}$$

where

$$\delta_t = r_{t+1} + \gamma \cdot \sum_{a \in A} \pi_t(s_{t+1}, a) \cdot P_{t+1}(a) - \sum_{a \in A} \pi_t(s_t, a) \cdot P_t(a) \tag{8}$$

Based on (7), the updating rule for learning policy parameters is determined as

$$\boldsymbol{\theta}_{t+1} \leftarrow \boldsymbol{\theta}_t + \lambda \cdot \delta_t \cdot \psi_{s,a} \tag{9}$$

The learning parameter λ in (9) will be set to the inverse of the maximum single-step reward in all experiments to be reported in Section 4. It can be verified that the computational complexity of the performance component in NXCS is $O\left(|[M]_t|\right)$, which is the same as XCS. Meanwhile, the complexity of the reinforcement component in NXCS is $O\left(Max(|[M]_t|, |[M]_{t+1}|)\right)$. Whereas in XCS, the corresponding complexity is $O\left(Max(|[A]_t|, |[M]_{t+1}|)\right)$, which should not appear significantly different in practice.

4   Experiment Results

Experiments on three reinforcement learning problems will be reported here. The Woods101 problem is a partially observable environment. It is used to understand whether NXCS can better cope with *perceptual aliasing* [7] than XCS and XCS_μ. The Woods14 problem is further used to study the performance of NXCS on benchmark deterministic multi-step problems with long-delayed reward. Finally, we will study the reliability of the learning system on stochastic maze problems, specifically the Maze5ϵ problem.

4.1   Experiments on the Woods101 Problem

The Woods101 problem, as described in [9], is a small grid environment that consists of 10 empty positions and one *terminal state* F (i.e. the goal). The agent, in any state, may choose to perform one of eight alternative actions. In the absence of an obstacle, each alternative action will bring the agent to a different adjacent position. To allow continued learning, whenever the agent reaches the terminal state, it will receive a maximum reward of 1000 and will be immediately relocated to a new state selected uniformly at random from the 10 empty positions.

Because an agent can only observe its adjacent positions in the grid, two states in the Woods101 problem are indistinguishable. Whenever a deterministic policy is followed, the same action will be performed in both states. There is hence a chance for the agent to be trapped in a local optima [9]. In comparison, an agent can achieve better performance by learning stochastic policies.

Fig. 1 presents the performance of NXCS, XCS, and XCS_μ on the Woods101 problem. Also included in this figure is the performance of another LCS named RXCS. RXCS is a learning system recently proposed by us for learning stochastic policies without using the natural gradient learning method. The maximum population size in our experiments is set to 300 classifiers. To reduce randomness, 30 independent tests have been conducted for each LCS. The average performance obtained is depicted in this figure. The same practice is also applied to build other result figures included in this paper.

As can be seen from Fig. 1, NXCS appears to perform better than XCS and XCS_μ throughout the whole learning process. In particular, at the end of the experiment (i.e. 8000 learning problems), the average performances achieved by NXCS, XCS, and XCS_μ are 4.39, 63.09, and 40.35 respectively. By using two-tailed t-test, it can be confirmed that the performance of NXCS is statistically better than that of XCS and XCS_μ. Specifically, the p-value is 1.5483×10^{-127} for the t-test between NXCS and XCS and it equals to 9.1783×10^{-146} for the t-test between NXCS and XCS_μ. Both the two p-values are far less than 0.05,

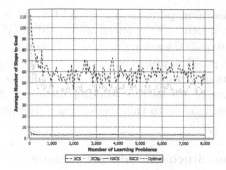

Fig. 1. Learning performance of NXCS, RXCS, XCS, and XCS$_\mu$ on the Woods101 problem. The performance is measured as the average number of actions to be performed by an agent in order to reach a goal. The theoretical optimal performance is also indicated in this figure.

which is commonly used as the standard statistical significance level for t-tests. Meanwhile, we found that, after about 2000 learning problems, NXCS achieved an average performance that is very close to the theoretical optimum of 4.3 (with a small difference less than 0.1).

4.2 Experiments on the Woods14 Problem

In this Subsection, the performance of NXCS is further tested on the Woods14 problem. The Woods14 is a difficult benchmark reinforcement learning problem [3]. In particular, due to the problem's long-delayed reward, XCS has been reported as failing to solve the problem properly [3].

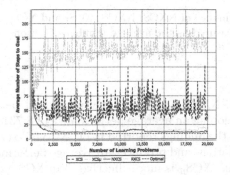

Fig. 2. Learning performance of NXCS, RXCS, XCS, and XCS$_\mu$ on the Woods14 problem. Performance is measured as the average number of actions an agent performs in order to reach a goal. The theoretical optimal performance is also indicated in this figure.

Fig.2 depicts the learning performance of NXCS, XCS and XCS_μ on the Woods14 problem. Evidently, NXCS successfully stabilized its performance at an average of 12.619 after 3000 learning problems. In comparison, both XCS and XCS_μ cannot solve the problem properly, eventually achieving averages of 33.684 and 162.088 in performance respectively. The best policy in theory can achieve an average performance of 9.5 on the Woods14 problem. The performance of NXCS appears to be quite close to this theoretical optimum.

4.3   Experiments on Stochastic Maze Problems

In this subsection, we investigate the reliability of the learning systems in stochastic environments. Particularly, we have tested NXCS, XCS, and XCS_μ on several stochastic maze problems, including the Maze4ϵ, Maze5ϵ and Maze6ϵ problems. All our results consistently show that NXCS is more effective at handling environmental randomness. Specifically, to support this claim, we present here the experiment results on the Maze5ϵ problem, which is a stochastic extension of the benchmark Maze5 problem, as described in [9].

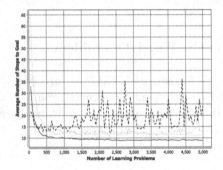

Fig. 3. Learning performance of NXCS, RXCS, XCS, and XCS_μ on the Maze5ϵ problem. Performance is measured as the average number of actions an agent performs in order to reach a goal.

As shown in Fig.3, NXCS has apparently performed better over the whole learning process, developing an average performance of 8.731 after 5000 learning problems. In line with the findings reported in [9], XCS failed to converge. Instead it exhibits large fluctuations and produces an average performance of 24.036 after 5000 learning problems. XCS_μ is more effective than XCS, achieving an average performances of 12.313 at the end of the experiment. If we perform a t-test between NXCS and XCS_μ, a p-value of 4.758×10^{-32} is obtained, confirming that the performance difference between the two learning systems is statistically significant.

5   Conclusions

Based on XCS, this paper successfully developed a new natural XCS (i.e. NXCS) classifier system. Our research was inspired by the natural gradient learning technology. To the best of our knowledge, this paper presented the first study of natural gradient learning in XCS. Our method is general and can potentially be applied to many other LCSs. Meanwhile, our experiments showed that NXCS performed competitively with XCS and XCS_μ.

Looking into the future, we would hope to see interesting applications of NXCS to real-world problems that require sequential and stochastic decision making. The potential usefulness of NXCS for a wide range of machine learning tasks, including data mining problems, may also deserve in-depth investigation.

References

1. Amari, S.: Natural gradient works efficiently in learning. Neural Computation 10(2), 251–276 (1998)
2. Bhatnagar, S., Sutton, R.S., Ghavamzadeh, M., Lee, M.: Natural actor-critic algorithms. Journal Automatica 45(11), 2471–2482 (2009)
3. Butz, M.V., Goldberg, D.E., Lanzi, P.L.: Gradient descent methods in learning classifier systems: improving xcs performance in multistep problems. IEEE Transactions on Evolutionary Computation (2005)
4. Butz, M.V., Wilson, S.W.: An Algorithmic Description of XCS. In: Lanzi, P.L., Stolzmann, W., Wilson, S.W. (eds.) IWLCS 2001. LNCS (LNAI), vol. 2321, pp. 253–272. Springer, Heidelberg (2002)
5. Holland, J.H.: Adaptation in Natural and Artificial Systems. University of Michigan Press (1975)
6. Holland, J.H.: Adaptation. In: Progress in Theoretical Biology, vol. 4, pp. 263–293. Academic Press (1976)
7. Lanzi, P.L.: An analysis of the memory mechanism of xcsm. In: Proceedings of the Third Genetic Programming Conference, pp. 643–651 (1998)
8. Lanzi, P.L.: Learning classifier systems: then and now. Evolutionary Intelligence (2008)
9. Lanzi, P.L., Colombetti, M.: An extension to the xcs classifier system for stochastic environments. In: Proceedings of the Genetic and Evolutionary Computation Conference, pp. 353–360 (2000)
10. Peters, J., Schaal, S.: Natural actor-critic. Neurocomputing, 1180–1190 (2008)
11. Sutton, R.S., Barto, A.G.: Reinforcement Learning: An Introduction. MIT Press (1998)
12. Sutton, R.S., McAllester, D., Singh, S., Mansour, Y.: Policy gradient methods for reinforcement learning with function approximation. In: Advances in Neural Information Processing Systems 12 (NIPS 1999), vol. 12, pp. 1057–1063. MIT Press (2000)
13. Wilson, S.W.: Classifier fitness based on accuracy. Evolutionary Computation 3(2), 149–175 (1995)

Quantum Inspired Evolutionary Algorithm by Representing Candidate Solution as Normal Distribution

Sreenivas Sremath Tirumala[1], Gang Chen[2], and Shaoning Pang[1]

[1] Unitec Institute of Technology, Auckland, New Zealand
stirumala@unitec.ac.nz
[2] Victoria University of Wellington, Wellington, New Zealand

Abstract. Application of Quantum principles on evolutionary algorithms was started as early as late 1990s and has witnessed continued improvements since then. Following the same quantization principle introduced by the Quantum inspired evolutionary algorithm (QEA) in 2003, most of the existing quantum inspired algorithms focused mainly on evolving a single set of homogeneous solutions. In this paper, we present a new quantization process. In particular, aimed at solving numerical optimization problems, the evolutionary selection procedure is quantified through a set of subsolution points that jointly define candidate solutions. Implementing this new method on competitive co-evolution algorithm (CCEA), a new Quantum inspired competitive coevolution algorithm (QCCEA) is proposed in this paper. QCCEA is experimentally compared with CCEA through 9 benchmark numerical optimization functions published in CEC 2013. The results confirmed that QCCEA is more effective than CCEA over a majority of benchmark problems.

Index Terms: Deep Architectures, Deep Learning, Evolutionary Algorithm, Deep Neural Networks.

1 Introduction

In recent years, many Quantum inspired algorithms were developed by applying quantum principles on evolutionary algorithms [1] [2] [3]. Among them, the Quantum inspired Evolutionary Algorithm (QEA) has received the highest attention [4]. Quantization is a crucial part of any quantum inspired algorithms. In QEA, a candidate solution to an optimization problem is quantized through "**qubits**", each of which represents a linear combination of the two binary bits "**0**" and "**1**". This method has triggered many similar approaches since then . The effectiveness of QEA in solving various types of optimization problems was extensively studied in [4] [5].

Despite of its prominent success, QEA and related research works focused only on evolving a single set of homogeneous solutions. No attempts have ever been made to apply quantum principles to co-evolutionary algorithms, especially the Competitive Co-evolution Algorithm (CCEA). In view of this limitation,

C.K. Loo et al. (Eds.): ICONIP 2014, Part III, LNCS 8836, pp. 308–316, 2014.

this paper proposes a new Quantum inspired Competitive Coevolution Algorithm (QCCEA). QCCEA follows a completely different way of quantization, in which each candidate solution is quantified through a series of subsolution points represented by normal probabilistic distributions. Meanwhile quantum inspired mechanism has also been employed to enhance QCCEA's capability of creating more unique solutions, in order to encourage vigorous competitions for better solutions and therefore the effectiveness of the evolutionary process.

To verify the effectiveness of our new algorithm, experimental comparison between QCCEA and CCEA has been conducted on 9 benchmark numerical optimization problems published in CEC2013. The experimental results show that QCCEA performed significantly better than CCEA for a majority of benchmark functions. The complete work of QCCEA is presented in a Master Thesis by Sreenivas [6]. We thereby believe that our quantum inspired method may open a new direction for studying co-evolutionary algorithms.

2   Related Work

While developing the QCCEA algorithm, we have studied two influential algorithms from the literature, i.e. QEA and QNN. QEA was introduced in [3] [7]. It uses quantum states to represent a candidate solution and the Q-gate [8] to diversify the candidate solution. The fundamental concept in QEA is "**qubits**", which are the smallest building blocks of a candidate solution. Each qubit is defined as a value pair, i.e. (α, β), where $|\alpha|^2 + |\beta|^2 = 1$, $|\alpha|^2$ gives the probability with which the qubit will be found in the "0" state, and $|\beta|^2$ gives the probability with which the qubit will be found in the "1" state. Accordingly, a candidate solution is represented as a string of n individual qubits,

$$\left\langle \begin{matrix} \alpha_1 \\ \beta_1 \end{matrix} \middle| \begin{matrix} \alpha_2 \\ \beta_2 \end{matrix} \middle| \begin{matrix} ... \\ ... \end{matrix} \middle| \begin{matrix} \alpha_n \\ \beta_n \end{matrix} \right\rangle \tag{1}$$

where $0 \le \alpha_i \le 1$, $0 \le \beta_i \le 1$, $|\alpha_i|^2 + |\beta_i|^2 = 1$, $i = 1, 2, ...n$ and $|\alpha_i|^2$, $|\beta_i|^2$ gives the probability with which the ith qubit will be found in state "0" and state "1" respectively.

The candidate solution is further diversified using Q-gate which is a variation operator as defined below

$$U(\Delta\theta_i) = \begin{bmatrix} \cos(\Delta\theta_i) - \sin(\Delta\theta_i) \\ \sin(\Delta\theta_i) \; \cos(\Delta\theta_i) \end{bmatrix} \tag{2}$$

where $\Delta\theta_i, i = 1, 2, ...n$, is the rotation angle of each qubit towards either 0 or 1 depending on its sign.

QEA can be applied to solve general purpose optimization problems as well as to evolve complex knowledge structures. One such implementation is Quantum based Neural Networks (QNN) which was developed to evolve neural networks. QNN is capable of optimizing network structures as well as connectivity weights [9]. For an arbitrary multilayer perceptron (MLP) model, its network connectivity C is defined as $\langle \alpha_1 | \alpha_2 | ...... | \alpha_n \rangle$ where each $\alpha_i, i = 1, 2, ..., c_{max}$

is a qubit. Here C_{max} refers to the maximum number of connections in the neural network. Accordingly, the connection weight configuration of the neural network, i.e. W, is further represented as $W = (Q_{w_1}, Q_{w_2}, ..., Q_{w_{cmax}})$, in which every $Q_{w_i}, i = 1, 2, ..., c_{max}$ gives the weight of a separate connection in the neural network. Q_{w_i} comprises k quantum bits or $Q_{w_i} = \langle \alpha_{i,1} | \alpha_{i,2} | ..... | \alpha_{c_{max},k} \rangle$, where k is an algorithm parameter. Similar to QEA, QNN also uses rotation gate to diversify candidate solutions. Rotation gate is updated according to the discrepancy between any solution and the pre-stored best. In the meantime, similar to the migration operation of QEA, QNN employs qubit swapping as an exchange operation in order to escape from local optima.

Traditional EAs represent a candidate solution as a single point, whereas QEA and QNN quantify the candidate solution through a linear combination of qubits. This change of representation may potentially increase the chances of identifying near optimal solutions and therefore expedite the evolutionary process. Specifically, we will propose a new QCCEA algorithm whose candidate solutions are obtained through a combination of subsolution points. Fig. 1(a) illustrates the proposed method that uses subsolution points and compares it with typical qubit quantization.

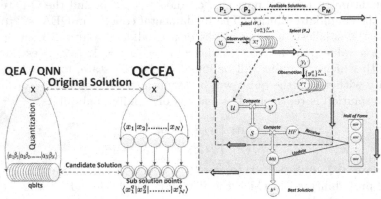

(a) qubit versus subsolution points   (b) Overall structure of QCCEA

Fig. 1. Results of Unimodal functions

3   Proposed QCCEA

In competitive co-evolution individuals of the population compete with each other resulting in a better species [10]. This competition can be of three types individuals, fitness sharing, shared sampling, and Hall of Fame (HF) that significant improvement in quality of the solution with each generation [10]. QCCEA is developed by applying quantum principle of superposition on Competitive Co-Evolution Algorithm (CCEA) Algorithm. The essence of QCCEA lies in the fact that it divides a candidate solution into a collection of solution points. Along the same vein as QEA, qubit in QCCEA is realized through these solution points. In another word, each such point assumes the form of a probability distribution (i.e. normal distribution) and functions similarly as a qubit. Using solution points that are stochastic in nature (i.e. stochastic points) may help to extend the search capability of CCEA. To be more specific, with stochastic points, more genetically diverse solutions can be produced by QCCEA during every generation, therefore effectively reducing the chance of premature convergence. Since QCCEA is developed based on CCEA and, as a result, it follows the basic evolutionary process of CCEA. The key distinguishing factor, as mentioned above, is marked by the representation of an individual. The overall structure of QCCEA is presented in Fig. 1(b). A detailed algorithmic description of QCCEA is further presented in Algorithm 1. As shown in Algorithm 1, QCCEA is initialized with a population of candidate solutions $P_1, P_2, ..., P_M$ (step 1) where M is the size of the population. HF and the best solution of the generation $b(t)$ are initialized by the first candidate solution P_1 in the population. A candidate solution P_m is represented as $\{x_n\}_{n=1}^N$, where N stands for the dimension of the search space. Each solution point of P_m, denoted as x_n, corresponds to a normal distribution and is evaluated at step 10 in Algorithm 1 in order to quantity solution P_m. Two quantified solutions u and v are specifically highlighted in Algorithm 1. Similar to CCEA, they will engage in a competition, resulting in s as the solution with the better fitness between u and v. This operation corresponds to step 18 of Algorithm 1. At step 19, the competition between s and the HF gives rise to b,

---

Algorithm 1. QCCEA: Quantum inspired Competitive coevolution (M, N, n, b^*)

---

1. Initialize $P_1, ..., P_M$; /*M solutions*/
2. Initialize b(1); /*current best solutions*/
3. $t \leftarrow 1$;
4. $HF(1) \leftarrow P(1)$;
5. repeat
6.   for $i = 1$ to M do
7. $X_i = \text{Select}(P_m), 1 \leq m \leq M$;
8.   Quantize $X_i = \{x_n\}_{n=1}^N$ to $X_i^q = \{x_n^q\}_{n=1}^N$; /*constant variable x_n is quantified as a normal distribution vector x_n^q*/
9.   for $k = 1$ to N do
10. $u \leftarrow u+$ Evaluate (x_k);
11.   end for
12.   for $j = 1$ to M and $j \neq i$ do
13. $Y_j = \text{Select}(P_m)$;
14.   Quantize $Y_j = \{y_n\}_{n=1}^N$ to $Y_i^q = \{y_n^q\}_{n=1}^N$;
15.   for $k = 1$ to N do
16. $v \leftarrow v+$ Evaluate (y_k);
17.   end for
18. $s \Leftarrow \text{Max}(\text{Evaluate}(u), \text{Evaluate}(v))$;
19. $b(t) \leftarrow \max(Evaluate(s), Evaluate(HF(t)))$ /*Compute the best solution with current Hall of Fame*/
20.   end for
21.   Add b(t) into $HF(t)$;
22.   end for
23. $t \leftarrow t + 1$;
24. $b^* \leftarrow max Evaluate(HF(t))$; /*select the best solution from current Hall of Fame*/
25. until enough solutions are evaluated

Table 1. COMPARISION BETWEEN CEEA AND QCCEA ON BENCH MARK FUNCTION $f_1 - f_9$. ALL RESULTS HAVE BEEN AVERAGED OVER 25 RUNS

No.	Functions	Benchmark	CCEA	QCCEA	T-Test Outcomes
f1	Sphere Function	-1400	-1241.50	-1327.20	5.82197×10^{-55}
f2	Rotated High Conditioned Elliptic Function	-1300	-1274.60	-1215.20	5.79889×10^{-49}
f3	Rotated Bent Cigar Function	-1200	-941.60	-1135.50	8.23198×10^{-75}
f4	Rotated Ackleys Function	-700	-592.0	-646.9	1.99006×10^{-45}
f5	Rotated Griewanks Function	-500	-357.2	-474.4	1.84575×10^{-63}
f6	Rotated Rastrigins Function	-300	-177.5	-224.3	2.15288×10^{-45}
f7	Lunacek Bi Rastrigin Function	300	222.6	253.4	1.15312×10^{-38}
f8	Rotated Lunacek Bi Rastrigin Function	400	290.9	309.1	4.1089×10^{-22}
f9	Expanded Griewanks plus Rosenbrocks Function	500	289.3	263.9	1.18431×10^{-34}

which is known as the best solution in a generation. HF is subsequently updated to include b. This process continues till the exist criteria are met. The solution b^* with the best fitness among all available solutions in the HF will be reported finally as the solution of the algorithm.

4 Experiment Results

10 benchmark numerical optimization functions from CEC2013 [11] were utilized in our experimental studies.All benchmark functions are defined on the same domain $[-100, 100]^D$, with the problem dimension $D = 10$. The experiment is performed for 3000 generations and each algorithm is tested independently on each function for 25 times.

4.1 Unimodal Funtions

The first set of experiments is on functions $f1 - f3$. For all experiments, the initial population size is set at 100. The obtained average results of 25 runs are presented in Table 1. Fig. 2 shows respectively the performance of CCEA and QCCEA on the benchmark functions. As seen from the results,QCCEA performs consistently closer to the optimum or near optimum than CCEA for four of the five unimodal functions. For function $f2$ QCCEA lags behind CCEA with a statistically negligible difference of 4.8%. Sphere Function $f1$ is a commonly used initial test function for performance evaluation of numerical optimization algorithms. For function $f1$ both CCEA and QCCEA exhibited similar performance at the beginning (till 721 generations) but at the later stages, QCCEA has improved progressively due to quantization. QCCEA has dominated in 2279 generations out of 3000 for function $f1$.

The biggest performance superiority of QCCEA occurs with Rotated Bent Cigar function $f3$ which is 17%. QCCEA achieved better convergence rate than CCEA due to its extensive search ability for the same number of generations while CCEA was caught within a small search space.

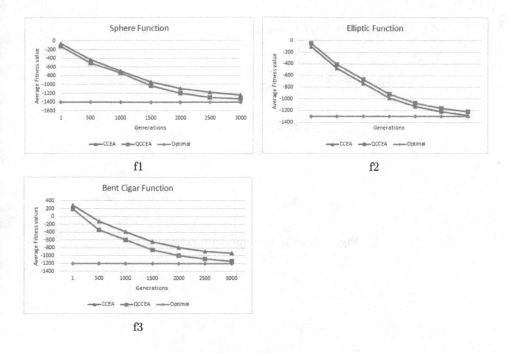

Fig. 2. Results of Unimodal functions

4.2   Multimodal Functions

The second set of experiments aimed at testing 6 multimodal functions $f4$ to $f9$. Multimodal functions are often considered more difficult to optimize because of their numerous local minima scattered in a huge search space. Functions $f4$ and $f6$ are extremely difficult to be optimized since their local minimum values are present everywhere inside the search space.

Table 1 summarizes the average fitness values obtained when applying QC-CEA on these multimodal functions. It also contains the average fitness values of CCEA for comparison. Fig. 3 further depict the performance of QCCEA and CCEA on functions f_4 to f_9 through the whole evolutionary process. As shown in Table 3, QCCEA outperformed CCEA on 5 out of the 6 multimodal functions. For the function $f9$ where QCCEA fall short of CCEA, the difference in the average fitness is 7.5% For the rest of functions where QCCEA dominated CCEA, the most prominent performance difference between QCCEA and CCEA is 24% , which occurs on function $f5$, i.e. the Rotated Griewank's function, as shown in Fig. 3

For Rotated Ackley's function $f4$ shown in Fig.4, both QCCEA and CCEA reaches the same fitness value at generation 2000. Afterwards QCCEA surpasses CCEA for the rest of evolutionary process.Both algorithms are unable to identify near optimal solutions. For Rotated Rastrigin's function $f6$, the same observation occurs as both algorithms cannot find near optimal solutions,

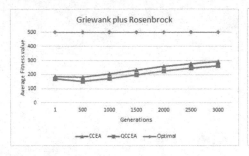
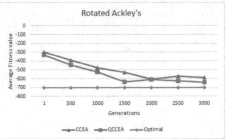

Fig. 3. Results Multimodal functions

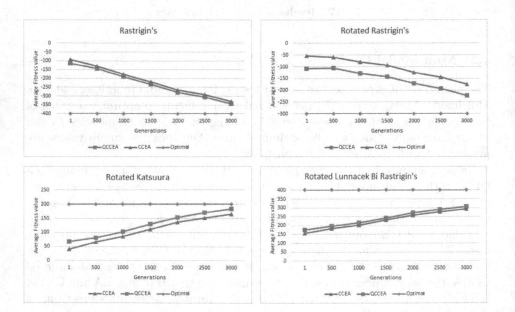

Fig. 4. Results Multimodal functions

as demonstrated in Fig.3. Nevertheless, QCCEA clearly outperformed CCEA, in that QCCEA reaches 25% away from the optima, which is 20% closer to the optima than the CCEA. For Lunacek Bi Rastrigin's function $f7$, both QCCEA and CCEA starts at the same point, but the performance of CCEA drops at around 500 generations. QCCEA performed consistently better than CCEA on this function, as witnessed in Fig.4. Based on the above discussion, it is evident that QCCEA tends to perform better than CCEA, irrespective of the shape, properties, and the number of local optima of the tested benchmark functions.

5 Conclusion

This paper presented QCCEA, a new quantum inspired competitive co-evolution algorithm. In QCCEA, candidate solution is represented through a combination of a set of solution points which are jointly described through normal distributions which is different from traditional quantization methods such as QEA [4] and QNN [9]. The performance of the QCCEA is evaluated on 9 benchmark functions and the obtained experiment results shows that QCCEA outperformed CCEA over 7 out of 9 test functions. The search speed of the QCCEA is also noticeably faster, since, in comparison with CCEA, QCCEA identifies near-optimal solutions in less number of generations.

References

1. Li, F., Zhou, M., Li, H.: A novel neural network optimized by quantum genetic algorithm for signal detection in mimo-ofdm systems. In: 2011 IEEE Symposium on Computational Intelligence in Control and Automation (CICA), pp. 170–177 (2011)
2. Li, F., Hong, L., Zheng, B.: Quantum genetic algorithm and its application to multi-user detection. In: 9th International Conference on Signal Processing, ICSP 2008, pp. 1951–1954 (2008)
3. Han, K.-H., Kim, J.-H.: Genetic quantum algorithm and its application to combinatorial optimization problem. In: Proceedings of the 2000 Congress on Evolutionary Computation, vol. 2, pp. 1354–1360 (2000)
4. Han, K.-H.: Quantum-inspired Evolutionary Algorithm. PhD thesis, Korea Advanced Institute of Science and Technology KAIST (2003)
5. Platel, M.D., Schliebs, S., Kasabov, N.: Quantum-inspired evolutionary algorithm: A multimodel eda. Trans. Evol. Comp. 13, 1218–1232 (2009)
6. Sreenivas, S.T.: A quantum inspired competitive coevolution evolutionary algorithm. Master's thesis, Unitec Institute of Technology, New Zealand (2013)
7. Han, K.-H., Hong Park, K., Ho Lee, C., Hwan Kim, J.: Parallel quantum-inspired genetic algorithm for combinatorial optimization problem. In: Proc. 2001 Congress on Evolutionary Computation, Piscataway, NJ, pp. 1422–1429. IEEE Press (2001)
8. Kim, Y., Kim, J.-H., Han, K.-H.: Quantum-inspired multiobjective evolutionary algorithm for multiobjective 0/1 knapsack problems. In: IEEE Congress on Evolutionary Computation, CEC 2006, pp. 2601–2606 (2006)
9. Lu, T.-C., Yu, G.-R., Juang, J.-C.: Quantum-based algorithm for optimizing artificial neural networks. IEEE Transactions on Neural Networks and Learning Systems 24(8), 1266–1278 (2013)

10. Stanley, K.O., Miikkulainen, R.: Competitive coevolution through evolutionary complexification. Journal of Artificial Intelligence Research 21, 63–100 (2004)
11. Liang, J., Qu, B., Suganthan, P.N., Hernndez-Daz, A.G.: Problem definitions and evaluation criteria for the cec 2013 special session on real-parameter optimization," tech. rep., Nanyang Technological University (2013)

Text Categorization with Diversity Random Forests

Chun Yang[1], Xu-Cheng Yin[1,*], and Kaizhu Huang[2]

[1] School of Computer and Communication Engineering, University of Science and Technology
Beijing, Beijing 100083, China
xuchengyin@ustb.edu.cn

[2] Department of Electrical and Electronic Engineering, Xi'an Jiaotong-Liverpool University,
Suzhou 215123, China

Abstract. Text categorization (TC), has many typical traits, such as large and difficult category taxonomies, noise and incremental data, etc. Random Forests, one of the most important but simple state-of-the-art ensemble methods, has been used to solve such type of subjects with good performance. most current Random Forests approaches with diversity-related issues focus on maximizing tree diversity while producing and training component trees. There are much diverse characteristics for component trees in TC trained on data of noise, huge categories and features. Consequently, given numerous component trees from the original Random Forests, we propose a novel method, Diversity Random Forests, which diversely and adaptively select and combine tree classifiers with diversity learning and sample weighting. Diversity Random Forests includes two key issues. First, by designing a matrix for the data distribution creatively, we formulate a unified optimization model for learning and selecting diverse trees, where tree weights are learned through a convex quadratic programming problem with given sample weights. Second, we propose a new self-training algorithm to iteratively run the convex optimization and automatically learn the sample weights. Extensive experiments on a variety of text categorization benchmark data sets show that the proposed approach consistently outperforms state-of-the-art methods.

1 Introduction

Classification techniques, especially text categorization, have many applications in Data Mining (DM) and Information Retrieval (IR), e.g., spam detection, sentiment detection, personal email sorting and document ranking [1]. Typical issues in text categorization and recommendation systems are large and difficult category taxonomies, huge samples, noise and incremental data, and various features. Classifier ensemble is a potential solution for such type of subjects. Many research efforts demonstrated that the Random Forests approach [2] is the most important but simple state-of-the-art ensemble for classification, consequently, for text categorization.

Random Forests can exploit implicit and explicit diversities together. The method combines the "Bagging" idea for instance sampling with the implicit diversity and the random selection of variables for feature selection with the explicit diversity. Generally, the performance of a classifier ensemble (including Random Forests) relies on not only the accuracy but also the diversity of component trees. Consequently, how to diversely generate and combine diverse classifiers plays an important role in Random Forests.

---

* Corresponding author.

C.K. Loo et al. (Eds.): ICONIP 2014, Part III, LNCS 8836, pp. 317–324, 2014.

On the field of Random Forests research, there are many researches for improving Random Forests with diversity-related issues, most of which focus on maximizing tree diversity while producing and training component trees. Liu et.al.[3] proposed Max-diverse Ensemble method, which has the maximum diversity and uses only simple probability averaging without any feature selection criterion or other random elements. Later, Liu et.al.[4] proposed Coalescence method, which coalesces a number of points in the random-half of the spectrum and is found to perform better than any single operating point in the spectrum, without the need to tune to a specific level of randomness.

Obviously, In TC, there are a lot of diverse characteristics for component trees which are trained on data of noise, large categories and huge features, i.e., some trees or a subset of trees by properly selecting will be much diverse from each other. Alternatively, we improve Random Forests with diversity from pruning ensemble, as ensemble of the partial available component trees may be better than that of the whole [5]. Given numerous component trees from the original Random Forests, we want to diversely and adaptively select and combine tree classifiers with diversity learning.

Moreover, in classifier ensemble, all existing diversity measures are calculated on the training set, which means the performance of optimization relies on the samples of training set besides the diversity learning itself [6,7,8]. In some relative fields, researchers suggest sample weighting is needed to correct for imperfections in the samples that might lead to bias and other departures between the sample and the reference population. Adaboost[9] is one of the most famous sample weighting models.

Consequently, given numerous component trees from the original Random Forests, we propose a novel method, **D**iversity **R**andom **F**orests (DRF), which diversely and adaptively select and combine tree classifiers with diversity learning and sample weighting. Diversity Random Forests uses a self-training algorithm to iteratively run the convex optimization and automatically learn the sample weights. Each iteration of this self-training algorithm consists of two main steps: (1) calculate tree weights by solving an optimization problem with sample weights known, and then (2) update sample weights. In the first step, diversity learning with sample weights is converted into a unified convex quadratic programming optimization model, by creatively setting the sample distribution as a diagonal matrix. In the second step, sample weights are automatically and adaptively updated with a dynamically damped learning trick. Therefore, the whole self-training algorithm has a good convergence performance. Moreover, experimental results on a variety of text categorization benchmark data sets definitely show that our proposed approach has very promising performance.

The rest of the paper is organized as follows. The DRF model is presented in Section 2, and more details on the learning algorithm is described in Section 3. Section 4 shows extensive experimental results. Finally, conclusion is drawn in Section 5.

2   Diversity Random Forests

2.1   Random Forests

Random Forests [2] are an ensemble learning method for classification. It generates a multitude of decision trees based on bootstrap samples of the training data and outputs the class that is the mode of the classes output by individual trees. For each node of a tree, m variables are randomly chosen and the best split based on these m variables is calculated based on the bootstrap data. Traditionally, m is set to $\lceil \sqrt{u} \rceil$, where u stands for the number of variable. Each decision tree results in a classification and is said to cast a weighted vote for that classification, and Random Forests returns the class that received the most votes.

As various theoretical and empirical studies shows[10,11,12], Random Forests are fast and easy to implement, produce highly accurate predictions and can handle a very large number of input variables without overfitting. In fact, they are considered to be one of the most accurate general-purpose learning techniques available.

In the paper, we formulate an optimization model based on the original Random Forests model [2]. Moreover, instead of the original output, the oracle output $\mathbf{O}$ of Random Forests is used for the optimization. Let the number of samples set be N, and the number of component trees L. $\mathbf{O}$ is a $N \times L$ matrix, and element

$$O_{ij} = \begin{cases} 1 & \text{the } j^{th} \text{ tree classified the } i^{th} \text{ sample correctly} \\ -1 & \text{otherwise} \end{cases} \tag{1}$$

2.2 Diversity Random Forests Model

As an ensemble approach, Random Forests can be improved by pruning component trees. Specially, for weighted-vote Random Forests, the improvement is equivalent to a mathematical optimization problem with tree weights. Define tree weights vector $\mathbf{w} = [w_1, w_2, ..., w_L]$, where $\sum_{j=1}^{L} w_j = 1$, $w_j \geq 0$. Traditionally, $\mathbf{w}$ is learned by

$$\mathbf{w_{opt}} = argmin_{\mathbf{w}} f_1(\mathbf{w}, \mathbf{P})$$
$$s.t. \ \mathbf{w_{opt}} \succeq 0, \ \mathbf{1^T w_{opt}} = 1. \tag{2}$$

where $\mathbf{P}$ is the accuracy of each tree on training set. $\mathbf{P} = [P_1, P_2, ..., P_L]^T$, where $P_j = \sum_{i=1}^{N} O_{ij}$. The optimization function in Equation (2) usually has functional relationship f_1 with the accuracy $\mathbf{P}$.

Previous works show that a multi-criteria searching for an ensemble that maximizes both accuracy and diversity leads to more accurate ensembles than a single optimization criterion. Thus, consider diversity in component trees of Random Forests and add a regularization term about diversity to expand Equation (2) as,

$$\mathbf{w_{opt}} = argmin_{\mathbf{w}} f_1(\mathbf{w}, \mathbf{P}) + \lambda div(\mathbf{w})$$
$$s.t. \ \mathbf{w_{opt}} \succeq 0, \ \mathbf{1^T w_{opt}} = 1. \tag{3}$$

In Equation (3), $div(\mathbf{w})$ is the diversity of ensemble with classifier weights $\mathbf{w}$. If use pairwise diversity method, $div(\mathbf{w})$ can be calculated as an average,

$$div(\mathbf{w}) = \mathbf{w}^T \mathbf{D} \mathbf{w}$$
$$\mathbf{D} = f_D(\mathbf{O}^T \mathbf{O}, \mathbf{1}_{N \times 1}^T \mathbf{O}) \tag{4}$$

where $\mathbf{D}$ is the diversity matrix of component trees, which has functional relationship f_D with $\mathbf{O}^T \mathbf{O}$ and $\mathbf{1}^T \mathbf{O}$.

In the paper, the Disagreement(dis) [13] is chosen to measure diversity, which is calculated by,

$$\mathbf{D_{dis}} = \frac{1}{2N}(N\mathbf{1}_{L \times L} - \mathbf{O}^T\mathbf{O}) \tag{5}$$

If use the average accuracy to calculate $f_1(\mathbf{w}, \mathbf{P})$, and the pairwise diversity Disagreement to calculate $div(\mathbf{w})$, then Equation (3) equals,

$$\mathbf{w_{opt}} = argmin_{\mathbf{w}} - \lambda\mathbf{w}^T\mathbf{D_{dis}}\mathbf{w} - \mathbf{Pw}$$
$$s.t. \ \mathbf{w_{opt}} \succeq 0, \ \mathbf{1}^T\mathbf{w_{opt}} = 1. \tag{6}$$

One issue of the optimization is how to determine the parameter λ. However, empirical analysis shows that the recognition rate has a very little change when the value λ changes.

More importantly, the performance of optimization function is totally different because of different training set selection. Considering the influence of training set, we expand Equation (6) as,

$$\mathbf{w_{opt}} = argmin_{\mathbf{w}} - \lambda\mathbf{w}^T\mathbf{D_{dis,\Omega}}\mathbf{w} - \mathbf{P_{\Omega}w}$$
$$s.t. \ \mathbf{w_{opt}} \succeq 0, \ \mathbf{1}^T\mathbf{w_{opt}} = 1. \tag{7}$$

where Ω is a parameter of the data distribution (sample weights). This (Equation (7)) is the model of our Diversity Random Forests.

To simplify calculation and remain the optimization as a convex problem, we creatively set Ω as a $N \times N$ diagonal matrix, and $diag(\Omega)_i = \Omega_{ii}$ stands for the weight of sample x_i, where $diag(\Omega)_i \geq 0$, $\mathbf{1}^T diag(\Omega) = 1$. Thus, $\mathbf{P_{\Omega}}$ and $\mathbf{D_{dis,\Omega}}$ can be calculated by,

$$\mathbf{P_{\Omega}} = \mathbf{1}^T\Omega\mathbf{O}$$
$$\mathbf{D_{dis,\Omega}} = \tfrac{1}{2}(\mathbf{1}_{L \times L} - \mathbf{O}^T\Omega\mathbf{O}) \tag{8}$$

Consequently, the optimization (7) can be simplified to a convex quadratic programming problem with a given Ω.

3   DRF Algorithm

It is difficult to find the solution for the optimization in Equation (7) without both $\mathbf{w}$ and Ω. However, with known Ω, the optimization is simplified to a quadratic programming problem. Thus, we propose an iterative learning algorithm, Diversity Random Forests (DRF) Algorithm, which is shown in Algorithm 1.

In Algorithm 1, the validation set is bootstrapped from the original training set of Random Forests. We assume the sample weights parameter Ω_{t+1} has a relationship with Ω_t, and use a dynamically damped trick, i.e., the damped factor $\beta_t \in [0, 1]$ and $\beta_t \leq \beta_{t+1}$. In the paper, we set β_t as,

$$\beta_t = \frac{1}{t} \tag{9}$$

Algorithm 1: DRF Algorithm

Input:
    Tr: the validation set. $|Tr| = N$
    $H = \{h_1, h_2, ..., h_L\}$: the component tree set, $|H| = L$.
    M: pairwise diversity method.

Output:
    w: the component tree weights.

Parameter:
    T: the max epoch.
    Ω_t: a diagonal matrix, and $diag(\Omega_t)_i$ is the weight of
        sample x_i used to calculate **w** on the tth turn.
    Ω_t^*: a diagonal matrix, and $diag(\Omega_t^*)_i$ is the updated
        weight of sample x_i on the tth turn.
    ϵ_t: the error rate on the tth turn.
    β_t: a parameter that $\beta_t \in [0, 1]$, and $\beta_t \leq \beta_{t+1}$.

Procedure:
1:    Set $diag(\Omega_1)_i = 1/N$.
2: **For** $t = 1, 2, ..., T$;
3:    Use Equation (7) and (8) to calculate **w**.
4:    Calculate ϵ_t by **w** and Tr.
5:    Use ϵ_t to calculate updated weight Ω_t^*.
6:    $\Omega_{t+1} = \beta_t \Omega_t^* + (1 - \beta_t)\Omega_t$
7: **End**

The updated weight matrix Ω_t^* increases the weights of easily wrong-classified samples. We update Ω_t^* by DRF-Exp, which gets the idea from the adaptive reweighting step in Boosting [9]. In Boosting, a distribution of weights over training samples is adaptively maintained, and component trees are created sequentially with each tree concentrating on instances that are not well learnt by previous ones. With this mechanism, the learning process is more efficient. Similarly, DSWL-Exp updates Ω_t^* by,

$$\alpha = \frac{1}{2} ln \frac{1 - \epsilon_t}{\epsilon_t}$$
$$diag(\Omega_{t+1}^*)_i = \frac{diag(\Omega_t^*)_i exp(-\alpha m_i)}{Z_{t+1}} \tag{10}$$

where Z_{t+1} is a normalization factor, then $diag(\Omega_{t+1}^*)$ is a valid distribution.

4   Experiments

We evaluated the performance of DRF by comparing against some state-of-art methods, such as Multinomial Naive Bayesian, J48, Support Vector Machines and Random Forests, on a variety of document collections.

4.1 Experimental Data

The detailed characteristics of the various document collections used in our experiments are available in [14].[1] More information for the data sets is presented in Table 1.

---

[1] http://sourceforge.net/projects/weka/files/datasets/
  text-datasets/19MclassTextWc.zip

Table 1. Benchmark Datasets

DataSet	Source	Docs	Words	Classes	DataSet	Source	Docs	Words	Classes
fbis	TREC	2463	2000	17	re1	Reuters	1657	3758	25
la1s	TREC	3204	13472	6	tr11	TREC	414	6429	9
la2s	TREC	3075	13472	6	tr12	TREC	313	5804	8
oh0	OHSUMED	1003	3182	10	tr21	TREC	336	7902	6
oh10	OHSUMED	918	3012	10	tr23	TREC	204	5832	6
oh15	OHSUMED	1050	3238	10	tr31	TREC	927	10128	7
oh5	OHSUMED	913	3100	10	tr41	TREC	878	7454	10
ohscal	OHSUMED	11162	11465	10	tr45	TREC	690	8261	10
re0	Reuters	1504	2886	13	wap	WebACE	1560	8460	20

4.2 Experimental Setup

The experiment compares DRF with some state-of-art methods, e.g., Multinomial Naive Bayes(MNB), J48, Support Vector Machines(SVM,[15]), Random Forests(RF, [2]). Both Multinomial Naive Bayes and J48 classifier are generated by WEKA,[2] and Random Forests classifier is generated by Matlab toolbox.[3] For each method, all parameters are set by default. In SVM, the Linear kernel is used, and the best c and g parameter is selected by cross validation from $c = 2^{-5}, 2^{-4}, ..., 2^5$, $g = 2^{-5}, 2^{-4}, ..., 2^5$.

In the experiment, 5-fold cross validation is performed on each data set. We assign Ranks to evaluate the methods' performance on each data set [16]. Mark the best method Rank 1, and the worse, the larger. Then calculate the average Rank for each method. Moreover, we also calculated the average recognition rate (AVE).

4.3 Results

The experimental results are shown in Table 2. In addition, the highest recognition rate for each data set is highlighted in boldface. As shown in Table 2, we can observe:

- Among four state-of-art methods(J48, MNB, SVM, RF), the best rank corresponds to RF(2.4), followed by SVM(2.8), MNB(3.6) and J48(4.6). On most data sets, RF achieves the best recognition rate, and is slightly worse than SVM on 'fbis', 're1', 'tr11', 'tr21', 'tr41' and 'wap' data sets. These results show that RF is a powerful technique for text categorization.
- Moreover, DRF ranks 1.6, and is 0.9% higher than Random Forests for the average classification precision. On most data sets, DRF achieves an 1%-4% higher recognition rate than RF, except on 'la1s', 'la2s' and 'wap'. That is to say, in TC, our proposed method, DRF, can utilize diversity in component trees and select a proper subset of trees in RF for ensemble.
- Specifically, by selecting training sets (calculate the sample weights) carefully, DRF has the minimum Rank and largest average recognition rate, and outperforms

---

[2] http://www.cs.waikato.ac.nz/ml/weka/
[3] https://code.google.com/p/randomforest-matlab/

J48, MNB, SVM and RF. In most cases, DRF achieves the best performance when there are enough training data for learning component trees, tree weights and sample weights. Consequently, our methods obtain the best rank (1.6) in all experimental approaches.

Table 2. Comparison of recognition rate (%) (Average±Standard Deviation).

Datasets	J48	MNB	SVM	RF	DRF
fbis	72.03 ± 2.07	77.30 ± 1.84	82.79 ± 1.07	82.74 ± 1.17	$\mathbf{83.35 \pm 1.20}$
la1s	75.56 ± 1.93	87.45 ± 0.51	87.83 ± 1.11	$\mathbf{88.08 \pm 1.61}$	88.05 ± 1.55
la2s	76.33 ± 1.66	88.78 ± 1.03	88.85 ± 1.17	$\mathbf{88.93 \pm 1.60}$	88.80 ± 1.60
oh0	81.05 ± 4.99	$\mathbf{88.43 \pm 3.09}$	85.14 ± 2.85	88.03 ± 2.66	88.03 ± 2.66
oh10	68.38 ± 3.06	78.00 ± 3.80	76.29 ± 4.54	80.95 ± 6.79	$\mathbf{81.14 \pm 6.79}$
oh15	72.39 ± 5.08	$\mathbf{82.04 \pm 1.81}$	76.88 ± 3.74	80.49 ± 5.08	81.04 ± 5.10
oh5	80.71 ± 5.13	87.47 ± 3.01	85.84 ± 4.68	87.58 ± 2.74	$\mathbf{89.32 \pm 2.74}$
ohscal	70.23 ± 5.10	73.99 ± 1.14	76.63 ± 1.49	80.87 ± 1.21	$\mathbf{80.93 \pm 3.21}$
re0	70.68 ± 1.96	76.87 ± 4.32	81.25 ± 4.30	81.32 ± 5.30	$\mathbf{81.52 \pm 5.26}$
re1	77.43 ± 4.43	79.05 ± 6.16	81.83 ± 4.23	81.81 ± 5.86	$\mathbf{82.35 \pm 5.86}$
tr11	77.06 ± 3.24	84.07 ± 3.07	87.20 ± 1.58	84.53 ± 2.87	$\mathbf{88.41 \pm 2.87}$
tr12	79.21 ± 4.05	81.76 ± 7.43	85.93 ± 4.02	87.19 ± 5.33	$\mathbf{87.84 \pm 5.33}$
tr21	77.95 ± 7.25	60.09 ± 6.01	86.00 ± 4.21	85.31 ± 4.53	$\mathbf{86.28 \pm 4.46}$
tr23	$\mathbf{92.68 \pm 5.17}$	69.07 ± 9.13	83.34 ± 4.66	83.89 ± 8.54	86.30 ± 8.54
tr31	93.53 ± 1.37	95.04 ± 1.35	97.09 ± 0.82	97.19 ± 2.52	$\mathbf{97.52 \pm 2.52}$
tr41	92.03 ± 2.67	93.97 ± 2.94	$\mathbf{94.76 \pm 1.69}$	92.94 ± 2.35	93.96 ± 2.35
tr45	91.01 ± 1.50	82.46 ± 3.78	89.28 ± 4.36	90.29 ± 4.51	$\mathbf{92.75 \pm 4.51}$
wap	65.38 ± 2.58	79.94 ± 3.94	$\mathbf{84.49 \pm 1.98}$	82.71 ± 2.15	81.23 ± 2.15
AVE	78.54	81.43	85.08	85.83	**86.60**
Ranks	4.6	3.6	2.8	2.4	1.6

5 Conclusion

Random Forests approach is widely considered as an effective method to improve accuracy of various component trees, which has a variety of applications in information retrieval and data mining, e.g., text categorization, image retrieval, and recommendation systems. By improving Random Forests from ensemble pruning aspect, we propose a convex mathematical model for ensembling components in Random Forests, which takes into account both diversity learning and sample weighting. We also propose an iterative self-training algorithm for DRF, where the optimization problem is simplified as a convex quadratic programming problem at each iteration. In the experiments, DRF is compared with other state of art methods, e.g., J48, Multinomial Naive Bayes, Support Vector Machines and Random Forests. A series of experiments on benchmark data sets show that our proposed method achieves very encouraging results for text categorization.

Acknowledgments. The research was partly supported by the National Natural Science Foundation of China (61105018, 61175020).

References

1. Manning, C.D., Prabhakar, R., Hinrich, S.: Introduction to Information Retrieval. Cambridge University Press (2008)
2. Breiman, L.: Random forests. Machine Learning 45, 5–32 (2001)
3. Liu, F.T., Ting, K.M., Fan, W.: Maximizing tree diversity by building complete-random decision trees. In: Proceeding of PAKDD, pp. 605–610 (2005)
4. Liu, F.T., Ting, K.M., Yu, Y., Zhou, Z.H.: Spectrum of variable-random trees. J. Artif. Intell. Res. 32, 355–384 (2008)
5. Zhou, Z.H., Wu, J., Tang, W.: Ensembling neural networks: Many could be better than all. Artificial Intelligence 137, 239–263 (2002)
6. Yin, X.-C., Huang, K., Hao, H.-W., Iqbal, K., Wang, Z.-B.: Classifier ensemble using a heuristic learning with sparsity and diversity. In: Huang, T., Zeng, Z., Li, C., Leung, C.S. (eds.) ICONIP 2012, Part II. LNCS, vol. 7664, pp. 100–107. Springer, Heidelberg (2012)
7. Yin, X.C., Huang, K., Hao, H.W., Iqbal, K., Wang, Z.B.: A novel classifier ensemble method with sparsity and diversity. Neurocomputing 134, 214–221 (2014)
8. Yin, X.C., Huang, K., Yang, C., Hao, H.W.: Convex ensemble learning with sparsity and diversity. Information Fusion 20, 49–59 (2014)
9. Freund, Y., Schapire, R.: Experiments with a new boosting algorithm. In: Proceedings of ICML, pp. 148–156 (1996)
10. Biau, G.: Analysis of a random forests model. J. Mach. Learn. Res. 13, 1063–1095 (2012)
11. Genuer, R., Poggi, J.M., Tuleau-Malot, C.: Variable selection using random forests. Pattern Recognition Letters 31(14), 2225–2236 (2010)
12. Verikas, A., Gelzinis, A., Bacauskiene, M.: Mining data with random forests: A survey and results of new tests. Pattern Recognition 44(2), 330–349 (2011)
13. Skalak, D.B.: The sources of increased accuracy for two proposed boosting algorithms. In: Proceeding of AAAI, pp. 120–125 (1996)
14. Han, E.H., Karypis, G.: Centroid-based document classification: Analysis and experimental results. In: Proceedings of European PKDD, pp. 424–431 (2000)
15. Chang, C.C., Lin, C.J.: Libsvm: A library for support vector machines. ACM Trans. Intelligent Systems and Technology 2(3), 1–27 (2011), http://www.csie.ntu.edu.tw/cjlin/libsvm
16. Brazdil, P., Soares, C.: A comparison of ranking methods for classification algorithm selection. In: Proceedings of ECML, pp. 63–74 (2000)

Unknown Attack Detection
by Multistage One-Class SVM
Focusing on Communication Interval

Shohei Araki[1], Yukiko Yamaguchi[2], Hajime Shimada[2], and Hiroki Takakura[2]

[1] Nagoya University, Graduate School of Information Science, Aichi, Japan
araki@net.itc.nagoya-u.ac.jp
[2] Nagoya University, Information Technology Center, Aichi, Japan
{yamaguchi,shimada,takakura}@itc.nagoya-u.ac.jp

Abstract. Cyber attacks have been more sophisticated. Existing countermeasures, e.g, Intrusion Detection System (IDS), cannot work well for
detecting their existence. Although anomaly-based IDS is considered to
be promising approach to detect unknown attacks, it still lacks the ability
to distinguish sophisticated attacks from trivial known ones. Therefore,
we applied multistage one-class Support Vector Machine (OC-SVM) to
detect such serious attacks. At the first stage, two training data are retrieved from traffic archive. The one is used for training OC-SVM and
then, attacks are obtained from the another. Also testing data from real
network are examined by the same OC-SVM and attacks are extracted.
The attacks from the traffic archive are used for training OC-SVM at
the second stage and those from real network are analyzed. Finally, we
can obtain unknown attacks which are not stored in archive.

Keywords: Intrusion Detection System, anomaly detection, network
security.

1   Introduction

In recent years, a threat of cyber attacks over the Internet has become a serious
issue. An attacker attacks computer systems and networks in order to steal
confidential information for earning money in black market. These cyber attacks
become more varied and sophisticated year by year.

To detect cyber attacks, Intrusion Detection System (IDS) plays an important
role. There are two types of IDS by detection method: a signature-based IDS
and an anomaly-based IDS. The signature-based IDS detects attacks by pattern
matching of predefined signatures. It has high detection rate, but it cannot detect unknown attacks. On the other hand, an anomaly-based IDS learns normal
behavior of network traffic and extracts abnormal values of network traffic behavior as attacks. Although the anomaly-based IDS can detect unknown attacks,
it has some problems. Above anomaly-based methods cannot tell whether the
detected attacks are known attacks or not. Furthermore, it is impossible to show
the seriousness of each unknown attack which affects on the network. Especially,

C.K. Loo et al. (Eds.): ICONIP 2014, Part III, LNCS 8836, pp. 325–332, 2014.

in case of a targeted attack, specially-crafted tools are used along with other conventional tools which cause the barrage of trivial attacks. A network administrator has to search the most serious attack among huge amount of unknown attacks. Therefore we need some method to extract such serious attack.

In this paper, in order to solve above problems, we introduce 6 new features to Kyoto2006+ Dataset [2]. Also new method is proposed to detect serious attacks by using multistage one-class Support Vector Machine (OC-SVM [6]). The multistage OC-SVM uses three sets of traffic, two sets retrieved from a traffic archive and one extracted from real network. At the first stage, OC-SVM learns older archive set and then analyzes newer archive set and one from real network. At the second stage, OC-SVM learns outlier traffic from the newer archive set and analyzes that from the real network. As a result, extracted traffic from outlier of the real network which does not exist in the newer set can be extracted, and we should pay attention to it as possible serious attack.

We evaluated our method using Kyoto2006+ Dataset and 6 new features. The results show that our method detects attacks with higher accuracy than by using only conventional features and successfully extracted unknown attacks.

2   Related Works

Several researcher proposed anomaly-based IDS methods using OC-SVM. Eskin et al. [3] proposed a method of anomaly detection. They compared methods based on OC-SVM, clustering and k-nearest neighbor and show higher detection rate at OC-SVM than others. Prdrisci et al. [5] proposed a method that extracts features from payload and detects attacks using OC-SVM.

An approach of unknown attack detection, Song et al. [8] proposed a method that extracts new features from alerts of a signature-based IDS and detects unknown attacks. 0-day attack shows quite irregular behavior from known attacks concerning packet size and communication interval because an attacker confirms the effectiveness of the attack. These irregular characteristic often raises alerts of the signature-based IDS so that they can detect 0-day attack by analyzing alerts. This method uses alerts of a signature-based IDS, so it cannot detect unknown attacks that a signature-based IDS does not report any alert.

Above anomaly-based methods cannot detect whether the detected attacks are known attacks or not. In other words, these methods cannot extract only unknown attacks. Because various countermeasures are deployed against known attacks, a network administrator has to dedicate to finding the existence of unknown attacks. Therefore, we propose a multistage OC-SVM method to extract such unknown attacks.

3   Multistage OC-SVM

The overall process of proposed method is composed of following steps (Fig. 1).

1. Features Extraction from past traffic data and a real network.

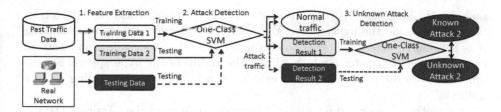

Fig. 1. Overall process

2. Attack detection by the first stage OC-SVM.
3. Classification to known attacks and unknown attacks by the second stage OC-SVM.

3.1 Feature Extraction

At first, we perform feature extraction. The features are extracted by each TCP sessions. Training Data 1 and Training Data 2 are extracted from past traffic data. Training Data 1 is extracted at older date than Training Data 2. Testing Data is extracted from a real network.

Kondo et al. [4] have shown that bot infected computers show different behavior from normal behavior about packet size and communication interval. By generating histograms for reflecting these characteristics, their method detects malicious sessions with Command & Control server by using SVM.

In addition to features of Kyoto2006+ Dataset, this paper introduces the following 6 new features from traffic data. Here, "duration" means the time length of a TCP session.

1. The number of received bytes divided by duration
2. The standard deviation of 1. concerning the past 100 sessions which have the same source IP addresses of the current session
3. The number of sent bytes divided by duration
4. The standard deviation of 3. concerning the past 100 sessions which have the same source IP addresses of the current session
5. The communication interval between the current and the previous sessions which have the same destination IP address.
6. The standard deviation of 5. in the past 100 times sessions

These features are intended to represent characteristics on communication interval. Our proposed method, therefore, analyzes 12 conventional and 6 new features to detect unknown attacks.

3.2 Attack Detection

The first stage classifier of OS-SVM is used to distinguish attack sessions from normal ones. By leaning normal traffic, the classifier creates a hypersphere that includes the large majority of normal sessions. If the classification assigns a testing session inside the hypersphere, the session is considered normal. Otherwise it is considered attack. In order to effectively use OC-SVM, it is mandatory to choose proper parameter ν which adjusts the radius of the hypersphere. For example, if ν is set to 0.1, OC-SVM calculates a hypersphere excluding 10% of data.

Our method requires three different time of traffic data, i.e., Training Data 1, Training Data 2 and Testing Data. Both training data are retrieved from traffic archive and the testing data is collected from the real network. Training data 1 is older than Training Data 2. By using Training Data 1, the first stage classifier learns normal traffic and then, evaluates Training Data 2 and Testing Data to obtain their outlier traffic. The outlier traffic from Training Data 2 and Testing Data is used for the second stage analysis describe in Section 3.3.

3.3 Extraction of Serious Attacks

As shown in Fig. 1, the second stage classifier is trained by using the outlier sessions from Training Data 2. After the training, the classifier obtains hypersphere that includes attacks observed by the time of Training Data 2.

Then the outlier sessions from Testing Data are examined. If the classifier assigns a session outside of the hypersphere, the session can be considered newly unknown attack. Because such a session has not been observed previously, it can be considered that a zero-day attack or a targeted attack causes the session.

By adopting our multi-stage OC-SVM, we can extract the most outlier sessions from the real network. By considering its degree of the outlier, the sessions can be considered the most hazardous and should be deeply analyzed as soon as possible. It means that our method can tell the network administrator the priority to take action to zero-day attacks.

4 Evaluation

4.1 Dataset

Because it is difficult to prepare labeled traffic data in a real network environment, we used Kyoto2006+ Dataset for evaluation. It is obtained from honeypot networks of Kyoto University. In this dataset, 24 features are generated by each TCP session. 24 features consist of 14 conventional features and 10 additional features. The former 14 features are based on features of KDDCup1999 [1] Dataset. These features consist of duration, service type, the number of connections whose source IP address and destination IP address are the same, a rate of "SYN" errors, and so on. The latter 10 features consist of signature-based IDS alerts, IP address, port number, start time of the session. We use 12 conventional features except 2 non-numeric attributes in the former 14 features. We do not use the latter 10 features for detection because these features are non-numeric attribute. In addition to these features, we use 6 features that mentioned in Sec. 3.1.

Table 1. Detection performance when $\nu = 0.06$

		Jan. 20th, 2008		Jan. 30th, 2008	
		True Classification			
		Attack	Normal	Attack	Normal
Detection	Attack	656	4,082	475	2,846
Result	Normal	40	63,168	221	70,257
Detection Rate		94.25%		68.22%	
False Positive Rate		6.07%		3.89%	

In this dataset, most of traffic data are attack traffic so that it does not represent the ratio of attack traffic in practical network. So, we adjusted attack rate to 1% because it is very few in general environment networks.

4.2 Evaluation of the First Stage Classifier

Overview. We selected November 1st in 2007 as training data. The number of normal sessions is 37,730. We selected January 20th and January 30th in 2008 as testing data. The former data have 67,250 normal sessions and 696 attack sessions and the latter data have 73,103 normal sessions and 834 attack sessions. Attack sessions are adjusted to approximately 1% of the normal sessions.

We use 12 conventional features and newly extracted 6 features explained in Section 3.1. For comparison, we also evaluate detection ratio by using only the 12 conventional features.

We define detection rate and false positive rate as following expressions.

$$\text{Detection Rate} = \frac{\#\text{ of sessions classified as an attack}}{\#\text{ of all attack sessions}}$$

$$\text{False Positive Rate} = \frac{\#\text{ of normal sessions classified as an attack}}{\#\text{ of all normal sessions}}$$

Results and Analysis. Fig. 2 shows that ROC curves under $\nu=\{0.01, 0.02, 0.04, 0.06, 0.08, 0.10\}$. This figure also shows the detection results of OC-SVM using only conventional 12 features in order to confirm the effectiveness of new features. The results show that the additional 6 features contributes to obtaining high detection rate and low false positive rate.

Table 1 shows the detailed results of detection on January 20th and 30th, 2008 under $\nu = 0.06$. The detection rate on January 30th is lower than the rate on January 20th. Although parameter ν is the same on two days, the detection rate is significantly different. Accordingly, it is necessary to estimate the appropriate parameter ν to perform high attack detection all dates.

There are some false positive sessions that cannot be found in the result by using only conventional features. Table 2 shows an example of false positive sessions in the result of proposed method. In Table 2, IP address indicates the

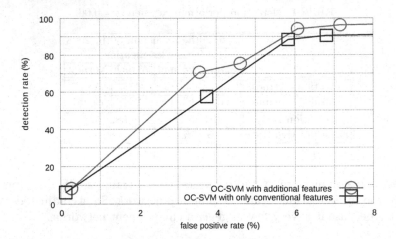

Fig. 2. ROC curve on January 20th, 2008

Table 2. Example of false positive sessions

Time	Duration	Received	Sent	SrcIP:port	DstIP:port
15:21:32	3.86	1,861,381,796	0	6403:9449	0f19:25
15:21:32	7.35	3,513	175	2133:3757	0f19:25
15:21:32	14.07	3,696	175	6110:49062	0f19:25
15:21:32	35.91	0	0	6028:33349	0f19:25
15:21:33	23.95	3,576	175	6110:48371	0f19:25

lower 16 bits of sanitized IPv6 address in Kyoto2006+. Since the 1st line in Table 2 shows that the received bytes of the session is over one million, it heavily affects on its following 100 sessions, i.e., their feature 2 described in Sec. 3.2. For this reason, sessions after huge traffic session were detected as attacks by mistake. To solve this problem, we need to filter out data such a noise before extracting features and training by OC-SVM. One of the filtering methods is grid-based data splitting algorithm [7]. By using such a method, we can exclude a noise data that has too extreme value.

4.3 Evaluation of the Second Stage Classifier

Overview. As similar to Fig. 1, we selected the attack sessions that were detected by the first stage classifier from January 20th to 27th in 2008 as Detection Result 1 for training data of the second stage classifier. Similarly, we selected from February 1st to 7th in 2008 as Detection Result 2 for testing data. We use training data and testing data for one week (longer than previous subsection) to obtain sufficient amount of data because the number of attacks is comparatively small. Although the training data includes normal session data as false positive, they are treated as attacks. Training Data 1 for the first stage classifier is November 1st in 2007 as same as in the previous subsection.

Table 3. Result of unknown attack detection

		True Classification	
		Unknown	Known
Detection	Unknown	107	18,559
Result	Known	43	30,952
Detection Rate		71.33%	
False Positive Rate		37.48%	

Table 4. Result of unknown attack detection by training only attacks

		True Classification	
		Unknown	Known
Detection	Unknown	120	10,370
Result	Known	30	39,141
Detection Rate		80.00%	
False Positive Rate		20.94%	

For the first stage classifier, we set the parameter ν to 0.05 in order to obtain moderate detection rate without extreme high false positive rate at any day. On the other hand, it is expected difference between known and unknown attacks is relatively small. In order to enhance the performance of the second stage classifier, we set larger value to v, i.e., 0.2.

As similar to the previous section, we define detection rate and false positive rate as follows. In this evaluation, normal sessions as false positive caused by the first stage classifier are also treated as known attack sessions.

$$\text{Detection Rate} = \frac{\text{\# of sessions classified as an unknown attack}}{\text{\# of all unknown attack sessions}}$$

$$\text{False Positive Rate} = \frac{\text{\# of attack sessions classified as an unknown attack}}{\text{\# of all known attack sessions}}$$

Results and Analysis. Table 3 shows a result of unknown attack detection. The result shows much higher false positive rate than the detection of the first stage classifier.

From these results, we suspected that the training data for the second stage classifier contain normal sessions as false positive caused by the first stage classifier. These noise session may affect the performance of pthe second stage classifier.

To prove this hypothesis, at first, we counted the number of sessions that were actual attack sessions and actual normal sessions in the training data for second stage classifier. The number of actual attack sessions are 3,547 and the number of actual normal sessions are 21,165. Although the first stage classifier can effectively detect attacks, there still remain huge amount of normal sessions in Detecting Result 1.

To prove effectiveness of our idea, next, we extracted 3,547 actual attack sessions from the Detection Result 1 and trained the second stage classifier by them. As same as the previous evaluation, we set the parameter ν to 0.2. Table 4 shows a result with new training data. The detection rate becomes 80.00% and the false positive rate becomes 20.94% which is much better result than that of Table 3. From this result, we can conclude the high possibility that the false positive of normal sessions becomes noise data. Therefore, if the first classifier has less false positive, we can improve unknown attack detection performance.

5   Conclusion

In this paper, we presented a method to detect unknown attacks using feature extraction and multistage OC-SVM. We added 6 new features based on communication interval, and applied them to OC-SVM.

We evaluated the proposed method with Kyoto2006+ Dataset and the features. By comparing with the first stage classifier without the new features, the classifier with these features shows higher precision rate in detecting attacks .c In the second stage classifier, our method detects unknown attacks, although there is a high false positive rate. A signature-based IDS cannot detect unknown attacks so that current network administrators take too much time and effort to find unknown attacks by analyzing all suspicious sessions. For these reasons, it can be said that our method has enough advantage by limiting on the number of suspicious unknown attacks.

For future works, we need to improve the detection rate of unknown attacks in the second stage classifier. Therefore, we have to perform filtering and clustering in order to reduce an affect of noise data. We also have to extract more effective features th at reflect unique characteristic of unknown attacks.

Acknowledgment. This work is supported by R&D of detective and analytical technology against advanced cyber-attack, administered by the Ministry of Internal Affairs and Communications.

References

1. KDD Cup 1999 Dataset,
   http://kdd.ics.uci.edu/databases/kddcup99/kddcup99.html
2. Kyoto2006+ Dataset, http://www.takakura.com/Kyoto_data/
3. Eskin, E., Arnold, A., Prerau, M., Portnoy, L., Stolfo, S.: A geometric framework for unsupervised anomaly detection. In: Applications of Data Mining in Computer Security, pp. 77–101. Springer (2002)
4. Kondo, S., Sato, N.: Botnet traffic detection techniques by C&C session classification using SVM. In: Miyaji, A., Kikuchi, H., Rannenberg, K. (eds.) IWSEC 2007. LNCS, vol. 4752, pp. 91–104. Springer, Heidelberg (2007)
5. Perdisci, R., Gu, G., Lee, W.: Using an ensemble of one-class svm classifiers to harden payload-based anomaly detection systems. In: Sixth International Conference on Data Mining, ICDM 2006, pp. 488–498. IEEE (2006)
6. Schölkopf, B., Platt, J., Shawe-Taylor, J., Smola, A., Williamson, R.: Estimating the support of a high-dimensional distribution. Neural Comput. 13(7), 1443–1471 (2001)
7. Song, J., Ohira, K., Takakura, H., Okabe, Y., Kwon, Y.: A clustering method for improving performance of anomaly-based intrusion detection system. IEICE - Trans. Inf. Syst. E91-D(5), 1282–1291 (2008)
8. Song, J., Takakura, H., Kwon, Y.: A generalized feature extraction scheme to detect 0-day attacks via IDS alerts. In: The 2008 International Symposium on Applications and the Internet (SAINT 2008), pp. 55–61 (2008)

Analysis and Configuration of Boundary Difference Calculations

Simon Dacey, Lei Song, Lei Zhu, and Shaoning Pang

Unitec Institute of Technology,
Carrington Road, Mt. Albert, Auckland, New Zealand

Abstract. In the field of land management, stakeholders (people) everywhere have many disputes over the location of boundaries between private land and public land. We find that the stakeholders disagree with each other over boundaries. We propose an approach that helps people to come to an agreement on position of boundaries (including pixel-based approach, polygon-based approach and middle boundary approach). The experiments are carried out on data relating to public parks in Auckland, New Zealand. The results of the experiments highlight the differences between different stakeholder's percieved boundaries.

Keywords: Boundary Negotiation, Boundary Disputes, Polygon-based Boundary Calcluation, GPS-based Boundary Detection.

1 Introduction

Land Management is the process of managing the use and development (in both urban and rural settings) of land resources. Land resources include organic agriculture, reforestation, water resource management and eco-tourism projects, is called public land. Private property is a legal designation of the ownership of property by non-governmental legal entities. The use of public land for private purposes, known as encroachment, has been identified as a problem affecting public parks in the Auckland region. Every boundary conflict contains a strong spatial component [1]. The spatial location of a public park is defined by its boundary. However there may be several different versions of the same boundary for a park. Perception of encroachment depends upon the viewpoint of the stakeholder. The stakeholders are residents living around the parks, non-residents who use the parks, organised sports groups and representatives of the Auckland City Council (including managers, councillors and surveyors). In land management, arguments occur over boundaries between stakeholders.

2 Related Work

This sections examines research work that has been carried out in related areas. The related areas are the mathematical formulas used to calculate the differences between two sets of data (boundary or area) and the magnitude of the

C.K. Loo et al. (Eds.): ICONIP 2014, Part III, LNCS 8836, pp. 333–340, 2014.

differences. A search of the literature reveals that people use the Hidden Markov Model [2] [3],the Boundary Element Method [4] [5] and point-set-based [6] [7] [8] to detect differences in boundaries. Hidden Markov models (HMM) are studied for the purpose of planar shape classification using curvature coefficients. A discrete-time HMM is a probabilistic model that describes a random sequence as the indirect observation of an underlying (hidden) random sequence where this hidden process in Markovian. The boundary element method (BEM) is a numerical computational method of solving linear partial differential equations which have been formulated as integral equations. A point-set-based model is developed for areal objects from a perspective that incorporates spatial cognition. This model is called with point-set-based regions (PSBR). Computing spatial relationships between two PSBRs using the derived areal objects consists of looking at topological relationships, directional relationships, metric relationships, distance between centroids, average distance and Hausdorff distance.

2.1 Motivation

After studying the methods listed above, we have found that they give an accurate detection of the differences. However, these works does not really solve the arguments among stakeholders, due to these methods not providing a possible solution for the stakeholders. We analysed the pixel-based calculation and proposed a polygon-based calculation to form a new point of view for the stakeholders, upon which they can base negotiations to solve the boundary dispute.

3   Methodology

Land use conflict occurs whenever land-use stakeholders have incompatible interests related to land areas that result in negative effects [9]. In order to resolve the arguments over boundaries, we examine two existing approaches: a) Pixel-based Approach and b) Polygon-based Approach and we then build upon the two approaches above to propose a new approach: Middle-boundary Approach.

3.1 Pixel-Based Approach

Given a set of n sequential GPS coordinate pairs $G = \{(Lo_1, La_1), \ldots, (Lo_n, La_n)\}$ for one area, we firstly transfer them into integer coordinates according to certain predefined precision, for example $C = \{(X_1, Y_1), \ldots, (X_n, Y_n) \mid X_i = round(Lo_i \times 1000), Y_i = round(La_i \times 1000) \; \forall i \in \{1, \ldots, n\}\}$. Then we shift the coordinate origin somehow to fit the coordination set as,

$$C_s = \{(x_1, y_1), \ldots, (x_n, y_n) \mid x_i = X_i - min(X) + 1, y_i = Y_i - min(Y) + 1 \\ \forall i \in \{1, \ldots, n\}\}$$

(1)

Now we have a bitmap with n positive pixels. Next we sequentially connect each neighboring pair (n-th point is the neighbor with $n-1$-th and 1-st). For example, if have (x_i, y_i) and (x_{i+1}, y_{i+1}), we need to compute a set of pixels approximately connecting (x_i, y_i) to (x_{i+1}, y_{i+1}) and also as a edge of closed polygon. To achieve such closed approximate pixel edge, we simply approximate the y coordination from a continuous series of x connecting x_i to x_{i+1} and do the reversed, as follows (assume $x_i \leq x_{i+1}$ and $y_i \leq y_{i+1}$)

$$
\begin{aligned}
x &= [x_i, x_i + 1, \ldots, x_{i+1}] \\
y &= y_i + x \times \tfrac{y_{i+1} - y_i}{x_{i+1} - x_i} \\
y &= round(y);
\end{aligned}
\tag{2}
$$

$$
\begin{aligned}
y &= [y_i, y_i + 1, \ldots, y_{i+1}] \\
x &= x_i + y \times \tfrac{x_{i+1} - x_i}{y_{i+1} - y_i} \\
x &= round(x).
\end{aligned}
\tag{3}
$$

The result with more points from (2) and (3) is taken as the edge point set between (x_i, y_i) and (x_{i+1}, y_{i+1}). Once all edges for one area if obtained, we can apply area fill algorithm to fill the edge graph and obtain a binary bitmap of that area. For area difference, we can simply do a matrix subtraction to compute following areas:

1. both A_1 and A_2 covers;
2. A_1 covers but A_2 not;
3. A_2 covers but A_1 not.

As any field area is represented by a binary matrix $M = \{0, 1\}^{m \times n}$, where m and n denote the number of rows and columns respectively of the bitmap, thus the area is simply calculated by

$$
S = sum(sum(M)).
\tag{4}
$$

3.2   Polygon-Based Approach

In the analytic geometry method, a boundary is seen as a polygon formed by connecting points sequentially. Given two boundaries $A = \{a_1, a_2 \ldots a_n\}$ and $B = \{b_1, b_2 \ldots b_m\}$, where the end points of A and B are clockwise distributed. There are four steps to find the difference $A - B$ (i.e. the area inside A but outside B) and $B - A$ (i.e. the area inside B but outside A).

1. Find all cross points C between any edge pairs, one from A and one from B. As each edge is a line segment, there are many existing algorithms for finding cross point between two given line segments. The result of this step is a set of points $C = c_1, c_2, \ldots c_o$, each cross point c_k is associated with one edge $a_i a_{i+1}$ from A and one edge $b_j b_{j+1}$ from B.
2. Form the difference polygons $D_1, D_2, \ldots, D_o$. As boundary A have o cross points with B, it is easy to imagine that there are o difference polygons. And the k-th difference polygon D_k is defined by cross point c_k, a sequence of points A_k from A, cross point c_{k+1} and a sequence of points B_k from B.

Note that when $k = o$ have $c_{k+1} = c_1$, as the 'next' for the last one in a circle is the first. Then, to determine the difference polygon D_k, we need to find out the sequences A_k and B_k. To compute the sequences A_k, we need firstly check if c_k and c_{k+1} are associated with the same edge $a_i a_{i+1}$ in A. If so, means the boundary B cross edge $a_i a_{i+1}$ at least twice and there is no end point from A between c_k and c_{k+1}, thus we have $A_k = \emptyset$. If not, say c_k associates with $a_i a_{i+1}$ and c_{k+1} associates with $a_l a_{l+1}$, then A_k is the sequences $[a_{i+1}, a_{i+2}, \ldots, a_l]$. Also note that if have $i + 1 > l$, the sequences A_k actually becomes $[a_{i+1}, a_{i+2}, \ldots, a_n, a_1, \ldots, a_l]$, always keep in mind that we are working on a 'circle'. Applying the same method, we can determine the sequences B_k say $B_k = [b_{j+1}, b_{j+2}, \ldots, b_p]$. Recall that both A and B are clockwise distributed, so are the A_k and B_k. Thus to define the difference polygon D_k, we need to use the inverse sequences $B_k' = [b_p, \ldots, b_{j+2}, b_{j+1}]$ with is counterclockwised. By now we have the difference polygon $D_k = c_k A_k c_{k+1} B_k'$ computed. And so for all difference polygons $D_1, D_2, \ldots, D_o$.

3. Determine which set of difference polygons belong to $A - B$ and which set to $B - A$. It is easy to conclude that for any neighboring difference polygon D_k and D_{k+1} should have the different identity, as if boundary of A is 'outside' of B between cross points c_k and c_{k+1} which indicates D_k belongs to $A - B$, after the cross in c_{k+1}, the boundary of A becomes 'inside' the boundary which indicates that D_{k+1} belongs to $B - A$, and vice versa. Thus, we can simply determine the difference polygons belonging by finding out the identity of any difference polygon. This can be done by detect whether a point from A (e.g. a_1) is inside polygon B. Make a ray from a_1 to any direction, if there is odd numbered cross point with polygon B, then a_1 is inside B; and if the cross number is even, then a_1 is outside B.

4. Compute the area of difference polygons. Given any polygon $A = a_1 a_2 \ldots a_n$ and its coordinate set $\{(x_1, y_1), (x_2, y_2), \ldots, (x_n, y_n)\}$, the area of A can be computed as

$$S_A = \tfrac{1}{2} \sum_{i=1}^{n} (x_i y_{i+1} - x_{i+1} y_i). \tag{5}$$

Note that when $i = n$, have $a_{n+1} = a_1$ and $x_{n+1} = x_1, y_{n+1} = y_1$.

3.3   Middle-Boundary Approach

To find the middle-boundary, we propose a nearest neighbor based algorithm. Let $A = a_1 a_2 \ldots a_n$ be the boundary with more endpoints than $B = b_1 b_2 \ldots b_m$, we calculated a middle boundary $M = m_1 m_2 \ldots m_n$ have the same number of end points with A, each m_i lays in the halfway of a_i to its nearest neighbor in $\{b_1, b_2, \ldots, b_m\}$. To find such nearest neighbor, we need firstly compute a distance vector $D = R^{m \times 1}$ have

$$D_j = dist(a_i, b_j) = \sqrt{(x_{ai} - x_{bj})^2 + (y_{ai} - y_{bj})^2}. \tag{6}$$

Then find the j with smallest D_j, and eventually compute the mean vector of $[x_{ai}, y_{ai}]$ and $[x_{bj}, y_{bj}]$.

Algorithm 1. Find the Middle Boundary

Require: Polygon $A = a_1 a_2 \ldots a_n$ and $B = b_1 b_2 \ldots b_m$, have $n >= m$; the coordinate
of each point $\{(x_{ai}, y_{ai}) | \forall i \in [1, n]\}$, $\{(x_{bj}, y_{bj}) | \forall j \in [1, m]\}$
Ensure: The middle boundary $M = m_1 m_2 \ldots m_n$ and its coordinate set
$\{(x_{mi}, y_{mi}) | \forall i \in [1, n]\}$.
  for $i \in [1, n]$ **do**
    Current point is a_i, (x_{ai}, y_{ai});
    Find the current point's nearest neighbor b_j (x_{bj}, y_{bj}) from B;
    Compute the m_i, have $x_{mi} = mean(x_{ai}, x_{bj})$, and $y_{mi} = mean(y_{ai}, y_{bj})$
  end for

Once M is determined, using algorithm mentioned in Section 3.2, we can compute the difference area between A to M and B to M.

4 Experiments

To demonstrate the boundary comparison techniques, for this paper, 20 parks from the Manukau and North Shore areas are selected as representative examples of the four categories of park discussed above.

4.1 Experimental Setup

A number of experiments were conducted to study the effectiveness of the new approach. In this section we look at the data used in the experiments, the setup and procedure of the experiments and we discuss the results of the experiments.

4.2 Location and Device

GPS Data Collection carried out at 20 sites across North Shore and Manukau areas in Auckland. One GPS devices used: Leica Viva TPS. The data collected, using the Leica Viva TPS device, uses the Mt Eden 2000 co-ordinates system.

4.3 Data

The data used in the experiments is boundary data of New Zealand specifically the Auckland area.

1 GPS Boundary Data. The GPS boundary data has been collected during field visits to the selected parks.
2 Land Boundary Data. The Council boundary data is used to compare against the observed boundary data.

(a) Area covered by both sets of data

(b) Positive differences between data sets

(c) Negative differences between data sets

(d) Composite image of data

Fig. 1. Pixel Based Approach

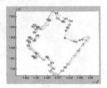 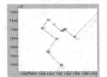  

(a) Boundary cross points

(b) Boundary cross points - zoomed in view

(c) Differences between polygons

(d) Differences between polygons - zoomed in view

Fig. 2. Polygon Based Approach

 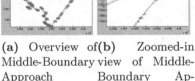

(a) Overview of Middle-Boundary Approach

(b) Zoomed-in view of Middle-Boundary Approach

Fig. 3. Middle-Boundary Approach

4.4 Discussion

The argument from people for the analyzing frames illuminates the underlying causes of Park boundries in the geographic analysis of public land use. Geographic information systems (GIS) have long been applied in resolving municipal/local boundary conflicts (e.g. US political redistricting) [10]. This approach utilises both GIS and GPS in presenting and resolving boundary disputes.

Table 1. Results of the polygon-based experiments

Name	Polygon-based				Middle-boundary	
	AreaOmG	AreaGmO	AreaOmM	AreaMmO	AreaGmM	AreaMmG
Agincourt Reserve	3050.970703	46.8359375	2566.961914	20.89550781	132.734375	590.8027344
Aorere Park	4635.09082	2241.937531	2224.094147	1173.392487	1138.012573	2480.46402
Auburn Reserve	12983.03125	0.544921875	7507.915039	111.6933594	0.13671875	5586.398438
Beaumont Park	351.6609802	854.4945984	291.2216797	363.8186035	596.5552368	166.3186951
Dale Reserve	0	0	0	0	13001.17609	3857.525452
Diana Reserve	234.1083984	179.5517578	118.3222656	93.94726563	88.72363281	118.9091797
Gallaher Park	1846.82312	2653.692627	1193.333954	1283.261169	1471.670105	754.7279663
Holland Reserve	2972.285156	319.7568359	1480.804688	9044.454102	502.2460938	10718.42871
James Watson Park	240.3355103	15420.63184	120.597168	6829.786682	9515.397919	1044.291077
Jellicoe Park	549.1130066	481.79776	483.4128418	275.4145508	272.3617249	131.6789246
Killarney Park	3160.15918	8609.833008	1402.926758	1428.557617	7316.314453	1892.266602
McFetridge Park	5461.837891	103.5576172	4819.770508	36.79882813	1404.405273	1979.706055
Normanton Reserve	544.5351563	1909.788086	354.5947266	373.9892578	1688.322266	342.4638672
Puhinui Domain	19511.99994	69455.7518	56116.37057	82893.59921	2093.553741	28110.97137
Robert White Park	0	0	0	0	0	0
Russell Road Reserve	1511.267151	9524.483978	1052.96701	7910.964752	1634.870209	479.6509399
Stancich Reserve	824200.4697	1560491.801	64279.37207	2873.81543	819018.793	50225.59277
Tadmoor Park	8388.599365	3037.15976	3787.05957	733.4524231	11932.83618	1146.519043
Taharoto Park	743.9765625	37836.19238	321.2314453	37842.42578	323.984375	38055.57227
Teviot Reserve	297.4169922	69.4296875	406.0068359	78.33398438	263.484375	163.7978516

The results of the pixel-based experiments show the differences between two views (sets of boundary data) for the same area. The first set of differences shows the encroachment of private land onto public land and the second set of differences shows the encroachment of public land onto private land. This approach highlights the differences between the boundaries but does not propose any possible solutions to the problem. The results of the polygon-based experiments in Table 1 show the plotting of the two sets of points and highlights the differences between the two views (sets of boundary data) for the same area. AreaOmG gives the area of the original boundary minus that of the GPS boundary. AreaGmO gives the area of the GPS boundary minus that of the original boundary. AreaOmM gives the area of the original boundary minus that of the middle boundary. AreaMmO gives the area of the middle boundary minus that of the original boundary. AreaGmM gives the area of the GPS boundary minus that of the Middle boundary. AreaMmG gives the area of the middle boundary minus that of the GPS boundary. The results of the experiments show the differences between two views (sets of boundary data) for the same area. We give a fair solution to people who have arguments on the measurement of boundary for Parks in Auckland, as seen in Fig. 3. The green line shows the boundary as perceived by the council and the blue line shows the boundary as perceived by the results of a field survey. The differences in the two boundaries show where the boundary is disputed. The red line shows a calculated middle boundary which may act as a starting point for resolving boundary disputes. Stakeholders have different views on encroachment. For example the council have a strict viewpoint and assume that the data they have is correct whereas some private residents have a relaxed viewpoint on the position of a boundary.

5   Conclusions and Future Work

In order to address the land encroachment problems in Auckland's parks, we firstly proposed two different approaches boundary calculations. Though both of them detect and highlight differences, in numerical terms, effectively, whereas neither approach offers a possbile solution to any boundary conflicts. We then proposed the middle-boundary approach, in which we address the boundary arguments from stakeholders by a nearest neighbor based algorithm. This solution could be a possible solution that disputes or at least a starting point for negotiations, because the (middle-boundary) have been addressed. The main limitation of the middle-boundary approach is that the proposed new boundary is based solely on mathematical calculations and does not take into account the stakeholders' views or motivations currently. Possible future work could involve adding weighting to the middle-boundary calculations so that different possible solutions could be generated.

References

1. Shmueli, D.F.: Framing in geographical analysis of environmental conflicts: Theory, methodology and three case studies. Geoforum 39(6), 2048–2061 (2008)
2. He, Y., Kundu, A.: 2-d shape classification using hidden markov model. IEEE Transactions on Pattern Analysis and Machine Intelligence 13(11), 1172–1184 (1991)
3. Bicego, M., Murino, V.: Investigating hidden markov models' capabilities in 2-d shape classification. IEEE Transactions on Pattern Analysis and Machine Intelligence 26(2), 281–287 (2004)
4. Deneme, I., Yerli, H., Severcan, M., Tanrikulu, A., Tanrikulu, A.: Use and comparison of different types of boundary elements for 2d soil-structure interaction problems. Advances in Engineering Software 40, 847–855 (2009)
5. Rom, H., Medioni, G.: Hierarchical decomposition and axial shape description. IEEE Transactions on Pattern Analysis and Machine Intelligence 15(10), 973–981 (1993)
6. Liu, Y., Yuan, Y., Xiao, D., Zhang, Y., Hu, J.: A point-set-based approximation for areal objects: A case study of representing localities. Computers, Environment and Urban Systems 34, 28–39 (2010)
7. Stehman, S.: Estimating area from an accuracy assessment error matrix. Remote Sensing of Environment 132, 202–211 (2013)
8. Baffetta, F., Fattorini, L., Franceschi, S., Corona, P.: Design-based approach to k-nearest neighbours technique for coupling field and remotely sensed data in forest surveys. Remote Sensing of Environment 113, 463–475 (2009)
9. von der Dunk, A., Grêt-Regamey, A., Dalang, T., Hersperger, A.M.: Defining a typology of peri-urban land-use conflicts–a case study from switzerland. Landscape and Urban Planning 101(2), 149–156 (2011)
10. Forest, B.: Information sovereignty and gis: the evolution of communities of interest in political redistricting. Political Geography 23(4), 425–451 (2004)

Morphological Associative Memory Employing a Split Store Method

Hakaru Tamukoh, Kensuke Koga, Hideaki Harada, and Takashi Morie

Graduate School of Life Science and Systems Engineering,
Kyushu Institute of Technology, Japan
tamukoh@brain.kyutech.ac.jp

Abstract. The morphological associative memory (MAM) has advantages of large memory capacity and high perfect recall rate in comparison with other associative memory models. However, the MAM cannot store a large data in memory matrices M and W because the space complexity of the ordinary method is $O(n^2)$ when the dimension of input data is n. In this paperCwe propose a MAM employing a split store method. The proposed method splits a given stored pattern into $\sqrt{n}$ sub-pattern, then memory matrices are independently generated for each sub-pattern. Experimental results show that the perfect recall rate and CPU time of the proposed method are nearly equal to the ordinary method while the proposed method reduces the space complexity to $O(n^{1.5})$.

Keywords: Morphological associative memory, split store, memory storage, space complexity.

1 Introduction

An associative memory is one of the important brain functions and has attracted many researchers [1]-[9]. Hopfield proposed Hopfield network which can be designed as an associative memory [1]. However, Hopfield network has drawbacks of low memory capacity and low convergence; the number of stored patterns can be no more than 15% of the number of neurons, and a convergence problem is caused by the local minimum.

As one of the associative memory models, Ritter proposed a morphological associative memory (MAM) [2] based on a morphological neural network [10]. In contrast with Hopfield network, the MAM has advantages; large memory capacity and high perfect recall rate. Especially in the auto associative memory mode such as (A → A), the MAM always provides the perfect recall for any number of patterns stored in its memory matrices where there is no noise in the input pattern. Therefore, it has unlimited memory capacity. In addition, the recall of MAM is one-shot process, thus the convergence problem does not exist.

Although the MAM has the above advantages, it also has drawbacks on a design of kernel image. The kernel image is used for the association and is obtained by trial and error in the ordinary MAM. It is hard to design as the total number of the stored patterns increases. To overcome the problem, several effective

C.K. Loo et al. (Eds.): ICONIP 2014, Part III, LNCS 8836, pp. 341–348, 2014.

methods to design the kernel image have been proposed [3]-[5]. However, those methods have a problem that the perfect recall cannot be achieved for an input pattern with a corrupted kernel image or the stored patterns with redundancy bits is necessary for the kernel image. As a MAM using no kernel image, the block splitting type MAM (BMAM) was proposed. The BMAM introduced a block splitting method and a majority logic approach to avoid spreading a noise over an image in the recall process and obtain a plausible recall pattern respectively [6]. However, the BMAM also has a problem that the perfect recall rate is inferior to the MAM using the kernel image. Therefore, a MAM using a stored pattern independent kernel image had been proposed [7]. This kernel design method offers advantages; the simple kernel design, good recall performance for the corrupted kernel image. However, the model cannot recall the correct pattern for a pattern completely included with other stored patterns (e.g., "C and G", "E and F"), as well as conventional MAMs. The existence of such inclusion patterns becomes serious when the number of the stored patterns increases. To solve the problem, we proposed a MAM employing a reverse recall [8][9]. In this model, we introduced the reverse recall method into the MAM using the stored pattern independent kernel image. By introducing these methods, the design of kernel image has been well improved.

However, the MAM still has a drawback on the space complexity. The recall process of the MAM is one-shot as mentioned before and its time complexity is $O(n)$ when dimension of the input pattern is n. On the other hand, the size of memory matrices requires $2n^2$ for association. The space complexity is $O(n^2)$ and it is difficult to develop a practical application when the dimension of the stored patterns increases. The BMAM reduces the size of memory matrices to $n^2 + \alpha, (\alpha \ll n^2)$ [6], but its space complexity is also $O(n^2)$.

In this paper, we propose a MAM employing a split store method to solve the space complexity of ordinary MAMs. The proposed method splits a given stored pattern into several sub-patterns, then memory matrices are independently generated for each sub-pattern. When a number of sub-patterns is $\sqrt{n}$, the proposed method reduces the space complexity to $O(n^{1.5})$. We evaluate a validity of the proposed method through experiments.

2   Morphological Associative Memory

2.1   Ritter's MAM

The MAM proposed by Ritter [2] has two memory matrices 'M' and 'W'. Here, let assume n-dimensional pattern pairs $X^r = (x_1^r, \cdots, x_n)$, $Y^r = (y_1^r, \cdots, y_n)$, $(r = 1, \cdots, p)$, the two memory matrices are given as;

$$m_{ij} = \bigvee_{r=1}^{p} (y_i^r - x_j^r), \tag{1}$$

$$w_{ij} = \bigwedge_{r=1}^{p} (y_i^r - x_j^r), \tag{2}$$

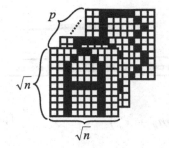

Fig. 1. Example of n-dimentional p-stored patterns

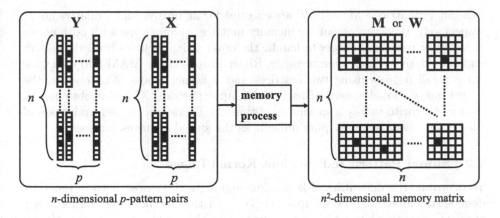

n-dimensional p-pattern pairs n^2-dimensional memory matrix

Fig. 2. Memory process of Ritter's MAM

where p is a number of pattern pairs, w_{ij} and m_{ij} represent (i, j)-th unit of memory matrices 'M' and 'W', respectively. The symbols $\bigwedge$ and $\bigvee$ denote minimum and maximum operators, respectively.

Figures 1 and 2 show an example of n-dimensional stored patterns and the memory process of Ritter's MAM to obtain the memory matrix, respectively. The size of two memory matrices is $2 \times n \times n$. Therefore, the space complexity of Ritter's MAM is $O(n^2)$.

In a recall process, the output Y^r is obtained by the given input X^r and memory matrices 'M' or 'W' as follows;

$$y_i^r = \bigwedge_{j=1}^{n} (m_{ij} + x_j^r). \tag{3}$$

or

$$y_i^r = \bigvee_{j=1}^{n} (w_{ij} + x_j^r), \tag{4}$$

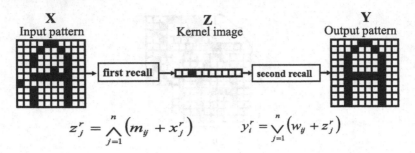

$$z_j^r = \bigwedge_{j=1}^{n} \left(m_{ij} + x_j^r\right) \qquad y_i^r = \bigvee_{j=1}^{n} \left(w_{ij} + z_j^r\right)$$

Fig. 3. Recall process using stored pattern independent kernel image

Memory matrices 'M' and 'W' are effective for an erosive and a dilative noise, respectively. However, one of the memory matrices cannot cope with both noises in the same pattern. In order to handle the noise which includes both the erosive and the dilative noise simultaneously, Ritter proposed the MAM adopts a two-stage recall process using two matrices and a kernel image 'Z'. However, the kernel image should consist of part of the input pattern 'X' and is obtained by trial and error to satisfy a specific condition [2]. Therefore, the determination of the kernel image becomes quite difficult as the stored patterns increases.

2.2 Stored Pattern Independent Kernel Image

To overcome the difficulties of kernel image design, the stored pattern independent kernel image has been proposed [7][8]. In this method, the kernel image consists of several bits which are equivalent to the number of stored patterns. Here, only one element of the kernel image is '1', the other elements are '0'. It is easy to design the kernel images for association. The memory matrices using the stored pattern independent kernel image are obtained as follows;

$$m_{ij} = \bigvee_{r=1}^{p} (z_i^r - x_j^r), \tag{5}$$

$$w_{ij} = \bigwedge_{r=1}^{p} (y_i^r - z_j^r), \tag{6}$$

The output is obtained by a two-stage recall process same as the Ritter's MAM given by follows;

$$z_i^r = \bigwedge_{j=1}^{n} (m_{ij} + x_j^r), \tag{7}$$

$$y_i^r = \bigvee_{j=1}^{n} (w_{ij} + z_j^r). \tag{8}$$

Figure 3 shows a shceme of the recall process of MAM employing stored pattern independent kernel image. In this paper, we employ this kernel image for experiments.

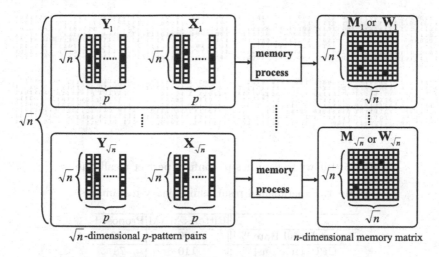

$\sqrt{n}$-dimensional p-pattern pairs n-dimensional memory matrix

Fig. 4. Memory process of MAM employing a split store method

3   MAM Employing a Split Store Method

The recall process of the MAM is one-shot and its time complexity is $O(n)$ when dimension of the input pattern is n. On the other hand, the size of memory matrices requires $2n^2$. The space complexity is $O(n^2)$. It takes huge memory space when the dimension of stored pattern increases. For instance, to realize an association between 256×256 pixel gray scale images (e.g. dimension of the stored pattern $n = 2^{16}$), the ordinary MAM requires 8 GByte for memory matrices 'M' and 'W'. It is hard to store the memory matrices into the main memory of general personal computer. In this section, we propose a split store method to reduce the space complexity of the ordinary MAMs.

Figure 4 shows a scheme of the proposed method. A memory process of the proposed method as follows;

1. Stored pattern pairs X^r and Y^r are divided in s sub-patterns.
2. Memory matrices M and W are generated for each sub-pattern.

The parameter s represents a number of sub-patterns. A recall process of the proposed method as follows;

1. A given input pattern is divided into s sub-patterns.
2. Recall processes are executed every sub-patterns.
3. Combining output images resulted by recall process.

The total size of memory matrices employing the proposed method is $2*s*(n/s)^2$. In Fig.4, we set $s = \sqrt{n}$ (e.g. the stored pattern shown in Fig.1 is divided in each row) in the step 1. In this case, the total size of memory matrices 'M' and 'W' is $2*n^{1.5}$. Therefore, the space complexity of the proposed method is $O(n^{1.5})$. For instance, the proposed method requires 34 MBytes for the association of 256×256 pixel gray scale images. It drastically reduces memory usage in comparison with the ordinary MAMs.

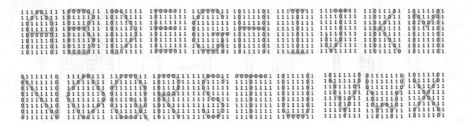

Fig. 5. Binary test images to confirm perfect recall rate

Table 1. Experimental resut using binary test images

	Ritter's MAM	Proposed
Perfect Recall Rate [%]	100	100
CPU Time [ms]	110	78

4   Experimental Results

In order to evaluate a validity of the split store method, we compare a recall rate of the proposed method with the conventional MAM.

4.1   Perfect Recall Rate

In the auto associative memory mode, Ritter's MAM always provides the perfect recall where there is no noise in the input pattern. It is one of the most important features of the conventional MAM.

In order to confirm the perfect recall rate of the proposed method, we used 20 test images which consisted of 64 binary elements as shown in Fig6. Here, we set $s = 8$ for the split store method. Table 1 shows experimental results which are average values of 320 trials. The proposed method achieved 100% of perfect recall rate same as the conventional MAM. In addition, CPU time of the proposed method was nearly equal to the conventional MAM. From these results, we confirmed that the proposed method achieved the same performance with the ordinary MAM while the proposed method reduced the space complexity.

4.2   Noise Tolerance

In order to evaluate noise tolerance of the proposed method, we randomly generated 1,000 stored patterns which consisted of 64 binary elements, and obtained memory matrices 'M' and 'W'. Here, we set $s = 8$ for the split store method. In recall process, we generated the erosive and dilative noises randomly on each element in an input pattern with a pre-assigned probability.

We evaluated four methods; (1) Ordinary MAM without kernel (Ritter's MAM), (2) Proposed MAM without kernel, (3) Ordinary MAM using kernel,

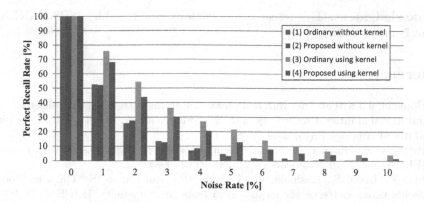

Fig. 6. Noise tolerace of the proposed and ordinary MAMs

(4) Proposed MAM using kernel. In experiments (3) and (4), we employed the stored pattern independent kernel image [7].

Figure 6 shows experimental results to evaluate noise tolerance of MAMs. Each perfect recall rate was an average of 1,000 trials. All MAMs (1)–(4) achieved 100% of perfect recall rate when noise rate was 0%. From the results, we confirmed again the MAM using the sprit store method achieved the same performance as the ordinary MAM.

From the results of noise rate were 1 to 10%, we confirmed that noise tolerance of the proposed method was well improved by introducing the kernel image. However, the perfect recall rate of the proposed method using kernel (4) was lower than the ordinary one (3). We supposed that it was caused by occurring inclusion patterns (e.g., "C and G", "E and F"). The number of pattern combination was reduced by splitting the stored pattern in the proposed method. In this case, a probability of occurring inclusion patterns was increasing. Therefore, we have to consider a tradeoff between the space complexity and the noise tolerance to set the parameter s in the proposed method. We also suppose that the performance degradation of the proposed method caused by inclusion patterns would be solved by introducing a reverse recall method [9].

5 Conclusion

In this paper, we proposed a MAM employing a split store method to reduce the space complexity of the ordinary MAMs. When we set a parameter $s = \sqrt{n}$, which represents number of sub-patterns, the proposed method reduced the space complexity from $O(n^2)$ to $O(n^{1.5})$. Experimental results showed that a perfect recall rate and CPU time of the proposed method were nearly equal to the ordinary method while the proposed method reduced the space complexity.

In future work, we will combine the split store method with a reverse recall method [9] to improve the noise tolerance of the proposed method.

Acknowledgment. This research was partially supported by JSPS KAKENHI Grant Number 26330279.

References

1. Hopfield, J.J.: Neural network and physical systems with emergent collective computational abilities. Proceedings of the National Academy of Sciences of the United States of America 79(8), 2554–2558 (1982)
2. Ritter, G.X., Sussner, P., Diaz-de-Leon, J.L.: Morphological associative memory. IEEE Trans. Neural Networks 9(2), 281–293 (1998)
3. Ida, T., Ueda, S., Kashima, M., Fuchida, T., Murashima, S.: On a method to decide kernel patterns of morphological associative memory. IEICE, D-II J83-D-II(5), 1372–1380 (2000)
4. Sussner, P.: Associative morphological memories based on variations of the kernel and dual kernel methods. Neural Networks 16(5-6), 625–632 (2003)
5. Hattori, M., Fukui, A., Ito, H.: A fast method to decide kernel patterns for morphological associative memory. Transactions of the Institute of Electrical Engineers of Japan 123(10), 1830–1838 (2003)
6. Saeki, T., Miki, T.: Block splitting type morphological associative memory and its recall rate improvement by majority logic approach. Int. Journal of Innovative Computing, Information and Control 4(9), 2195–2204 (2008)
7. Harada, H., Saeki, T., Miki, T.: A morphological associative memory employing a stored pattern independent kernel image and its hardware model. In: Proc. of 5th Int. Workshop on Computational Intelligence and Applications, pp. 219–224 (2009)
8. Harada, H., Miki, T.: A Hardware Model of a Morphological Associative Memory Employing a Simplified Reverse Recall. ICIC Express Letters 6(3), 833–838 (2012)
9. Harada, H., Miki, T.: A Morphological Associative Memory Employing Simplified Reverse Recall and Its Hardware Model. Int. Journal of Innovative Computing, Information and Control 10(5) (2013)
10. Davidson, J.L., Ritter, G.X.: A theory of morphological neural networks. SPIE Digital Optical Computing II 1215, 378–388 (1990)

A Novel Hybrid Approach for Combining Deep and Traditional Neural Networks

Rui Zhang, Shufei Zhang, and Kaizhu Huang

Xi'an Jiaotong-Liverpool University,
SIP, Suzhou, 215123, China
{Rui.Zhang02@,Shufei.zhang10@student.,Kaizhu.Huang@}xjtlu.edu.cn

Abstract. Over last fifty years, Neural Networks (NN) have been important and active models in machine learning and pattern recognition. Among different types of NNs, Back Propagation (BP) NN is one popular model, widely exploited in various applications. Recently, NNs attract even more attention in the community because a deep learning structure (if appropriately adopted) could significantly improve the learning performance. In this paper, based on a probabilistic assumption over the output neurons, we propose a hybrid strategy that manages to combine one typical deep NN, i.e., Convolutional NN (CNN) with the popular BP. We present the justification and describe the detailed learning formulations. A series of experiments validate that the hybrid approach could largely improve the accuracy for both CNN and BP on two large-scale benchmark data sets, i.e., MNIST and USPS. In particular, the proposed hybrid method significantly reduced the error rates of CNN and BP respectively by 11.72% and 28.89% on MNIST.

1 Introduction

In the last half century, Neural Networks (NN) have been actively researched and widely applied in most areas of pattern recognition and computer vision [1]. In the field, there are many famous NN models including Back Propagation (BP) [1], Self-organizing Map NNs [5], and Hopfield network [8]. In particular, BP is regarded as one of the state-of-the-art supervised learning approaches, which has been extensively exploited in a large number of applications such as classification and regression. A typical BP NN usually exploits a three-layer structure with the non-linear activation function. In training BP, the error could be propagated from latter to previous layers, which motivated its model name.

Recently, NNs even receive more attention in the community partly because a novel deep hierarchy structure proves to be able to improve the learning accuracy in many real tasks [4,6]. Such NNs, exploiting many layers rather than a shallow structure (e.g., 3 layers) are called as Deep NNs (DNN). There are two typical types of NNs in this line: (1) DNNs based on Restricted Boltzmann Machine (RBM) [3], and (2) Convolutional NNs (CNN) [6]. The first type of DNNs adopts RBMs to obtain high-level features gradually layer by layer using a fast training algorithm, namely a pre-training strategy combined with fine tuning [4], while

C.K. Loo et al. (Eds.): ICONIP 2014, Part III, LNCS 8836, pp. 349–356, 2014.
© Springer International Publishing Switzerland 2014

CNNs engage explicit convolutional masks to extract high-level features also in a layer-wise way [6]. Both types of DNNs achieve big success across the fields of object detection, image recognition and speech recognition.

Due to the success of both traditional NNs and DNNs, it remains natural yet interesting that if combination of them could further lift up the performance. To this end, in this paper, we try to investigate how to combine a traditional NN, i.e. BP, with the deep learning model CNN, and if such combination could indeed improve the accuracy. For this purpose, we propose a novel probabilistic hybrid approach assuming that the output values given by the output layer of both BP and CNN follow a Gaussian Mixture Model (GMM). Each class (i.e., each output neuron) is one component for GMM. Based on the GMM model, we could easily compute the posterior probability (confidence value) that a new sample belongs to a specific class in both BP and CNN. Consequently, a simple decision can be done by choosing the output class with the highest confidence value. Importantly, we have validated this approach in two benchmark datasets, i.e., MNIST and USPS data. Experimental results showed that the proposed strategy could significantly improve the accuracy. In particular, the proposed hybrid method largely reduced the error rates of CNN and BP respectively by 11.72% and 28.89% on MNIST.

The rest of this paper is organized as follows. In the next sections, we first briefly introduce BP, CNN, and then present our combination algorithm in details. In Section 5, we report the experimental results to validate the effectiveness of our algorithm. Finally, We set out concluding remarks in Section 6.

2   Back Propagation Neural Network

Back Propagation NN is actually a typical feed-forward NN [1]. In more details, each neuron receives a signal from the neurons in the previous layer, and each of those signals is multiplied by a separate weight value. The weighted inputs are summed, and passed through a so-called activation function which "squashes" the output to a fixed range of values. Such output is passed to all the neurons in the next layer. In this sense, we usually feed the input values to the first layer so that the signals are propagated through the network, and finally receive the output values. In order to make the networks output meaningful values for classification or regression, we need a method of adjusting the weights connected among different layers. For this purpose, one of the most common learning algorithms is called Back Propagation. Given a training set, a typical BP network could update the weights using a stochastic gradient algorithm per training sample.

The BP algorithm applies iteratively in the following steps: one sample is feeded to the network, and the network produces certain output based on the current values of the weights among different layers. This output is compared to the ground truth, and a mean-squared error can be computed. The error value is then propagated backwards through the network (which motivates the name of BP). Small updating is made to the weights in each layer. The whole process is repeated for each training instance until a stable solution was obtained for the weights. Details can be seen in [1].

3 Deep Convolutional Neural Network

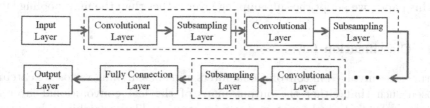

Fig. 1. Framework of Hybrid System

CNNs recently receive more and more attention because of their big accuracy improvement in various applications. CNNs adopt a deep hierarchy structure [2]. More specifically, CNNs usually alternate with convolutional layers and subsampling layers. This could be seen in Figure 1. Different CNNs vary in how convolutional and subsampling layers are constructed and how the nets are trained. Basically, a CNN can be regarded as one hierarchy or deep network where the previous layers (convolutional layers and subsampling layers) extract features gradually from low levels to high levels. The last classification layers, also referred to as fully-connected layers, could simply be considered as traditional BP networks. One typical setting for CNN can be seen in Figure 3, borrowed from [7]. We will describe each typical layer involved in CNN as follows.

3.1 Optional Image Processing Layer

The image processing layer is an optional pre-processing layer that is kept fixed during training. Such layer would provide potentials to enable additional information besides the raw input image into the network. In this paper, we just feed the raw image into CNN and hence has no this pre-processing layer.

3.2 Convolutional Layer

A specific convolutional layer is decided by the size and the number of the maps, kernel sizes, skipping factors, and the connection table. Each layer has M maps of equal size (M_x, M_y). A kernel of size (K_x, K_y), used to extract certain high-level features, is shifted over the valid region of the input image. Moreover, neurons of a given map share their weights but have different receptive fields.

3.3   Subsampling Layer

After the convolutional layer generates M maps to middle-level images, a subsampling layer will be followed to simply resize the map to a smaller one. Such operation could be done according to a specific subsampling ratio. In the literature, there are also some researchers changing the subsampling layer as a max-pooling layer [2], where the output of the max-pooling layer is given by the maximum activation over non-overlapping rectangular regions of size (K_x, K_y). In this paper, we adopt the subsampling layer rather than the max-pooling layer.

3.4   Classification Layer

Kernel sizes involved in convolutional filters and subsampling ratios are carefully chosen such that either the output maps of the last convolutional layer are downsampled to 1 pixel per map, or a fully connected layer combines the outputs of the topmost convolutional layer into a one-dimensional feature vector. The top layer is always fully connected, with one output unit per class label. Hence the classification layer is also called as the fully-connected layer.

4   Hybrid Approach Based on GMM

In this section, we focus on describing our hybrid approach in details.Note that both BP and CNN usually have an output layer where the number of neurons is the same as that of classes (denoted by c) involved in the data. When a test sample $\mathbf{z}$ of p-dimension is input , the class associated with the output neuron containing the largest response will be assigned to the sample. If we assume that $p(z, C)$ (C is the class variable) is modeled by a Gaussian Mixture Model with each component corresponding to the class $C = c_i$,, we could have

$$P(\mathbf{z}, C; \theta) = \sum_{i=1}^{c} \pi_i \mathcal{N}(\mathbf{z}; \mu_i, \sigma_i),$$

where the unknown parameter vector θ consists of the mixture weight π_i, the means of component μ_i, and the covariance of component matrices $\sigma_i (i = 1, \ldots, c)$.

Clearly, the posterior probability can be calculated by

$$P(c_j | \mathbf{z}; \theta) = \frac{\pi_j \mathcal{N}(\mathbf{z}; \mu_j, \sigma_j)}{\sum_{i=1}^{c} \pi_i \mathcal{N}(\mathbf{z}; \mu_i, \sigma_i)} .$$

Since $\mathbf{z}$ is usually a high-dimensional vector, it would be difficult to estimate $\mathcal{N}(\mathbf{z}; \mu_j, \sigma_j)$. Instead, we could simply estimate the class conditional probability using the actual output value of neuron o_i associated with class c_i, i.e.,

$$\mathcal{N}(\mathbf{z}; \mu_j, \sigma_i) \propto \mathcal{N}(o_j(\mathbf{z}); \mu_{o_j}, \sigma_{o_j}).$$

Here, $o_j(\mathbf{z})$ is the value of output neuron j when the input sample is $\mathbf{z}$. As o_j is one-dimensional variable, its mean and covariance can be much easier and more stable to be estimated from training samples. By combining all the above, we could finally get the simplified posterior probability equation:

$$P(c_j|\mathbf{z};\theta) = \frac{\pi_j \mathcal{N}(\mathbf{z};\mu_j,\sigma_j)}{\sum_{i=1}^{c} \pi_i \mathcal{N}(\mathbf{z};\mu_i,\sigma_i)} \tag{1}$$

$$= \frac{\pi_j \mathcal{N}(o_j(\mathbf{z});\mu_{o_j},\sigma_{o_j})}{\sum_{i=1}^{c} \pi_i \mathcal{N}(o_j(\mathbf{z});\mu_{o_i},\sigma_{o_i})} . \tag{2}$$

Consequently, given a test sample $\mathbf{z}$ and the trained BP (or CNN), one can easily calculate the posterior probability according to Eq. 1 and the output values of $\mathbf{z}$. In another word, the value of output neuron associated with $\mathbf{z}$ could be transformed to a probability. The class or neuron with the maximum posterior probability can then be assigned to $\mathbf{z}$.

Based on the probability transformation by GMM, we can easily get the maximum posterior probability obtained from BP and from CNN, when a specific test sample is input. Therefore, it is reasonable and obvious that we could obtain the hybrid result by comparing which output is more confident according to the two probabilities given by BP and CNN. Namely, we simply output the class judged by the more confident classifier (BP or CNN) by the maximum operation. An illustration of this hybrid system can be seen in Figure 4.

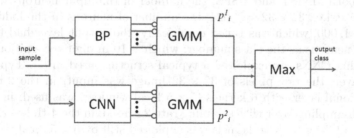

Fig. 2. Framework of Hybrid System

5   Experiments

In this section, we evaluate the performance of the hybrid approach on two benchmark real data sets, MNIST and USPS in comparison with BP and CNN.

5.1   Experimental Setup

In this section, we first introduce the two used data sets briefly. Then we provide the BP and CNN network structure. MNIST contains a training set of $60,000$ and

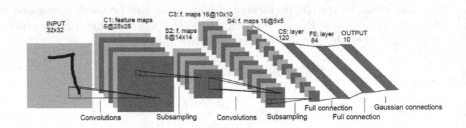

Fig. 3. Training structure used in CNN [7]

a test set of 10,000 handwritten numerals. This data set is a subset of a larger set available from NIST. Each sample has been size-normalized and centered in a fixed-size image of 28 × 28 pixels. USPS contains a training set of 7,291 handwritten digits and a test set of 2,007 digits. Both data sets have frequently been used in the literature and are regarded as the benchmark data sets in pattern recognition. Following many previous research, we simply exploited raw pixel features for both data sets. For MNIST data, we put each sample in the center of 32 × 32 white image. For the USPS data, we resize each sample from 16 × 16 to 28 × 28, and then center it in a white image of 32 × 32.

In training BP neural networks, we used a three-layer structure. In the experiments of both MNIST and USPS, the number of the input neurons was equal to image size, i.e., 32 × 32 = 1,024; the number of neurons in the hidden layer was set to 1,000, which was tuned empirically; the output layer had the same number of neurons as the class number, which is 10 in digit recognition.

In training CNNs, we exploited a typical structure used in [7]. Typically, in the 1-st layer, the raw pixels of 32 × 32 image was input; in the 2-nd layer, a convolutional layer with 6 kernels (5 × 5 local window) was used; in the 3-rd layer, a subsampling layer with sampling rate 2 is used; in the 4-th layer, another convolutional layer with 16 kernels was exploited still by 5 × 5 local window; in the 5-th layer, a further subsampling layer with sampling rate 2 was adopted; in the 6-th layer, a convolutional layer with 120 kernels was used (still with 5 × 5 local window); in the 7-th layer, a fully connected layer with 84 neurons were used; the 8-th layer is the output layer with 10 neurons. For clarity, we slightly modified the graph use in [7] and displayed it in Figure 3.

Table 1. Error Rate Comparison of Different Approaches on MNIST and USPS Data

Data	CNN (%)	BP (%)	Hybrid (%)	Err. Redu. to CNN (%)	Err. Redu. to BP (%)
MNIST	2.90	3.60	**2.56**	11.72 ↓	28.89 ↓
USPS	9.72	8.97	**8.57**	11.83 ↓	4.46 ↓

5.2 Experimental Results

We trained the BP and CNN in both the data sets and reported the test error rates in the test sets respectively. The hybrid system was also tested on the same test sets. The experimental results were shown in Table 1. For a better visualization, we also plotted the error rates in Figure 4. In order to clearly inspect the difference among different models, we also reported the relative error rate reduction of the hybrid approach to CNN and BP respectively in the last two columns of Table 1. It is evident that the hybrid approach significantly reduced the classification error rates involved in BP and CNN. More specifically, on MNIST, its relative error rate reductions were respectively 11.72% and 28.89% to CNN and BP, while on USPS, the relative error rate reductions were respectively 11.83% and 4.46% to CNN and BP. Such results clearly demonstrated the usefulness of our proposed hybrid methods.

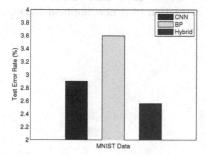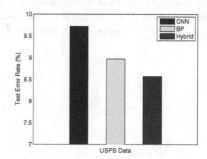

Fig. 4. Test error rate comparison among different approaches. Clearly, the proposed hybrid approach significantly outperformed BP and CNN.

6 Conclusion

In this paper, we have proposed a novel hybrid approach to combine one typical deep NN, i.e. CNN with one popular traditional NN, i.e. BP. Specifically, the hybrid approach assumes a Gaussian Mixture Model over the output neurons for both models. Based on the probabilistic calculation, the confidence of each model, namely, CNN and BP can be provided for the input instance. We have presented the justification and described the detailed learning formulation. A series of experiments validate that the combined approach could significantly improve the accuracy for both CNN and BP on two large-scale benchmark data sets.

Acknowledgement. The research was partly supported by the National Basic Research Program of China (2012CB316301) and Jiangsu University Natural Science Research Programme (14KJB520037).

References

1. Bishop, C.M.: Neural Networks for Pattern Recognition. Oxford University Press, London (1995)
2. Ciresan, D.C., Meier, U., Masci, J., Gambardella, L.M., Schmidhuber, J.: High-performance neural networks for visual object classification. In: Proceedings of Internet Joint Conference on Artificial Intelligence, IJCAI (2011)
3. Hinton, G.E., Osindero, S., Teh, Y.: A fast learning algorithm for deep belief nets. Neural Computation 18, 1527–1554 (2006)
4. Hinton, G.E., Salakhutdinov, R.R.: Reducing the dimensionality of data with neural networks. Science 33(5786), 504–507 (2006)
5. Kohonen, T., Honkela, T.: Kohonen Network. Scholarpedia (2007)
6. Krizhevsky, A., Sutskever, I., Hinton, G.E.: Imagenet classification with deep convolutional neural networks. In: Advances in Neural Information Processing Systems (NIPS) 25 (2012)
7. LeCun, Y., Bottou, L., Bengio, Y., Hakner, P.: Gradient-based learning applied to document recognition. Proceedings of the IEEE 86(11), 2278–2324 (1998)
8. Storkey, A.J., Valabregue, R.: The basins of attraction of a new hopfield learning rule. Neural Networks 12(6), 869–876 (1999)

A Classification Method of Darknet Traffic for Advanced Security Monitoring and Response

Sangjun Ko[1,2], Kyuil Kim[1], Younsu Lee[1], and Jungsuk Song[1,2,*]

[1] Korea Institute of Science and Technology Information, Daejeon, Korea
[2] Korea University of Science and Technology, Daejeon, Korea
{ksj87,kisados,zizeaz,song}@kisti.re.kr

Abstract. Most organizations or CERTs deploy and operate Intrusion Detection Systems (IDSs) to carry out the security monitoring and response service. Although IDSs can contribute for defending our information property and crucial systems, they have a fatal drawback in that they are able to detect only known attacks that were matched to the predefined signatures. In our previous work, we proposed a security monitoring and response framework based on not only IDS alerts, but also darknet traffic. The proposed framework regards all incoming darknet packets that were not detected by IDSs as unknown attacks. In our further analysis, we recognized that not all of darknet traffic is related to the real attacks. In this paper, we propose an advanced classification method of darknet packets to effectively identify whether they were caused by the real attacks or not. With the proposed method, the security analyst can ignore the darknet packets that were not related to the real attacks. In fact, the experimental results show that it succeeded in removing 23.45% of unsuspicious darknet packets.

Keywords: Security Monitoring and Response, IDS alerts, Darknet, Classification Method.

1   Introduction

The security monitoring and response, which mainly consists of three phases, i.e., detection of potential cyber attacks, analysis of them and response to the real attacks, is one of the most powerful services in order to fight against cyber threats happening on the Internet. Most organizations or CERTs deploy and operate Intrusion Detection Systems (IDSs) to carry out the security monitoring and response service [1]. Although IDSs can contribute to defending our information property and crucial systems, they have two fatal drawbacks.

Firstly, their detection accuracy is very low. In fact, more than 99% of IDSs alerts are false positive [2,3]. Second, they are able to detect only known attacks that were matched to the predefined signatures. In order to cope with the second weakness, in our previous work [5,6,7], we proposed a security monitoring and response framework based on not only IDS alerts, but also darknet traffic. Since

---

* Corresponding author.

C.K. Loo et al. (Eds.): ICONIP 2014, Part III, LNCS 8836, pp. 357–364, 2014.

darknet is a set of unused IP addresses(i.e., no real server,system,etc), packets that were observed on the darknet can be regarded as malicious activities. Based on the concept of the darknet, the proposed framework regards all incoming darknet packets that were not detected by IDSs as unknown attacks. In our further analysis, however, we recognized that not all of darknet traffic is related to the real attacks.

In this paper, we propose an advanced classification method of darknet packets to effectively identify whether they were caused by the real attacks or not. With the proposed method, the security analyst can ignore the darknet packets that were not related to the real attacks For example, backscatter packets of DDoS attack, reply packets by third-party victims, torrent packets, etc. In order to verify the effectiveness of the proposed method, we used real darknet traffic of three months (Sep. 1st, 2013 ~ Nov. 30th, 2013) that was obtained from Science and Technology Security Center (S&T-SEC) which provides the security monitoring and response service to 51 Korea government-funded research institutes. The number of the source hosts within the 51 research institutes that sent packets to the darknet was 661. Among the 661 source hosts, the proposed method succeeded in identifying the 155 source hosts (i.e., 23.45%) that sent unsuspicious darknet packets. This means that it is able to dramatically improve the performance of the security monitoring and response service.

The rest of the paper is organized as follows. Section 2 gives a brief description of the existing approaches related to darknets. Section 3 presents the proposed method in detail and the experimental results are given in Section 4. Finally, we make conclusions and suggestions for future research in Section 5.

2   Related Work

Researchers have done many efforts on the reduction of meaningless IDS alerts [2,3,4]. They are mainly based on data mining and machine learning techniques to deal with the IDS alerts in an automated manner. Darknet based approaches have been also proposed to develop countermeasures against malicious activities on the Internet [8,9,10,11]. For example, Nakao et al. introduced a network incident analysis center for tactical emergency response (nicter) and its main purpose is to carry out correlation analysis between the network threats observed in the darknet and malwares captured in the various types of honeypots. [5,6,7].

The experimental results showed that the proposed framework could detect unknown attacks that were not detected by IDSs. This framework has a problem; it regards all the incoming darknet packets as real attacks. In practice, however, darknet packets can be sent by backscatter packets of DDoS attack, reply packets by third-party victims, torrent packets, whose source hosts are not the original attackers or attack hosts. Therefore, we propose a classification method of darknet packets so as to overcome this problem of our previous framework.

3   Proposed Method

3.1   Overall Architecture

Figure 1 shows the overall architecture of the proposed method. The method is divided in three main phases : Monitoring, Extraction and Classification. During the Monitoring phase, it collects all the incoming packets to the darknet space and feed them into the Extracting phase. The Extraction phase extracts the specific darknet packets whose source IP addresses belong to the target organizations for the security monitoring and response service. Finally, the Classification phase consists of three modules : ICMP Module, Torrent Module, TCP Module. The three modules inspect the payload or the header of the darknet packets in order to identify whether their source hosts are related to the original hosts or not. The detailed description of the three modules is shown in Section 3.2, 3.3 and 3.4, respectively.

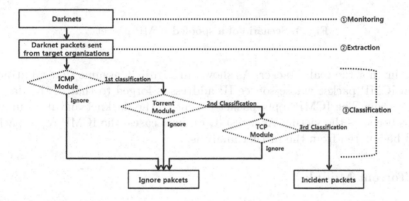

Fig. 1. Overall architecture of the proposed method

3.2   ICMP Module

Figure 2 shows the procedure of ICMP module, and it consists of three main steps. Firstly, it inspects the header of the darknet packets to extract only ICMP packets. Second, if the protocol of the darknet packets is 'ICMP', then it extracts 'TYPE' value (e.g., echo reply, echo request, time exceed, etc) from their header. Finally, if the 'TYPE' value of the darknet packets is relevant to 'reply' types (e.g., echo reply, timestamp reply and so on), they are ignored for the security monitoring and response service. Otherwise, they are regarded as the real attacks.

The reason why we ignore the ICMP reply packets is as follows. In normal cases, the ICMP reply packets must be sent to their original source hosts. If the ICMP reply packets were observed on the darknet, this means that the original source hosts spoofed their IP addresses to those of the darknet. Therefore, source hosts of the ICMP reply packets that sent to the darknet should be regarded as

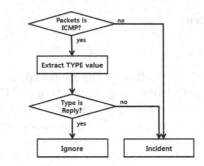

Fig. 2. Procedure of ICMP module

Fig. 3. Scenario of a spoofed ICMP packet

the victim, not the real attacker. As shown in Figure 3, assume that an attacker sent an ICMP packet whose source IP address is forged to that of the darknet to the victim. The ICMP reply packet is sent to the darknet and its source IP address becomes the victim. As a result, in most cases, the ICMP reply packets should be ignored from the further analysis.

3.3 Torrent Module

Figure 4 shows the procedure of the Torrent module. The main purpose of this module is to filter out the specific darknet packets that were sent by the Torrent protocol. Since the torrent client carries out scanning activities to the other peers, the scanning packets can be observed on the darknet. Therefore, we have to ignore the specific darknet packets that were sent by the torrent software.

To this end, the Torrent module first extracts the payload from each of the darknet packets. And then compares the payload with the three strings, i.e., 'Bittorrent Protocol', 'info_hash' and 'd1:ad2:id20', that are generally represented for using the torrent protocol. Finally, if the payload of the darknet packets contain one of three strings, they are ignored for the security monitoring and response service.

3.4 TCP Module

Figure 5 shows the procedure of the TCP module and it consists of three main steps. First, inspecting the header of the darknet packets to extract only TCP packets. Second, if the protocol of the darknet packets is 'TCP', then it extracts 'flag' value (e.g., SYN, ACK, RST, etc) from their header. Finally, if the 'flag'

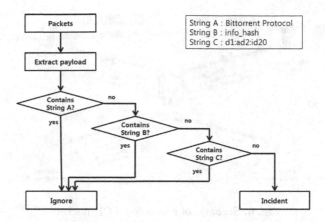

Fig. 4. Procedure of Torrent Module

value of the darknet packets is related to reply packets(e.g., 'SYN + ACK', 'RST + ACK', etc), they are ignored for the security monitoring and response service. Otherwise, they are regarded as the real attacks.

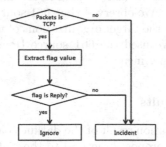

Fig. 5. Procedure of TCP Module

Similar to the ICMP module, in normal cases, the TCP reply packets must be sent to their original source hosts. If the TCP reply packets were observed on the darknet, this means that the original source hosts spoofed their IP addresses to those of the darknet. In many cases, this situation is caused by the backscatter, i.e., reflection of DDoS attack. Therefore, source hosts of the TCP reply packets that sent to the darknet should be regarded as the victim, not the real attacker. As shown in Figure 6, assume that an attacker sent an TCP packet whose source IP address is forged to that of the darknet to the victim. The TCP reply packet is sent to the darknet and its source IP address becomes the victim. As a result, in most cases, the TCP reply packets should be ignored from the further analysis.

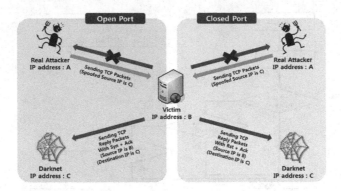

Fig. 6. Scenario of a spoofed TCP packet

4   Evaluation

4.1   Experimental Environment

In order to verify the effectiveness of the proposed method, we prepared 32 /24 darknet (i.e., 8,160 IP addresses) in S&T-SEC and collected all the darknet packets during 3 months (Sep. 1st, 2013 ∼ Nov. 30th, 2013). The total number of the darknet packets and the unique source IP addresses were 237,599,930 and 9,053,688, respectively. We observed that 661 source IP addresses of the 51 research institutes that are the target organizations of the S&T-SEC sent 49,945 packets to our darknet. We used the 661 source IP addresses and the 49,945 darknet packets for the experiment.

4.2   Experimental Results

Figure 7 shows the classification result of the darknet packets and their source IP addresses. We applied the proposed method to the 49,945 darknet packets that were sent from the 661 source IP addresses. Among the 661 source IP addresses, the three modules extracted 190 source IP addresses (28.75%) that sent the ICMP, Torrent and TCP packets to the darknet.

The ICMP module extracted the 101 source IP addresses that sent the 1,855 ICMP packets to the darknet and it succeeded in ignoring 96 source IP addresses (14.52%) that sent the ICMP reply packets to the darknet. The Torrent module extracted 59 source IP addresses that sent the 199 torrent packets to the darknet. It succeeded in ignoring all the source IP addresses (8.93%) that sent the torrent packets whose payload contain one of the predefined three strings. In case of the TCP module, it extracted the 30 source IP addresses that sent the 313 TCP packets to the darknet. However, we observed that all the source IP addresses did not sent the TCP packets with 'SYN+ACK' or 'RST+ACK'. As a result, it could be concluded that the proposed method contributed to removing of 23.45% of unsuspicious darknet packets.

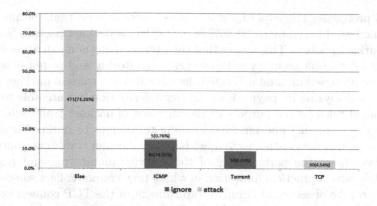

Fig. 7. Classification result of the proposed method

Although the TCP module could not remove the TCP packets with 'SYN+ACK' or 'RST+ACK', we recognized that the original darknet packets include such TCP packets. Fig 8 show the number of TCP packets according to their flag. From Fig 8, we can easily see that about 34% of the TCP packets were set by 'SYN+ACK' or 'RST+ACK'. This means that if we expand the proposed method to the all source IP addresses, not only those of the 51 research institutes, The TCP module can contribute to filtering out the unsuspicious TCP packets.

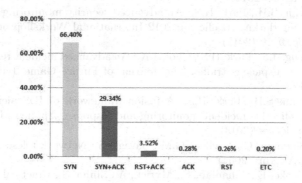

Fig. 8. Number of TCP packets according to their flag

5   Conclusion

In this paper, we have proposed an advanced classification method of the darknet packets. The proposed method consists of three main modules : ICMP, Torrent and TCP. The main purpose of the three modules is to filter out unsuspicious darknet packets based on the communication characteristics of ICMP, Torrent and TCP protocols. The experimental results demonstrated that the proposed method succeeded in removing 23.45% of unsuspicious darknet packets from the original darknet packets. This means that the proposed method can dramatically improve the detection accuracy of the security monitoring and response service.

Since the proposed method extracted the classification rules of darknet traffic by manually analyzing its payload and header information, we are able to make the additional rules for the purpose of classification of darknet traffic effectively if we apply data mining and machine learning techniques to darknet traffic. In addition, although the existing techniques based on darknet traffic can contribute to improvement of the performance of the security monitoring and response service, they have a practical limitation in which they cannot collect some attack codes that can be observed after only establishment of the TCP connection. Our future work will try to overcome this limitation by integrating with honeypot techniques or constructing a reply server for the darknet traffic.

References

1. Denning, D.E.: An intrusion detection model. IEEE Transactions on Software Engineering SE–13, 222–232 (1987)
2. Julisch, K.: Clustering intrusion detection alarms to support root cause analysis. ACM Transactions on Information and System Security 6(4), 443–471 (2003)
3. Manganaris, S., Christensen, M., Zerkle, D., Hermiz, K.: A Data Mining Analysis of RTID Alarms. Computer Networks 34(4), 571–577 (2000)
4. Humphrey, W.N., Luo, J.: Using alert cluster to reduce IDS alerts. In: ICCIT 2010, pp. 467–471. IEEE (2010)
5. Choi, S.S., Kim, S.H., Park, H.S.: An advanced security monitoring and response framework using darknet traffic. In: 2012 International Workshop on Information & Security, pp. 9–10 (2012)
6. Choi, S.S., Song, J.S., Park, H.S., Choi, J.K.: An advanced incident response framework based on suspicious traffic. The Journal of Future Game Technology 2(2), 171–176 (2012)
7. Choi, S.S., Kim, S.H., Park, H.S.: A fusion framework of IDS alerts and darknet traffic for effective incident monitoring and response. Applied Mathematics & Information Sciences (2013)
8. Moore, D., Shannon, C., Voelker, G.M., Savage, S.: Network telescopes, technical report. CAIDA (April 2004)
9. Bailey, M., Cooke, E., Jahanian, F., Myrick, A., Sinha, S.: Practical darknet measurement. In: 2006 40th Annual Conference on Information Sciences and Systems, pp. 1496–1501. IEEE (2007)
10. Nakao, K., Inoue, D., Eto, M., Yoshioka, K.: Practical correlation analysis between scan and malware profiles against zero-day attacks based on darknet monitoring. IEICE Transactions on Information and Systems 92(5), 787–798 (2009)
11. Eto, M., Inoue, D., Song, J., Junji, N., Kazuhiro, O., Nakao, K.: Nicter: A large-scale network incident analysis system. In: Workshop on Development of Large Scale Security-Related Data Collection and Analysis Initiatives (BADGERS 2011), pp. 37–45. ACM, Salzburg (2011)

Detecting Malicious Spam Mails:
An Online Machine Learning Approach

Yuli Dai[1], Shunsuke Tada, Tao Ban[2], Junji Nakazato[2], Jumpei Shimamura[3],
and Seiichi Ozawa[1]

[1] Graduate School of Engineering, Kobe University,
1-1 Rokko-dai, Nada-ku, Kobe, 657-8501, Japan
`131t272t@stu.kobe-u.ac.jp`,
`ozawasei@kobe-u.ac.jp`
[2] National Institute of Information and Communications Technology (NICT),
Koganei, Tokyo, 184-8795, Japan
`{bantao,nakazato}@nict.go.jp`
[3] Clwit Inc., Tokyo, Japan
`shimamura@clwit.co.jp`

Abstract. Malicious spam is one of the major problems of the Internet
nowadays. It brings financial damage to companies and security threat to
governments and organizations. Most recent spam emails contain URLs
that redirect spam receivers to malicious Web servers. In this paper,
we propose an online machine learning based malicious spam email detection system. The term-weighting scheme represents each spam email.
These feature vectors are then used as the input of the classifier. The
learning is periodically performed to update the classifier so that the
system provides increased adaptability to take account of spam emails
whose contents change from time to time. A real data set is labeled by
the *SPIKE* system which is developed by *NICT*. Evaluation experiments
show that the detection system is efficient and accurate to identify malicious spam emails.

Keywords: malicious spam detection, online learning, tf-idf, vector space
model.

1 Introduction

E-mail is an important and efficient communication technique in today's life.
Because of its convenience, it is abused by spammers for commercial, political
and other purposes. As a result, the emailboxes of people get cluttered with
unsolicited bulk emails, i.e., spams. According to the 2012 statistics, 68.8% of
all email traffic was spam[2].

On the other hand, recent spam emails are possibly sent by various malware
(e.g., bots, worms). Such kind of spam emails always contain URLs linked to
a Web server for the purpose of diverse cyber attacks such as malware infection, user information theft and phishing attacks, etc. We call such spam emails
malicious spam emails.

C.K. Loo et al. (Eds.): ICONIP 2014, Part III, LNCS 8836, pp. 365–372, 2014.

There are many strategies to detect malicious spam emails. One of these techniques is to download HTML contents and malware by crawling the URLs within spam emails, and then analyse the obtained contents. Despite of its high accuracy in malicious spam email detection, the above approach costs a good deal of time and the analyses result cannot be shown immediately. *SPIKE* is a system detecting malicious emails in this way. It is part of Network Incident analysis Center for Tactical Emergency Response (*nicter*)[3][4] developed by National Institute of Information and Communications Technology (*NICT*).

In this paper, we propose an online machine learning based malicious spam email detection system. We use double bounce emails, which are collected by *nicer*[3][4], as our data. Text-based features are extracted from the body of these emails because malicious spam emails are likely to have identical content. This online system periodically updates the classifier since spammers change the contents of emails frequently. Our system can identify malicious spam emails from unmalicious ones as soon as the spam emails arrive at the SMTP server. The experimental results show that our system is highly accurate on malicious spam email detection and robust against the change of spam emails as well.

The rest of the paper is organized as follows. In section 2, we present the process of our system. Section 3.1 gives an introduction of double bounce email and why we use them as our data set. The experimental results are given in section 3.2. Finally, we present our concluding remarks.

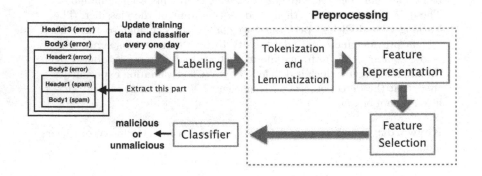

Fig. 1. An illustration of the system process

2   Malicious Spam Email Detection System

Figure 1 shows the overall structure of the proposed online system. Valid header (we just use the subject information) and body (the actual contents of the message) are extracted from the double bounce emails. Preprocessing steps include labeling, tokenization, lemmatization, feature representation and feature selection. All the emails are then represented by the *vector space model*[5]. Linear $l1$-norm SVM is used for feature selection, and linear $l2$-norm SVM is used for classification. The everyday update to the classifier makes our system robust against spam emails whose contents change frequently.

SPIKE takes a long time to identify a malicious spam email. This kind of inefficiency leads to an anxious waiting of email receivers. This drawback can be solved availably by our system, i.e., the category of an arrived spam email can be known instantaneously. Malicious spam emails will be deleted to prevent cyber attack.

2.1 Preprocessing

Before the formulation of a classifier, appropriate preprocessing steps are required as the classifier expects numerical feature vectors with a fixed size rather than the raw text data with variable length. The steps are illustrated in Figure 1, and can be grouped into labeling, tokenization and lemmatization, feature representation and feature selection.

Labeling: The data need to be labeled on account of supervised learning and the labels of those mails are dependent on *SPIKE*, which is a malware analysis system. It is part of Network Incident analysis Center for Tactical Emergency Response (*nicter*)[3][4] developed by *NICT*.

SPIKE analyzes and detects malware by crawling URLs embedded in the emails. Spam mails are then labelled with Black or White based on the results obtained by *SPIKE*. Here Black and White represent malicious and unmalicious, respectively. The process of crawling URLs is as follows:

(1) extract URLs from the body of the email.
(2) regard the URLs as entrance and start crawling.
(3) analyze the destination web pages linked with the URLs by downloading their HTML contents and malware if possible.
(4) extract URLs embedded in the web pages obtained from (3), and repeat (3) and (4).
(5) label the email as Black if there exists at least one web page being judged with malicious, otherwise White.

In addition, we label the email as Grey in case of invalid entrance URL and as None if there is no URL existing in the body of the email.

Tokenization and Lemmatization: Tokenization is a process of extracting the words in the extracted contents and lemmatization is a process of reducing words to their possible root forms (e.g., "forms" to "form")[6]. After this step, a text corpus of training data is created.

Feature Representation: *Tf-idf* (term frequency-inverse document frequency) is applied in our research for feature representation. It is a numerical statistic that is intended to reflect how important a word is to a document in a collection or corpus[7]. The calculation of the *tf-idf* term weight of each word is given by

$$tf\text{-}idf_{j,k} = tf_{j,k} \times idf_k .$$
(1)

where the $tf_{j,k}$ is the term frequency of word k in document j(i.e., spam mail in our research). Given the document frequency df_k as the frequency of documents in the database that contains word k, the inverse document frequency idf_k is defined as $\log(\frac{N}{df_k})$, where N is the total number of documents.

The text corpus of mails can thus be represented as a row of column vectors $A_{j,k}$, with $a_{j,k}$ representing the weight of word k in document j. This specific strategy is called the *bag of words*, also known as the *vector space model*. As most emails will typically use a very subset of the words used in the corpus, the resulting matrix will have many feature values that are zeros.

Feature Selection: Feature selection is a process of selecting the most effective and "representative" features from the original features. It also boosts estimators' performance on high-dimensional datasets. In our research, a linear *l1* SVM is applied to reduce the dimensionality of the data. We replace the *l2*-norm $\|\mathbf{w}\|_2^2$ with a *l1*-norm $\|\mathbf{w}\|_1$ that will promote feature selection[8]. Coefficients for the weakest features are set to zero, i.e., sparse solution. This approach evaluates feature importances and selects the most relevant features. The optimal value of the parameter C can be found by cross-validation.

Classifier: We apply different classification methods, such as decision tree, support vector machines, naive bayes and k-nearest neighbors. Among these algorithms, SVMs achieves the best performance.

Support vector machines (SVMs) developed by V. Vapnik is one of the most successful classification methods for many applications including text classification. For its capability of dealing with high dimensional datasets and efficiency for training, we apply linear *l2*-norm SVM as the classifier. The goal of SVM classification is to find the separating hyperplane with maximal margin, for which the distance to the closest training sample is maximal[9]. In standard two-class classification problems, we are given a set of training data $(\mathbf{x}_1, y_1), ..., (\mathbf{x}_n, y_n), \mathbf{x}_i \in R^n$ and the output $y_i \in \{1, -1\}$ is binary. The standard *l2*-norm SVM is equivalent to fit a model that

$$\min_{b,\mathbf{w}} C \sum_{i=1}^{n} [1 - y_i(b + \mathbf{w}^T \phi(\mathbf{x}_i))]_+ + \|\mathbf{w}\|_2^2 .$$
(2)

where C is a penalty parameter on the training error and ϕ is a function that mapped training data into higher dimensional space. Practically, a kernel function $K(\mathbf{x}_i, \mathbf{x}_j) = \phi(\mathbf{x}_i)^T \phi(\mathbf{x}_j)$ may be used to train the SVM. A linear SVM has $\phi(\mathbf{x}) = \mathbf{x}$, so the kernel function is $K(\mathbf{x}_i, \mathbf{x}_j) = \mathbf{x}_i^T \mathbf{x}_j$[10].

3   Experiments

3.1   Data Set

As we know, a normal email contains at least one return-path address in its header field, even if a sender mistyped the recipient address to his or her email. In contrast, double bounce emails have no valid recipient address and return-path address. Therefore, if such an email is sent to the external SMTP server, a user unknown error message will be returned to the internal SMTP server which sent the original email. After that, the internal SMTP server sends back the same error message to the external SMTP server since the original email has no valid return-path address. The process is shown in Figure 2. In this situation, spammers fabricate inexistent recipient addresses and conceal their detrimental activities. Therefore, double bounce emails can be regarded as pure spam. For this reason, we use them as our raw data.

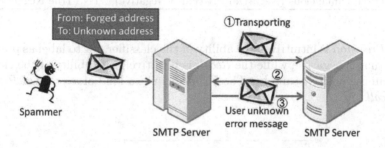

Fig. 2. The generating process of double bounce email

Our data set was collected by *nicter*[3][4]. Considering the structure of double bounce email as shown in Figure 1, we ignore the error messages and extract the spam part consisting of header 1 (i.e., subject) and body 1. The preprocessing of the raw data has been explained in Section 2.1. We collected the data set from March 1st to July 9th 2013. The total number of collected emails is 19924. It is worth mentioning that the number of double bounce emails we received in each day subjects to a large variance. This characteristic is also reflected in the ratio of malicious double bounce emails to unmalicious ones.

3.2   Evaluation

It is different from the general classification problem that in our research a malicious spam email misclassified as unmalicious can be unacceptable. For this reason, describing the performance in terms of the classification accuracy is not adequate. Moreover, in a highly unbalanced scenario, a classifier can attain a high accuracy. Therefore *Precision, Recall* and *F1* [11] are applied as the evaluation measures. They are defined as follows

$$Precision = \frac{TP}{TP + FP}, \tag{3}$$

$$Recall = \frac{TP}{TP + FN}, \tag{4}$$

$$F1 = 2 \cdot \frac{Precision \cdot Recall}{Precision + Recall}. \tag{5}$$

where *TP, TN, FP, FN* are defined in Table 1.

Table 1. The definitions of *TP, TN, FP, FN*

	Condition Malicious (Positive)	Condition Unmalicious (Negative)
Prediction Malicious (Positive)	TP (True Positive)	FP (False Positive)
Prediction Unmalicious (Negative)	FN (False Negative)	TN (True Negative)

The *Precision* is intuitively the ability of the classifier not to label as positive a sample that is negative, while the *Recall* is intuitively the ability of the classifier to find all the positive samples. The *F1 measure* is a kind of average of *Precision* and *Recall*.

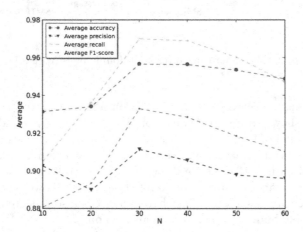

Fig. 3. Average results for N

Fist, we need to select the number of the training data and the parameter C of *l1*-norm that can achieve the best performance. We use the data of Day x as test data and the data of the N days ($N = 10, 20, 30, 40, 50, 60$) before Day x

as training data, e.g., if the test data is collected on March 31st, the training data should be from March 21st to March 30th when $N = 10$. We update the classifier and do testing every one day from Day 61 to Day 100 (Totally 40 days). The reason why it starts from Day 61 is that the maximal N is 60, moreover, the test data must be same. The averages (the average of all the testing results in 40 days) for *Accuracy, Precision, Recall* and *F1-score* are shown in Figure 3. It indicates that the best performance is achieved when $N = 30$.

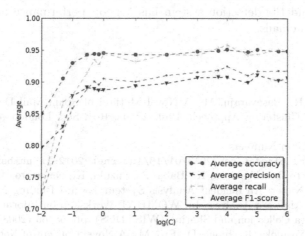

Fig. 4. Average results for parameter C

We set $N = 30$ and fix the penalty parameter of $l2$-norm in classification, then step into the next stage (i.e., selection of parameter C). The result shown in Figure 4 indicates that the average for each evaluation measure generally becomes larger along with the increase of parameter C and the curves become flat after $log(C) = 0$. Consequently we set $C = 10000$ in which case best performance can be attained.

After feature selection, we apply different classifiers. As we receive emails everyday, it is necessary to update the classifier. We predict the malicious spam emails received in Day x after updating the training data and classifier.

The average results of the classifiers are shown in Table 2. The linear $l2$-norm SVM has the best overall performance compared with other algorithms.

Table 2. Classification Evaluation

Classifier	Accuracy	Precision	Recall	F1-score
Decision Tree	0.930	0.866	0.901	0.864
Naive Bayesian	0.945	0.895	0.910	0.889
linear $l2$-norm SVM	0.953	0.909	0.957	0.925
kNN	0.893	0.843	0.819	0.811

4   Conclusion

In this paper, an online machine learning based malicious spam email detection system is proposed. The classifier is updated every one day in order to catch up with the change of contents of malicious spam emails. The dataset was labeled by *SPIKE*. Different performance measures such as the *Precision, Recall* and the *F1 measure* were observed. Several popular classification algorithms are studied and evaluated. The results show that the linear *l2*-norm SVM has a better overall performance and the detection system has a good performance on identifying malicious spam emails.

References

1. Prabhakar, R., Basavaraju, M.: A Novel Method of Spam Mail Detection Using Text Based Clustering Approach. Phil. Trans. Roy. Soc. London A247, 529–551 (2010)
2. Internet 2012 in Numbers, http://royal.pingdom.com/2013/01/16/internet-2012-in-numbers
3. Inoue, D., Eto, M., Yoshioka, K., Baba, S., Suzuki, K., Nakazato, J., Ohtaka, K., Nakao, K.: Nicter: An Incident Analysis System Toward Binding Network Monitoring with Malware Analysis. In: WOMBAT Workshop on Information Security Threats Data Collection and Sharing (WISTDCS), pp. 58–66 (2008)
4. Nakao, K., Yoshioka, K., Inoue, D., Eto, M.: A Novel Concept of Network Incident Analysis based on Multi-layer Observations of Malware Activities. In: The 2nd Joint Workshop on Information Security (JWIS 2007), pp. 267–279 (2007)
5. Salton, G., Wong, A., Yang, C.S.: A Vector Space Model for Automatic Indexing. Communications of the ACM 18(11), 613–620 (1975)
6. Guzella, T.S., Caminhas, W.M.: A review of machine learning approaches to spam filtering. Expert Syst. Appl. 36, 10206–10222 (2009)
7. Rajaraman, A., Ullman, J.D.: Mining of Massive Datasets, pp. 1–17 (2011)
8. Zhu, J., Rosset, S., Hastie, T., Tibshirani, R.: 1-norm support vector machines. Advances in Neural Information Processing Systems 16, 49–56 (2004)
9. Tretyakov, K.: Machine Learning Techniques in Spam Filtering. Technical Report, Institute of Computer Science, University of Tartu (2004)
10. Chang, Y.W., Lin, C.J.: Feature Ranking using linear SVM. In: JMLR Workshop and Conference Proceedings, vol. 3, pp. 53–64 (2008)
11. Lewis, D.D.: Evaluating and optimizing autonomous text classification systems. In: Fox, E.A., Ingwersen, P., Fidel, R. (eds.) Proceedings of the 18th Annual International ACM SIGIR Conference on Research and Development in Information Retrieval, New York, pp. 246–254 (1995)

Freshness-Aware Thompson Sampling

Djallel Bouneffouf

Orange Labs, 2, avenue Pierre Marzin, 22307 Lannion, France
Djallel.Bouneffouf}@Orange.com

Abstract. To follow the dynamicity of the user's content, researchers have recently started to model interactions between users and the Context-Aware Recommender Systems (CARS) as a bandit problem where the system needs to deal with exploration and exploitation dilemma. In this sense, we propose to study the freshness of the user's content in CARS through the bandit problem. We introduce in this paper an algorithm named Freshness-Aware Thompson Sampling (FA-TS) that manages the recommendation of fresh document according to the user's risk of the situation. The intensive evaluation and the detailed analysis of the experimental results reveals several important discoveries in the exploration/exploitation (exr/exp) behaviour.

Keywords: CARS, Thompson Sampling, Contextual bandits.

1 Introduction

Mobile technologies have made access to a huge collection of information, anywhere and any-time. In this sense, recommender systems must promptly identify the importance of documents to recommend in the great location and moment. Recently, CARS tackle this problem by relating the user's interest to the user's situation (time, location, friends). However, they cannot avoid to recommend the same document under the same situations. As a result, a small set of documents are recommended again and again and then are seen as favourite documents, however recommend the same set of documents many times in a short period makes the users feel bored. Works found in literature [9, 8, 1] tackle this problem by addressing the recommendation as a need for balancing exr/exp studied in the "bandit algorithm". Actually the greatest result in exr/exp is performed by the Thompson Sampling (TS), but its drawback is in the none consideration of the freshness of document in the recommendation. The Freshness can be considered as the strength of strangeness or the amount of forgotten experience [6], and it leads the system to recommend some documents that have not been clicked for a long time because these documents are fresh to users even though they do not click to them multiple times. To this effect, we introduce in this paper an algorithm named Freshness-Aware Thompson Sampling (FATS) that achieves this goal by balancing adaptively the exr/exp trade-off according to the user's situation and the document's freshness. This algorithm extends the TS strategy by exploring fresh documents in suitable user's situations.

C.K. Loo et al. (Eds.): ICONIP 2014, Part III, LNCS 8836, pp. 373–380, 2014.

The remaining of the paper is organized as follows. Section 2 reviews related works. Section 3 gives key notion used in the paper. Section 4 describes the algorithms involved in the proposed approach. The experimental evaluation is illustrated in Section 5. The last section concludes the paper and points out possible directions for future work.

2   Related Work

We refer, in the following, techniques that study the different dimensions of our problem.

Multi-Armed Bandit Problem in RS. Recently, research works are dedicated to study the multi-armed bandit problem in RS, considering the user's behaviour as the context. In [3], authors model CARS as a contextual bandit problem. The authors propose an algorithm called Contextual-ϵ-greedy which a perform recommendation sequentially recommends documents based on contextual information about the users' documents. In [1], authors analyse the TS in contextual bandit problem. The study demonstrate that it has better empirical performance compared to the state-of-art methods. The authors in [3, 1] describe a smart way to balance exr/exp, but do not consider the user's context and document freshness during the recommendation.

User's Content Dynamicity in RS. To follow the dinamicity of the user's content, the authors in [5] formulate and study a new variant of the k-armed bandit problem, motivated by e-commerce applications. In their model, arms have (stochastic) lifetime after which they expire. In this setting an algorithm needs to continuously explore new arms, contrarily to the standard k-armed bandit model in which arms are available indefinitely and exploration is reduced once an optimal arm is identified. In this work the dynamicity of the content is considered but the authors do not address the notion of freshness. A notion of freshness of document is used in [7], where the authors propose an RS that considers the freshness of music in recommendation. However they neither consider the freshness in CARS nor in multi-armed bandit problem.

The Risk-Aware Decision. The risk-aware decision has been studied for a long time in reinforcement learning, where the risk is defined as the reward criteria that not only takes into account the expected reward, but also some additional statistics of the total reward, such as its variance or standard deviation [10]. In RS the risk is recently studied. The authors in [4] consider the risk of the situations in the recommendation process, and the study yields to the conclusion that considering the risk level of the situation on the exr/exp strategy significantly increases the performance of the recommender system.

Contribution. From this state of the art we observe that none of the existing works have studied the correlation between the user's situation risk and the freshness document recommendation. This is precisely what we intend to do with Freshness-Aware Thompson Sampling (FATS), the proposing algorithm exploits the following new features: (1) The algorithm takes into consideration the document's freshness in its exr/exp trade-off by considering the "Forgetting Curve" to assess freshness and evaluate favouredness. (2) The algorithm manages the recommendation of fresh documents according to the user's situation, where the fresh documents are more explored in non-risky situation (the user is at home the user may be interested by a freshness documents) rather than risky or critical situation (the user is at the office, in a meeting or with a client) the system has to do less exploration to avoid disturbing the user.

3   Key Notion

This section focuses on introducing the key notions used in this paper.

Situation: A situation is an external semantic interpretation of low-level context data, enabling a higher-level specification of human behaviour. More formally, a situation S is a n-dimensional vector, $S = (O_{\delta_1}.c_1, O_{\delta_2}.c_2, ..., O_{\delta_n}.c_n)$ where each c_i is a concept of an ontology O_{δ_i} representing a context data dimension. According to our need, we consider a situation as a 3-dimensional vector $S = (O_{Location}.c_i, O_{Time}.c_j, O_{Social}.c_k)$ where c_i, c_j, c_k are concepts of Location, Time and Social ontologies.

User Preferences: User preferences UP are deduced during the user navigation activities. $UP \subseteq D \times A \times V$ where D is a set of documents, A is a set of preference attributes and V a set of values. We focus on the following preference attributes: *click, fail , time* and *recom* which respectively correspond to the number of clicks for a document, number of failure (recommended and not clicked), the time spent on a document and the number of times it was recommended.

The User Model: The user model is structured as a case base composed of a set of situations with their corresponding UP, denoted $UM = \{(S^i; UP^i)\}$, where $S^i \in S$ is the user situation and $UP^i \in UP$ its user preferences.

Definition of Risk: "The risk in recommender systems is the possibility to disturb or to upset the user (which leads to a bad answer of the user)".

From the precedent definition of the risk, we have proposed to consider in our system Critical Situations (CS) which is a set of situations where the user needs the best information that can be recommended by the system, because he can not be disturbed. This is the case, for instance, of a professional meeting. In such a situation, the system must exclusively perform exploitation rather than exploration-oriented learning. In other cases where the risk of the situation is less important (like for example when the user is using his information system at home, or he is on holiday with friends), the system can make some exploration by recommending information without taking into account his interest.

To consider the risk level of the situation in RS, we go further in the definition of situation by adding it a risk level R, as well as one to each concept: $S[R]=(O_{\delta_1}.c_1[cv_1], O_{\delta_2}.c_2[cv_2], ..., O_{\delta_n}.c_n[cv_n])$ where $CV=\{cv_1, cv_2, ..., cv_n\}$ is the set of risk levels assigned to concepts, $cv_i \in [0, 1]$. $R \in [0, 1]$ is the risk level of situation S, and the set of situations with $R = 1$ are considered as CS.

Definition (Situation Bandit Problem). In a situation bandits problem, there is a distribution P over $(S^i, r(d_1), ..., r(d_k))$, where S is the situation,

$d_i \in D$ is one of the k document to be recommended, and $r(d) \in [0,1]$ is the reward for document d. The problem is a repeated game: on each round, a sample $(S^i, r(d_1), ..., r(d_k))$ is drawn from P, the situation S is announced, and then for one document chosen by the system, its reward $r(d)$ is revealed.

Definition (Thompson Sampling). The Thompson Sampling (TS) is a randomized algorithm based on Bayesian ideas. Using Beta prior and considering the Bernoulli bandit problem (the rewards are either 0 or 1), TS initially assumes document d to have prior $Beta(1,1)$ on μ_d (the probability of success). At time t, having observed $SU_d(t)$ successes (reward = 1) and $FU_d(t)$ failures (reward = 0) in $\theta_d(t) = SU_d(t) + FU_d(t)$ selects of document d, the algorithm updates the distribution on μ_d as $Beta(SU_d(t)+1, FU_d(t)+1)$. The algorithm then generates independent samples from these posterior distributions of the μ_d, and selects the document with the largest sample value.

4 FA-TS

To adapt the FA-TS algorithm to consider freshness document in context aware environment, we propose to compute the similarity between the present situation and each one in the situation base; if there is a situation that can be reused; the algorithm retrieves it, and then applies the TS algorithm. The proposed FA-TS algorithm is described in Algorithm 1 and involves for each trial $t = 1...T$ the following tasks. **Task 1:** Let S^t be the current user's situation, and PS the set of past situations. The system compares S^t with the situations in PS in order to choose the most similar S^p using the $RetrieveCase()$ method. **Task 2:** Let D be the document collection and $D^p \in D$ the set of documents recommended in situation S^p. After retrieving S^p, the system observes the user's behaviour when reading each document $d^i \in D^p$. Based on observed rewards, the algorithm chooses the document d^p with the greater expected reward r^t using the $RecommendDocuments()$ method. To have the appropriate exploration at each situation, the $RecommendDocuments()$ method include a module $R(S^t)$ that computes the risk of the situation. **Task 3:** The algorithm improves its document-selection strategy with the new observation (S^t, d^t, r^t). The updating of the case base is done using the $Auto\_improvement()$ method.

Algorithm 1. The FA-TS algorithm

---
1. **Require:** $d \in D$ set UP, PS, N
2. Foreach $t = 1, 2, ..., T$ do
3. $(S^p, UP^p) = RetrieveCase(S^t, PS, UP, D)$ // Retrieve the most similar case
4. $SelectDocuments(UP^p, S^t, S^p, D, N)$ // Recommend N documents
5. **Receive a feedback** UP^t from the user
6. $Autoimprovement(UP^p, UP^t, S^t, S^p, N)$ // Update user's profile

---

RetrieveCase(): The system compares S^t with the situations in PS in order to choose the most similar one, $S^p = argmax_{S^i \in PS} sim(S^t, S^i)$. The semantic similarity metric is computed by:

$$sim(S^t, S^i) = \sum_{\delta \in \Delta} \alpha_\delta sim_\delta(c_\delta^t, c_\delta^i) \tag{1}$$

In Eq. 1, sim_δ is the similarity metric related to dimension δ between two concepts c_δ^t and c_δ^i, and Δ is the set of dimensions (in our case Location, Time and Social); α_δ is the weight associated to dimension δ and it is set out by using an arithmetic mean as follows: $\alpha_\delta = \frac{1}{t-1}(\sum_{k=1}^{t-1} y_\delta^k)$,where $y_\delta^k = sim_\delta(c_\delta^K, c_\delta^p)$ at trial $k \in \{1, ..., t-1\}$ from the $t-1$ previous recommendations, and $c_\delta^p \in S^p$. The idea here is to augment the importance of a dimension with the previously corresponding computed similarity values, reflecting the impact of the dimension when computing the most similar situation in Eq.1. The similarity between two concepts of a dimension δ depends on how closely c_δ^t and c_δ^i are related in the corresponding ontology. To compute sim_δ, we use the same similarity measure as [11]:

$$sim_\delta(c_\delta^t, c_\delta^i) = 2 * \frac{depth(LCS)}{depth(c_\delta^t) + depth(c_\delta^i)} \tag{2}$$

In Eq. 2, LCS is the Least Common Subsumer of c_δ^t and c_δ^i, and $depth$ is the number of nodes in the path from the current node to the ontology root.

SelectDocuments(): The algorithm chooses the document d^P with the greatest index P computed as follows:

$$P(d) = (1 - \epsilon) * \theta(d, S^p) - \epsilon * Mr(d) \tag{3}$$

In Eq. 3, $\theta(d, S^p) = SU_d(S^p, t) + FU_d(S^p, t)$. The idea here is to consider the sampling for each user's situation rather than all over the situations.

$Mr(d)$ is the strength of strangeness or the amount of experience forgotten. We apply Forgetting Curve [6] to evaluate the freshness of a document to a user. The Forgetting Curve is shown as follows:

$$Mr(d) = e^{-\frac{t(d)}{rsm(d)}} \tag{4}$$

In Eq. 4, Mr is memory retention, rsm is the relative strength of memory and t is time. The least the amount of memory retention of a document is in a user's mind, the freshest is the document to the user. In our work, rsm is defined as the number of times the document has been clicked and t is the distance from present time to the last time the document has been clicked.

To adapt the impact of the user's memory retention to context-aware environment, we consider an ϵ that manage the weight of the Mr in computing the pertinence of documents. With the assumption that more the situation is risky more the user does not forget the document related to this situation, we propose to reduce recommending fresh document according the risk of the situation. More the situation is risky less fresh document is explored. Concretely, the algorithm computes the weight of ϵ, by using the situation risk level $R(S^t)$, as indicated in Eq. 5.

$$\epsilon = \epsilon_{max} - R(S^t) * (\epsilon_{max} - \epsilon_{min}) \tag{5}$$

A strict exploitation (ϵ=0) leads to a non optimal documents selection strategy, this is why R is multiplied by $(1-\epsilon_{min})$, where ϵ_{min} is the minimum exploration allowed in CS and ϵ_{max} is the maximum exploration allowed in all situations (these metrics are fixed to $\epsilon_{max} = 0.5 \wedge \epsilon_{min} = 0.05$ using an off-line simulation).

Autoimprovement(): Depending on the similarity between the current situation S^t and its most similar situation S^p, two scenarios are possible: (1) If $sim(S^t, S^p) \neq 1$ then $PS = PS \cup S^t \wedge UP = UP \cup UP^t$: the current situation does not exist in the case base; the system adds this new case composed of the current situation S^t and the current user preferences UP^t; (2) If $sim(S^t, S^p) = 1$ then $S^p = S^p \cup S^t \wedge UP^p = UP^p \cup UP^t$: the situation exists in the case base; the system updates the case having premise the situation S^p with the current user preferences UP^t.

Computing the Risk Level of the Situation: The risk complete level $R(S^t)$ of the current situation is computed by aggregating three approaches R^c, R^v and R^m as follows:

$$R(S^t) = \sum_{j \in J} \lambda_j R_j(S^t) \tag{6}$$

In Eq. 6, R_j is the risk metric related to dimension $j \in J$, where $J = \{m, c, v\}$; λ_j is the weight associated to dimension j and it is set out using an off-line evaluation. R_c compute the risk using concepts, R_m compute the risk using the semantic similarity between the current situation and situations stocked in the system and R_v compute the risk using the variance of the reward. The three approaches and their aggregation are described in [2].

5   Evaluation of FA-TS

In order to empirically evaluate the performance of our approach in on-line environment, we conduct our experiment with 3500 users of mobile application. We have randomly split users on five groups, and we assign to each group the mobile application with different recommendation algorithms (the algorithms are described below). Each time the user opens his software he gets 10 documents recommended by the system. To evaluate the impact of the risk we compare FA-TS to a variant with a fixed ϵ exploration of freshness like: **FA-TS-1:** In FA-TS, the risk is fixed to 1 ($\epsilon = 0$), which means that the algorithm does not consider the freshness in its recommendation. **FA-TS-0.5:** In FA-TS, the risk is fixed to 0.5 ($\epsilon = 0.5$), which means that the algorithm considers the freshness of the documents and the probability computed by the TS to recommend document. **FA-TS-0:** In FA-TS, the risk is fixed to 0 ($\epsilon = 1$), which means that the algorithm considers just the freshness to recommend document (no consideration of the risk of the situation) and **TS:** The TS uses the algorithm described in [1] to recommend document without consideration of freshness documents.

Average Precision on Top 10 Documents. We compare the algorithms regarding the precision which is the number of user's clicks on the 10 recommended documents during a navigation session. The average precision (AP) is the mean of the system's precision for all session during one day, a navigation session is the interval between the time when the user opens the mobile application and the time when he closes it. Note that we do not compute the recall because we can not know a priori all pertinent documents. In Fig. 5, the horizontal axis represents the day of the month and the vertical axis is the performance metric.

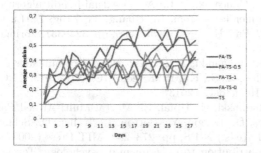

Algorithms	AP	ATSD
FA-TS	0,6542	1,3705
FA-TS-0.5	0,6187	1,3701
FA-TS-1	0,5450	1,3585
FA-TS-0	0,5109	1,3605
TS	0,4950	1,3714

Fig. 1. Average Precision on top 10 documents for each algorithm

We have displayed in the Table. 5 the average number of clicks per recommendation and the average time spent on documents (ATSD) for all the 28 days. We have several observations regarding the different algorithms. From the Fig. 5 we can observe that the FA-TS algorithm has effectively the best average precision during this month. We have also observed that $FA - TS - 1$ gives better results than TS in term of average clicks, which shows that considering the user's situation awareness in the TS approach improves its result. $FA - TS - 0.5$ gives better result than $FA - TS - 1$, which is explained by the consideration of the documents freshness in the TS. $FA - TS$ outperforms $FA - TS - 0.5$, which shows that managing the freshness of the document according to the situation's risk gives better result than a fixed approaches. An other interesting observation is in the fact that $FA - TS - 0$ outperform TS, which shows the impotence of considering the freshness which is not done by the TS. From the Table. 5 we can say that the ATSD does not significantly change from an algorithm to an other, which means that the exr/exp trade-off does not impact the user's time spent on documents and let us say that FA-TS gives better result on precision without reducing the quality of the recommended documents.

6 Conclusion

In this paper, we have studied the problem of document freshness in CARS and have proposed a new approach that considers the freshness of the document in recommendation regarding the user's situation. The experimental results demonstrate that considering the freshness on CARS significantly increases their performance. Moreover, this study yields to the conclusion that managing the recommendation of fresh document according to the risk of the situation gives a real add-value in recommendation performance.

References

[1] Agrawal, S., Goyal, N.: Thompson sampling for contextual bandits with linear payoffs. CoRR, abs/1209.3352 (2012)

[2] Bouneffouf, D.: DRARS, A Dynamic Risk-Aware Recommender System. PhD thesis, Institut National des Télécommunications (2013)

[3] Bouneffouf, D., Bouzeghoub, A., Gançarski, A.L.: A contextual-bandit algorithm for mobile context-aware recommender system. In: Huang, T., Zeng, Z., Li, C., Leung, C.S. (eds.) ICONIP 2012, Part III. LNCS, vol. 7665, pp. 324–331. Springer, Heidelberg (2012)

[4] Bouneffouf, D., Bouzeghoub, A., Ganarski, A.L.: Risk-aware recommender systems. In: Lee, M., Hirose, A., Hou, Z.-G., Kil, R.M. (eds.) ICONIP 2013. LNCS, vol. 8226, pp. 57–65. Springer, Heidelberg (2013)

[5] Chakrabarti, D., Kumar, R., Radlinski, F., Upfal, E.: Mortal Multi-Armed Bandits. In: Koller, D., Schuurmans, D., Bengio, Y., Bottou, L., Koller, D., Schuurmans, D., Bengio, Y., Bottou, L. (eds.) NIPS, pp. 273–280. MIT Press (2008)

[6] Ebbinghaus, H.: Memory: A contribution to experimental psychology. Teachers college, Columbia university (1913)

[7] Hu, Y., Ogihara, M.: Nextone player: A music recommendation system based on user behavior. In: Proceedings of the 12th International Society for Music Information Retrieval Conference, Miami (Florida), USA, October 24-28, pp. 103–108 (2011), http://ismir2011.ismir.net/papers/PS1-11.pdf

[8] Li, L., Chu, W., Langford, J., Schapire, R.E.: A contextual-bandit approach to personalized news article recommendation. In: Proceedings of the 19th International Conference on World Wide Web, WWW 2010, pp. 661–670. ACM, USA (2010)

[9] Li, W., Wang, X., Zhang, R., Cui, Y.: Exploitation and exploration in a performance based contextual advertising system. In: Proceedings of the 16th ACM SIGKDD International Conference on Knowledge Discovery and Data Mining, KDD 2010, pp. 27–36. ACM, USA (2010)

[10] Luenberger, D.: Investment Science. Oxford University Press (1998)

[11] Mladenic, D.: Text-learning and related intelligent agents: A survey. IEEE Intelligent Systems 14(4), 44–54 (1999)

Condition Monitoring of Broken Rotor Bars Using a Hybrid FMM-GA Model

Manjeevan Seera[1,*], Chee Peng Lim[2], and Chu Kiong Loo[1]

[1] Faculty of Computer Science and Information Technology, University of Malaya, Malaysia
mseera@um.edu.my
[2] Centre for Intelligent Systems Research, Deakin University, Australia

Abstract. A condition monitoring system for induction motors using a hybrid Fuzzy Min-Max (FMM) neural network and Genetic Algorithm (GA) is presented in this paper. Two types of experiments, one from the finite element method and another from real laboratory tests of broken rotor bars in an induction motor are conducted. The induction motor with broken rotor bars is operated under different load conditions. FMM is first used for learning and distinguishing between a healthy motor and one with broken rotor bars. The GA is then utilized for extracting fuzzy if-then rules using the *don't care* approach in minimizing the number of rules. The results clearly demonstrate the effectiveness of the hybrid FMM-GA model in condition monitoring of broken rotor bars in induction motors.

Keywords: Condition monitoring, fault diagnosis, fuzzy min-max neural network, genetic algorithms, induction motor.

1 Introduction

Condition monitoring is vital in machine maintenance, especially in the manufacturing environment for safe-guarding the efficiency and reliability of manufacturing machinery [1]. It is important to have a proper maintenance strategy in order to avoid machine or process failures; therefore minimizing the overall production time and cost [2]. The task of detection and isolation of faults can be challenging, especially in operations where dependent failures occur [3]. In this regards, the output loss owing to unplanned shutdown caused by process or machine failures cannot be recovered without incurring additional time and cost, such as wages for workers in overtime periods [4]. As such, condition monitoring has become a vital part in production operations and planning.

Typically, maintenance of machines can be accomplished either in a reactive, preventive, or predictive manner [5]. In reactive maintenance, the fix-upon-failure strategy is used while in preventive maintenance, the pre-planned strategy is utilized. Predictive maintenance, or better known as condition-based maintenance uses a

---

* Corresponding author.

C.K. Loo et al. (Eds.): ICONIP 2014, Part III, LNCS 8836, pp. 381–389, 2014.

forecasting strategy. Based on the practical implications of condition-based maintenance, the focus in this study is on developing a hybrid intelligent model for detecting broken rotor bars in induction motors. The main goal of condition-based maintenance is not only preventing machine failures, but also minimizing redundant maintenance activities [6]. On the other hand, induction motors are commonly used in various processes in production facilities such as in manufacturing machines, cranes, compressors, trolleys, cranes, and fans [7]. It is, therefore, important to minimize the running cost of induction motors. A useful way is employing an effective condition-based monitoring tool in reducing unforeseen failures in induction motors, as well as reducing unscheduled downtimes.

Among various neural network-based models, the Fuzzy Min-Max (FMM) network [8] is useful for data classification problems. Some salient features of FMM include online learning ability and a rapid learning process. One key limitation of FMM, however, is the inability to provide explanation for its predictions. This limitation, commonly known as the black-box phenomenon [9], exists in many neural network models. One effective way in solving the black-box phenomenon is through rule extraction. Here, we use a Genetic Algorithm (GA) for rule extraction from FMM. While the GA can be slow in its execution, it is able to perform global search and arrive at the optimal solution. Specifically, the GA's capability is exploited to minimize the input features in the extracted rules using the *don't care* approach in this study. We then evaluate the applicability of the hybrid FMM-GA model to condition monitoring of broken rotor bars in induction motors using simulated and real data samples, which forms the main contribution of this work.

The organization of this paper is as follows. The FMM-GA model is first presented in Section 2. The experimental setup, results, and discussion using the finite element method and real laboratory experiments are detailed in Sections 3 and 4, respectively. Conclusions and suggestions for further work are presented in section 5.

2   Hybrid FMM-GA Model

In this section, a hybrid FMM-GA model, which comprises FMM and the GA, is explained, as follows.

2.1   The Fuzzy Min-Max Network

FMM consists of three layers of nodes, i.e., F_A as the input layer, F_B as the hidden layer, and F_C as the output layer. The hidden layer is also known as the hyperbox layer. The input and output layers comprise nodes equal in numbers of the dimension of input patterns and the target classes, respectively. The connections between F_A and F_B encode the minimum (V) and maximum (W) points of the hyperboxes. The connections between F_B and F_C are binary-valued, whereby each F_C node represents one target class. The output from each F_C node represents the degree to which the input pattern fits within a target class.

Each hidden layer node in F_B is represented by a hyperbox fuzzy set. For an n-dimensional input pattern, X, a unit cube, I^n, is defined, where the membership value is from 0 and 1. Each hyperbox fuzzy set B_j is defined by [8]:

$$B_j = \left\{ X, V_j, W_j, f\left(X, V_j, W_j \right) \right\} \quad \forall X \in I^n \tag{1}$$

where V_j is the minimum point and W_j is the maximum point. The FMM learning process finds and fine-tunes the boundaries of the classes formed by the hyperboxes. Fig. 1 shows the decision boundary of a two-class problem.

The number of hyperboxes created in FMM is reduced when θ is increased from a small to a large value as the size of each hyperbox is increased. The jth hyperbox membership function is used to measure the extent the input pattern falls outside hyperbox B_j. This serves as a measurement on the extent that each component is lesser (or greater) than the minimum (or maximum) point along each dimension that falls outside the maximum and minimum boundary of the hyperbox. When the membership function is close to 1, the point is said to be "more contained" by the hyperbox. A sensitivity parameter, γ, manages how swiftly the membership value reduces when the distance between the input pattern and the hyperbox increases.

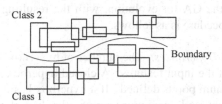

Fig. 1. An example of the FMM decision boundary of a two-class problem

The learning algorithm in FMM comprises a series of expansion and contraction processes for the hyperboxes. The training set, D, consists of M ordered pairs $\{X_h, C_h\}$, where, $X_h = (x_{h1}, x_{h2}, \ldots, x_{hn}) \in I^n$ is the hth input pattern, and $C_h \in \{1, 2, \ldots, m\}$ is the index of one of the m target classes. The process of learning starts with the selection of an ordered pair from D and finding of a hyperbox from the same class that can be expanded. The expansion criterion has a constraint to be met, as follows.

$$n\theta \geq \sum_{i=1}^{n} \left(\max\left(w_{ji}, x_{hi} \right) - \min\left(v_{ji}, x_{hi} \right) \right), \tag{2}$$

where θ is the hyperbox size. When the expansion criterion cannot be satisfied, a new hyperbox is formed in the network. Online learning is realized whereby new hyperboxes are added without the need of retraining.

During hyperbox expansion, there is a possibility of an overlap between existing hyperboxes to occur. As such, an overlap test is introduced. For all dimensions, provided one of the following cases (eq. 3 to 6) is met, it is said that an overlap between two hyperboxes from different classes exists. If an overlap is found during the search process, the index of the dimension and the smallest overlap value is used during contraction. Given an assumption of $\delta^{old} = 1$ initially, the four test cases and their corresponding minimum overlap value for the i-th dimension are as follows.

$$Case\ 1:\ v_{ji} < v_{ki} < w_{ji} < w_{ki},\ \delta^{new} = \min(w_{ji} - v_{ki}, \delta^{old}) \tag{3}$$

$$Case\ 2:\ v_{ki} < v_{ji} < w_{ki} < w_{ji},\ \delta^{new} = \min(w_{ki} - v_{ji}, \delta^{old}) \tag{4}$$

$$Case\ 3:\ v_{ji} < v_{ki} < w_{ki} < w_{ji},$$
$$\delta^{new} = \min(\min(w_{ki} - v_{ji}, w_{ji} - v_{ki}), \delta^{old}) \tag{5}$$

$$Case\ 4:\ v_{ki} < v_{ji} < w_{ji} < w_{ki},$$
$$\delta^{new} = \min(\min(w_{ji} - v_{ki}, w_{ki} - v_{ji}), \delta^{old}) \tag{6}$$

where j=hyperbox B_j that has been expanded in the previous step and k=hyperbox B_k representing another class and is being tested for possible overlap. If $\delta^{old} - \delta^{new} > 0$, then $\Delta = i$ and $\delta^{old} = \delta^{new}$, whereby if this is met, there is an overlap in the Δth dimension, and the overlap test proceeds to the next dimension. When the overlap test stops, the minimum overlap index variable is set to indicate that the next contraction step is not necessary, *i.e.*, Δ=−1. More details of FMM is available in [8].

2.2   Genetic Algorithm-Based Rule Extractor

In this section, the GA-based rule extractor is described. The hyperboxes generated from FMM are fed to the GA for evolution, with the resulting hyperboxes used for rule extraction. The procedure is as follows.

1) Generating Open Hyperboxes: The number of hyperbox dimensions created by FMM is equal to that of the input features. A closed hyperbox is a hyperbox with all its minimum and maximum points defined. If a hyperbox has dimensions that are not defined by its minimum and maximum points, the hyperbox is called an *open* hyperbox, and the non-declared dimension is designated as the *don't care* dimension. The *don't care* dimension is said to fully cover the particular *don't care* feature of the input space. To satisfy this requirement, the minimum and maximum points of the *don't care* dimension are set to zero and one, respectively. All possible combinations of open hyperboxes that can be generated by a hyperbox are examined. Note that the number of possible open hyperboxes (except one where all dimensions are designated as *don't care*) is $(2^d - 2)$, where d is the dimension of the input space.

2) Extracting Fuzzy If-Then Rules: For rule extraction, each hyperbox is transformed into one fuzzy rule. The rule extraction procedure starts by quantizing the minimum and maximum values of each input feature. The quantization level (Q) equals the number of fuzzy partitions in the quantized rules [11]. As an example, with $Q = 5$ as used in this paper, an input feature, A_q, is quantized to "very low", "low", "medium", "high", or "very high" in a fuzzy rule. For quantization, the round-off method [11] is used. As such, interval [0, 1] is divided into Q intervals, and the input feature is assigned to the quantization points evenly with one at each of the end points using [11]:

$$A_q = \frac{(q-1)}{(Q-1)} \tag{7}$$

where $q = 1, \ldots, Q$.

3 Simulation Study

The motor harmonics is first detailed in this section. This is followed by the experimental setup, results, and discussion.

3.1 Motor Harmonics

The rotor magnetic field in an ideal induction motor has only one space harmonic. However, owing to rotor slotting, the rotor magnetic field has harmonics in addition to the fundamental component [13]. The harmonics induced in the stator winding are typically known as the Rotor Slot Harmonics (RSHs), and can be calculated using [14]:

$$f_{\text{RSH}} = \left[k \frac{N_r}{p}(1-s) \pm 1 \right] F_s, \quad k = 1, 2, 3, \ldots \tag{8}$$

where p is the number of pole-pairs, N_r is the number of rotor slots, F_s the supply frequency, and s is the slip. The frequency of RSH is dependent on the numbers of rotor slots and pole pairs. For every k, there is a pair of harmonics. As such, the frequency between the harmonics is twice the supply frequency. The order of RSH can be determined by [15]:

$$H_{\text{RSH}} = k \frac{N_r}{p} \pm 1, \quad k = 1, 2, 3, \ldots \tag{9}$$

3.2 Simulation Using the Finite Element Method

The Finite Element Method (FEM) is a useful way for modelling and simulating motor designs, in either two or three dimensions. In addition, FEM brings the advantage of modeling motors with a high degree of accuracy, which represents the most viable way in solving problems of electromagnetic fields in electrical machines [12].

We used the Vector Fields Opera-2d 13.0 FEM software in this paper. The induction motor simulation was performed using the time-stepping finite element analysis. The simulated motor model was a three-phase induction motor, 0.5Hp, 415V, 4-poles. The models of one and two broken rotor bars created in Opera-2d are shown in Fig. 2.

Fig. 2. (*a*) One broken rotor bar (*b*) Two broken rotor bars

The aim of using the simulation model in Opera-2d was to observe the differences in stator currents when the motor operated under normal and faulty conditions; therefore allowing a fast, inexpensive, and accurate evaluation of different motor faults to be carried out.

3.3   Results and Discussion

In the experiments, a total of 21 features comprising of the 1st, 5th, 7th, 11th, 13th, 17th and 19th harmonics from the three motor phases (A, B, C) of an induction motor was used. The output was the predicted motor condition, either a fault-free motor or one with broken rotor bars. The 5-fold cross-validation method was used, where four data sub-sets were used for learning, while the remaining for testing. A total of 20,000 data samples were recorded for each motor condition, at different load conditions. The experiments were repeated five times, and the averages and standard deviations (StdDev) were computed with the bootstrap method using 5,000 resamplings. The simulation results from FEM are shown in Table 1.

Table 1. Simulation results

Network	Accuracy	StdDev	Complexity
MLP [16]	91.82%	5.67	20 Hidden Nodes
FMM [16]	98.62%	2.28	7 Hyperboxes
FMM-GA	98.85%	1.87	6 Hyperboxes

FMM-GA acquired the highest accuracy rate when compared with the previous results reported in [16], which used the same data set and the same test set-up. FMM-GA also had the least complex network, when compared with FMM and Multi-Layer Perceptron (MLP) neural network.

4   Laboratory-Based Study

A series of real experiments was conducted using a laboratory-scale test rig. The details are as follows.

4.1   Experimental Set-Up

The experimental set-up consisted of an induction motor, three current probes connected to a digital oscilloscope, a load controller, and a load inducer. The induction motor used was a 1 Hp, 4-pole, 415 V, 50 Hz. It operated under different load conditions, i.e. quarter, half, three quarter, and full load. The load control unit was used to electronically control the motor load. Three current probes were employed to measure the stator currents. For each load condition of the induction motor, a total of 20,000 data samples were recorded. Fig. 3 shows the induction motors with one and two broken rotor bars.

(a)   (b)

Fig. 3. (a) One broken rotor bar (b) Two broken rotor bars

4.2   Results and Discussion

The experimental results are shown in Table 2. Again, FMM-GA produced the highest accuracy rate as compared with those from FMM and MLP reported in [16]. Similarly, FMM-GA created the least complex network as compared with those from FMM and MLP. The standard deviation of FMM-GA was the lowest as well, as compared with those from FMM and MLP.

Table 2. Real experiments results

Network	Accuracy	StdDev	Complexity
MLP [16]	90.14%	6.86	20 Hidden Nodes
FMM [16]	98.41%	2.56	8 Hyperboxes
FMM-GA	98.92%	1.92	6 Hyperboxes

Rule extraction plays an important part in condition monitoring as it provides a useful way for understanding the predictions from a system. Unlike [16], not only FMM-GA is able to achieve improved accuracy rates compared to FMM, a number of if-then rules are extracted for explaining its predictions. Based on the experiments, an extracted rule set is form, and one exemplar of the rules is as follows.

IF harmonics 1 (A, B, C) is *don't care*
 harmonics 5 (A) is medium and (B, C) is *don't care*
 harmonics 7 (C) is medium and (A, B) is *don't care*
 harmonics 11 (A, B, C) is *don't care*
 harmonics 13 (A, B, C) is *don't care*
 harmonics 17 (A, B, C) is *don't care*
 harmonics 19 (A, B, C) is *don't care*
THEN Output is broken rotor bars

The above rule can be simplified by omitting the *don't care* antecedents. The new rule is as follows.

IF harmonics 5 (A) is medium
harmonics 7 (C) is medium
THEN Output is broken rotor bars

As can be seen, the advantage of using the *don't care* approach is evident, whereby the number of rule antecedents is reduced from 7 to 2, making the rule easy to understand. According to the rule, the 5th and 7th harmonics at a "medium" stage indicates that a motor with broken rotor bars. This is in line with the findings in [17], where it has been showed that the 5th and 7th harmonic amplitudes in the power spectral density are different between a fault-free motor and one with static eccentricity and broken rotor bars. As a result, the rule set extracted using the GA serves as important knowledge to provide justification to the domain users (e.g. maintenance engineer) with respect to the prediction yielded from FMM-GA.

5   Conclusions

A hybrid FMM-GA model for condition monitoring of broken rotor bars in induction motors has been presented in this paper. Two sets of experiments on broken rotor bars, comprising simulation studies using the FEM and real laboratory tests, have been conducted. In both experiments, the current signals are converted into their frequency spectrum, which is then used to form the input features to FMM-GA.

The results of the experiments indicate that FMM-GA is able to correctly differentiate motors with broken rotor bars from fault-free motors with higher accuracy rates, as compared with FMM or MLP. More importantly, useful if-then rules are extracted from FMM-GA. The rules extracted by FMM-GA are also in line with findings reported in the literature. In the real world environment, these extracted rules are important as they provide justifications to domain users with respect to the predictions generated from FMM-GA.

For further work, the hybrid FMM-GA model will be applied to other types of motor faults. In addition, hardware implementation of the system will be conducted in order to allow real-time condition monitoring of motors with FMM-GA.

Acknowledgment. This project is supported by UMRG Research Subprogram (Project Number RP003D-13ICT).

References

[1] Venugopal, S., Wagstaff, R.A., Sharma, J.: Exploiting Phase Fluctuations to Improve Machine Performance Monitoring. IEEE Transactions Automation Science and Engineering 4(2), 153–166 (2007)

[2] Portioli-Staudacher, A., Tantardini, M.: Integrated maintenance and production planning: a model to include rescheduling costs. Journal Quality in Maintenance Engineering 18(1), 42–59 (2012)

[3] Weber, J., Wotawa, F.: Diagnosis and repair of dependent failures in the control system of a mobile autonomous robot. Applied Intelligence 36(3), 511–528 (2012)

[4] Alsyouf, I.: The role of maintenance in improving companies' productivity and profitability. International Journal of Production Economics 105(1), 70–78 (2007)

[5] Chen, Y., Ding, T., Jin, J., Ceglarek, D.: Integration of Process-Oriented Tolerancing and Maintenance Planning in Design of Multistation Manufacturing Processes. IEEE Transactions Automation Science and Engineering 3(4), 440–453 (2006)

[6] Camci, F., Chinnam, R.: Health-State Estimation and Prognostics in Machining Processes. IEEE Transactions Automation Science and Engineering 7(3), 581–597 (2010)

[7] Montanari, M., Peresada, S.M., Rossi, C., Tilli, A.: Speed sensorless control of induction motors based on a reduced-order adaptive observer. IEEE Transactions Control Systems Tech. 15(6), 1049–1064 (2007)

[8] Simpson, P.: Fuzzy Min-Max Neural Networks-Part 1: Classification. IEEE Transactions Neural Networks 3(5), 776–786 (1992)

[9] Kolman, E., Margaliot, M.: Are artificial neural networks white boxes? IEEE Transactions Neural Networks 16(4), 844–852 (2005)

[10] Ishibuchi, H., Murata, T., Turksen, I.: Single-objective and twoobjective genetic algorithms for selecting linguistic rules for pattern classification problems. Fuzzy Sets Syst. 89(2), 135–150 (1997)

[11] Carpenter, G., Tan, A.: Rule extraction: From neural architecture to symbolic representation. Connection Sci. 7(1), 3–27 (1995)

[12] Greconici, M., Koch, C., Madescu, G.: Advantages of FEM analysis in electrical machines optimization used in wind energy conversion systems. In: IEEE 3rd International Exploitation of Renewable Energy Sources, pp. 91–94 (2011)

[13] Sharifi, R., Ebrahimi, M.: Detection of stator winding faults in induction motors using three-phase current monitoring. ISA Transactions 50(1), 14–20 (2011)

[14] Joksimovic, G.M., Penman, J.: The detection of inter-turn short circuits in the stator windings of operating Motors. IEEE Transactions on Industrial Electronics 47(5), 1078–1084 (2000)

[15] Nandi, S., Toliyat, H.: Novel frequency-domain-based technique to detect stator interturn faults in induction machines using stator-induced voltages after switch-off. IEEE Transactions on Industrial Applications 38(1), 101–109 (2002)

[16] Seera, M., Lim, C.P., Ishak, D.: Detection and Diagnosis of Broken Rotor Bars in Induction Motors Using the Fuzzy Min-Max Neural Network. International Journal of Natural Computing Research 3(1), 44–55 (2012)

[17] Faiz, J., Ebrahimi, B.M., Toliyat, H.A., Abu-Elhaija, W.: Mixed-fault diagnosis in induction motors considering varying load and broken bars location. Energy Conv. and Mgmt. 51(7), 1432–1441 (2010)

Employing Genetic Algorithm to Construct Epigenetic Tree-Based Features for Enhancer Region Prediction

Pui Kwan Fong[1], Nung Kion Lee[1,*], and Mohd Tajuddin Abdullah[2]

[1] Department of Cognitive Sciences, Universiti Malaysia Sarawak,
Kota Samarahan, Malaysia
[2] Center of Tasik Kenyir EcoSystem, Universiti Malaysia Terengganu,
Kuala Terengganu, Malaysia
nklee@fcs.unimas.my

Abstract. This paper presents a GA-based method to generate novel logical-based features, represented by parse trees, from DNA sequences enriched with H3K4me1 histone signatures. Current methods which mostly utilize k-mers content features are not able to represent the possible complex interaction of various DNA segments in H3K4me1 regions. We hypothesize that such complex interaction modeling is significant towards recognition of H3K4me1 marks. Our propose method employ the tree structure to model the logical relationship between k-mers from the marks. To benchmark our generated features, we compare it to the typically used k-mer content features using the mouse (mm9) genome dataset. Our results show that the logical rule features improve the performance in terms of f-measure for all the datasets tested.

Keywords: Genetic Algorithm, Feature extraction, Histone modifications.

1 Introduction

Initiation of gene transcription involves variants of regulatory elements whereby locating cis-regulatory elements enlighten the comprehension of complex gene regulation. One of the essential cis-regulatory elements known as enhancer comprises clusters of transcription factor binding sites (TFBS), each spans about 6 to 20 base pair(bp). Enhancer is capable of regulating gene expressions locating ten to hundred thousand bp away regardless of its location [1]. Locating enhancer regions remain a challenging task due to the unusual characteristics of distant-acting and short DNA sequences. In addition,the binding sites of enhancer degenerates easily yet retaining the original function [2]. Thus, it is difficult to find a general pattern of sequence to represent a specific type of enhancer.

Pioneer computational methods focus on implementing motif profiles searching to discover TFBS. Review by [3] highlights that these methods achieve high prediction accuracy for lower organisms only and often produce high false positive hits for complex organisms. Recently, the advancement of chIP-chip and

---

* Corresponding author.

C.K. Loo et al. (Eds.): ICONIP 2014, Part III, LNCS 8836, pp. 390–397, 2014.

chIP-seq techniques on genome-wide mapping of epigenetic marks [4,5] facilitates the use of these features for enhancer prediction. Epigenetic marks such as histone modification is prominent as a landmark for enhancer identification and this features are widely used [6,7] with different representation approaches which can give high impact on the prediction results.

In this paper, we hypothesize that DNA features co-exist in which their interaction are complex and need to be represented in higher order features as oppose to using only content information such as k-mer frequency [7]. A framework for modelling complex parse tree features using Genetic Algorithms (GA) [8] is proposed. Tree features generated from this framework are scrutinize for the competence in discriminating enrichment of H3K4me1 in DNA sequences.

2 Related Works

Early enhancer prediction approaches employ motif profile search and comparative genomic. Motif profile search utilizes annotated motif databases such as JASPAR or TRANSFAC to construct statistical or supervised learning model for predicting associated sites. While comparative genomic approach identify evolutionary conserved region using multiple sequence alignment algorithm. Supervised/statistical model has the limitation of returning many false positives because of its representation model which is non-specificity and the difficulties to determine matching cut-off value. While conservation analysis is useful, it can only detect evolutionary conserved sites. Therefore, machine learning method which utilizes different features related to binding sites has been proposed [6,7]. These methods employ features associated with enhancer site or region for supervised algorithm training. Ultimately, the set of features use in training determine the prediction accuracy rates.

Significant findings revealed that enrichment of H4K3me1 have striking correlations with enhancer whereby the distance between them are approximately 100 bp to 1500 or 2000 bp [5], [7]. Thus, there is an increasing need to locate DNA sequences enriched with these modified histones whereby experimentally, ChIP-chip or ChIP-seq is used to produce high resolution mapping and profiling [4,5]. However, these experimental techniques are tedious and expensive thus histone modification information is not easily accessible and available for all organisms. Therefore, the key approach of this paper is to propose a computational method for determining and characterizing the DNA locations of histone modified marks.

Characterizing histone modification marks using computational methods are proven successful using different features representation methods. Combination of content (k-mer frequency) and context (distance from gene) based features are used by [9] to predict H3K4me1 of yeast genome. Study showed that utilizing both features could achieve high H3K4me1 prediction accuracy of 90.86% while only 72.61% and less is achieved when small k-mer frequency (less than 9-mer feature) is used. Clearly, simple k-mer features are insufficient to represent DNA sequences with histone modification enrichment. Thus, a new method to represent H3K4me1 sequences with complex combination of nucleotides is proposed instead of just employing fixed k-mer frequency.

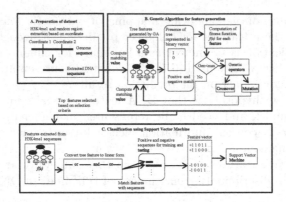

Fig. 1. Framework of GA-based tree feature generation

3 Methods and Materials

The framework of the proposed method is illustrated in Fig. 1. Generally, it consists of 3 main components labeled by A, B and C. Fig. 1A depicts extraction of DNA sequences to form positive and negative sequence sets based on coordinate of sequences with and without H3K4me1 enrichment. Discriminating complex tree features are extracted from these sets of sequences using Genetic Algorithm [10] as shown in Fig. 1B. Subsequently, the generated tree features which made up of logical interaction between short DNA segment (i.e.,k-mers) are selected and converted to feature vectors which function as an input to Support Vector Machine (SVM). SVM is used to perform classification for identification of sequences with and without H3K4me1 enrichment. Finally, results from SVM are analyzed using a few performance measurements as shown in Fig. 1C.

3.1 Feature Generation Using Genetic Algorithm

Fig. 1B detailed some of the main steps involved in generating features that consist of logical interaction 'AND' and 'OR' with continuous k-mers and k-mers without gap. Renowned for its heuristic search using survival of fittest theory [10], Genetic Algorithm is used in this framework to generate and select good candidate tree features from large feature search space.

Processes in Fig. 1B are repeated based on the predefined number of GA generation to produce a set of complex tree feature. For each generation, tree features generated are matched with sequences enriched with H3K4me1 (positive) and sequences without H3K4me1 enrichment (negative) to find out the presence and absence of each tree in each sequence. This binary value 1 or 0 is then used to score the fitness value for each tree feature whereby Roulette wheel selection is employed to choose feature with higher fitness value. Selected tree features undergo genetic operations to generated a set of new tree features which will have higher fitness value as compared to the previous set of features.

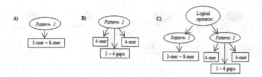

Fig. 2. Parse tree representation of basic feature units in general form: (A) *pattern-1*; (B) *pattern-2*; and (C) conjunctive or disjunctive features

Feature Representation. Features generated from GA are represented using parse tree which aims to capture distinct interactive features which are not possible by a single k-mer features with fixed length. This novel element is motivated by [11] in which this approach is implemented using different subunit of features and search strategy for DNA splice site prediction. In our approach, each root node and internal node is made up of logical operator AND or OR while the leaf node can be of any combination between 3-mer to 8-mer without gap or two 4-mer with gaps less than 5. A summary of each node with its respective possible elements and connection with child node is shown in Table 1.

Basic Units. Fig. 2A and Fig. 2B illustrate the general form of basic unit which made up of leaf nodes. Unlike k-mer features which can only capture either one of this unit, this approach is capable of modelling both motif with and without gaps labeled as *pattern-1* and *pattern-2* which are useful for prediction.

Conjunctive and Disjunctive Features. These complex features consist of a minimal two basic units joined by one logical operator either 'AND' known as conjunctive feature or 'OR' known as disjunctive feature. Various combination of logical operators and basic units are also permitted in which it can capture motifs which presence simultaneously or optionally in target sequences. The general form of conjunctive and disjunctive feature are shown in Fig. 2C.

Depth of Tree Features. In this approach, the minimum level of tree depth is predefined to three while the maximum is five. Since the aim of this method is to capture the interactive component of motifs, the minimum number of logical operators in each tree is one. Simultaneously, the maximum complexity of the tree is kept to depth five with seven logical operator.

Table 1. Characteristic of elements in parse tree

Element	Type of node	Child node
AND	Root or internal	AND, OR, pattern 1 or pattern 2
OR	Root or internal	AND, OR, pattern 1 or pattern 2
Pattern 1	Leaf	K-mer
Pattern 2	Leaf	K-mer and gaps

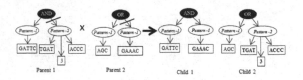

Fig. 3. Example of crossover operations for tree feature with depth value 3

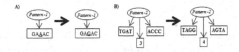

Fig. 4. Examples of mutation operations for tree *pattern-1* (A) and *pattern-2* (B)

Genetic Operators. Genetic operators consisting of one point crossover and random mutation performed on each selected tree feature based on random probability are crucial to generate good candidate features. These are performed by evolving the parse tree features through combination and exchange of basic tree units. Two parent tree features are selected for crossover to generate new child tree features with better fitness value. Whereby only one tree feature is needed for mutation to increase the diversity of feature set which can represent majority of the search space of H3K4me1 sequences.

Crossover. Crossover between two selected tree features are performed when the random probability is less then the predefined crossover probability. Two point of crossover are randomly picked for each of the tree indicated by the red line. Subsequently, information in that branch are exchanged and two child tree features are produced as shown in Fig. 3.

Mutation. Since both *pattern-1* and *pattern-2* basic unit possess different characteristic and matching ability, different mutation strategies are implemented as illustrated in Fig. 4. Mutation on *pattern-2* changes the whole element to increase the chance of matching while mutation on *pattern-1* only changes a randomly selected nucleotide as it is easier to match.

Fitness Function. Each tree feature is evaluated using this fitness function and tree with larger value will have higher chances to be selected for genetic operations. Fitness function is formulated as Equation 1 where T represents a tree feature; N, a and b represent the total number of patterns in T. First part of the equation aims to capture individual pattern with highest discriminative value between sequences enriched with and without H3K4me1 where $f(p_j, s_i^+)$ is a binary indicator function which return 1 if motif p_j is presence in s_i^+, ith positive sequence or s_k^-, kth negative sequence. The second part of this equation functions to evaluate the tree feature as a whole which takes into account of its logical operators.

$$Fitness(T) = \arg\max_{j=1...N} \left\{ \frac{\sum_{i=1}^{a} f(p_j, s_i^+)}{\sum_{k=1}^{b} f(p_j, s_k^-)} \right\} + \frac{\sum_{i=1}^{a} f(T, s_i^+)}{\sum_{k=1}^{b} f(T, s_k^-)} \tag{1}$$

3.2   Classification Using Support Vector Machine

Prediction of enhancer regions inferred by H3K4me1 enrichment is performed using LibSVM [12] with different input of feature vectors based on the type of features used. Feature vector for each sequences in training and testing set are computed based on the operation of logical operators on each of the matched (1) or unmatched (0) basic unit of selected tree feature to each of the positive and negative sequence involved. Number of rows in feature vector represent total number of sequences while number of columns store binary value to indicate the presence and absence of the particular tree feature. LibSVM [12] is first trained with feature vector from training sequences using default parameters and subsequently used to classify testing sequences enriched with or without H3K4me1.

3.3   Datasets

Sequences enriched with H3K4me1 from melan-a cells of mouse (mm9) genome are obtained from the study by [7]. Study by [7] provides coordinate of sequences enriched with and without H3K4me1 for each chromosome. In this paper, 1000 sequences enriched with H3K4me1 (positive sequences) from chromosome 1 and 1000 sequences without H3K4me1 (negative sequences) are chosen randomly for complex tree feature generation using GA. In addition another 1000 negative sequences and five set of positive sequences from chromosome 2 to 6 are prepared for testing purposes.

Table 2. H3K4me1 prediction performance using different features

Features	Chr2		Chr 3		Chr 4		Chr 5		Chr 6	
	P [1]	R [2]	P	R	P	R	P	R	P	R
Top 50 tree [3]	0.220	0.010	0.286	0.014	0.222	0.010	0.239	0.011	0.407	0.02
Top 500 tree	**0.821**	0.582	**0.835**	0.645	**0.825**	0.597	**0.841**	0.673	**0.826**	0.603
Top 50 4-mer	0.675	0.567	0.765	0.651	0.695	0.621	0.695	0.621	0.608	0.608
Top 50 5-mer	0.719	**0.600**	0.741	**0.674**	0.730	**0.635**	0.743	**0.680**	0.726	**0.622**

---

[1] Precision.
[2] Recall.
[3] Histone tree feature.

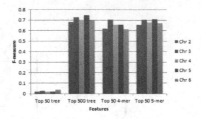

Fig. 5. F–measure value for each feature used in classification

4   Results and Discussion

In this study, the discriminative power of complex tree features in identifying H3K4me1 enriched sequences are evaluated by comparing to the widely used k-mer frequency features. Complex tree features are generated using default GA procedure with 15 generations, 1000 populations, 0.9 crossover rate and 0.2 mutation rate. Selected top tree features from chromosome 1 are used to train SVM while prediction is carried out on different chromosomes to discover the generality of these features. Frequency of 4-mer and 5-mer features are used as a benchmark in which frequencies of the top 50 k-mers are normalized to form feature vectors that acts as input to the SVM.

Table 2 shows the results of comparisons between tree and k-mer feature in which the best result for each test case is highlighted in bold. It can be seen that our approach (top 500 tree features) achieved higher precision rates in comparison to both k-mer features. In addition, it is noted that when the number of tree features is small (e.g. 50), it performed poorly on the evaluation sets. This can be explained by the fact that each tree feature only represents a small number of histone marks features and possibly there are overlaps between the tree feature.Therefore, large number of tree features are needed to achieve good coverage on the feature space.

On the contrary, it is observed that the average recall value of 5 chromosomes for top 50 5-mer is slightly higher than top 500 tree feature by 0.022. This implies that the capability of tree feature in extracting positive sequence from search space is slightly lesser than 5-mer feature. However, our approach attained better balance between precision and recall rates, given by the F-measure as depicted in Fig. 5. In all cases our method achieve higher F-measure value in comparison to using k-mer features. For example, the average F-measure using 500 histone features is 0.709 which significantly outperformed 4-mer and 5-mer features by 0.061 and 0.026 respectively. This indicates that top 500 histone features are more capable in predicting both positive and negative sequences correctly.

In addition, it is found that top 500 tree features generated from 30, 45 and 50 generations also perform equally well and better than the k-mer features alone (results not shown). From these performance evaluations, it can be concluded that all features except top 50 histone feature perform reasonably well in predicting H3K4me1 sequences. Consistent precision and recall rates shown in

Table 2 highlights that tree features generated based on one chromosome can be generalized to predict histone modifications enrichment in other chromosomes.

In this paper, a method for generating complex feature representation for histone sequences is proposed. The features are constructed from the ungapped and the gapped k-mer pattern, represented by hierarchical trees. GA is used to generate the tree features with customized genetic operators. Each tree feature is hypothesized to represent a small subset of salient features in the histone regions thus large number of feature is needed for effective representation (i.e, 500). The proposed representation is shown to perform well on the tested datasets.

Acknowledgments. PK is supported by the Malaysian MyBrain scholarship.

References

1. Lettice, L.A.: A Long-range Shh Enhancer Regulates Expression in the Developing Limb and Fin and is Associated with Preaxial Polydactyly. Human Molecular Genetics 12, 1725–1735 (2003)
2. Wittkopp, P.J., Kalay, G.: Cis-regulatory elements: Molecular Mechanisms and Evolutionary Processes Underlying Divergence. Nature Review Genetics 13, 56–69 (2012)
3. Das, M.K., Dai, H.: A Survey of DNA Motif Finding Algorithms. BMC Bioinformatics 8, S21 (2007)
4. Barski, A., Cuddapah, S., Cui, K., Roh, T., Schones, D.E., Wang, Z., Wei, G., Chepelev, L., Zhao, K.: High-Resolution Profiling of Histone Methylations in the Human Genome. Cell 129, 823–837 (2007)
5. Heintzman, N.D., Stuart, R.K., Hon, G., Fu, Y., Ching, W.C., Barrera, L.O., Van Calcar, S., Qu, C., Ching, K., Wang, W., Weng, Z., Green, R.D., Crawford, G.E.: Distinct and Predictive Chromatin Signatures of Transcriptional Promoters and Enhancers in the Human Genome. Nature Genetics 39, 311–318 (2007)
6. Firpi, H.A., Ucar, D., Tan, K.: Discover Regulatory DNA Elements Using Chromatin Signatures and Artificial Neural Network. Bioinformatics 26, 1579–1586 (2010)
7. Gorkin, D.U., Lee, D., Reed, X., Fletez-Brant, C., Bessling, S.L., Loftus, S.K., Beer, M.A., Pavan, W.J., Mccallion, A.S.: Integration of ChIP-seq and Machine Learning Reveals Enhancers and a Predictive Regulatory Sequence Vocabulary in Melanocytes. Genome Research 22, 2290–2301 (2012)
8. Holland, J.: Adaptation in Natural and Artificial Systems. The University of Michigan Press, Ann Arbor (1975)
9. Pham, T.H., Ho, T.B., Tran, D.H., Satou, K.: Prediction of Histone Modifications in DNA Sequences. Bioinformatics and Bioengineering, 959–966 (2007)
10. Mitchell, M.: An Introduction to Genetic Algorithms. The MIT Press, London (2001)
11. Kamath, U., Compton, J., Islamaj-Dogan, R., De Jong, K.A., Shehu, A.: An Evolutionary Algorithm Approach for feature Generation from Sequence Data and Its Application to DNA Splice Site Prediction. IEEE/ACM Transactions on Computational Biology and Informatics 9, 1387–1397 (2012)
12. Chang, C., Lin, C.: A Library for Support Vector Machines. ACM Transactions on Intelligent Systems and Technology 27, 27 (2001)

Model and Algorithm for Multi-follower Tri-level Hierarchical Decision-Making

Jialin Han[1,2], Guangquan Zhang[2,*], Jie Lu[2], Yaoguang Hu[1], and Shuyuan Ma[1]

[1] School of Mechanical Engineering, Beijing Institute of Technology, China
[2] Faculty of Engineering and Information Technology, University of Technology, Sydney, Australia
{Hjl,hyg,bitmc}@bit.edu.cn, {Guangquan.Zhang,Jie.Lu}@uts.edu.au

Abstract. Tri-level decision-making addresses compromises among interacting decision entities that are distributed throughout a three-level hierarchy. Decision entities at the three hierarchical levels are respectively termed as the top-level leader, the middle-level follower and the bottom-level follower. When multiple followers are involved at the middle and bottom levels, the leader's decision will be affected not only by reactions of the followers but also by various relationships among them. To support such a multi-follower tri-level (MFTL) decision-making process, this study first proposes a general MFTL decision model for the situation involving both cooperative and uncooperative relationships among multiple followers. It then develops a MFTL *K*th-Best algorithm to find an optimal solution to the model. Lastly, we use the proposed MFTL decision techniques to deal with a supply chain management problem in applications.

Keywords: Hierarchical decision-making, multilevel programming, tri-level decision-making, *K*th-Best algorithm, supply chain.

1 Introduction

Tri-level decision-making (also known as tri-level programming) technique is proposed to solve decentralized decision problems involving interacting decision entities distributed throughout a three-level hierarchy, which is a subfamily of multilevel programming [1] motivated by Stackelberg game theory [2]. Decision entities at the three hierarchical levels are respectively termed as the leader, the middle-level follower and the bottom-level follower, and make their decisions in sequence from the top level to the middle level and then to the bottom level seeking to optimize their individual objectives. The decision process means the higher level has the priority to make its decision and the lower-level decision entity reacts after and in view of a decision made by the higher level. However, the decision of each entity is affected by the actions of the others. The decision process is repeatedly executed until the Stackelberg equilibrium among them is achieved. This category of hierarchical

---

* This work is supported by the Australian Research Council (ARC) under discovery grant DP140101366, and the National High Technology Research and Development Program of China (NO. 2013AA040402).

C.K. Loo et al. (Eds.): ICONIP 2014, Part III, LNCS 8836, pp. 398–406, 2014.

decision-making process often appears in many decentralized management problems, such as supply chains management [3], resource allocation optimization [4,5] and hierarchical production operations [6].

Take the three-stage supply chain composing of a manufacturer, a distributor and a vendor as an example. The manufacturer, distributor and vendor are distributed throughout three hierarchical levels, respectively called the leader, the middle-level follower and the bottom-level follower. Within a marketing circle, each of them has to keep a certain amount of inventory to fulfill exceeded market requirements and meanwhile aims to minimize its own inventory cost. However, their total inventories must satisfy the exceeded market requirements, which means that one decision entity has to increase its holding inventory if the others reduce their inventories. Furthermore, the leader has the priority to determine its own inventory by considering market requirements and the implicit reactions of both followers. The middle-level follower then adjusts its holding inventory to respond to the leader likewise considering the implicit reactions of the bottom-level follower. In the light of the decisions of the top and middle levels, the bottom-level follower determines its inventory to optimize its own objective at last. The example describes a typical tri-level decision problem in which the execution of decisions is sequential, interactive and iterative among the three decision entities seeking to optimize their individual objectives until the Stackelberg equilibrium is achieved.

In general, there are two fundamental issues in supporting the tri-level decision-making process in applications. One is how to model a real-world tri-level decision problem, which may manifest different characteristics at the three decision levels, and the other is how to find an optimal solution to the problem. Although tri-level decision-making has been attracting numerous investigations on models [7,8], solution algorithms [7,8] and applications [3,4],[6],[8], the existing research has been mainly limited to the situation that a single decision entity is involved at each level. Actually, two or more decision entities are often involved at the middle and bottom levels in real-world cases called multi-follower tri-level (MFTL) decision-making [9]. In the three-stage supply chain example, the manufacturer (the leader) may have multiple subordinate distributors (middle-level followers), and simultaneously, there may also be several vendors (bottom-level followers) attached to each distributor. Moreover, multiple followers at the same level may have a variety of relationships with one another, such as cooperative and uncooperative relationships. Such situations will make the MFTL decision complex and generate different decision processes, which need to be described and solved using different decision models and solution methods.

This study considers a special situation that a cooperative and an uncooperative relationship appear at the middle and bottom levels respectively based on related definitions in our previous research [9]. The situation means that multiple followers at the middle level have the same decision variables but have individual optimization objectives and separate constraints, while multiple bottom-level followers attached to the same middle-level follower optimize their own objectives by controlling individual decision variables under their separate constraints. For example, the multiple distributers or vendors may collaborate with other counterparts by making joint decisions to enhance market competitiveness, called a cooperative relationship among them, or may consider their counterparts as competitors and make their decisions independently, known as an uncooperative relationship among them.

The main contribution of this paper is the provision of a general model and a solution method to describe and solve the proposed MFTL decision process. The paper first presents a linear MFTL decision model for the above situation. To find an optimal solution to the model, a MFTL Kth-Best algorithm is proposed. Lastly, a case study on three-stage supply chain decision illustrates the proposed MFTL decision techniques.

2   A MFTL Decision Model and Related Theoretical Properties

The organizational structure among decision entities in the three-level hierarchy that is studied in this paper is shown as Fig. 1. Let $x \in X \subset R^k$, $y \in Y_i \subset R^{k_0}$, $z_{ij} \in Z_{ij} \subset R^{k_{ij}}$ denote the decision variables of the leader, the middle-level follower i, and the bottom-level follower ij respectively where $j = 1, 2, \cdots, m_i, i = 1, 2, \cdots, n$. We give detailed definitions of the cooperative and uncooperative relationships proposed in our previous research [9].

Definition 1. [9] If both the objective function and constraint conditions of the middle-level follower i only involve the shared decision variable y controlled by all of them in common apart from the decision variables $x, z_{i1}, \cdots, z_{im_i}$ determined by the leader and the bottom-level followers, this means a cooperative relationship among multiple followers at the middle level.

Definition 2. [9] If both the objective function and constraint conditions of the bottom-level follower ij only involve its own decision variable z_{ij} apart from the decision variables x and y respectively determined by the leader and the middle-level follower i, this can be called a uncooperative relationship among multiple bottom-level followers attached to the same middle-level follower i.

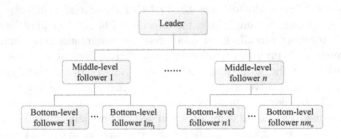

Fig. 1. The organizational structure of the three-level hierarchy

For $x \in X \subset R^k$, $y \in Y_i \subset R^{k_0}$, $Y = Y_1 \cap \cdots \cap Y_n$, $y \in Y \subset R^{k_0}$, $z_{ij} \in Z_{ij} \subset R^{k_{ij}}$, $F: X \times Y \times Z_1 \times \cdots \times Z_{1m_1} \times \cdots \times Z_{n1} \cdots \times Z_{nm_n} \to R^1$, $f_i^{(2)}: X \times Y_i \times Z_{i1} \times \cdots \times Z_{im_i} \to R^1$, $f_{ij}^{(3)}: X \times Y_i \times Z_{ij} \to R^1$ and $j = 1, 2, \cdots, m_i, i = 1, 2, \cdots, n$, a linear MFTL decision model in which one leader, n cooperative middle-level followers and m_i uncooperative bottom-level followers attached to the middle-level follower i are involved is defined as follows:

$$\min_{x \in X} F(x, y, z_{11}, \cdots, z_{1m_1}, \cdots, z_{n1}, \cdots, z_{nm_n}) = cx + dy + \sum_{i=1}^{n} \sum_{j=1}^{m_i} e_{ij} z_{ij} \qquad \text{(Leader)} \quad (1a)$$

$$\text{s.t.} \quad Ax + By + \sum_{i=1}^{n} \sum_{j=1}^{m_i} C_{ij} z_{ij} \le b, \qquad (1b)$$

where $(y, z_{i1}, \cdots, z_{im_i})$ $(i = 1, 2, \cdots, n)$, for each given value x, solves (1c-1f):

$$\min_{y \in Y_i} f_i^{(2)}(x, y, z_{i1}, \cdots, z_{im_i}) = c_i x + d_i y + \sum_{j=1}^{m_i} g_{ij} z_{ij} \qquad \text{(Middle-level follower } i \text{)} \quad (1c)$$

$$\text{s.t.} \quad A_i x + B_i y + \sum_{j=1}^{m_i} D_{ij} z_{ij} \le b_i, \qquad (1d)$$

where z_{ij} $(j = 1, 2, \cdots, m_i)$, for the given value (x, y), solves (1e-1f):

$$\min_{z_{ij} \in Z_{ij}} f_{ij}^{(3)}(x, y, z_{ij}) = c_{ij} x + d_{ij} y + h_{ij} z_{ij} \qquad \text{(Bottom-level follower } ij \text{)} \quad (1e)$$

$$\text{s.t.} \quad A_{ij} x + B_{ij} y + E_{ij} z_{ij} \le b_{ij}, \qquad (1f)$$

where $c, c_i, c_{ij} \in R^k$, $d_i, d_i, d_{ij} \in R^{k_0}$, $e_{ij}, g_{ij}, h_{ij} \in R^{k_{ij}}$, $A \in R^{s \times k}$, $A_i \in R^{s_i \times k}$, $A_{ij} \in R^{s_{ij} \times k}$, $B \in R^{s \times k_0}$, $B_i \in R^{s_i \times k_0}$, $B_{ij} \in R^{s_{ij} \times k_0}$, $C_{ij} \in R^{s \times k_{ij}}$, $D_{ij} \in R^{s_i \times k_{ij}}$, $E_{ij} \in R^{s_{ij} \times k_{ij}}$, $b \in R^s$, $b_i \in R^{s_i}$, $b_{ij} \in R^{s_{ij}}$, $j = 1, 2, \cdots, m_i, i = 1, 2, \cdots, n$.

Concepts related to the solutions to the model (1) are defined as follows.

Definition 3.

1） Constraint region of the MFTL decision model (1):

$$S = \{(x, y, z_{11}, \cdots, z_{1m_1}, \cdots, z_{n1}, \cdots, z_{nm_n}) \in X \times Y \times \prod_{i=1}^{n} \prod_{j=1}^{m_i} Z_{ij} : Ax + By + \sum_{i=1}^{n} \sum_{j=1}^{m_i} C_{ij} z_{ij} \le b,$$

$$A_i x + B_i y + \sum_{j=1}^{m_i} D_{ij} z_{ij} \le b_i, A_{ij} x + B_{ij} y + E_{ij} z_{ij} \le b_{ij}, j = 1, 2, \cdots, m_i, i = 1, 2, \cdots, n\}.$$

2） Feasible set of the ith middle-level follower and its bottom-level followers for each fixed $x \in X$:

$$S_i(x) = \{(y, z_{i1}, \cdots, z_{im_i}) \in Y_i \times \prod_{j=1}^{m_i} Z_{ij} : A_i x + B_i y + \sum_{j=1}^{m_i} D_{ij} z_{ij} \le b_i, A_{ij} x + B_{ij} y + E_{ij} z_{ij} \le b_{ij}, j = 1, 2, \cdots, m_i\}, i = 1, 2, \cdots, n.$$

3） Feasible set of the ith middle-level follower's jth bottom-level follower for each fixed $(x, y) \in X \times Y_i$:

$$S_{ij}(x, y) = \{z_{ij} \in Z_{ij} : A_{ij} x + B_{ij} y + E_{ij} z_{ij} \le b_{ij}\}, j = 1, 2, \cdots, m_i, i = 1, 2, \cdots, n.$$

4） Rational reaction set of the ith middle-level follower's jth bottom-level follower:

$$P_{ij}(x, y) = \{z_{ij} \in Z_{ij} : z_{ij} \in \arg\min[f_{ij}^{(3)}(x, y, \hat{z}_{ij}), \hat{z}_{ij} \in S_{ij}(x, y)]\}, j = 1, 2, \cdots, m_i, i = 1, 2, \cdots, n.$$

5） Rational reaction set of the ith middle-level follower and its bottom-level followers:

$$P_i(x) = \{(y, z_{i1}, \cdots, z_{im_i}) \in Y_i \times \prod_{j=1}^{m_i} Z_{ij} : (y, z_{i1}, \cdots, z_{im_i}) \in \arg\min\langle f_i^{(2)}(x, \hat{y}, \hat{z}_{i1}, \cdots, \hat{z}_{im_i}), (\hat{y}, \hat{z}_{i1}, \cdots, \hat{z}_{im_i}) \in S_i(x),$$

$$\hat{z}_{ij} \in P_{ij}(x, \hat{y}), j = 1, 2, \cdots, m_i\rangle\}, i = 1, 2, \cdots, n.$$

6） Inducible region:

$$IR = \{(x, y, z_{11}, \cdots, z_{1m_1}, \cdots, z_{nm_1}, \cdots, z_{nm_n}) : (x, y, z_{11}, \cdots, z_{1m_1}, \cdots, z_{nm_1}, \cdots, z_{nm_n}) \in S,$$

$$(y, z_{i1}, \cdots, z_{im_i}) \in P_i(x), i = 1, 2, \cdots, n\}.$$

3   A MFTL Kth-Best Algorithm

The inducible region of the linear bi-level programming problem is composed of connected faces of S so that a vertex of the original polyhedron will provide the solution, which can be extended to the tri-level programming problem [7],[9].

Consider the following linear programming problem:

$$\min\{F(x, y, z_{11}, \cdots, z_{1m_1}, \cdots, z_{n1}, \cdots, z_{nm_n}) : (x, y, z_{11}, \cdots, z_{1m_1}, \cdots, z_{n1}, \cdots, z_{nm_n}) \in S\} \tag{2}$$

and let $(x^1, y^1, z_{11}^1, \cdots, z_{1m_1}^1, \cdots, z_{n1}^1, \cdots, z_{nm_n}^1), \cdots, (x^N, y^N, z_{11}^N, \cdots, z_{1m_1}^N, \cdots, z_{n1}^N, \cdots, z_{nm_n}^N)$ denote the

N-ranked basic feasible solution to (2), such that $cx^K + dy^K + \sum_{i=1}^{n}\sum_{j=1}^{m_i} e_{ij} z_{ij}^K \le$

$cx^{K+1} + dy^{K+1} \sum_{i=1}^{n}\sum_{j=1}^{m_i} e_{ij} z_{ij}^{K+1}$, $K = 1,2, \cdots N-1$. Then solving the problem (1) is equivalent to

searching the index $K^* = \min\{K \in \{1,2, \cdots, N\} : (x^K, y^K, z_{11}^K, \cdots, z_{1m_1}^K, \cdots, z_{n1}^K, \cdots, z_{nm_n}^K) \in IR\}$, which ensures

that $(x^{K^*}, y^{K^*}, z_{11}^{K^*}, \cdots, z_{1m_1}^{K^*}, \cdots, z_{n1}^{K^*}, \cdots, z_{nm_n}^{K^*})$ is an optimal solution to the model (1). As

this requires finding the K^*th best vertex solution to problem (2) to obtain an optimal solution to model (1), the algorithm therefore is named Kth-Best algorithm. Procedures of the MFTL Kth-Best algorithm are developed as follows.

The MFTL Kth-Best algorithm

[Begin]

Step 1: Set $k=1$, adopt the simplex method to obtain an optimal solution $(x^1, y^1, z_{11}^1, \cdots, z_{1m_1}^1, \cdots, z_{n1}^1, \cdots, z_{nm_n}^1)$ to the linear programming problem (2). Let T be a set of feasible vertices of problem (2) that has been searched and W be a set of feasible vertices to be searched. Let $T = \varnothing$ and $W = \{(x^1, y^1, z_{11}^1, \cdots, z_{1m_1}^1, \cdots, z_{n1}^1, \cdots, z_{nm_n}^1)\}$. Set $i=1$ and go to Step 2.

Step 2: Put $x = x^k$ and solve the problem (1c-1f) and obtain an optimal solution $(\hat{y}, \hat{z}_{i1}, \cdots, \hat{z}_{im_i})$ using the bi-level Kth-Best algorithm [10]. Then go to Step 3.

Step 3: If $(\hat{y}, \hat{z}_{i1}, \cdots, \hat{z}_{im_i}) \ne (y^k, z_{i1}^k, \cdots, z_{im_i}^k)$, go to Step 4. If $(\hat{y}, \hat{z}_{i1}, \cdots, \hat{z}_{im_i}) = (y^k, z_{i1}^k, \cdots, z_{im_i}^k)$ and $i \ne n$, set $i=i+1$ and go to Step 2. If $(\hat{y}, \hat{z}_{i1}, \cdots, \hat{z}_{im_i}) = (y^k, z_{i1}^k, \cdots, z_{im_i}^k)$ and $i=n$, stop and $(x^k, y^k, z_{11}^k, \cdots, z_{1m_1}^k, \cdots, z_{n1}^k, \cdots, z_{nm_n}^k)$ is an optimal solution to the MFTL decision model (1) and $K^* = k$.

Step 4: Let W_k denote the set of adjacent extreme points of $(x^k, y^k, z_{11}^k, \cdots, z_{1m_1}^k, \cdots, z_{n1}^k, \cdots, z_{nm_n}^k)$ such that $(x, y, z_{11}, \cdots, z_{1m_1}, \cdots, z_{n1}, \cdots, z_{nm_n}) \in W_k$ implies

$cx + dy + \sum_{i=1}^{n}\sum_{j=1}^{m_i} e_{ij} z_{ij} \ge cx^k + dy^k + \sum_{i=1}^{n}\sum_{j=1}^{m_i} e_{ij} z_{ij}^k$. Let $T = T \cup \{(x^k, y^k, z_{11}^k, \cdots, z_{1m_1}^k, \cdots, z_{n1}^k, \cdots, z_{nm_n}^k)\}$

and $W = (W \cup W_k) \setminus T$. Go to Step 5.

Step 5: Set $k=k+1$ and choose $(x^k, y^k, z_{11}^k, \cdots, z_{1m_1}^k, \cdots, z_{n1}^k, \cdots, z_{nm_n}^k)$ such that $cx^k + dy^k +$

$\sum_{i=1}^{n}\sum_{j=1}^{m_i} e_{ij} z_{ij}^k = \min\{cx + dy + \sum_{i=1}^{n}\sum_{j=1}^{m_i} e_{ij} z_{ij} : (x, y, z_{11}, \cdots, z_{1m_1}, \cdots, z_{n1}, \cdots, z_{nm_n}) \in W\}$. Set $i=1$ and go to

Step 2.

[End]

4   A Case Study

Following the three-stage supply chain decision example proposed in the introduction of the paper, we give more detailed decision information as follows. In the MFTL decision case, the manufacturer (leader) and the vendors (bottom-level followers) keep their individual inventories using their respective warehouses and determine their inventories independently to optimize their individual objective under separate constraints. Clearly, there exists the uncooperative relationship among multiple bottom-level followers. However, to reduce the inventory cost, all the distributers at the middle level share a common warehouse and determine the inventory in common, which means they have the same decision variable even though they may have individual objectives and separate constraints called a cooperative relationship among them. Therefore, the leader seeks to minimize its inventory cost $F(x, y, z_{11}, \cdots, z_{1m_1}, \cdots, z_{n1}, \cdots, z_{nm_n})$ by controlling its own decision variable x (its inventory). The middle-level followers want to optimize their individual objectives $f_i^{(2)}(x, y, z_{i1}, \cdots, z_{im_i})$ ($i = 1,2, \cdots, n$) by determining the shared decision variable y (the common inventory) for the given x determined by the leader. The bottom-level follower ij attached to the middle-level follower i ($i = 1,2, \cdots, n$) optimize its objective function $f_{ij}^{(3)}(x, y, z_{ij})$ ($j = 1,2, \cdots, m_i$) by choosing its individual decision variable z_{ij} in view of the given x and y. In this paper, we simplify the three-stage supply chain decision problem as the following numerical model in the format of model (1).

For $x \in X \subset R^1$, $y \in Y_i \subset R^1$, $z_{ij} \in Z_{ij} \subset R^1$ and $X = \{x : x \geq 0\}$, $Y_i = \{y : y \geq 0\}$,

$Z_{ij} = \{z_{ij} : z_{ij} \geq 0\}$, $j = 1,2, \cdots, m_i$, $i = 1,2$, $m_i = 2$.

$$\min_{x \in X} F(x, y, z_{11}, z_{12}, z_{21}, z_{22}) = -1.5x - y + 2z_{11} + z_{12} - z_{21} - 0.5z_{22}$$

s.t.
$\quad x + y + z_{11} + z_{12} + z_{21} + z_{22} \geq 10,$

$\quad x \leq 1.5,$

$\displaystyle \min_{y \in Y_i} f_1^{(2)}(x, y, z_{11}, z_{12}) = x + y + z_{11} + z_{12} \qquad \min_{y \in Y_2} f_2^{(2)}(x, y, z_{21}, z_{22}) = x - y + 2z_{21} + 3z_{22}$

s.t. $x + y + z_{11} + z_{12} \geq 6.5,$ s.t. $x + y + z_{21} + z_{22} \geq 5.5,$

$\qquad x + y \leq 2,$ $\qquad x + y \leq 2,$

$\displaystyle \min_{z_{11} \in Z_{11}} f_{11}^{(3)}(x, y, z_{11}) = x + y + 3z_{11} \qquad \min_{z_{21} \in Z_{21}} f_{21}^{(3)}(x, y, z_{21}) = x + y + 2z_{21}$

s.t. $x + y + z_{11} \geq 3.5,$ s.t. $x + y + z_{21} \geq 3,$

$\qquad z_{11} \leq 2,$ $\qquad z_{21} \leq 2,$

$\displaystyle \min_{z_{12} \in Z_{12}} f_{12}^{(3)}(x, y_1, y_0, z_{12}) = x + y + 2z_{12} \qquad \min_{z_{22} \in Z_{22}} f_{22}^{(3)}(x, y, z_{21}) = x + y + z_{22}$

s.t. $x + y + z_{12} \geq 5,$ s.t. $x + y + z_{22} \geq 4.5,$

$\qquad z_{12} \leq 4,$ $\qquad z_{22} \leq 3.$

We can adopt the MFTL Kth-Best algorithm to solve the MFTL decision problem. First, we have to solve a linear programming problem in the format of (2).
Step 1: Set $k=1$ and adopt the simplex method to obtain an optimal solution to the problem (2). The optimal solution to (2) is $(x^1, y^1, z_{11}^1, z_{12}^1, z_{21}^1, z_{22}^1) = (1.5, 0.5, 1.5, 3, 2, 3)$ and

now $W = \{(1.5,0.5,1.5,3,2,3)\}$, $T = \varnothing$. Set $i=1$ and go to Step 2 and the iteration 1 will begin.

Step 2: Put $x = x^1 = 1.5$, and solve the problem in the form of (1c-1f). We can get the optimal solution $(\hat{y}, \hat{z}_{11}, \hat{z}_{12}) = (0.5,1.5,3)$ to (1c-1f) by bi-level Kth-Best algorithm and go to Step 3.

Step 3: Obviously, $(\hat{y}, \hat{z}_{11}, \hat{z}_{12}) = (y^1, z_{11}^1, z_{12}^1) = (0.5,1.5,3)$, $i \neq n$, set $i=2$ and go to Step 2.

Step 2: Put $x = x^1 = 1.5$ and $i=2$, and solve the problem in the form of (1c-1f). We can get the optimal solution $(\hat{y}, \hat{z}_{21}, \hat{z}_{22}) = (0.5,1,2.5)$ to (1c-1f) by bi-level Kth-Best algorithm and go to Step 3.

Step 3: Now, $(\hat{y}, \hat{z}_{21}, \hat{z}_{22}) \neq (y^1, z_{21}^1, z_{22}^1)$ and go to Step 4.

Step 4: Find the adjacent extreme points of $s^1 = (x^1, y^1, z_{11}^1, z_{12}^1, z_{21}^1, z_{22}^1)$ $=(1.5,0.5,1.5,3,2,3)$ and now the set of adjacent extreme points $W_1 = \{(0,2,1.5,3,2,3), (1.5,0.5,1.5,3,1,3), (1.5,0.5, \quad 1.5,3,2,2.5)\}$, $T = \{s^1\} = \{(1.5,0.5,1.5,3,2,3)\}$, $W = W_1$. Go to Step 5.

Step 5: Set $k=k+1=2$ and choose $(x^2, y^2, z_{11}^2, z_{12}^2, z_{21}^2, z_{22}^2) = (1.5,0.5,1.5,3,2,2.5)$ from the vertices set W such that $F(x^2, y^2, z_{11}^2, z_{12}^2, z_{21}^2, z_{22}^2) = \min\{F(x, y, z_{11}, z_{12}, z_{21}, z_{22}):$ $(x, y, z_{11}, z_{12}, z_{21}, z_{22}) \in W\}$, set $i=1$ and go to Step 2. This step means the iteration 1 has stopped and we cannot get an optimal solution through the iteration so the second iteration will then be executed.

Table 1. The detailed computing process by the MFTL Kth-Best algorithm

Iteration k	$s^k = (x^k, y^k, z_{11}^k,$ $z_{12}^k, z_{21}^k, z_{22}^k)$	W_k	T	W
2	(1.5,0.5,1.5,3,2,2.5)	{(1.5,0.5,1.5,3,1,2.5)}	$\{s^1, s^2\}$	{(0,2,1.5,3,2,3), (1.5,0.5,1.5,3,1,3), (1.5,0.5,1.5,3,1,2.5)}
3	(0,2,1.5,3,2,3)	{(0,2,1.5,3,1,3), (0,2,1.5,3,2,2.5)}	$\{s^1, s^2, s^3\}$	{(1.5,0.5,1.5,3,1,3), (1.5,0.5,1.5,3,1,2.5), (0,2,1.5,3,1,3), (0,2,1.5,3,2,2.5)}
4	(1.5,0.5,1.5,3,1,3)	$\varnothing$	$\{s^1, s^2, s^3, s^4\}$	{(1.5,0.5,1.5,3,1,2.5), (0,2,1.5,3,1,3), (0,2,1.5,3,2,2.5)}
5	(0,2,1.5,3,2,2.5)	{(0,2,1.5,3,1,2.5)}	$\{s^1, s^2, s^3, s^4, s^5\}$	{(1.5,0.5,1.5,3,1,2.5), (0,2,1.5,3,1,3), (0,2,1.5,3,1,2.5)}
6	(0,2,1.5,3,1,3)	$\varnothing$	$\{s^1, s^2, s^3, s^4, s^5, s^6\}$	{(1.5,0.5,1.5,3,1,2.5), (0,2,1.5,3,1,2.5)}
7	(1.5,0.5,1.5,3,1,2.5)	---	---	---

In this way, we finally get an optimal solution through seven iterations. The searched extreme points and the detailed computing process of iterations 2-7 are shown as Table 1. In iteration 7, $(x^7, y^7, z_{11}^7, z_{12}^7, z_{21}^7, z_{22}^7) = (1.5,0.5,1.5,3,1,2.5)$ is an optimal solution to the MFTL decision problem, which implies that all decision entities achieve the equilibrium at the vertex solution. The objective function values of all decision entities are $F(x, y, z_{11}, z_{12}, z_{21}, z_{22}) = 1.0$, $f_1^{(2)}(x, y, z_{11}, z_{12}) = 6.5$, $f_2^{(2)}(x, y, z_{21}, z_{22}) = 10.5$, $f_{11}^{(3)}(x, y, z_{11}) = 6.5$, $f_{12}^{(3)}(x, y, z_{12}) = 8.0$, $f_{21}^{(3)}(x, y, z_{21}) = 4.0$, $f_{22}^{(3)}(x, y, z_{22}) = 4.5$. It is noticeable that $W_4 = \varnothing$ and $W_6 = \varnothing$ in Table 1 do not mean there does not exist adjacent extreme points of $(x^4, y^4, z_{11}^4, z_{12}^4, z_{21}^4, z_{22}^4)$ and $(x^6, y^6, z_{11}^6, z_{12}^6, z_{21}^6, z_{22}^6)$ but may imply their adjacent extreme points have been found in previous iterations and have been involved in W. However, when plenty of followers are involved or a large number of decision variables and constraints exist, the execution efficiency of the algorithm may experience a steep decline as superabundant vertices are needed to complete the search.

5   Conclusions and Further Study

In a tri-level decision problem, multiple followers are often involved at the middle and bottom levels. Various relationships among multiple followers at the same level can result in different decision processes in a three-level hierarchical system. To support MFTL decision problems involving both cooperative and uncooperative relationships among multiple followers, this paper proposes a decision model and then develops a Kth-Best algorithm to find an optimal solution to the model. Lastly, a case study on three-stage supply chain decision illustrates the effectiveness of the proposed MFTL decision techniques. The results show that the MFTL decision model and Kth-Best algorithm provide an available way in describing and solving the proposed MFTL decision process. Our future research will develop a decision support system driven by the proposed decision techniques to explore the performance of the MFTL Kth-Best algorithm through sufficient numerical experiments and to handle large-scale or more complex MFTL decision problems in applications.

References

1. Vicente, L., Calamai, P.: Bilevel and multilevel programming: A bibliography review. Journal of Global Optimization 5(3), 291–306 (1994)
2. Stackelberg, H.V.: The Theory of Market Economy. Oxford University Press, Oxford (1952)
3. Xu, X., Meng, Z., Shen, R.: A tri-level programming model based on Conditional Value-at-Risk for three-stage supply chain management. Computers & Industrial Engineering 66(2), 470–475 (2013)
4. Yao, Y., Edmunds, T., Papageorgiou, D., Alvarez, R.: Trilevel optimization in power network defense. IEEE Transactions on Systems, Man, and Cybernetics 37(4), 712–718 (2007)
5. Mitiku, S.: A multilevel programming approach to decentralized (or hierarchical) resource allocation systems. Applied Mathematics and Mechanics 7(1), 2060003–2060004 (2007)
6. Torabi, S.A., Ebadian, M., Tanha, R.: Fuzzy hierarchical production planning (with a case study). Fuzzy Sets and Systems 161(11), 1511–1529 (2010)
7. Zhang, G., Lu, J., Montero, J., Zeng, Y.: Model, Solution concept and the Kth-best algorithm for linear tri-level programming. Information Sciences 180(4), 481–492 (2010)

8. Alguacil, N., Delgadillo, A., Arroyo, J.M.: A trilevel programming approach for electric grid defense planning. Computers & Operations Research 41, 282–290 (2014)
9. Lu, J., Zhang, G., Montero, J., Garmendia, L.: Multifollower trilevel decision making models and system. IEEE Transactions on Industrial Informatics 8(4), 974–985 (2012)
10. Shi, C., Zhang, G., Lu, J.: The Kth-Best approach for linear bilevel multi-follower programming. Journal of Global Optimization 33(4), 563–578 (2005)

A Fuzzy ART-Based Approach for Estimation of High Performance Concrete Mix Proportion

Fei Ha Chiew[1,2], Kok Chin Chai[2], Chee Khoon Ng[2], and Kai Meng Tay[2]

[1] Faculty of Civil Engineering, Universiti Teknologi MARA, Malaysia
[2] Faculty of Engineering, Universiti Malaysia Sarawak
chiewfa@sarawak.uitm.edu.my, kcchai@live.com,
{ckng,kmtay}@feng.unimas.my

Abstract. Mix design of high performance concrete (HPC) is a complicated procedure as the mixtures usually consist of many possible combinations. Conventionally, numerous laboratory works are required. In this paper, the use of Fuzzy Adaptive Resonance Theory (ART) in estimating HPC mix proportion from experimental data is proposed. The proposed Fuzzy ART–based approach attempts to search for a set of suitable mix proportions which is near to a desired compressive strength, and deduce a mix proportion. Such approach is useful as experimental data is subjected to experimental errors or even potentially outliers. The applicability of the proposed approach is demonstrated with a set of benchmark data. This paper contributes to a new methodology for estimating the mix proportion of HPC.

Keywords: Mix proportion, Mix design, Fuzzy ART, High performance concrete, Civil engineering.

1 Introduction

High performance concrete (HPC) is a concrete with better strength, low permeability or better durability [1]. Besides water, Portland cement, fine aggregates and coarse aggregates, HPC usually needs additional cementitious materials, e.g., fly ash, silica fume and blast furnace slag, and chemical admixtures. The production of HPC in construction usually involves seven stages i.e., material selections, mix proportioning, batching, mixing, transporting, placing and curing [2]. The focus of this research is on the second stage, i.e., mix proportioning. Mix proportioning (also known as mix design) is the process of selecting the proportion of raw materials based on a set of desired properties in the most economical solution [3].

Regardless of the importance of mix proportioning, a generalized systematic approach to the selection of mix proportion of HPC is still lacking [1],[4]. In practice, the mix design of HPC is based on existing mix design standards for normal concrete [5]. However, modifications need to be made to the mix proportion obtained, and large numbers of trial mixes are required in order to obtain the desired combination of materials [6]. Later, empirical models of concrete properties were proposed [6],[7], where mix proportioning methods were developed with aid of mathematical models.

C.K. Loo et al. (Eds.): ICONIP 2014, Part III, LNCS 8836, pp. 407–414, 2014.
© Springer International Publishing Switzerland 2014

Again, these approaches require experimental works to be conducted and functions relating materials to concrete properties are derived based on experimental results.

Since studies [4]-[7] were done separately by different researchers, the use of experimental results from each study was limited to the purpose of each study and materials used. Many possible mix proportions may lead to similar desired compressive strength [8]. In this case, if experimental results from various studies were combined in a database, further analyses can be conducted in order to choose the most suitable mix proportion. These analyses require clustering approach [9] to organize experimental data from various studies, with multi-dimensional attributes into separate groups, according to their similarities. As experimental data is subjected to experimental errors i.e., prolonged mixing, unthorough mixing and existence of silt on coarse aggregates, it is important that experimental data with inaccurate results can be identified.

Recent literatures reveal that soft computing approaches can be used as a decision support tool to HPC mix design, based on available experimental data. The works in [10]-[13] proposed to analyze experimental data with soft computing approaches, and further estimate compressive strength of concrete for a given mix proportion. These models are beneficial, yet they do not solve the mix design problem in a reverse way i.e., to estimate the mix proportion for a given compressive strength. Studies by Lim *et al.* [4], and Lee and Yoon [14] proposed soft computing approaches to formulate fitness functions relationships between materials and strength for the experimental data, and further propose mix proportion for targeted strength and slump. However, modeling of HPC properties is difficult due to high nonlinear relationships between concrete properties and materials [10]. Therefore, this research proposes to use clustering approach, which can be used to classify experimental data according to their similarities, and further decide a suitable mix proportion. In clustering, data within the same cluster share some similar features, compared with other clusters [9].

Motivated by the above issues, a mix proportion estimation model based on Fuzzy ART is developed in this study. Fuzzy ART approach is chosen due to the simple nature of its architecture. This approach gives easy explanation of the responses of neural network to input patterns [15]. The proposed method is beneficial as it can search for data with compressive strengths near the desired strength, identifies outlier data and optimizes the mix proportion estimation based on similarity between data with desired strength. A benchmark database by Yeh [12] was used to evaluate the usefulness of the proposed model. The rest of this paper is organized as follows. In Section 2, a Fuzzy ART-based approach in estimating mix proportion is proposed. In Section 3, an experimental study and the associated results of the proposed model are presented with a benchmark database. Finally, concluding remarks and suggestions for further work are given in Section 4.

2   The Proposed Methodology

To illustrate the general idea of the proposed methodology, blast furnace slag concrete data from Yeh [12] is exampled. Six raw materials (i.e., cement, blast furnace slag, water, superplasticizer, coarse aggregate and fine aggregate) are used to produce HPC. A database consists of 128 different proportions of the six raw materials used to produce 1 m³ of HPC and their respective compressive strengths is gathered.

The proposed Fuzzy ART-based approach for estimating mix design of HPC is depicted in Fig. 1. Each of the steps is explained in detail as follows:

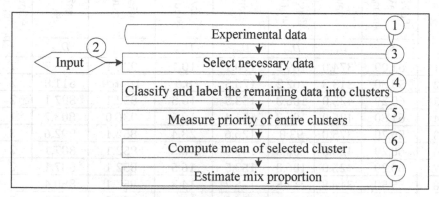

Fig. 1. The proposed Fuzzy ART-based approach

Step 1. Consider 128 experimental data from Yeh [12].

Step 2. Consider a desired compressive strength (in MPa) (i.e., I_{DCS}) and the acceptable tolerance in % (i.e., ε). In this study, $I_{DCS} = 60$MPa and $\varepsilon = 10\%$, are considered.

Step 3. Select necessary data from 128 data points.

Consider that the concrete compressive strength of a data in the database, S_i where $i = 1,2,3,...,128$. If S_i falls in the range of $I_{DCS} \pm \varepsilon\%$ where $i = 1,2,3,...,128$, then that data point is selected. The selected data points are denoted as $D_{r,x}$ where $r = 1,2,...,R$ and R denotes the total selected data. The selected data are stored into a $R \times 6$ matrix i.e., I as follows:

$$I = \begin{bmatrix} D_{11} & \cdots & D_{16} \\ \vdots & \ddots & \vdots \\ D_{R1} & \cdots & D_{R6} \end{bmatrix} \tag{1}$$

The selected concrete compressive strength is denoted as S_r where $r = 1,2,...,R$. Table 1 presents the selected data with their respective compressive strengths. In this paper, concrete compressive strength is denoted as S_r and the proportion of each raw material, is denoted as $D_{r,x}$, where r and x are number of data point and and type of raw material, respectively. The "Cluster" column indicates the cluster of each data point, using fuzzy ART (obtained from Step 4). As an example, in Table 1, S_1 and $D_{1,1}$ refer to the concrete compressive strength and cement content for the first data point i.e., 61.09 MPa and 374.0 kgm⁻³ respectively. This mix proportion is categorized in Cluster 1, using Fuzzy ART.

Table 1. Selected data from database and their respective clusters (from Yeh [12])

Concrete compressive strength (MPa)	Raw materials (in kgm^{-3})							Cluster(obtained from Step 4)
	Cement	Blast Furnace Slag	Water	Superplasticizer	Coarse Aggregate	Fine Aggregate	
r	S_r	D_{r1}	D_{r2}	D_{r3}	D_{r4}	D_{r5}	D_{r6}	
1	61.09	374.0	189.2	170.1	10.1	926.1	756.7	1
2	59.80	313.3	262.2	175.5	8.6	1046.9	611.8	2
3	60.29	425.0	106.3	153.5	16.5	852.1	887.1	2
4	61.80	425.0	106.3	151.4	18.6	936.0	803.7	1
5	56.70	375.0	93.8	126.6	23.4	852.1	992.6	3
6	60.29	425.0	106.3	153.5	16.5	852.1	887.1	2
7	60.29	425.0	106.3	153.5	16.5	852.1	887.1	2
8	55.50	318.8	212.5	155.7	14.3	852.1	880.4	2
9	66.00	439.0	177.0	186.0	11.1	884.9	707.9	3
10	59.00	356.0	119.0	160.0	9.0	1061.0	657.0	3
11	57.21	321.0	164.0	190.0	5.0	870.0	774.0	1
12	65.91	366.0	187.0	191.0	7.0	824.0	757.0	1
13	61.23	326.0	166.0	174.0	9.0	882.0	790.0	1
14	56.61	331.0	170.0	195.0	8.0	811.0	802.0	1
15	56.62	330.5	169.6	194.9	8.1	811.0	802.3	1
16	57.22	321.3	164.2	190.5	4.6	870.0	774.0	1
17	65.91	366.0	187.0	191.3	6.6	824.3	756.9	1
18	61.24	325.6	166.4	174.0	8.9	881.6	790.0	1

Step 4. Apply Fuzzy ART algorithm [16]-[17] to group the selected data into several clusters.

The architecture of Fuzzy ART is depicted in Fig. 2. In layer 1 (or input layer), there are 12 nodes, i.e., NI_{r1}, NI_{r2},..., NI_{r6}, NI_{r1}^c, NI_{r2}^c,..., NI_{r6}^c. In layer 2 (or recognition layer), there are s cluster prototypes, $s > 0$, and s can be increased over time depending on the availability of data samples. Each cluster prototype is labeled as C_z, where $z = 1,2,3,...,s$. The weight connecting C_1 and NI_{r1} is denoted as $w_{NI_{r1},1}$, and that connecting C_s and NI_{r6}^c is denoted as $w_{NI_{r6}^c,s}$. $\overline{NI^c}$ is a generalized matrix of two matrices i.e., NI and NI^c. All weights are contained in a matrix, i.e., $w_{\overline{NI^c},z}$. Each component of $w_{\overline{NI^c},z}$ is labeled as $w_{\overline{NI^c},z}(v)$, where $v = 1,2,3...,12$. The complete algorithm is summarized into Steps 4(a)-(f) as follows:

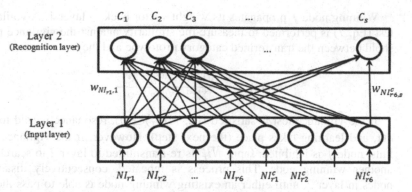

Fig. 2. Fuzzy ART architecture, in which Layer 2 is an incremental layer

a. *Normalization*: Selected data I from database are normalized into $[0,1]$ using Eq. (2) as follows:

$$NI_{r,x} = \frac{(I_{r,x} - min_{r=1,2,...,R}\, I_{r,x})}{(max_{r=1,2,...,R}\, I_{r,x} - min_{r=1,2,...,R}\, I_{r,x})} \tag{2}$$

b. *Performing complement coding*: The complement of matrix NI is computed using Eq. (3) as follows:

$$NI^c = 1 - NI \tag{3}$$

c. *Combining two matrices NI and NI^c*: Two matrices are combined and formed an individual matrix i.e., $\overline{NI^c}$ as denoted as follows:

$$\overline{NI^c} = \begin{bmatrix} NI_{11} & \cdots & NI_{16} & NI_{11}^c & \cdots & NI_{16}^c \\ \vdots & \ddots & \vdots & \vdots & \ddots & \vdots \\ NI_{R1} & \cdots & NI_{R6} & NI_{R1}^c & \cdots & NI_{R6}^c \end{bmatrix} \tag{4}$$

d. *Initialization and parameter setting*: The number of cluster is set to 1 (i.e., $s = 1$) and it is incremental as learning progresses. The parameters for Fuzzy ART algorithm are initialized as follows: The weights connecting C_z and $\overline{NI^c}$ are initialized to 1, choice parameter (i.e., $\alpha = 0.001$), learning rate (i.e., $\beta = 1$) and vigilance parameter (i.e., $\rho = 0.8$).

e. *Category choice, test, and search*: Each input vector (i.e., $\overline{NI^c}$) is transmitted from layer 1 to layer 2. The response of each node in layer 2 is computed using the *choice function* (Eq. 5). The node that has the highest response, denoted as node $J \in (1,2,3,...,s)$, is selected as the winning node (Eq. 6). If there is a tie on $T_{r,z}$, the node with the smallest index is chosen.

$$T_{r,z} = \frac{\sum_{v=1}^{v=12}\left(\overline{NI_r^c}(v) \wedge w_{\overline{NI^c},z}(v)\right)}{\alpha + \sum_{v=1}^{v=12}\left(w_{\overline{NI^c},z}(v)\right)} \tag{5}$$

$$T_J = \max\left(T_{r,z} : z = 1,2,3,...,s\right) \tag{6}$$

Winning node J propagates its weight vector back to layer 1. A vigilance test (Eq. 7) is performed to measure the similarity against the vigilance threshold between the transformed category prototype and the input vector.

$$M(T_J) = \frac{\sum_{v=1}^{v=12}\left(\overline{NI_r^c}(v) \wedge w_{\overline{NI^c},J}(v)\right)}{\sum_{v=1}^{v=12}\left(\overline{NI_r^c}(v)\right)} \tag{7}$$

If the vigilance test is satisfied (i.e, $M(T_J) \geq \rho$), resonance is said to occur, and learning takes place (the next step). However, if the vigilance test fails, node J is inhibited. Input $\overline{NI_r^c}$ is re-transmitted to layer 2 to search for another winning node. This process is repeated, consecutively disabling nodes in layer 2, until either an existing winning node is able to pass the vigilance test. If no such node is available, a new node is created to encode the input vector.

f. *Learning*:   Once the search process ends, learning takes place by adjusting $w_{\overline{NI^c},J}$ using Eq. (8).

$$w_{\overline{NI^c},J,new}(v) = \beta\left(\overline{NI_r^c}(v) \wedge w_{\overline{NI^c},J,old}(v)\right) + (1-\beta)w_{\overline{NI^c},J,old}(v), \ v = 1,2,\dots,12 \tag{8}$$

Step 5. Prioritize the concrete compressive strength in each cluster using similarity measure. The highest similarity measure between C_z and I_{DCS} indicates the highest priority or known as the winning cluster.

From Table 1, it is observed that S_r is clustered into three clusters. As an example, C_2 consists of five data points with their respective concrete compressive strengths i.e., 59.80 MPa, 60.29 MPa, 60.29 MPa, 60.29 MPa and 55.50 MPa. The concrete compressive strength of each cluster is generalized and represented as $C_z(t)$ where $z = 1,2,\dots,s$ and $t = 1,2,\dots,T$. $C_z(t)$ and I_{DCS} are initially normalized using Eqs. (9) and (10) as follows:

$$NC_z(t) = \frac{C_z(t) - min_{r=1,2,\dots,R}\,S_r}{max_{r=1,2,\dots,R}\,S_r - min_{r=1,2,\dots,R}\,S_r} \tag{9}$$

$$NI_{DCS} = \frac{I_{DCS} - min_{r=1,2,\dots,R}\,S_r}{max_{r=1,2,\dots,R}\,S_r - min_{r=1,2,\dots,R}\,S_r} \tag{10}$$

For instance, $C_2(1)$ refers to the first component in Cluster 2 i.e., 59.80MPa.

$$NC_2(1) = \frac{59.80 - 55.50}{66.00 - 55.50} = 0.410$$

A correlation coefficient is adopted to measure the similarity between NC_z and NI_{DCS}, as written as Eq. (11).

$$sim(NC_z, NI_{DCS}) = \frac{\sum_{t=1}^{T}(NC_z(t) \times NI_{DCS})}{\sqrt{\sum_{t=1}^{T}(NC_z(t))^2}\sqrt{NI_{DCS}^2 \times T}} \tag{11}$$

Step 6. Compute the arithmetic mean of the winning cluster.

Consider that the winning cluster is denoted as Z, then the mix proportions in cluster Z are considered. Arithmetic mean of the mix proportions
in cluster Z is computed.

Step 7. Estimate the suitable mix proportion for the desired concrete compressive
strength.

3   An Experimental Study

Consider the desired compressive strength, I_{DCS} = 60MPa with tolerance, ε = 10%.
Using such inputs, the data selected from Yeh [12] is presented in Table 1. Table 1
shows selected data with compressive strengths in the range (60 $\pm$ 6MPa) and their
respective mix proportions.

The priority selection of each cluster is decided based on the similarity measure between the desired compressive strength (i.e., I_{DCS}) and C_z, as expressed in Eqs. (9)-
(11). The results are tabulated in Table 2. It is observed that Cluster 2 indicates the
highest similarity to I_{DCS} i.e., 0.8935 and Cluster 3 has the smallest similarity, i.e.,
0.7883. This indicates that Cluster 2 is the winning cluster. The arithmetic means of
the data in Cluster 2 are presented in Table 3. Based on our proposed methodology,
59.23 MPa is a result of 1 m^3 concrete with mix proportion of 381.42 kg of cement,
158.72 kg of blast furnace slag, 158.34 kg of water, 14.48 kg of superplasticizer,
891.06 kg of coarse aggregates and 830.70 kg of fine aggregates.

Table 2. Similarity measure between Cs_z and I_{DCS}

Cluster, z	1	2	3
$sim(NC_z, NI_{DCS})$	0.8286	0.8935	0.7883

Table 3. Arithmetic means of data in cluster 2 and estimated mix proportion for concrete
compressive strength of 60 MPa

Concrete compressive strength (MPa)	Cement (kgm^{-3}),	Blast Furnace Slag (kgm^{-3})	Water (kgm^{-3})	Superplasticizer (kgm^{-3})	Coarse Aggregate (kgm^{-3})	Fine Aggregate (kgm^{-3})
59.23	381.42	158.72	158.34	14.48	891.06	830.70

It is worth-mentioning that this simulated study was carried out with a laptop, with
the following specifications: Intel® Core i7 2670QM, 2.2GHz with 4GB of RAM
and MATLAB® 7.10.0 (R2010a). The time complexity is about 0.146 seconds.

4   Concluding Remarks and Recommendations

In this paper, the use of a Fuzzy ART-based approach to estimate mix proportion of
HPC for a desired compressive strength is proposed. This approach is beneficial in
analyzing multi-dimensional experimental data. This approach also has the advantage
to identify potential outlier data, which do not share some similarities with the rest of

the database. The developed model can be used as a mix proportioning decision making tool, and reduce the number of trial mix required. In this paper, a benchmark database was adopted to demonstrate the usefulness of the proposed method, and positive results were obtained.

For future works, the results from the proposed method and real experimental results will be compared. Apart from that, a further study incorporating slump values as another input parameter is to be conducted.

References

1. Neville, A.M.: Properties of Concrete. 4th ed. Wiley, England (1995)
2. ACI 363R-92:State-of- the art report on high strength concrete, Manual of Concrete Practice, Part 1, American Concrete Institute (1992)
3. Jackson, N., Dhir, R.K.: Civil Engineering Materials, 5th ed. Palgrave, New York (1996)
4. Lim, C.H., Yoon, Y.S., Kim, J.H.: Genetic algorithm in mix proportioning of high performance concrete. Cement and Concrete Research 34, 409–420 (2004)
5. Alves, M.F., Cremonini, R.A., Dal Molin, D.C.C.: A comparison of mix proportioning methods for high-strength concrete. Cement and Concrete Composites 26, 613–621 (2004)
6. Bharatkumar, B.H., Narayanan, R., Raghuprasad, B.K., Ramachandramurthy, D.S.: Mix proportioning of high performance concrete. Cement and Concrete Composites 23, 71–80 (2001)
7. Sobolev, K.: The development of a new method for proportioning of high performance concrete mixtures. Cement and Concrete Composites 26, 901–907 (2004)
8. Aïtcin, P.C.: High performance concrete. E & FN Spon, New York (1998)
9. Xu, R., Wunsch, D.C.: Clustering. John Wiley & Sons, New Jersey (2009)
10. Chou, J.S., Chiu, C.K., Farfoura, M., Al-Taharwa, I.: Optimizing the prediction accuracy of concrete compressive strength based on a comparison of data-mining techniques. Journal of Computing in Civil Engineering 25, 242–253 (2011)
11. Yeh, I.C., Lien, L.C.: Knowledge discovery of concrete material using genetic operation trees. Expert Systems with Applications 36, 5807–5812 (2009)
12. Yeh, I.C.: Modeling of strength of high performance concrete using artificial neural networks. Cement and Concrete Research 28, 1797–1808 (1998)
13. Ahmadi-Nedushan, B.: An optimized instance based learning algorithm for estimation of compressive strength of concrete. Engineering Applications of Artificial Intelligence 25, 1073–1081 (2012)
14. Lee, J.H., Yoon, Y.S.: Modified harmony search algorithm and neural networks for concrete mix proportion design. Journal of Computing in Civil Engineering 23, 57–61 (2009)
15. Pacella, M., Semeraro, Q., Anglani, A.: Manufacturing quality control by means of a Fuzzy ART network trained on natural process data. Engineering Applications of Artificial Intelligence 17, 83–96 (2004)
16. Carpenter, G.A., Grossberg, S., Rosen, D.B.: Fuzzy ART: Fast stable learning and categorization of analog patterns by an adaptive resonance system. Neural Networks 4, 759–771 (1991)
17. Carpenter, G.A., Grossberg, S., Rosen, D.B.: Fuzzy ART: An adaptive resonance algorithm for rapid, stable classification of analog patterns. In: IJCNN-91-Seattle International Joint Conference on Neural Networks 2, pp. 411–416. IEEE (1991)

A New Application of an Evolving Tree to Failure Mode and Effect Analysis Methodology

Wui Lee Chang[1], Kai Meng Tay[1,*], and Chee Peng Lim[2]

[1] Faculty of Engineering, Universiti Malaysia Sarawak, Sarawak, Malaysia
[2] Centre for Intelligent Systems Research, Deakin University, Melbourne, Australia
kmtay@feng.unimas.my

Abstract. Failure Mode and Effect Analysis (FMEA) is a popular safety and re-liability analysis methodology for examining potential failure modes of prod-ucts, process, designs, or services, in a wide range of industries. Despite its popularity, there are a number of limitations of FMEA, and two highlighted is-sues are the bulky FMEA form and its intricacy of use. To overcome these shortcomings, we introduce the idea of visualisation pertaining to the failure modes or control actions in FMEA. A visualisation model with an incremental learning feature, i.e., the evolving tree (ETree), is adopted to allow the failure modes or control actions in FMEA to be clustered and visualized. The failure modes or control actions are grouped and visualized with consideration of their Severity, Occurrence, and Detection scores. Our proposed approach allows the failure modes or control actions to be mapped into a tree structure for visualisa-tion. The devised approach is evaluated with a benchmark problem. The ex-periments show that the control actions of FMEA can be visualised through the tree structure, which provides a quick and easily understandable platform of the FMEA spreadsheet to facilitate decision making tasks.

Keywords: Failure mode and effect analysis, clustering, visualisation, evolving tree .

1 Introduction

Failure Mode and Effect Analysis (FMEA) is a popular and effective problem preven-tion methodology for defining, identifying, and eliminating potential failure modes and errors of a system, design, process, or service [1]. FMEA attempts to identify the potential failure modes of a system, understand the causes and effects of each potential failure mode, and determine a series of appropriate actions to eliminate or reduce the risk of failure modes [1]. Traditionally, the risk of potential failure modes is determined by computing the Risk Priority Number (RPN) [1]. The RPN model considers three factors as its inputs, i.e. Severity (S), Occurrence (O), and Detection (D), and produces an RPN score (i.e. multiplication of S, O, and D) as the output [1].

---

[*] Corresponding author.

C.K. Loo et al. (Eds.): ICONIP 2014, Part III, LNCS 8836, pp. 415–422, 2014.
© Springer International Publishing Switzerland 2014

S and O are seriousness and frequency of a failure mode, respectively, while D is the effectiveness of the existing measures in detecting a failure before the failure effect reaches the customer [1].

Despite the popularity of FMEA, many investigations to improve its shortcomings have been reported. Generally, three main issues in FMEA are (1) the risk prioritisation issues [1, 2], (2) the bulky FMEA form [3, 4], and (3) the intricacy of use [3, 4]. To tackle the first issue, a number of multi-criteria decision making (MCDM) methods [2], mathematical programming (MP) methods [2], artificial intelligence (AI) methods [2] have been proposed. It is worth mentioning that in our previous works, a fuzzy rule-base system [5], and an fuzzy adaptive resonance theory (ART) neural network model [6] were investigated. However, little attention has been paid to the second and third issues. The second issue indicates that the FMEA spreadsheet is voluminous [4]. The third issue highlights that the overall procedure of FMEA is time consuming [4] and hard to maintain [3, 4].

For tackling the second and third issues, we suggest the use of a visualisation tool, which allows the failure modes, or control actions, to be visualized. Note that the use of visualisation in risk analysis is not new, for example, in [7], the failure risks are visualized in the likelihood and impact (a.k.a. severity) dimensions, i.e., a two-dimensional space. Such visualisation is useful to understand the distribution of risk. Unfortunately, such approach is hardly applicable to FMEA, which involves a three-dimensional input (i.e., S, O, and D) space. Motivated by this challenge, the aim of this paper is to use an evolving tree (ETree) [8] to allow the failure modes or control actions in FMEA to be visualized as a tree structure (i.e., a tree model). ETree is an evolving clustering model which is adaptive to new data, for feasible visualisation through a tree structure [8]. We attempt to depict the behaviours and relationship of the failure modes or action in a tree structure.

To the best of our knowledge, there is no investigation reported in the literature pertaining to visualisation of the failure modes or actions in FMEA. Nevertheless, visualisation in FMEA is important because (1) it conveys the failure modes or actions into a structure which can be grasped and understood more quickly than that of the raw failure modes or actions in FMEA spreadsheet, (2) it allows FMEA users to access or analyse FMEA with a large number of failure modes or actions, that might not be otherwise possible, and (3) it enables FMEA users to have an effective and efficient insight of the failure modes or actions in FMEA as well as to facilitate a quick decision making process through good visualisation, in which good understanding and insight of FMEA usually need to come from a relatively long duration of effort and experience. In short, this paper contributes to a new ETree-based approach to visualisation of the failure modes or control actions in FMEA. It is also a new application of ETree. A benchmark case study with data and information from [1] (see pages 231-242) is employed to evaluate the usefulness of the proposed approach.

2   Proposed Approach

2.1   Background of the Evolving Tree

The Evolving Tree (ETree) is a neural network for clustering and visualisation tasks. An example of a tree model is shown in Figure 1. In this sub-section, the terminologies related to ETree are summarized, as follows.

- Nodes are denoted as $N_{l,p}$, where $l = 1,2,3,...$ is the identity of the node, and $p = 0,1,2,...$ is its parent node, $p \neq l$.
- A unique weight vector, W_l, is associated with each node, i.e., $N_{l,p} = [w_{l,s}, w_{l,o}, w_{l,d}]$, as the centroid of the failure modes.
- The total number of nodes is denoted as n_{node}.
- Each node is attributed with a BMU (Best-Matched Unit) hit counter, i.e., b_l.

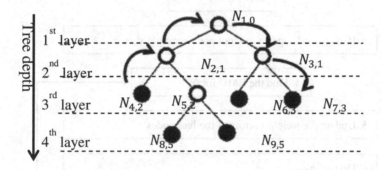

Fig. 1. A schematic diagram of the ETree

There are three types of nodes, i.e., root node, trunk node, and leaf node. The root node is the first created node, denoted as $N_{1,0}$. The trunk nodes are in white, and are located below the root node. The leaf nodes are in black, and are located below the trunk nodes. The leaf nodes are also known as the clusters.

The nodes arrangement produces a number of layers of the tree. The tree depth is defined as the maximum layer of the tree, while the tree size is defined as the total number of nodes in a tree. A tree distance is used to describe the relationship of the clusters. As an example, the tree distance between $N_{4,2}$ and $N_{7,3}$ are determined by the number of nodes (non-leaf nodes) that are connected to form the simplest pathway from $N_{4,2}$ towards $N_{7,3}$, i.e., denoted as $d_T(N_{4,2}, N_{7,3})$. As shown in Figure. 1, $d_T(N_{4,2}, N_{7,3}) = 3$.

2.2   The ETree-Based FMEA Model

In this study, the failure modes or control actions are denoted as $\overline{x_k}$, where $k = 1, 2, \ldots, m$, and $\overline{x_k} = [s_k, o_k, d_k]$. The number of failure modes or control action are m. Besides that, s_k, o_k, and d_k are the elements of S, O, and D, respectively, i.e., $s_k \in$ S, $o_k \in$ O, and $d_k \in$ D. There are three pre-defined parameters for the ETree-based FMEA model, i.e., the number of child nodes, θ_{child}, the splitting threshold, $\theta_{splitting}$, and the number of iteration, $epoch$. Figure 2 shows the learning procedure for the ETree-based FMEA model.

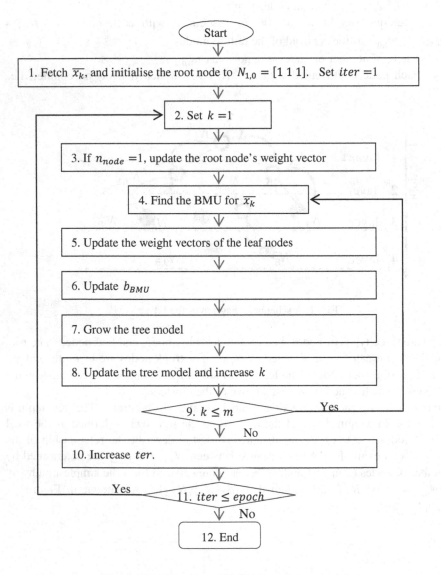

Fig. 2. The ETree-based FMEA learning methodology

Each of the steps is explained, as follows.

Step 1: Fetch $\overline{x_k}$, and initialise the root node to $N_{1,0} = [1\ 1\ 1]$. Set $iter =1$.
Step 2: Set $k =1$ by initialising k to 1.
Step 3: If $n_{node} =1$, update the root node's weight vector with Equation (1).

$$W_1(new) = W_1(old) + \left[\propto (iter) \times exp\left(\frac{-d_T(N_{1,0},N_{leaf})^2}{2\sigma^2(iter)}\right)\right] \times (\overline{x_k} - W_1(old)) \quad (1)$$

Step 4: Find the BMU for $\overline{x_k}$. If $n_{node} =1$, the root node is the BMU. Otherwise, based on the Euclidean similarity measurement, as in Equation (2), the child nodes of the root node (if any) with the minimum *Euclidean similarity* is selected as the winner. If the winner is not a leaf node, the child nodes of the winner are considered. The child nodes of the winner with the minimum *Euclidean similarity* is selected as the winner. The winner is the BMU, if it is a leaf node. If there are more than one winners, the BMU is randomly selected from the winners.

$$Euclidean\ Similarity(N_{l,p},\overline{x_k}) = \sqrt[2]{(w_{l,s} - s_k)^2 + (w_{l,o)} - o_k)^2 + (w_{l,d} - d_k)^2}\,(2)$$

Step 5: Update the weight vectors of the leaf nodes. The weight vectors of the leaf nodes are updated using the Kohonen learning rule, as in Equation (3). Note that $\propto$ and σ are monotonically reduced by the number of iterations $(iter)$. N_{leaf} is the leaf node, where $leaf \in l$ and $leaf \notin p$.

$$W_l(new) = W_l(old) + \left[\propto (iter) \times exp\left(\frac{-d_T(N_{BMU},N_{leaf})^2}{2\sigma^2(iter)}\right)\right] \times (\overline{x_k} - W_l(old)) \quad (3)$$

Step 6: Update b_{BMU} with $b_{BMU}(new) = b_{BMU}(old) + 1$.
Step 7: Grow the tree model. If $b_{BMU} = \theta_{splitting}$, the BMU is split into θ_{child} child nodes. The weight vectors of the child nodes are cloned from their parent's, i.e., the BMU's.
Step 8: Update the tree model and increase k to $k = k + 1$.
Step 9: If $k \leq m$, goto Step 4.
Step 10: Increase $iter$ to $iter = iter + 1$.
Step 11: If $iter \leq epoch$, goto Step 2.
Step 12: End

3   A Case Study

A case study with the data and information in a FMEA spreadsheet in [1] (pages 231-242) was considered. There are total of 76 control actions in the FMEA spreadsheet. Some of the control actions contain the same s, o, and d values. In this paper, the control actions are visualized using the proposed ETree approach with the following settings, i.e., $\theta_{child} =2$, $epoch =5$, and $\theta_{splitting} =30, 20$, and 10.

3.1   Visualisation

Figure 3 shows the tree structure of the FMEA spreadsheet from [1], with $\theta_{splitting}$ = 30. The nodes are indexed with l, e.g., $N_{1,0}$ is indexed as 1. $N_{1,0}$ has two child nodes, i.e., $N_{2,1}$ and $N_{3,1}$, and are indexed as 2 and 3, respectively.

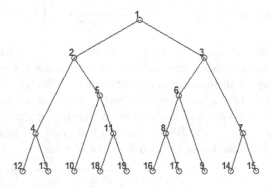

Fig. 3. Tree model for $\theta_{splitting}$ =30

Comparing with the FMEA spreadsheet, which is lengthy (i.e., 11 pages), the control actions can also be represented as a tree structure. This tree structure allows the control actions to be visualised and understood quickly, as compared with the raw control actions in the FMEA spreadsheet. Besides that, through the tree structure, a large number of control actions can be analyzed and assessed at one glance. As an example, an S, O, and D combination of [1 3 9], is considered, and $N_{12,4}$ with weight vector of [1.3711 3.8228 9.0001] is retrieved. Among the related control actions are those with S, O, and D combinations of [1 3 9], [1 4 9], [2 3 9], [4 3 9], and [1 5 8]. Such approach is useful as an effective and efficient means to understand the control actions of FMEA can be realised. In addition, it can be seen that $N_{12,4}$ is close to $N_{13,4}$(with a weight vector of [1.0015 2.0012 9.9992]), and share the same parent node, i.e., $N_{4,2}$ (with a weight vector of [1.3036 3.0707 9.4564]), whereby the control actions associated with $N_{12,4}$ and $N_{13,4}$ are part of $N_{4,2}$.

Table 1 shows a comparison of the resulting tree models with $\theta_{splitting}$ =30, 20, and 10. As expected, the smaller the $\theta_{splitting}$ setting, the more complex the tree becomes.

Table 1. Comparison of the tree model at $\theta_{splitting}$ =10, 20, and 30

$\theta_{splitting}$	Tree size	Tree depth	Number of leaf nodes
30	19	5	10
20	27	6	14
10	57	9	29

3.2   Discussion

Figure 4 depicts a three-dimensional plot of the control actions in terms of their s, o, and d scores. Identical control actions are labelled and indexed with l. $N_{12,4}$ is associated with the control actions labelled with 12, while $N_{13,4}$ is associated with control actions labelled with 13, and both are highlighted in a dotted circle. Note that $N_{12,4}$ and $N_{13,4}$ are located close to each other, and they share the same parent node. Visualisation in a three-dimensional plot can be confusing, as shown in Figure 4. However, the control actions can be summarized and visualised as a tree structure conveniently, as shown in Figure.3.

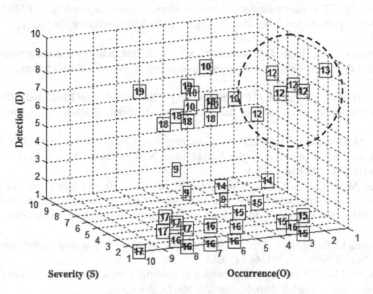

Fig. 4. The failure modes in the S, O, and D space

4   Concluding Remarks

The usefulness of ETree for analysing the control actions of FMEA has been demonstrated with a set of benchmark information. With such approach, the control actions are clustered and visualised as a tree structure effectively. The tree structure allows the control actions to be understood quickly as an entirety, as compared with the   raw control actions in the FMEA spreadsheet. Besides that, a large number of control actions can be analyzed and assessed easily through the tree structure.

For future works, application of the proposed approach to other domains, e.g., agriculture [5-6] and semiconductor [9], will be studied. In addition, the use of other visualisation tools, e.g., self-organizing maps [10], self-organizing trees [11] and etc., for visualisation of failure modes or control actions, will be investigated.

Acknowledgements. The financial support of the FRGS grant (i.e., FRGS/1/2013/ICT02/UNIMAS/02/1), ERGS grant (i.e., ERGS/02(01)/807/2011(02)) and RACE grant (i.e., RACE/F2/TK/UNIMAS/5) is gratefully acknowledged.

References

1. Stamatis, D.H.: Failure Mode and Effect Analysis: FMEA from Theory to Execution. ASQ Press (2003)
2. Liu, H.C., Liu, L., Liu, N.: Risk evaluation approaches in failure mode and effects analysis: A literature review. Expert Syst. Appl. 40, 828–838 (2013)
3. Signor, M.C.: The failure-analysis matrix; a kinder, gentler alternative to FMEA for information systems. In: Proc. Ann. Reliability & Maintainability Symp., pp. 173–177 (1996)
4. Montgomery, T.A., Pugh, D.R., Leedham, S.T., Twitchett, S.R.: FMEA automation for the complete design process. In: Proc. Ann. Reliability & Maintainability Symp., pp. 30–36 (1996)
5. Jong, C.H., Tay, K.M., Lim, C.P.: Application of the fuzzy Failure Mode and Effect Analysis methodology to edible bird nest processing. Comput. Electron. Agric. 96, 90–108 (2013)
6. Tay, K.M., Jong, C.H., Lim, C.P.: A Clustering-based Failure Mode and Effect Analysis Model and Its Application to the Edible Bird Nest Industry. Neural Comput. Appl., doi:10.1007/s00521-014-1647-4
7. Feather, M.S., Cornford, S.L., Kiper, J.D., Menzies, T.: Experiences using visualisation technique to present requirements, risk to them, and options for risk mitigation. In: REV 2006, First International Workshop on Requirement Engineering Visualisation, p. 10 (2006)
8. Pakkanen, J., Iivarinen, J., Oja, E.: The evolving tree—analysis and applications. IEEE Trans. Neural Netw. 17, 591–603 (2006)
9. Tay, K.M., Lim, C.P.: Fuzzy FMEA with a guided rules reduction system for prioritization of failures. Int. J. Qual. Reliab. Manag. 23, 1047–1066 (2006)
10. Kohonen, T.: Self-organizing maps, vol. 30. Springer (2001)
11. Campos, M.M., Carpenter, G.A.: S-TREE: self-organizing trees for data clustering and on-line vector quantization. Neural Networks 14, 505–525 (2001)

An Application of Fuzzy Adaptive Resonance Theory to Engineering Education

See Hung Lau[1], Kai Meng Tay[2], and Chee Khoon Ng[3]

Faculty of Engineering, Universiti Malaysia Sarawak,
94300 Kota Samarahan, Sarawak, Malaysia
lauseehung@gmail.com, {kmtay,ckng}@feng.unimas.my

Abstract. In this paper, the use of Fuzzy Adaptive Resonance Theory (ART) in education data mining is demonstrated. Criterion-referenced assessment (CRA) attempts to determine students' score by comparing their achievements with a clearly stated criterion for learning outcomes. Scoring rubrics are usually used in CRA. The aim of this paper is on the use of Fuzzy ART to group students with scores from CRA, via scoring rubrics. Such approach is useful to assist instructors to establish a personalized learning system, to promote effective group learning, and to provide adaptive contents, for engineering education. In this paper, the applicability of Fuzzy ART-based approach is demonstrated with a real case study relating laboratory project assessment in Universiti Malaysia Sarawak, with positive results obtained. This paper contributes to a new application of an incremental learning neural network with no prefixed number of clusters required, i.e., Fuzzy ART, to engineering education.

Keywords: Criterion-referenced Assessment (CRA), Engineering Education, Fuzzy Adaptive Resonance Theory (ART), Grouping Students.

1 Introduction

The use of data mining techniques in education (i.e., traditional classroom and distance education) is not new [1,2]. Data mining techniques can be used to discover valuable information that is useful in formative evaluation to assist educators establish a pedagogical basis for decision when designing or modifying an environment or teaching approach [2]. Romero & Ventura (2010) [2] reviewed recent works relating to the use of data mining techniques (e.g., clustering, classification, pattern matching, regression and etc.) in education and classified these works to eleven categories. The focus of this paper is on one of these categories, i.e., grouping students.

The category of works aims to group students according to their customized features, for building a personalized learning system and promoting effective group learning [2]. It attempts to create groups of students in such a way that students in the same cluster share some similarities than those in other groups [3] (i.e., according to their customized features and personal characteristics). The clusters/groups of students obtained can then be used by instructors to build a personalized learning system, to promote effective group learning and to provide adaptive contents to the

C.K. Loo et al. (Eds.): ICONIP 2014, Part III, LNCS 8836, pp. 423–430, 2014.

syllabus. Among the popular tools for grouping students are classification and clustering tools. Among the clustering techniques are: (i) hierarchical agglomerative clustering, (ii) K-means and (iii) model-based clustering. A search in literature reveals that the use of Fuzzy Adaptive Resonance Theory (ART) in education is new.

With respect to assessment in education, its purposes are three folds: (i) it derives a standards-referenced judgment; (ii) it supports selection decision, potentially determine the next teaching and learning strategies; and (iii) it intends to ensure that the students remain motivated [4,5]. It is also worth mentioning that assessment in education could be classified into two: (i) norm-reference assessment (i.e., a measure of performance that is interpretable in terms of an individual's relative position held in some known group); and (ii) criterion-referenced assessment (CRA) (i.e., a measure of performance that is interpretable in terms of a clearly defined and delimited domain of learning task). In CRA, scores are given to students by comparing their achievements with a clearly stated criterion for learning outcomes and the standards for particular levels of performance are clearly stated [6]. Scoring rubrics are usually used in CRA. It is worth mentioning that the use of fuzzy inference system in CRA has been reported in our previous work [7].

In this paper, the focus is on the use of Fuzzy ART in engineering education, for grouping students. Scoring rubrics are used. A real case study related to laboratory project assessment for electronic engineering program in Universiti Malaysia Sarawak (UNIMAS) is considered. Students are requested to complete a series of learning tasks, and scores are given to each of the learning tasks, by instructors. Instead of giving total score, which is an aggregation of scores from learning tasks [7], students are grouped using Fuzzy ART. Such approach is important: (i) to build a personalized learning system; (ii) to promote effective group learning; (iii) to provide adaptive content that may contribute to overall learning outcomes. In this paper, Fuzzy ART is chosen due to: (i) its incremental learning features; (ii) no prefixed number of clusters required, compared with most of the approaches in [2]; and (iii) its simple architecture and algorithm which can be explained to most engineering instructors.

To evaluate the proposed method, a real-world data and information from laboratory project [7] is used. The experimental results are discussed and analyzed. This paper contributes to a new application of Fuzzy ART to engineering education. This paper is organized as follows. In Section 2, the background of the case study and the proposed method are described. In Section 3, the experimental results are presented and discussed. Finally, concluding remarks are provided in Section 4.

2   Proposed Methodology

2.1   Background of the Case Study and the Scoring Rubrics

A case study from [7] is considered. Students are required to perform three test items: (i) to design an electronics system based on the knowledge learnt from their digital system subject, and their creativity and technical skills (i.e., *System Design*, (Dg)); (ii) to develop the system either using printed circuit board or on breadboard (i.e., *System Development*, (Dv)); and (iii) to present and demonstrate their works (i.e., *Oral Presentation*, (*Pr*)).   Tables 1 and 2 show the scoring rubrics used for the *System Design* and *System Development*, respectively [7].

Table 1. Scoring rubric for *System Design*, Dg

Rank	Linguistic Terms	Criteria
10	Excellent	The circuit is complex ($\geq$ 10 necessary ICs). All necessary components are included. Able to apply all learned knowledge in circuit design. Able to simulate and clearly explain the operation of designed circuit.
9-8	Very good	The circuit is moderate (7-9 necessary ICs). Some components are not included. Able to apply most of the learned knowledge. Able to simulate and clearly explain the operation of the circuit.
7-6	Good	The circuit is moderate (5-6 necessary ICs). Some unnecessary components are included. Able to apply most of the learned knowledge. Able to simulate the circuit and briefly explain circuit operation.
5-3	Satisfactory	The circuit is simple (3-4 necessary ICs). Some unnecessary components are included. Apply moderate of the learned knowledge. Simulate only parts of circuit and briefly explain the circuit operation.
2-1	Unsatisfactory	The circuit is simple (1-2 necessary ICs). Some components are not included and unnecessary components are added. Only apply some of the learned knowledge. Unable to simulate and explain the operation of designed circuit.

Table 2. Scoring rubric for *System Development*, Dv

Rank	Linguistic Terms	Criteria
10-9	Excellent	PCB: Demonstrated excellent solder techniques (No cold solder joints, no bridge joints and all components leads were soldered to the pad). Components are installed on the PCB correctly. Circuit fully operated as expected. Project board: All the components, jumpers and cables are well-arranged and tidy. Circuit fully operated as expected.
8-7	Very good	PCB: Demonstrated good solder techniques (Some cold solder and bridge joints, some components leads were not soldered to the pad). Components are installed on the PCB correctly. Circuit operated as expected. Project board: Most of the components, jumpers and cables are well-arranged and tidy. Circuit operated as expected.
6-5	Good	PCB: Demonstrated good solder techniques. (Some cold solder and bridge joints, some components lead were not soldered to the pad). Some components are not installed correctly. Some parts of circuit malfunction. Project board: The components are well-arranged but jumpers and cables are messy. Some parts of the circuit malfunction.

Table 2. (*Continued*)

4-3	Satisfactory	PCB: Demonstrated poor solder techniques (Many cold solder and bridge joints and many components leads were not soldered to the pad). Some components are not installed correctly. Most parts of circuit not function. Project board: The arrangement of components, jumpers and cables are messy. Most parts of the circuit malfunction.
2-1	Unsatisfactory	PCB: Demonstrated poor solder techniques. (Many cold solder and bridge joints and many components leads were not soldered to the pad). Most of the components are not installed correctly. The circuit totally not functions. Project board: The arrangement of components, jumpers and cables are very messy. The circuit totally not functions.

2.2   The Proposed Fuzzy ART-Based Approach

Fuzzy ART is adopted for grouping the students according to their scores. Set of scores given to the student, $\# k$, are denoted as $\bar{x}_k = \left[Dg_k, Dv_k, \Pr_k \right]$, where $k = 1, 2, \cdots, m$. The architecture of Fuzzy ART is illustrated in Fig. 1. In layer 1, there are six nodes, i.e., Dg_{nor}, Dv_{nor}, Pr_{nor}, Dg_{nor}^c, Dv_{nor}^c and Pr_{nor}^c. In layer 2, there are s cluster prototypes, $s > 0$, and s can be increased over the time depending on the availability of the data samples. Each cluster prototype is labeled as C_z, where $z = 1, 2, 3, .., s$. The weight connecting C_z and x_{nor} is denoted as $w_{x,nor,z}$. For example, the weight connecting C_1 and Dg_{nor} is denoted as $w_{Dg_{nor,1}}$, and the one connecting C_s and Pr_{nor}^c is denoted as $w_{Pr_{nor,s}^c}$. All the weights are contained in a matrix, $w_{x,nor,z}$. Each component of $w_{x,nor,z}$ is labeled as $w_{x,nor,z}(v)$, where $v = 1, 2, 3, ..., 6$.

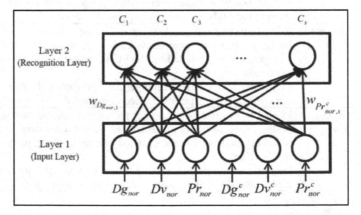

Fig. 1. Fuzzy ART architecture, in which Layer 2 is an incremental layer

The Fuzzy ART algorithm [8] is used to group the students' score into several clusters. The stepwise explanation of the Fuzzy ART for students' score can be explained as follow:

Step 1 – Normalization: Each input Dg_k, Dv_k, Pr_k is normalized by Eq. (1) where $\underline{x}$ is the minimum score (i.e., 1) and $\bar{x}$ is the maximum score (i.e, 10) considered. The normalized input, $\bar{x}_k$ is denoted as Eq. (2)

$$x_{k,nor} = \frac{x_k - \underline{x}}{\bar{x} - \underline{x}}, \text{ where } x \in \left[Dg, Dv, Pr\right] \tag{1}$$

$$\bar{x}_{k,nor} = \left[Dg_{k,nor}, Dv_{k,nor}, Pr_{k,nor}\right], \text{ where } x \in \left[Dg, Dv, Pr\right] \tag{2}$$

Step 2 – Perform complement coding: The complement of $Dg_{k,nor}, Dv_{k,nor}, Pr_{k,nor}$, i.e., $Dg_{k,nor}^c$, $Dv_{k,nor}^c$ and $Pr_{k,nor}^c$ is computed using Eq. (3).

$$\bar{x}_{k,nor} = 1 - x_{k,nor} \text{ where } x \in \left[Dg, Dv, Pr\right] \tag{3}$$

Form the complement-coded score of $\bar{x}_{k,nor}$, i.e., $\bar{x}_{k,nor}^c = \left[Dg_{k,nor}, Dv_{k,nor}, Pr_{k,nor}, Dg_{k,nor}^c, Dv_{k,nor}^c, Pr_{k,nor}^c\right]$.

Step 3 – Parameter setting: Set the value for the choice (α), vigilance (ρ) and learning rate (β). The vigilance threshold ρ (i.e., $0 < \rho < 1$) is responsible for the number of categories in regulating the granularity of the cluster structures formed. Choice parameter α is effective in category selection while learning rate β controls the pace of categorization [8]. In this paper, α = 0.000001 and β = 1 (i.e., fast learning) are adopted while varying ρ with its effect examined.

Step 4 – Initialization: Initial weights are taken as 1 and the number of cluster is set to 1 (i.e., $s = 1$).

Step 5 – Category choice, test and search: Each input is transmitted from layer 1 to layer 2. The response of each layer is determined using the *choice function* (Eq. 4). The node that has the highest response, denoted as node $J \in \left(1, 2, 3, .., s\right)$, is selected as the winning node (Eq. (5)). If there is a tie on $T_{k,z}$, the node with the smallest index is chosen.

$$T_{k,z} = \frac{\sum_{v=1}^{v=6}\left(\bar{x}_{k,nor}^c(v) \wedge w_{xnor,z}(v)\right)}{\alpha + \sum_{v=1}^{v=6} w_{xnor,z}(v)} \tag{4}$$

where " $\wedge$ " is fuzzy AND operator and $(x \wedge y) = \min(x, y)$

$$T_J = \max\left(T_{k,z} : z = 1,2,3,...s\right) \tag{5}$$

The winning node J propagates its weight vector back to layer 1. A vigilance test (Eq. (6)) is performed to measure the similarity against the vigilance threshold between the transformed category prototype and the input vector.

$$M\left(T_J\right) = \frac{\sum\limits_{v=1}^{v=6}\left(\overline{x}_{k,nor}^{c}(v) \wedge w_{xnor,z}(v)\right)}{\sum\limits_{v=1}^{v=6}\left(\overline{x}_{k,nor}^{c}(v)\right)} \tag{6}$$

If $M\left(T_J\right) \geq \rho$ then T_J is passing the test (i.e., resonance is said to occur) and learning will takes place. However, if $M\left(T_J\right) \leq \rho$ then T_J is not passing the test (i.e., resonance is not occurring). It prohibits node J from participating in the subsequent competitions. Input $\overline{x}_{k,nor}^{c}$ is retransmitted to layer 2 to search for new winner. The process is repeated, consecutively disabling nodes in layer 2, until either an existing winning node is able to pass the vigilance test, or, if no such node is available, a new node is created to encode the input vector.

Step 6 – Learning: Once the searching process ends, the learning takes place by adjusting $w_{xnor,j}$ using Eq. (7).

$$w_{xnor,j,new}(v) = \beta\left(\overline{x}_{k,nor}^{c}(v) \wedge w_{xnor,j,old}(v)\right) + \left(1 - \beta\right)w_{xnor,j,old}(v) \tag{7}$$

where $v = 1,2,3,...,6$.

Step 7 – Repeat: The algorithm continues with the next input at step 4. Stop of all data is allocated to s different categories.

Step 8 – Prioritization of cluster: Acquired students' score should be prioritized using similarity measure. The highest similarity measure indicates the highest priority i.e., winning cluster. Arithmetic mean of the input values in winning cluster is computed.

3   Results and Discussion

Table 3 summarizes the assessment results from Fuzzy ART based CRA model. The column " k " shows the label of each student's project. Columns " Dg ", " Dv " and " Pr " list the scores for each test item respectively. Column " C_k " is the clustering results using Fuzzy ART with three different vigilance parameter setting (i.e., $\rho = 0.75$, 0.85 and 0.95. As an example, student with $k = 1$ was awarded Dg , Dv and

Pr scores 4, 4, and 6, respectively. With Fuzzy ART, the student belongs to the first cluster for ρ = 0.75, 0.85 and 0.95.

Table 3. Grouping of students

k	Score of each task			C_k		
	Dg	Dv	Pr	0.75	0.85	0.95
1	4	4	6	1	1	1
2	5	4	6	1	1	1
3	5	4	7	1	1	1
4	7	4	6	1	1	2
5	5	5	7	1	3	4
6	6	8	5	2	2	3
7	5	7	7	2	3	4
8	7	6	7	3	4	5
9	8	6	6	3	4	5
10	7	7	6	3	4	6
11	7	7	8	3	4	6
12	7	9	8	3	5	7
13	8	8	10	4	5	8
14	10	8	8	4	6	9
15	10	9	8	4	6	9

For ρ = 0.75, there are four clusters as shown in Table 3. Students k = 1 to 5 belong to the first cluster; students k = 6 and 7 belong to the second cluster; students k = 8 and 12 belong to the third cluster; and students k = 13 to 15 belong to the fourth cluster. Based on the clustering results, students belonging to the first cluster need intensive guidance in all three aspects i.e., designing of electronic system, development of electronic and presentation skills. Students in the fourth cluster on the other hand need very minimal guidance. Such clustering outcome is useful for designing a personalized learning system with adaptive contents. It is viewed as an useful technique for formative assessment [9].

4   Concluding Remarks

In this paper, the use of Fuzzy ART in engineering education is demonstrated. The proposed approach is able to group students according to their learning profile. Such approach provides useful information to instructors for designing a personalized learning system with adaptive contents. The usefulness of the proposed approach is evaluated with a real-world data from UNIMAS.

As future work, the usefulness of the proposed tool to facilitate formative assessment will be investigated. Besides, the use of advanced visualization tools to visualize students considering their scores will be conducted.

References

[1] Baker, R., Yacef, K.: The state of educational data mining in 2009: A review and future visions. J. Educ. Data Mining 1(1), 3–17 (2009)

[2] Romero, C., Ventura, S.: Educational Data Mining : A review of the state of the art. IEEE Transactions on System, Man, and Cybernetics - Part C: Applications and Reviews 40(6), 601–618 (2010)

[3] Rui, X., Donald, C.W.I.: Clustering. John Wiley & Sons, Inc., Hoboken (2009)

[4] Newton, P.E.: Clarifying the purposes of educational assessment. Assessment in Education 14(2), 149–170 (2007)

[5] Miller, M.D., Linn, R.L., Gronlund, N.E.: Measurement and assessment in teaching, 9th edn. Pearson (1999)

[6] Sadler, D.R.: Interpretations of criteria-based assessment and grading in higher education. Assessment & Evaluation in Higher Education 30(2), 175–194 (2005)

[7] Jee, T.L., Tay, K.M., Ng, C.K.: A new fuzzy criterion-referenced assessment with a fuzzy rule selection technique and a monotonicity-preserving similarity reasoning scheme. Journal of Intelligent & Fuzzy Systems 24, 261–279 (2013)

[8] Carpenter, G.A., Grossberg, S., Rosen, D.B.: Fuzzy ART: Fast stable learning and categorization of analog patterns by an adaptive resonance system. Neural Networks 4, 759–771 (1991)

[9] Stiggins, R.: From formative assessment to assessment FOR learning. Phi Delta Kappan 87(4), 324–328 (2005)

Augmented Query Strategies for Active Learning in Stream Data Mining

Mustafa Amir Faisal[1], Zeyar Aung[2,*], Wei Lee Woon[2], and Davor Svetinovic[2]

[1] Department of Computer Science, University of Texas at Dallas, Richardson,
TX 75080, USA
[2] Institute Center for Smart and Sustainable Systems (iSmart), Masdar Institute of
Science and Technology, Abu Dhabi 54224, UAE
mustafa.faisal@utdallas.edu, {zaung,wwoon,dsvetinovic}@masdar.ac.ae

Abstract. Active learning is used in situations where the amount of
unlabeled data is abundant but it is costly to manually label the data.
So, depending on our available budget, from all unlabeled instances we
are to select only a subset of them to ask the oracle for manual labeling.
Thus, the query strategy, i.e., how relevant instances are selected to be
sent to the oracle, plays an important role in active learning. Though
active learning is a very established research area, only a few research
works have been done on it in the context of stream data mining. Active learning for stream data is more challenging than for static data
because the repetition of queries is not feasible as revisiting the data is
almost impossible. In this paper, we propose two augmented query strategies for active learning in stream data mining, namely, Margin Sampling
with Variable Uncertainty (MSVU) and Entropy Sampling with Uncertainty using Randomization (ESUR). These two strategies are derived
and improved from the existing methods of Variable Uncertainty (VU)
and Uncertainty using Randomization (UR) respectively. We evaluate
the effectiveness of our proposed MSVU and ESUR strategies by comparing them against the original VU and UR on 6 different datasets
using two base classifiers: Leveraging Bagging (LB) and Single Classifier
Drift (SCD). Experimental results show that our proposed strategies offer promising outcomes for various datasets and detecting concept drift
in the data.

Keywords: Stream data mining, Active learning, Query strategy.

1 Introduction

Active learning has been a popular area of research since the 1990s. It is very
useful in machine learning applications in which the amount of unlabeled data is
abundant but manually labeling the data is costly. An example is spam filtering
where it would be very difficult to manually label all instances in a training set.
Thus, we need to select right instances to ask the oracle for manual labeling
because of the limited labeling "budget". In the case of a large mail server only
a relatively small subset, say hundreds to thousands (from amongst millions of

---

* Corresponding author.

C.K. Loo et al. (Eds.): ICONIP 2014, Part III, LNCS 8836, pp. 431–438, 2014.

available emails) need to be manually labeled (it is assumed that our oracle, i.e., the manual labeling process, is not noisy and always provides correct labels.) Then, these selected instances and their manually assigned labels can be used to build an automatic classifier. The main motivation of active learning is to use very small amounts of manually labeled data to train classifiers while at the same time keeping accuracy high. Therefore, the query strategy, i.e., the procedure for selecting instances to be sent to the oracle, plays an important role in active learning.

Though active learning is a very established area of research, very little research have been done on it in the context of stream data mining. Active learning in stream data mining imposes more challenges than active learning in static data mining. In stream mining, repetition of a query is not feasible as revisiting data is almost impossible. Moreover, historical data cannot be stored because of limited functional memory. Hence, with a limited amount of labeled data, maintaining high accuracy is a crucial challenge.

In this paper, the focus is on active learning query strategies with stream data. Instead of pool-based sampling we assume that data cannot be buffered and a decision should be made for each data instance. Thus, we are interested in stream-based selective sampling using different query strategies. The motivation for this is to use active learning for stream data mining in small devices like active RFID tags [1], wireless sensor nodes [2], and smart meters [3], etc. in which the amount of memory is very limited. In addition, we also consider evolving nature of data where concept drift can happen.

The main objective of this research is to investigate the existing query strategies and to develop new ones that outperform the existing approaches — particularly in the context of stream data mining. Consequently, our main contributions in this paper are that: (1) We have proposed two new query strategies for stream data mining, namely MSVU and ESUR, by enhancing the existing state-of-the-art ones, namely VU and UR respectively. We have shown that the proposed MSVU and ESUR strategies outperform their original counterparts in a majority of test cases. (2) We have made some important observations regarding various query strategies that their performances vary greatly depending on the base classifier or the change detection technique used.

2   Relevant Background

Research on active learning with stream data is comparatively new while more work has been done with static data. Only a brief overview is provided in this section. For more detailed information, readers are referred to [4], which provides a comprehensive survey of existing query strategies for active learning in both static and stream settings.

The Random Strategy is a very basic one in which the learner selects random instances which are then presented to the oracle for labeling [5]. Every incoming instance is presented to the oracle with probability β, which is the pre-defined budget.

The Uncertainty Sampling Strategy is a general strategy first introduced by [6] where the learner asks the oracle about the instances about which it is the least certain. There are a number of subcategories of this strategy, which are described below.

2.1   Least Confidence

- Fixed Uncertainty: In this strategy for online learning, as described in [5], instances for which the certainty is below a fixed threshold are flagged for labeling, where certainty is based on the posterior probability estimates provided by a classifier or learner. However, this is not practical in stream data mining, where data evolves quickly and with constant concept drift.
- Variable Uncertainty (VU): To overcome the limitations of the fixed uncertainty strategy, Žliobaitė et al. [5] introduced a variable threshold which adapts to the changing characteristics of the data.
- Uncertainty using Randomization (UR): In data stream variation or concept drift can occur anywhere in the input space. However, uncertainty strategy labels the instances that are near to the decision boundary. To mitigate this problem, the labeling threshold is randomized by multiplying by a normally distributed random variable that is within $\mathcal{N}(1, \sigma^2)$ [5].

In Section 3, we propose two augmented strategies based on VU and UR respectively and compare the performances of the augmented strategies with those of the original ones.

2.2   Margin Sampling

In the Least Confidence strategy, the most possible label is considered. This may lead to information about the remaining label distribution being ignored. An attempt has been made to correct this shortcoming using a different multiclass uncertainty sampling variant named Margin Sampling [7]. The definition of margin sampling is $X_C = \arg\min_{D_t} P_C(\hat{y}_1|D_t) - P_C(\hat{y}_2|D_t)$, where $\hat{y}_1$ and $\hat{y}_2$ are the first and second most possible class labels respectively under model C, and D_t is an incoming instance at time t.

In this paper, we amalgamate the idea of margin sampling and the Variable Uncertainty (VU) strategy [5] to come up with a better method of Margin Sampling with Variable Uncertainty (MSVU) as described below in Section 3.

2.3   Entropy

Entropy [8], an information-theoretic measure, is an uncertainty gauge presenting the amount of information required to encode a distribution. The definition of entropy is: $X_H^* = \arg\max_x \left(-\sum_i P_C(y_i|x) \log P_C(y_i|x)\right)$, where y_i ranges over all possible labelings under model C. Entropy is usually regarded as a measure of uncertainty or impurity in machine learning. The entropy-based approach generalizes well to probabilistic multi-label classifiers and probabilistic models for more complex data like sequences [4].

In this paper, we enhance the Uncertainty using Randomization (UR) strategy [5] by incorporating sequence entropy in order to develop a better strategy named

Entropy Sampling with Uncertainty using Randomization (ESUR) as descried below in Section 3.

3 Proposed Query Strategies

Let us consider $D_1, D_2, D_3, \ldots, D_t, \ldots$ as a data stream, where D_t as an instance at time t. The budget β indicates that in incoming data, $\beta\%$ of data are expected to be labeled by the oracle. The assumption is that labeling cost is same for every instance. Each query strategy takes an instance D_t, budget β, and other necessary parameters and decides whether to ask for labeling or not. After getting the label, the classifier is trained with this instance until budget, β is not exhausted.

In [5], the authors proposed two basic strategies with least confidence for uncertainty sampling, namely, Variable Uncertainty (VU) and Uncertainty using Randomization (UR). In our work, we augment those strategies by incorporating the ideas of Margin Sampling [7] to VU and Entropy Sampling [4] to UR respectively. This results in two new query strategies, namely *Margin Sampling with Variable Uncertainty* (MSVU) and *Entropy Sampling with Uncertainty using Randomization* (ESUR), which perform better particularly in steam mining context. These two augmented strategies are described in detail below.

3.1 Margin Sampling with Variable Uncertainty (MSVU)

The MSVU algorithm is presented below. The main difference with VU is in lines 3 and 4. In line 3, we calculate the minimum margin for two promising class labels determined by the classifier. And in line 4, this minimum margin is compared with the threshold θ.

These modifications help improve the performance of the algorithm over the original VU. The reason is that while VU considers the most possible label ignoring the information about the remaining label distribution, MSVU tackles this shortcoming by considering the difference between the two most possible labels or classes by the model.

```
Algorithm MSVU (D_t, C, β, a)
Input:   (1) Incoming instance, D_t
Input:   (2) Trained classifier, C
Input:   (3) Budget, β
Input:   (4) Adjusting step, a
Output: (1) label ∈ {true, false} implies whether to ask the true label y_t
Initialization: Total labeling cost u = 0, initial labeling threshold, θ = 1.0
1.        if (u/t < β)
2.            then budget is not exceeded,
3.                X_C = arg min_{D_t} P_C(ŷ_1|D_t) − P_C(ŷ_2|D_t) where ŷ_1 and ŷ_2 are the
                      first and second possible class labels respectively under model, C
4.                if(X_C < θ)
5.                    then margin difference is below the threshold
6.                        u = u + 1 labeling costs increase,
7.                        θ = θ(1 − a) the threshold decreases,
8.                        return true
9.                    else margin region is wider
10.                       θ = θ(1 + a) make the threshold wider,
11.                       return false
12.           else budget is exceeded
13.               return false
```

3.2 Entropy Sampling with Uncertainty using Randomization (ESUR)

ESUR is a variant of entropy sampling in stream data mining context. It utilizes **sequence entropy** (SE) [4], which is defined as:

$$\Gamma^{SE}(x) = -\sum_{\hat{y}} P(\hat{y}|x : C) \log P(\hat{y}|x : C) \tag{1}$$

where $\hat{y}$ ranges over all possible label sequences for input sequence x under model C.

An algorithmic description of ESUR is presented below. ESUR is very similar to the original UR except in lines 3–5. Using Equation 1 in line 3, sequence entropy is calculated. Sequence entropy is multiplied by a random multiplier in line 4 and this resultant value is compared with threshold value in line 5. If current threshold is big enough than randomized entropy, the strategy would ask for labeling the instance. The rest of the strategy is same as UR.

These enhancements help improve the algorithm's performance over the original UR because ESUR considers the disturbance in data and also provides a more balanced coverage of the whole input space.

Algorithm **ESUR** (D_t, C, β)
Input: (1) Incoming instance, D_t
Input: (2) Trained classifier, C
Input: (3) Budget, β
Output: (1) $label \in \{\textbf{true}, \textbf{false}\}$ implies whether to ask the true label y_t
Initialization: Total labeling cost $u = 0$, initial labeling threshold, $\theta = 1$.
1.   if $(u/t < \beta)$
2.   **then** budget is not exceeded,
3. $\Gamma^{SE}(x) =$ Compute entropy using equation
4. $\Gamma^{SE}(x)_{randomized} = \Gamma^{SE}(x) \times \eta,$
                    where $\eta \in \mathcal{N}(1, \sigma^2)$ is a random multiplier
5.   **if**$(\Gamma^{SE}(x)_{randomized} < \theta)$
6.   **then** entropy is less than threshold
7. $u = u + 1$ labeling cost increases,
8. $\theta = \theta(1 - a)$ the threshold decreases,
9.   **return true**
10.   **else** certainty is good
11. $\theta = \theta(1 + a)$ make the uncertainty region wider
12.   **return false**
13.   **else** budget is exceeded
14.   **return false**

4   Experimental Results

We compare the performances of five query strategies: (1) Variable Uncertainty (VU), (2) Uncertainty with randomization (UR), (3) Random, (4) Margin sampling with variable uncertainty (MSVU), and (5) Entropy sampling with variable uncertainty using randomization (ESUR). Among them, Random strategy is used as a baseline method.

For each of the five strategies, two different classifiers were used: (1) Leveraging Bagging (LB) [9] and (2) Single Classifier Drift (SCD) [10,11]. SCD can be configured either with drift detection method (DDM) [10] or early drift detection method (EDDM) [11]. Since EDDM generally offers better outcomes, we choose to use it in our experiments.

Two sets of experiments were conducted, namely (1) Experiment on 3 prediction datasets and (2) Experiments on 3 textual datasets. All experiments are done in *Massive Online Analysis* (MOA) platform [12]. For each method, only the budget is changed for each testing instance, and default values are used for all the remaining parameters.

4.1 Experiment I: On Prediction Datasets

The three prediction datasets used are: Electricity, Forest, and Airlines [13]. These datasets are also used in the experiments of [5]. Electricity dataset is about predicting a rise or a fall in electricity demands and prices in New South Wales, Australia, provided immediate consumptions and prices in the same and neighboring regions. In Forest dataset, the task is to predict forest cover type from cartographic variables. In Airlines dataset, the task is to predict whether a given flight will be delayed or not by using supplied information of the scheduled departures.

Normal accuracy is used to evaluate the performances on the prediction datasets. The performances 5 query strategies each using 2 classifiers are summarized in Table 1.

For Airlines dataset, the performances of all strategies fluctuate almost between 65% to 50% for both classifiers. Variants of variable uncertainty strategy (VU and MSVU) outperform the variants of randomization strategy (UR and ESUR). In particular, our proposed strategy, MSVU outperforms the other strategies for both classifiers.

In the case of the Electricity dataset, MSVU outperforms other strategies for LB. With SCD as budget increases, UR shows better performance than other variable uncertainty variants. There is an accuracy fluctuation among the strategies for both of the classifiers.

All the strategies show good performance for Forest dataset. After a small budget (around 0.1), all the strategies with LB and SCD achieve just below 100% accuracy and remain stable throughout the budget change.

4.2 Experiment II: On Textual Datasets

The three textual datasets used are IMDB-E, IMDB-D, and Reuters [13]. They are also used in the experiments of [5]. IMDB (Internet Movie Database) dataset is divided into two categories. For IMDB-E (easy), only one category is considered as interesting at a time and for IMDB-D (difficult), five associated categories are interesting at a time. With the purpose of deliberately initiating concept drifts, the authors of [5] introduce three changes after 25, 50, and 75 thousand instances. In Reuters dataset, the first half of the data stream legal or judicial is considered to be relevant and in second half the share listings category was considered to be relevant. For these textual data, the labels were assigned by authors of [5].

Geometric accuracy is used to measure the performances on textual datasets. It is defined as $GA = (A_1 \times A_2 \times \ldots \times A_c)^{\frac{1}{c}}$. Here A_i is the accuracy on class i and c is the number of classes. The performances 5 query strategies each using 2 classifiers are summarized in Table 1.

Table 1. Summary of results on 3 prediction datasets (Experiment I) and 3 textual datasets (Experiment II). Highest scores are highlighted in red for LB and blue for SCD. Our proposed methods (MSVU and ESUR) provide better results than the original VU and UR methods [5] do in 9 out of 12 test cases.

	Prediction Datasets Average Accuracy (%)						Textual Datasets Average Geometric Accuracy (%)					
	Airlines		Electricity		Forest		IMDB-D		IMDB-E		Reuters	
	LB	SCD	LB	SCD	LB	SCD	LB	SCD	LB	SCD	LB	SCD
MSVU	63.77	64.52	76.45	80.62	95.26	96.36	36.61	33.70	46.96	47.80	93.27	64.56
ESUR	59.24	58.3	66.29	73.50	94.55	96.23	45.87	45.71	50.37	52.10	82.99	44.98
Random	61.58	60.11	73.18	79.63	94.67	96.10	43.51	41.56	49.25	51.37	88.70	54.47
VU	63.20	63.74	75.81	80.42	95.24	96.41	36.74	34.41	46.85	48.00	93.21	65.77
UR	61.70	61.89	74.87	78.87	95.35	96.38	42.29	41.14	48.75	50.08	89.24	56.26

The proposed strategies exhibit both high and low accuracies. For both IMDB-D and IMDB-E, ESUR attains the highest accuracy level, while VU as well as MSVU show bad accuracies. However, opposite behavior is observed in the case of Reuters dataset: MSVU achieves the highest geometric accuracy, while ESUR receives the lowest accuracy.

The variants randomization strategy present dominating performances. Among them, ESUR shows highest accuracy in all datasets. At the change in 50 thousand instances, all strategies receive their respective lowest geometric accuracies and the change in 75 thousand instances, there is a rising tendency in accuracy in the case of IMDB-E dataset while a falling tendency in accuracy is shown in the case of IMDB-D dataset at this change.

Both the classifiers, LB and SCD, show almost same behavior. The variants of variable uncertainty strategy outperform the variants of randomization strategy. In the case of LB classifier, VU shows slight better performance than MSVU. On the other hand, for SCD, MSVU shows slightly better result than VU.

5 Conclusion and Future Works

The results of the experiments described here show that ESUR perform well with the remote changes and IMDB datasets. The same is also true for MSVU for close changes as well as Airlines, Electricity, and Reuters datasets. In comparison with VU and UR [5], the proposed augmented MSVU and EUSR strategies outperform them in the majority of cases. Future research directions include designing new strategies which have the ability to tackle both close and remote changes. A more comprehensive study could also be conducted in which a larger number of base classifiers are deployed.

References

1. Bin, S., Yuan, L., Xiaoyi, W.: Research on data mining models for the internet of things. In: Proc. 2010 International Conference on Image Analysis and Signal Processing (IASP), pp. 127–132 (2010)

2. Tripathy, A.K., Adinarayana, J., Merchant, S.N., Desai, U.B., Ninomiya, S., Hirafuji, M., Kiura, T.: Data mining and wireless sensor network for groundnut pest/disease precision protection. In: Proc. 2013 National Conference on Parallel Computing Technologies (PARCOMPTECH), pp. 1–8 (2013)

3. Faisal, M.A., Aung, Z., Williams, J., Sanchez, A.: Data-stream-based intrusion detection system for advanced metering infrastructure in smart grid: A feasibility study. IEEE Systems Journal (in press, 2014)

4. Settles, B., Craven, M.: An analysis of active learning strategies for sequence labeling tasks. In: Proc. 2008 Conference on Empirical Methods on Natural Language Processing (EMNLP), pp. 1070–1079 (2008)

5. Žliobaitė, I.e., Bifet, A., Pfahringer, B., Holmes, G.: Active learning with evolving streaming data. In: Gunopulos, D., Hofmann, T., Malerba, D., Vazirgiannis, M. (eds.) ECML PKDD 2011, Part III. LNCS, vol. 6913, pp. 597–612. Springer, Heidelberg (2011)

6. Lewis, D., Gale, W.: A sequential algorithm for training text classifiers. In: Proc. 17th ACM SIGIR Conference on Research and Development in Information Retrieval (SIGIR), pp. 3–12 (1994)

7. Scheffer, T., Decomain, C., Wrobel, S.: Active Hidden Markov Models for information extraction. In: Proc. 4th International Conference on Advances in Intelligent Data Analysis (IDA), pp. 309–318 (2001)

8. Shannon, C.E.: A mathematical theory of communication. Bell System Technical Journal 27, 379–423 (1948)

9. Bifet, A., Holmes, G., Pfahringer, B.: Leveraging bagging for evolving data streams. In: Balcázar, J.L., Bonchi, F., Gionis, A., Sebag, M. (eds.) ECML PKDD 2010, Part I. LNCS, vol. 6321, pp. 135–150. Springer, Heidelberg (2010)

10. Gama, J., Medas, P., Castillo, G., Rodrigues, P.: Learning with drift detection. In: Proc. 17th Brazilian Symposium on Artificial Intelligence (SBIA), pp. 286–295 (2004)

11. Baena-García, M., Campo-Avila, J.D., Fidalgo, R., Bifet, A., Gavaldà, R., Morales-Bueno, R.: Early drift detection method. In: Proc. 4th International Workshop on Knowledge Discovery from Data Streams (IWKDDS), pp. 77–86 (2006)

12. Bifet, A., et al.: Massive Online Analysis (2012), http://moa.cs.waikato.ac.nz (release March 2012)

13. Bifet, A., Kirkby, R.: MOA (Massive Online Analysis) datastream (2012), http://sourceforge.net/projects/moa-datastream/files/Datasets/Classification/

Feature Selection and Mass Classification Using Particle Swarm Optimization and Support Vector Machine

Man To Wong[1], Xiangjian He[1], Wei-Chang Yeh[2], Zaidah Ibrahim[1,3],
and Yuk Ying Chung[4]

[1] Faculty of Engineering and Information Technology,
University of Technology Sydney, NSW 2007, Australia
[2] Department of Industrial Engineering and Engineering Management,
National Tsing Hua University, Hsinchu 300, Taiwan
[3] Faculty of Computer and Mathematical Sciences,
Universiti Teknologi MARA, Malaysia
[4] School of Information Technologies,
University of Sydney, NSW, Australia
eemtwong@gmail.com,
Xiangjian.He@uts.edu.au

Abstract. This paper proposes an effective technique to classify regions of interests (ROIs) of digitized mammograms into mass and normal breast tissue regions by using particle swarm optimization (PSO) based feature selection and Support Vector Machine (SVM). Twenty-three texture features were derived from the gray level co-occurrence matrix (GLCM) and gray level histogram of each ROI. PSO is used to search for the gamma and C parameters of SVM with RBF kernel which will give the best classification accuracy, using all the 23 features. Using the parameters of SVM found by PSO, PSO based feature selection is used to determine the significant features. Experimental results show that the proposed PSO based feature selection technique can find the significant features that can improve the classification accuracy of SVM. The proposed classification approach using PSO and SVM has better specificity and sensitivity when compared to other mass classification techniques.

Keywords: mass classification, support vector machine, particle swarm optimization, feature selection.

1 Introduction

Breast cancer is the most common cancer of women in America [1]. Mammography is the most effective method for early detection of breast cancers [2]. Masses are important early signs of breast cancer [3]. Mass detection in mammogram is difficult because the features of masses can be obscured and can be similar to normal breast parenchyma [4]. The results from a computer aided detection system can be used as a second opinion to a radiologist and improve the detection accuracy.

Many mass detection algorithms have the following two steps. In the first step, suspicious regions of interest (ROIs) are detected on the mammogram images by

C.K. Loo et al. (Eds.): ICONIP 2014, Part III, LNCS 8836, pp. 439–446, 2014.
© Springer International Publishing Switzerland 2014

using some image processing techniques such as segmentation or thresholding. In the second step, one typical approach is to extract features from the suspicious regions. Classifiers can then be applied on these features to classify the regions as mass or normal tissue. This will reduce the number of false positives. Sahiner et al. [6] used texture features and convolution neural networks in mass classification. He obtained 90% sensitivity and 69% specificity. Tourassi et al. [7] had applied template matching scheme based on the mutual information and obtained 90% sensitivity and 65% specificity. Christoyianni [8] used the GLCM [9] texture features and MLP and obtained 85% sensitivity and 83% specificity. Petrosian et al. [10] used the GLCM texture features and a modified decision tree classifier and obtained 76% sensitivity and 64% specificity. Angelini et al. [11] had tested and compared the performance of different image representations for mass classification. Instead of extracting features from the suspicious regions, the features are embodied by the image representation used to encode the suspicious regions. The best result was given by the pixel image representation, using SVM as classifier, with 90% sensitivity and 94% specificity.

The objective of this paper is to propose a novel feature selection and mass classification technique using SVM and PSO. The regions of interests (ROIs) are manually extracted from the MIAS Mini-Mammographic database [12]. The ROIs can contain mass or normal tissue. The ROIs will be classified as mass or non-mass regions using texture features calculated from the gray level co-occurrence matrix (GLCM) and statistical features from the gray level histogram. A PSO-based feature selection technique is proposed to select a smaller subset of significant features which can provide comparable or even better performance when compared to the full set of features.

2   Feature Selection Using PSO and SVM

Support Vector Machine (SVM) [13] is a classifier that has robust and accurate classification performance in many different applications. SVM finds the best hyperplane that separates the data by maximizing the margin between the hyperplane and the support vectors. The performance of SVM depends on the selection of kernel, the kernel's parameters, and cost parameter C. The RBF kernel is used in this paper. This kernel nonlinearly maps samples into a higher dimensional space and can handle the case when the relation between class labels and attributes is nonlinear. When RBF kernel is used, two parameters have to be properly chosen for good classification performance: the gamma (γ) parameter of the RBF kernel and the C parameter.

In this paper, the SVM software implementation in OpenCV [15] software library is used. The SVM in OpenCV is based on LIBSVM [16]. The C-Support Vector Classification (C-SVC) type and the RBF kernel of LIBSVM are used. According to the recommendation of [14], the feature values are linearly scaled to the range of [0,1]. The parameters C and γ (gamma) of SVM (using RBF kernel) are chosen by using PSO to search for C and gamma (γ) that can provide the best fitness function value of PSO. The fitness function used is the classification accuracy of SVM in the training set, using leave one out (LOO) cross validation.

PSO is a population based stochastic optimization technique modelled after the social behavior of bird flocks [17]. In PSO each particle represents a potential solution to the optimization problem. Initially each particle is assigned a randomized velocity. Then the particles are flown through the problem space [17, 18]. The aim of PSO is to find the particle position with the best fitness function value.

Each particle keeps track of the following information in the problem space: x_i, the current position of the particle; v_i, the current velocity of the particle; and y_i, the personal best position of the particle which is the best position that it has achieved so far. This position yields the best fitness value for that particle. The fitness value of this position, called *pbest*, is also stored. In this paper, the *gbest* model of PSO is used. The best particle is determined from the entire swarm. The overall best value (*gbest*) obtained so far by any particle in the population and its location y_g are also tracked.

The velocity and position of the particle are given by equations (1) and (2) [18].

$$v_i(t+1) = wv_i(t) + c_1 r_1(t)\big(y_i(t) - x_i(t)\big) + c_2 r_2(t)\big(y_g(t) - x_i(t)\big) \qquad (1)$$
$$x_i(t+1) = x_i(t) + v_i(t+1) \qquad (2)$$

where w is the inertia weight, c_1 and c_2 are the acceleration constants, and $r_1(t)$ and $r_2(t)$ are random numbers generated in the range between 0 and 1.

Before feature selection, the parameters C and gamma (γ) of SVM, using the RBF kernel, are chosen by using PSO to search for values of C and γ that can provide the best fitness function value, using all the available features. The classification accuracy of SVM is used as the fitness function for PSO. In the training set, leave-one-out (LOO) cross validation is used. The LOO cross validation is especially suitable for small training set as it can maximize the use of training data. The two values $\log_2 C$ and $\log_2 \gamma$ are searched by PSO within the range from -10 to 10. Hence the actual range of C and γ that can be found in the search is from 2^{-10} to 2^{10}.

The original version of PSO described above operated in continuous space. The binary version of PSO (BPSO) has been developed for discrete problems [19] which can be used in feature selection. The velocity in BPSO represents the probability of an element in the position taking value 1. Equation (1) is used to update the velocity while x_i, y_i and y_g are restricted to 1 or 0. A sigmoid function $s(v_i)$ is used to transform v_i to the range of (0,1). BPSO updates the position of each particle according to the following formulae:

$$x_i = 1 \ if \ rand() < s(v_i), else \ 0 \ ; \quad s(v_i) = \frac{1}{1+e^{-v_i}} \qquad (3)$$

rand( ) is a random number selected from a uniform distribution in [0,1].

In this paper, binary PSO (BPSO) is used to search for the feature subset in the training set. When x_i is 1, the feature corresponding to this bit position will be selected. When x_i is 0, the feature will not be selected. SVM classifier is used to evaluate the feature subset using LOO cross validation. The fitness function used in the proposed BPSO based approach is to maximize classification accuracy.

3   Texture Features

In Gray Level Co-occurrence matrix (GLCM), the texture context information is specified by the matrix of relative frequencies $P(i, j, d, \theta)$ with which two neighboring pixels separated by distance d and along direction θ occur on the image; one pixel with gray level i and the other with gray level j [4,9]. After the number of neighboring pixel pairs R used in computing a particular GLCM matrix is obtained, the matrix is normalized by dividing each entry by R, the normalizing constant [9]. For each ROI, eight texture features were derived from each GLCM [5, 9, 10]. The notation $p(i,j)$ is used to represent the $(i, j)th$ entry in a normalized GLCM matrix and $p(i,j)$ is obtained by dividing each entry of the matrix $P(i, j)$ by R [9]. $\sum_{i,j}$ represents $\sum_{i=0}^{n-1} \sum_{j=0}^{n-1}$ where n is the number of gray levels per pixel.

$$Energy = \sum_{i,j} p(i,j)^2 . \tag{4}$$

$$Inertia = \sum_{i,j} (i-j)^2 p(i,j) . \tag{5}$$

$$Entropy = - \sum_{i,j} p(i,j) \log (p(i,j)) . \tag{6}$$

$$Homogeneity = \sum_{i,j} \frac{1}{1 + (i-j)^2} p(i,j) . \tag{7}$$

$$Max. probability = maximum\ of\ p(i,j) . \tag{8}$$

$$Cluster\ Shade = \sum_{i,j} (i+j - \mu_x - \mu_y)^3 p(i,j) . \tag{9}$$

$$Cluster\ Tendency = \sum_{i,j} (i+j - \mu_x - \mu_y)^2 p(i,j) . \tag{10}$$

$$Correlation = \frac{\sum_{i,j}(i - \mu_x)(j - \mu_y)p(i,j)}{\sigma_x \sigma_y} . \tag{11}$$

where μ_x , μ_y, σ_x and σ_y are the means and standard deviations of the marginal distributions associated with $P(i, j) / R$, and R is the normalizing constant [5, 9, 10].

In finding the GLCM, d is set to 1. Four directions are used for θ : 0, 45, 90 and 135 degrees. Then the average and range of the four values of each feature are calculated. The range is defined as the difference between the maximum and minimum of the four values. Hence a total of sixteen texture features are found for each ROI.

In addition to the GLCM features, seven statistical features are also derived from the gray level histogram of each ROI [8, 20]. The seven features are mean, standard deviation, skew, entropy, smoothness, uniformity and kurtosis [8,20]. The equations for these seven features can be found in [20].

4   Experimental Result and Discussion

4.1   Mammogram Database and Test Method

The MIAS MiniMammographic Database is provided by the Mammographic Image Analysis Society in UK [12]. The mammograms are digitized at 200 micron pixel edge and have a resolution 1024 x 1024. The types of abnormality in the database include calcification, masses, architectural distortion and asymmetry. Mammograms which do not contain any abnormality (classified as normal) are also provided.

One hundred and twenty ROIs were manually extracted from the images in the MIAS database. The approach of extracting ROIs from the mammogram database is based on [11]. In the ground truth file of the MIAS database, the location of the center of the mass (if it exists) is given, together with the radius of circle which completely encloses the mass. A square crop centered on the location of each annotated mass is selected. The size of square crop is chosen so that the ratio between the crop area and the area of the annotated mass is approximately 1.3 .   All the crops containing a mass are then resized to a fixed size of 128 x 128 pixels. The resizing of variable size ROI to a fixed size region has been used in other research paper on mass classification [11]. For the non-mass class (normal tissue), the 128x128 pixel regions are extracted randomly from the normal mammograms. 44 of the 120 ROIs contain mass and 76 of them contain normal tissue only. For ROIs which contain mass, the mass can be benign or malignant. Three types of masses were used in this paper: circumscribed, spiculated and ill-defined masses. For ROIs which contain normal tissue only, the ROIs are randomly chosen inside the breast body.   Five-fold stratified cross validation is used in testing. The 120 ROIs are divided into five equal sets. Four sets are used as a training set and the remaining set as a test set. Hence there are 96 ROIs in the training set and 24 ROIs in the test set. Feature selection by BPSO-SVM is done using the training set only. Then only the significant features obtained from feature selection are used to train the classifier, using the training set only. The trained classifier is then used to classify the test set using the significant features. The above process is repeated by using another set of data as a test set and the other four sets as a training set. Every ROI is used in the test set once only. The average classification accuracy of the five test sets is calculated.

In BPSO-SVM based feature selection, SVM is used to evaluate the feature subset in the training set. The classification accuracy of the feature subset on the training set is evaluated using SVM and LOO cross validation. Once the significant features have been found by the BPSO-SVM technique, only the significant features are used in the training set to train the classifier. Note that 5-fold cross validation is used to calculate the classification accuracy of the SVM on the test set while LOO cross validation is used to evaluate the feature subset found by BPSO-SVM in the training set. The PSO based parameters tuning for SVM and the BPSO-SVM feature selection method were implemented using C++ language and OpenCV software library [15]. The BPSO based feature selection method is compared with other wrapper based feature selection methods which are all available in the WEKA machine learning workbench [13]. The wrapper subset evaluation technique used is SVM. The three different search techniques in WEKA library used to find feature subsets include stepwise forward selection, stepwise backward selection and best first search [13].

4.2   Experimental Result and Discussion

In Table 1, 2 and 3, the values of specificity, sensitivity and overall accuracy are all measured in the test set, using 5-fold cross validation. The notation "BPSO-SVM" refers to the proposed method in this paper. Sensitivity, specificity and overall accuracy are defined as follows [13]:

$$Sensitivity = \frac{TP}{TP + FN} \tag{12}$$

$$Specificity = \frac{TN}{FP + TN} \tag{13}$$

$$Accuracy = \frac{TP + TN}{TP + FP + TN + FN} \tag{14}$$

where TP is the number of true positives, FN is the number of false negatives, TN is the number of true negatives and FP is the number of false positives. In Table 1, the proposed BPSO-SVM feature selection method has the best sensitivity, specificity and overall classification accuracy. In Table 2, except the proposed method, all the other classifiers shown were used to classify the test set without using feature selection. For the MLP, J48 and KNN classifiers, their implementations in the WEKA machine learning software library [13] are used. From Table 2, the proposed method BPSO-SVM gives the highest sensitivity and overall accuracy while its specificity performance is very close to KNN.

Table 1. Comparison of feature selection methods using SVM as classifier

Feature Selection Method	Specificity (%)	Sensitivity(%)	Accuracy (%)
BPSO-SVM	97.33	97.78	97.50
All Features	96.05	88.64	93.33
Stepwise forward search	96.10	85.84	92.50
Stepwise backward search	94.76	88.34	92.50
Best first search	96.10	88.06	93.32

Table 3 compares the performance of the proposed BPSO-SVM method with other existing mammogram mass classification techniques. The specificity and sensitivity of the proposed method in this paper are better than other existing methods.

Table 2. Comparison of classification methods using BPSO-SVM (with feature selection) and other classifiers without feature selection

Classifier	Specificity (%)	Sensitivity(%)	Accuracy (%)
BPSO-SVM + SVM	97.33	97.78	97.50
SVM (all features)	96.05	88.64	93.33
MLP	94.76	83.12	90.82
J48 (decision tree)	89.58	88.34	89.16
KNN (K=3)	97.42	86.40	93.34

Table 3. Comparison of proposed BPSO-SVM based classification and other existing mammogram mass classification techniques

Classification method	Specificity (%)	Sensitivity(%)
BPSO-SVM + SVM	97.33	97.78
Angelini et al. [11]	94.00	90.00
Christoyianni et al. [8]	83.05	86.66
Sahiner et al. [6]	69.00	90.00
Petrosian et al. [10]	64.00	76.00
Tourassi et al. [7]	65.00	90.00

5   Conclusion

The objective of this paper is to demonstrate the good performance of the proposed feature selection and mass classification approach using BPSO and SVM. PSO is used to search for the optimal parameters C and gamma of SVM, using the RBF kernel. Then BPSO-SVM feature selection technique is used to find the significant features in the training set. Finally SVM is used to classify the test set, using the significant features only. The experimental results show that the proposed BPSO-SVM feature selection method can have better result than other widely used feature selection methods when it is applied to mammogram mass classification. By using features from GLCM and gray level histogram, a small number of significant features found by BPSO-SVM can have better performance in classification accuracy than the full set of features in mass classification. Also the proposed mass classification approach has better performance when compared to other existing mass classification techniques. The proposed classification approach using PSO and SVM can achieve 97.78% sensitivity and 97.33% specificity on the test set using 5-fold cross validation.

References

1. Garfinkel, L., Catherind, M., Boring, C., Heath, C.: Change trends: an overview of breast cancer incidence and mortality. Cancer 74(1), 222–227 (1997)
2. Bovis, K., Singh, S., Fieldsend, J., Pinder, C.: Identification of masses in digital mammograms with MLP and RBF nets. In: Proc. of the IEEE-INNS-ENNS International Joint Conference in Neural Networks, pp. 342–347 (2000)

3. Cheng, H., Cai, X., Chen, X., Hu, X., Lou, X.: Computer aided detection and classification of microcalcifications in mammograms:a survey. Pattern Recog. 36, 2967–2991 (2003)
4. Cheng, H., Shi, X., Min, R., Hu, L., Cai, X., Du, H.: Approaches for automated detection and classification of masses in mammograms. Pattern Recog. 39, 646–668 (2006)
5. Eisa, M.M., Ewees, A.A., Refaat, M.M., Elgamal, A.F.: Effective medical image retrieval technique based on texture features. International Journal of Intelligent Computing and Information Science 13(2), 19–33 (2013)
6. Sahiner, B., Chan, H.P., Petrick, N., Wei, D., Helvie, M.A., Adler, D.D., Goodsitt, M.M.: Classification of mass and normal breast tissue: a convolution neural network classifier with spatial domain and texture images. IEEE Trans. Med. Imaging 15, 598–610 (1996)
7. Tourassi, G.D., Vargas-Voracek, R., Catarious, D.M., Floyd, C.E.: Computer-assisted detection of mammographic masses: a template matching scheme based on mutual information. Med. Phys. 30, 2123–2130 (2003)
8. Christoyianni, I., Dermatas, E., Kokkinakis, G.: Neural classification of abnormal tissue in digital mammography using statistical features of the texture. In: Proc. of the 6th IEEE Int'l Conf. on Electronics, Circuits & Systems, vol. 1, pp. 117–120 (1999)
9. Haralick, R.M., Shanmugam, K., Dinstein, I.: Texture features for image classification. IEEE Trans. Syst. Man Cybernet, SMC 3(6), 610–621 (1973)
10. Petrosian, A., Chan, H.P., Helvie, M.A., Goodsitt, M.M., Adler, D.D.: Computer-aided diagnosis in mammography: classification of mass and normal tissue by texture analysis. Physics in Medicine and Biology 39(12), 2273–2288 (1994)
11. Angelini, E., Campanini, R., Iampieri, E., Lanconelli, N., Masotti, M.: Testing the performances of Different Image Representations for Mass Classification in Digital mammograms. Int'l Journal of Modern Phys. C 17(1), 113–131 (2006)
12. The Mini-MIAS Database of Mammograms, http://peipa.essex.ac.uk
13. Witten, I.H., Frank, E.: Data mining: practical machine learning tools and techniques, 2nd edn. Morgan Kaufmann (2005)
14. Hsu, C.W., Chang, C.C., Lin, C.J.: A practical guide to support vector classification, http://www.csie.ntu.edu.tw/~cjlin (last updated on April 15, 2010)
15. Bradski, G., Kaehler, A.: Learning OpenCV, 1st edn. O'Reilly (September 2008)
16. Chang, C.C., Lin, C.J.: LIBSVM: a library for support vector machines. ACM Trans. on Intelligent Systems and Technology 2(3), Article No. 27 (2011)
17. Kennedy, J., Eberhart, R.: Particle swarm optimization. In: Proc. of the IEEE International Joint Conf. on Neural Networks, Australia, vol. 4, pp. 1942–1948 (1995)
18. Eberhart, R., Shi, Y.: Particle swarm optimization: developments, applications and resources. In: Proc. of the Congress on Evolutionary Computation, pp. 81–86 (2001)
19. Kennedy, J., Eberhart, R.: A discrete binary version of the particle swarm algorithm. In: IEEE Int'l Conf. on Syst., Man, and Cybernetics, vol. 5, pp. 4104–4108 (1997)
20. Islam, M.J., Ahmadi, M., Sid-Ahmed, M.A.: An efficient automatic mass classification method in digitized mammograms using artificial neural network. International Journal of Artificial Intelligence and Applications 1(3), 1–13 (2010)

Strategic Decision Support in Waste Management Systems by State Reduction in FCM Models

Miklós F. Hatwágner[1], Adrienn Buruzs[2], Péter Földesi[3], and László T. Kóczy[4]

[1]Department of Information Technology, Széchenyi István University, Győr, Hungary
miklos.hatwagner@sze.hu
[2]Department of Environmental Engineering, Széchenyi István University, Győr, Hungary
buruzs@sze.hu
[3]Department of Logistics, Széchenyi István University, Győr, Hungary
foldesi@sze.hu
[4]Department of Telecommunications and Media Informatics,
Budapest University of Technology and Economics, Hungary and Department of Automation,
Széchenyi István University, Győr, Hungary
koczy@tmit.bme.hu, koczy@sze.hu

Abstract. In this paper, we introduce a new design for modeling sustainable waste management systems. By its complexity, this model is much more precise in describing the real systems than those found in the relevant literature. We set up a model with six factors and then decomposed the constituting factors up to around thirty subcomponents, thereby established an extremely complex and completely novel model of the Integrated Waste Management System (IWMS) using the system-of-system (SoS) approach with the help of experts. After the investigation of the basic and detailed model and their connection matrices, the following idea arises. The two models differ conceptually and so greatly that less than thirty-three factors should be enough to approximately describe the mechanism of action of the real IWMS. In the following, a new state reduction method is proposed. It can be considered as a generalization of the state reduction procedure of sequential systems and finite state machines. The essence of the proposal is to create clusters of factors and to build a new model using these clusters as factors. This way the number of factors can be decreased to make the model easier to understand and use. Our main goal with this method is to support the strategic decision making process of the stakeholder in order to ensure the long-term sustainability of IWMS.

Keywords: fuzzy cognitive maps, integrated waste management system, cycles, state reduction.

1 Introduction

Fuzzy Cognitive Maps (FCM) offers a very convenient and simple tool for modeling complex systems. According to [1], human experts are generally rather subjective and can handle only relatively simple networks therefore there is an urgent need to develop methods for automated generation of FCM models.

C.K. Loo et al. (Eds.): ICONIP 2014, Part III, LNCS 8836, pp. 447–457, 2014.
© Springer International Publishing Switzerland 2014

A FCM is a fuzzy graph structure representing causal reasoning by using nodes corresponding to the factors. During the simulation, these factors interact by implementing the dynamics of the original systems [3]. The FCM is a combination of neural networks and fuzzy logic [3].

Modern technological systems such as waste management systems are often comprised of a large number of interacting and coupled entities called subsystems and/or components. Such systems have nonlinear behavior and thus cannot be derived as the summation of the individual components [2]. The involved feedback loops are essential in the analysis of the vulnerability and resilience of such complex systems.

FCMs represented by directed signed fuzzy graphs were introduced by Kosko [4]. This tool allows quantitative simulation of the system consisting of factors and relationships. This model may be used for the analysis, simulation and testing of the dynamic behavior of the parameters, and for the prediction of the long term behavior of the system [5].

The design of a FCM is a modeling process that heavily relies on the input from experts and/or stakeholders and starts by the extraction of the knowledge of the latter exploiting their professional experience. This knowledge extraction procedure ensures the inclusion of the various interests and points of view in order to build up synergies and partnerships and to find sustainable solutions for the problem on hand [6].

While the cognitive elements in the FCM model inevitably involve subjectivity, the goal is to build a model as independently as possible from the subjective elements by carefully filtering and cumulating the input from the stakeholders [7].

The first step of the design process is to determine the number and features of constituting factors with the help of a group of experts. They also describe the existence, the type and the strength of the causal relationships among these factors. The strengths are then normalized in the [-1; 1] double unit interval.

In the FCM, factors have time variant states (corresponding to the available time series) while edges w_{ij} represent causality between concepts C_i and C_j. In our model the states of the concepts are also normalized between [0; 1]. From the interrelations among the concepts in the FCM, a corresponding adjacency matrix may be formed.

To run the simulation of the modeled system using the connection matrix, from the initial states of concepts and the transition functions, the subsequent states of the concepts (time series) can be calculated until the system reaches a steady state.

2   The Approach Applied

During the process of our research [8] we used the FCM to simulate a waste management system described by six factors. As a validation of the simulation results [9] we collected data based on the relevant literature to set up a relevant time series. This time series served as an input to the Bacterial Evolutionary Algorithm (BEA) which generated a connection matrix and defined the value of λ (the parameter of the transition function) that produce the most similar time series to the original one. With other words, the validation was the 'inverse' of the above mentioned FCM modeling process.

To resolve the contradiction experienced between the two connection matrices and to go below the level of generally recognized components, decomposing the factors up to around thirty subcomponents we established an extremely complex and completely novel model of the Integrated Waste Management System (IWMS) using the system-of-system (SoS) approach with the help of experts. **Table 1** introduces the factors and the sub-factors, while **Table 2** and **Table 3** describe the detailed connection matrix.

Table 1. The identified sub-factors of the main factors and the concept IDs (CID) of them

Main factor	Sub-factor	CID	Main factor	Sub-factor	CID
Technology (C1)	Engineering knowledge	C1.1	Society (C4)	Public opinion	C4.1
	Technological system and its coherence	C1.2		Public health	C4.2
	Local geographical and infrastructural conditions	C1.3		Political and power factors	C4.3
	Technical requirements in the EU and national policy	C1.4		Education	C4.4
	Technical level of equipment	C1.5		Culture	C4.5
Environment (C2)	Impact on environmental elements	C2.1		Social environment	C4.6
	Waste recovery	C2.2		Employment	C4.7
	Geographical factor	C2.3	Law (C5)	Monitoring and sanctioning	C5.1
	Resource use	C2.4		Internal and external legal coherence (domestic law)	C5.2
	Wildlife (social acceptance)	C2.5		General waste management regulation in the EU	C5.3
	Environmental feedback	C2.6		Policy strategy and method of implementation	C5.4
Economy (C3)	Composition and income level of the population	C3.1	Institution (C6)	Publicity, transparency (data management)	C6.1
	Changes in public service fees	C3.2		Elimination of duplicate authority	C6.2
	Depreciation and resource development	C3.3		Fast and flexible administration	C6.3
	Economic interest of operators	C3.4		Cooperation among institutions	C6.4
	Financing	C3.5		Improvement of professional standards	C6.5
	Structure of industry	C3.6			

Table 2. Refined connection matrix created by experts as a result of the workshop, Part 1

CID	C1.1	C1.2	C1.3	C1.4	C1.5	C2.1	C2.2	C2.3	C2.4	C2.5	C2.6	C3.1	C3.2	C3.3	C3.4	C3.5	C3.6
C1.1	0	0.2	0	0.6	0.4	0.6	0.2	0	0.8	0.2	0.6	0.4	0.8	0.4	0.8	0.4	0.4
C1.2	0.4	0	0.4	0.4	0.6	0.2	0.2	0	0.4	0.2	0.4	0.6	0.8	0.6	0.6	0.6	0.6
C1.3	0	0.2	0	0.2	0	0	0	0	0.2	0	0.4	0.6	0.6	0.6	0.6	0.4	0.4
C1.4	0.2	0	0	0	0	0.6	0.2	0	0.6	0.6	0.8	0.8	0.8	0.4	0.8	0.8	0.8
C1.5	0.8	0.2	0	0.8	0	0.4	0.2	0	0.4	0.4	0.6	0.6	0.8	0.6	0.6	0.6	0.6
C2.1	0	0	0.6	0.2	0	0	0	0	0.2	0.4	-0.6	0	0.2	0	0	0	0
C2.2	0	0.2	0	0	0.2	0.4	0	0.6	-1	0	-0.6	0	-0.4	0.4	0.8	0.6	1
C2.3	0	0	0.6	0	0	0.4	0.4	0	0.4	0	0	0	0.2	0	0	0	0.6
C2.4	0	0.2	0.4	0	0.6	-0.6	-0.8	-0.6	0	-0.4	-0.6	0	-0.2	0	0	-0.2	0.2
C2.5	0	0	0	0.6	0	0.4	0	0.4	0	0	0.4	0	0	0	0.2	0	0
C2.6	0	0.6	-0.8	0.6	0.6	-0.8	0.6	0	0.6	-0.8	0	-0.6	0.2	0	0	0	0.2
C3.1	0	0.2	0	0	0.2	-0.8	0.4	0	0	0.2	0.2	0	0.8	0.8	0.6	0.6	0
C3.2	0	0.6	0	0	0.6	-0.6	0.4	0	0.6	0	0.4	0	0	0.6	0.8	0.8	1
C3.3	0	0.6	0	0.2	0.4	0.4	0.4	0	0.2	0.2	0.2	0	0.6	0	0.4	0.8	0.8
C3.4	0.8	0.8	0	0.2	0.8	-0.6	0.8	0	-0.2	0.2	0.2	0	1	0.6	0	0.6	0.4
C3.5	0	0.4	0	0	0.6	0.4	0.8	0	0.6	0	0	0	0.6	0.6	0.6	0	0.8

CID	C1.1	C1.2	C1.3	C1.4	C1.5	C2.1	C2.2	C2.3	C2.4	C2.5	C2.6	C3.1	C3.2	C3.3	C3.4	C3.5	C3.6
C3.6	0	0.6	0	0	0.8	0.6	1	0.8	-0.8	0.4	0.4	0	0.4	0.2	0.6	0.4	0
C4.1	0.2	0.2	0	0.6	0.6	0.8	0.6	0.4	0.8	1	0.6	0.2	0.6	0.4	0.6	0.4	0.4
C4.2	0.4	0.2	0.2	0.6	0.6	0.6	-0.2	0.2	0.8	0.8	1	0.6	0.4	0.4	0.4	0.4	0.4
C4.3	0	0.8	0	0.4	0	0	0	0	-0.2	0.4	-0.2	0.6	1	0.8	0.6	0.8	0.4
C4.4	0.2	0	0	0.2	0.2	0.4	0.2	0	0.6	0.6	0.6	0.8	0.2	0.2	0.2	0.2	0.2
C4.5	0.2	0	0.4	0.6	0.8	-0.2	0.6	0.2	0.4	0.8	0.6	0.2	0.2	0.2	0.2	0.2	0.2
C4.6	0	0	0.4	0.6	0.4	0.2	0.6	0.2	0.4	0.6	0.4	0.2	0.2	0.2	0.2	0.2	0.2
C4.7	0	0	0	0.2	0	0	0.4	0	0.6	0.4	0.4	0.6	0.2	0.6	0.4	0.2	0.4
C5.1	0	0.4	0	0	0.4	0.2	0.2	0	0.2	0.2	0.2	0	0	0.6	0.2	0	-0.4
C5.2	0.4	0.6	0	0	0.4	0.8	0.8	0.6	0.6	0.6	0.8	0	1	0.6	0.6	1	0.6
C5.3	0.2	0.4	0	0.4	0.4	0.8	0.8	0.6	0.8	0.6	0.8	0	0.4	0	0.2	0.8	0.6
C5.4	0.2	0.6	0	0	0.8	0.8	0.6	0	0.6	0.6	0.6	0	0.8	0.2	0.2	0.2	0.4
C6.1	0	0.6	0	0.4	0	0.2	0	0	0.4	0.2	0.4	0.4	0.6	0.6	0.6	0.8	0.2
C6.2	0	0.4	0	0	0	0	0	0	-0.4	0	-0.2	0.4	0.6	0.8	0.8	0.6	0.4
C6.3	0	0.4	0	0	0	0	0	0	0	0	0.4	0.8	0.8	0.6	0.8	0.6	0.6
C6.4	0	0.4	0	0.4	0	0.2	0	0.2	0.2	0	0.6	0.6	0.8	0.4	0.4	0.4	0.8
C6.5	0.4	0.2	0	0.6	0.2	0.2	-0.2	0	0.6	0.4	0.8	0.6	0.6	0.8	1	1	1

Table 3. Refined connection matrix created by experts as a result of the workshop, Part 2

CID	C4.1	C4.2	C4.3	C4.4	C4.5	C4.6	C4.7	C5.1	C5.2	C5.3	C5.4	C6.1	C6.2	C6.3	C6.4	C6.5
C1.1	0	0	0	0.4	0	0	-0.6	0	0.4	0.4	0.4	0	0	0	0	0.2
C1.2	0	0.2	0.2	0	0	0.2	-0.2	-0.6	0.6	0.2	0.6	0.8	0.4	0	0.6	0.4
C1.3	0	0.2	0	0	0.4	0.2	0	0	0	0.2	0	0.2	0	0	0	0.2
C1.4	0.2	0.6	0.4	0.2	0.2	0	0	0.2	0.2	0.6	0	0.4	0	0.2	0	0.6
C1.5	0	0.2	0	0	0.4	0.2	-0.2	0.6	0	0.2	0.4	0	0	0	0.2	0.8
C2.1	0.4	0.8	0	0.2	0	0	0	0.6	0.4	0.4	0.2	0	0	0	0.2	0
C2.2	-0.6	0.4	0	0.2	0	-0.2	0.4	0	0	1	0	0	0	0	0.4	0.4
C2.3	-0.6	-0.6	0	0	0	0.2	0.4	0	0	0	0	0	0	0	0	0
C2.4	0.2	-0.6	0	0.2	0.2	0.4	0.4	0	0	0	0.4	0	0	0	0	0.2
C2.5	0.6	0.4	0	0.4	0.2	0	0	0	0	0	0	0	0	0	0	0
C2.6	0.8	0.4	0	0.2	0	0.4	0	0.6	0.2	0.8	0.8	0	0	0	0.2	0
C3.1	1	0.4	0	0.6	0.6	0.8	0	0	0	0.2	0	0.2	0	0.6	0	0
C3.2	1	0	0	0.2	0	-0.4	-0.4	0	0	0	0	0	0	0	0	0
C3.3	0	0	0	0.6	0	0	0.4	0	0	0	0	0	0	0	0	0.2
C3.4	0.4	0	0	0.2	0	0	-0.4	0	0	0	0	0.2	0	0	0.6	0
C3.5	0	0	0	0.2	0	0	0.4	0	0	0	0	0.2	0	0	0	0.2
C3.6	-0.4	0	0	0.4	0	0	-0.6	0	0.6	0.6	0.6	0	0	0	0	0.8
C4.1	0	0.8	0.4	0.6	0.8	0.8	0	0	0.8	0.4	0.4	0.6	0	0.2	0.4	0.2
C4.2	0.8	0	0.2	0.4	0.8	0.8	0	0.6	0.8	0.6	0.4	0.4	0	0	0	0
C4.3	0	0	0	0	0	0	0	0	0.2	0	0.6	0.6	0.8	0.2	0.8	0
C4.4	0.4	1	0.2	0	0.6	0.6	0	0.2	0.4	0.4	0.4	0	0	0	0	0
C4.5	0.6	0.4	0.2	0.8	0	0.6	0	0.4	0.4	0.4	0.4	0.2	0	0	0	0
C4.6	0.8	0.4	0.2	0.8	1	0	0	0.2	0.2	0.2	0.2	0.2	0	0	0	0
C4.7	0	0	0.4	0	0.2	0	0	0.2	0	0.2	0.2	0.4	0.4	0.6	0.4	0
C5.1	0	0.4	0	0	0	0	0	0	0.2	0.2	0.2	0	0	0	0	0
C5.2	0.6	1	0	0.6	0.4	0	0.2	0.8	0	0	0.6	0.8	0	0	0	0
C5.3	0.4	0	0	0.4	0.2	0	0	0.4	1	0	0.4	0.2	0	0	0	0
C5.4	0	0.4	0	0.4	0	0	0.2	0.8	0.8	0	0	0	0	0	0	0.4
C6.1	0.2	0.8	0.4	0.2	0	0.6	0	-0.4	0.4	0	0.8	0	0.8	0.4	0.6	0
C6.2	0	0	0.2	0	0	0	0	-0.2	0.4	0	0.8	1	0	0	0.4	0
C6.3	0	0	0.2	0	0	0.4	0	-0.6	0.4	0	0.8	0.8	1	0	0.4	0

CID	C4.1	C4.2	C4.3	C4.4	C4.5	C4.6	C4.7	C5.1	C5.2	C5.3	C5.4	C6.1	C6.2	C6.3	C6.4	C6.5
C6.4	0	0	0.6	0	0	0	0	-0.2	-0.2	0	0.6	1	0.6	0.2	0	0
C6.5	0	0	0.2	0.4	0	0.4	0	0.4	0.6	0.6	1	0.4	0.6	0.2	0.2	0

Using the experts' connection matrix, the initial state of concepts (t_0), the parameter λ of the transition function the time series of the concept states can be calculated. **Table 4** and **Table 5** show the two time series generated by using different λ values. It may be noticed that the order of the concepts in the last steps is very similar. While the qualitative behavior of the simulation result is virtually independent from the steepness (λ), the actual constant values, where the concept influence state converges to, are more or less similar to each other thus after normalization the results are very consistent.

Table 4. Time steps generated by the connection matrix ($\lambda=1$)

CID	t_0	t_1	t_2	t_3	t_4	CID	t_0	t_1	t_2	t_3	t_4
C1.1	0,0	0,8	1,0	1,0	1,0	C4.1	0,0	0,9	1,0	1,0	1,0
C1.2	0,2	0,8	1,0	1,0	1,0	C4.2	0,1	0,8	1,0	1,0	1,0
C1.3	0,0	0,6	1,0	1,0	1,0	C4.3	0,0	0,7	1,0	1,0	1,0
C1.4	0,0	0,7	1,0	1,0	1,0	C4.4	0,0	0,7	1,0	1,0	1,0
C1.5	0,7	0,7	1,0	1,0	1,0	C4.5	0,0	0,8	1,0	1,0	1,0
C2.1	0,4	0,6	1,0	1,0	1,0	C4.6	0,0	0,7	1,0	1,0	1,0
C2.2	0,4	0,7	1,0	1,0	1,0	C4.7	0,0	0,7	1,0	1,0	1,0
C2.3	0,0	0,6	0,7	0,9	0,9	C5.1	0,0	0,7	0,9	1,0	1,0
C2.4	0,0	0,4	0,5	0,4	0,4	C5.2	0,6	0,8	1,0	1,0	1,0
C2.5	0,0	0,6	0,9	1,0	1,0	C5.3	0,2	0,9	1,0	1,0	1,0
C2.6	0,1	0,7	1,0	1,0	1,0	C5.4	0,0	0,9	1,0	1,0	1,0
C3.1	0,1	0,6	1,0	1,0	1,0	C6.1	0,2	0,7	1,0	1,0	1,0
C3.2	0,0	0,7	1,0	1,0	1,0	C6.2	0,0	0,7	1,0	1,0	1,0
C3.3	0,0	0,7	1,0	1,0	1,0	C6.3	0,1	0,7	1,0	1,0	1,0
C3.4	0,1	0,7	1,0	1,0	1,0	C6.4	0,2	0,6	1,0	1,0	1,0
C3.5	0,0	0,8	1,0	1,0	1,0	C6.5	0,0	0,8	1,0	1,0	1,0
C3.6	0,1	0,9	1,0	1,0	1,0						

Table 5. Time steps generated by the connection matrix ($\lambda=0.2$)

CID	t0	t1	t2	t3	t4	CID	t0	t1	t2	t3	t4
C1.1	0,0	0,6	0,7	0,7	0,8	C4.1	0,0	0,6	0,8	0,9	0,9
C1.2	0,2	0,6	0,8	0,8	0,8	C4.2	0,1	0,6	0,8	0,9	0,9
C1.3	0,0	0,5	0,6	0,7	0,7	C4.3	0,0	0,5	0,7	0,8	0,8
C1.4	0,0	0,6	0,8	0,8	0,8	C4.4	0,0	0,6	0,7	0,8	0,8
C1.5	0,7	0,5	0,8	0,8	0,8	C4.5	0,0	0,6	0,8	0,8	0,8
C2.1	0,4	0,5	0,6	0,6	0,7	C4.6	0,0	0,6	0,7	0,8	0,8
C2.2	0,4	0,5	0,6	0,7	0,7	C4.7	0,0	0,5	0,7	0,7	0,7
C2.3	0,0	0,5	0,6	0,6	0,6	C5.1	0,0	0,5	0,6	0,6	0,6
C2.4	0,0	0,5	0,5	0,5	0,5	C5.2	0,6	0,6	0,8	0,9	0,9
C2.5	0,0	0,5	0,6	0,6	0,6	C5.3	0,2	0,6	0,8	0,8	0,8
C2.6	0,1	0,5	0,6	0,7	0,7	C5.4	0,0	0,6	0,7	0,8	0,8
C3.1	0,1	0,5	0,7	0,8	0,8	C6.1	0,2	0,5	0,8	0,8	0,8
C3.2	0,0	0,5	0,7	0,7	0,7	C6.2	0,0	0,5	0,7	0,7	0,7
C3.3	0,0	0,5	0,7	0,7	0,7	C6.3	0,1	0,5	0,7	0,8	0,8
C3.4	0,1	0,6	0,7	0,7	0,7	C6.4	0,2	0,5	0,7	0,8	0,8
C3.5	0,0	0,6	0,7	0,7	0,7	C6.5	0,0	0,6	0,8	0,9	0,9
C3.6	0,1	0,6	0,7	0,7	0,8						

3   Loops and Self-loops in the Connection Matrix

Usually it is accepted that causality is not self-reflexive, i.e., a factor cannot have a caused effect on itself. This means that the connection matrix has '0-s' in its diagonal [10]. It has been assumed that otherwise the component would grow without limits. This is, however not exactly true, as FCM uses discrete time series and also, because these are negative, causal effects involved. The time steps can have various scales, depending on the specific problem. In a sense the operation of FCM can be considered as a synchronous sequential network, and so, a direct feedback in the system is unable to induce excitation, because the effect of the positive feedback is compensated by the negative excitation coming from other concepts. Nevertheless, direct feedback loops were not used in the paper regarding to the wide consensus in the literature, but in the future we plan to make investigations into the study of the effects of a possible removal of this boundary condition. Our hypothesis is that a self-loop (an edge that connects a vertex to itself) could be permitted because anyway a large number of positive cycles can be found in most connection matrices and thus the effect of such cycles could be essentially the same as the effect of the self-loops. (Due to the compensation of cycles by other effects, we think that self-loops could be permitted also, but the detailed examination of it is going to be the subject of a future research.)

While evaluating the connection matrix of the thirty-three factors, we have found over 70,000 positive and negative cycles (see **Table 6** for some examples).

Table 6. Some examples of cycles in the connection matrix

C3.4 → C3.2 → C3.4
C2.2 → C3.4 → C2.2
C1.5 → C1.4 → C3.6→ C1.5
C4.2 → C5.2 → C2.1 → C4.2
C6.1 → C5.4 → C5.2 → C6.1
C3.6 → C1.5 →C1.4 → C2.6 →C5.3 → C2.1 → C4.2 → C5.2 → C2.2 → C3.6

4   State Reduction of the FCM Model

Six factors should be enough to describe the behavior of waste management systems regarding to the wide-ranging consensus which can be experienced in the literature and between the experts [11, 12]. Nevertheless an FCM model containing six factors (called 'concepts' in FCM theory) was very inaccurate [13] and a more detailed model (thirty-three factors) was developed. After the thorough investigation of the basic and the detailed models and their connection matrices, the following idea arises. The two models differ conceptually and so greatly that less than thirty-three factors should be enough to approximately describe the mechanism and the action of a real IWMS.

In the following a new state reduction method is proposed. It can be considered as a generalization of the state reduction procedure of the sequential systems and the finite state machines. The essence of our proposal is to create clusters of factors and to build a new model using these clusters as factors. This way the number of factors can be decreased to make the model easier to understand and to use.

Initially, the clusters are disjoint sets of factors, each of them containing one factor only. $K_i = \{C_i\}$ for every $i = 1 \ldots n$ where K_i is the ith cluster, C_i is the ith factor (concept) and n is the number of factors in the model (thirty-three in our case). Next, an agglomerating strategy is applied for all of the clusters. For those factors not included in the current cluster, the 'distance' between the factor and the cluster (that is, all members of the cluster) is measured with an appropriate metric.

The used metric measures the difference between the connections starting from two factors (C_i and C_j) to the third (C_k), where $i \neq j \neq k$, $i = 1 \ldots n, j = 1 \ldots n, k = 1 \ldots n$. If this difference is less than a specified ε value ($|w_{ik} - w_{jk}| < \varepsilon$ and $|w_{ki} - w_{kj}| < \varepsilon$, where w specifies the sign and magnitude of connections between factors) in a predefined proportion (p) of the cases, the current factor is added to the cluster. This process can be described more exactly with the following C-style pseudo-code (see **Fig. 1**).

```
function isNear(i, j, eps, p)  // i, j = factor indexes
  near = 0                     // eps = ε, p = p
  far = 0
  for(k=0; k<n; k++)                // n = number of factors
    if(k!=i and k!=j)
      if(abs(w(i, k) - w(j, k)) < eps) // w(i, k) = w_ik
        near = near + 1
      else
        far = far + 1
      if(abs(w(k, i) - w(k, j)) < eps)
        near = near + 1
      else
        far = far + 1
  if(near==0 or far/near >= p)
    return false
  else
    return true

function buildCluster(initialFactor, eps, p)
  c = {initialFactor}
  for(i=0; i<n; i++)
    if(i != initialFactor)
      member = true
      while(member and hasNextElement(c))
        j = nextElement(c)
        member = isNear(j, i, eps, p)
      if(member)
        c = c + {i}
  return c
```

```
function buildAllClusters(eps, p)
  clusters = {}
  for(i=0; i<n; i++)
    k = buildCluster(i, eps, p)
    if(!isElementOf(k, clusters))
      clusters = clusters + {k}
  return clusters
```

Fig. 1. Pseudo-code of the state reduction algorithm, Part 1

In **Fig. 1** the function `buildAllClusters` initiates the state reduction process building all the clusters and providing the uniqueness of clusters. The `buildCluster` creates a new cluster using the specified initial factor. It investigates the additional factors and makes a decision using the `isNear` function about to add the factor to the cluster or not. The `isNear` function implements the metric of the state reduction.

The next step is the creation of the connection matrix of the new, reduced FCM model. Function `getWeight` describes the details of the calculation. The function returns the weight of the connection between clusters a and b. The weight is the average weight of the connections between the factors of cluster a to cluster b. The zero-weight connections are ignored (see **Fig. 2**).

```
function getWeight(a, b)
  count = 0
  sum = 0
  while(hasNextElement(a))
    i = nextElement(a)
    while(hasNextElement(b))
      j = nextElement(b)
      w = w(i, j)
      if(w != 0)
        count = count + 1
        sum = sum + w
  if(count == 0)
    return 0
  else
    return sum/count
```

Fig. 2. Pseudo-code of the state reduction algorithm, part 2

The ε and p are the critical parameters of the state reduction. If ε is too small, no or only a few factor-merge can be made, and it does not make the model simple enough. On the other hand, if the value of ε is too high, e.g. 2, the whole sensitivity matrix 'collapses' into a 1 by 1 matrix. The useful value of ε is somewhere in the [0; 2] interval, but the specific value is different in every case. The value of p can be in the interval [0; 1]. Small p values can help to extend the clusters even if in some insignificant number of cases the 'distance' measured by the applied metric is greater than the allowed ε value. However, high value of p would enable to merge almost every factor into the same cluster regardless of the value of ε, therefore this case must be avoided. The state reduction described above is similar to classification, where the number of classes is estimated in an appropriate way.

Considering the properties of the specific engineering problem, the number of states in the reduced FCM need to be six (the consensus of the experts) or at least much less than the number of concepts in the detailed model.

Several attempts were made to find the suitable ε and p values. The definition of the values need some experimenting: if only a few factors can be merged, the desired goal of merging cannot be achieved. On the other hand, if too many factors are merged, the interpretation of the merged factors can be difficult or impossible. The results of experiments are collected in **Table 7**. We have found that an acceptable value for ε could be 0.6 and for p could be 0.2 thus reducing the detailed matrix into a matrix containing only fifteen factors. The merged factors in this new matrix are presented in **Table 8**. The aggregation of these sub-factors (representing mainly the factors such as 'economy', 'law', 'technology' and sub-factors such as 'cooperation among institutions' and 'social environment') indicates that the main driving elements belong strongly together and have a similar but very strong role in ensuring the sustainability of the IWMS.

Table 7. The number of factors after merging with different ε and p values

ε	p	Number of factors after merging
0.2	0.2	30
0.4	0.05	30
0.4	0.1	26
0.4	0.2	23
0.4	0.45	18
0.4	0.5	14
0.6	0.05	23
0.6	0.1	21
0.6	0.2	15
0.8	0.05	19
0.8	0.1	10

Table 8. The merged factors ($\varepsilon = 0.6$, $p = 0.2$; for legend see **Table 1**)

New CIDs	Merged factors (concepts)
Q1	C3.1+C3.2+C3.3+C3.4+C3.5+C5.2+C5.3+C5.4+C1.1+C1.2+C1.4+C6.4+C4.6
Q2	C3.2+C3.3+C3.4+C3.5+C3.6+C5.3+C5.4+C1.1+C1.2+C1.4+C1.5
Q3	C3.1+C3.5+C2.1+C2.3+C2.5+C5.2+C5.3+C5.4+C1.1+C1.2+C1.3+C4.4+C4.5+C4.6
Q4	C3.2+C3.3+C3.5+C3.6+C2.2+C2.5+C1.1+C1.2+C1.5
Q5	C3.1+C3.3+C3.5+C2.3+C2.5+C5.2+C5.3+C5.4+C1.1+C1.2+C1.3+C1.4+C6.1+C6.3+C6.4+C6.5+C4.4+C4.5+C4.6
Q6	C3.5+C3.6+C2.3+C2.4+C2.5+C5.3+C1.1+C4.4
Q7	C3.1+C3.2+C3.3+C3.5+C2.5+C5.2+C5.3+C5.4+C1.1+C1.2+C1.4+C6.4+C4.6
Q8	C3.1+C3.4+C3.5+C2.6+C5.2+C5.3+C5.4+C1.1+C1.4+C6.1+C4.4+C4.5+C4.6
Q9	C3.3+C3.5+C2.2+C2.3+C2.5+C5.1+C1.1+C1.2+C1.5+C6.5+C4.4
Q10	C3.1+C3.3+C3.4+C3.5+C5.2+C5.3+C5.4+C1.1+C1.2+C1.4+C6.1+C6.4+C6.5+C4.4+C4.5+C4.6
Q11	C3.1+C3.3+C3.4+C3.5+C5.4+C1.1+C1.2+C1.4+C6.1+C6.2+C6.4+C6.5+C4.3+C4.4+C4.6
Q12	C3.1+C3.3+C2.3+C2.5+C5.2+C5.3+C5.4+C1.1+C1.2+C1.3+C1.4+C6.1+C6.3+C4.1+C4.2+C4.4+C4.5+C4.6
Q13	C3.1+C3.3+C2.3+C2.5+C5.2+C5.3+C5.4+C1.1+C1.2+C1.3+C1.4+C6.1+C6.3+C6.4+C6.5+C4.2+C4.4+C4.5+C4.6
Q14	C3.1+C3.2+C3.3+C3.4+C3.5+C5.3+C5.4+C1.1+C1.2+C1.4+C6.4+C4.3+C4.6
Q15	C3.1+C3.3+C2.3+C2.5+C5.2+C5.3+C1.1+C1.2+C1.3+C1.4+C6.1+C6.3+C6.4+C6.5+C4.2+C4.4+C4.5+C4.6+C4.7

5   Conclusions

We introduced a new design for modeling sustainable waste management systems. First, we set up a model with six factors and then decomposed the constituting factors up to around thirty subcomponents, thereby established an extremely complex and completely novel model of the Integrated Waste Management System (IWMS).

Realizing the excessive largeness of the detailed model, a new fuzzy state reduction method was suggested in the paper. It can be considered as a generalization of the reduction of finite state machines and it is based on fuzzy tolerance relations [14]. The rules and preconditions of cluster building were defined and made easier to understand by using pseudo codes. On the basis of this method, we might be able to support the strategic decision making process of the stakeholder in order to ensure the long-term sustainability of IWMS.

On the one hand the validation of the presented method by experts, on the other hand the examination of some details, e.g. to study the effect of other metrics and the effect of self-loops are the target of our future research.

Acknowledgement. The authors would like to thank to TÁMOP-4.2.2.A-11/1/KONV-2012-0012, TÁMOP-4.1.1.C-12/1/KONV-2012-0017, to the Hungarian Scientific Research Fund (OTKA) K105529 and K108405 for the support of the research.

References

1. Stach, W., Kurgan, L., Pedrycz, W., Reformat, M.: Genetic Learning of Fuzzy Cognitive Maps. J. of Fuzzy Sets and Systems 153, 371–401 (2005)
2. Stylos, D., Groumpos, P.P.: Modeling Complex Systems Using Fuzzy Cognitive Maps. IEEE Transactions on Systems, Man, and Cybernetics – Part A: Systems and Humans 34(1) (2004)
3. Stylos, C.D., Georgopoulos, V.C., Groumpos, P.P.: The Use of Fuzzy Cognitive Maps in Modeling Systems. In: Proceedings of 5th IEEE Mediterranean Conf. on Control and Systems, Paphos, Cyprus (1997)
4. Kosko, B.: Fuzzy Cognitive Maps. Int. J. of Man–Machine Studies 24(1), 65–75 (1986)
5. Papageorgiou, E., Kontogianni, A.: Using Fuzzy Cognitive Mapping in Environmental Decision Making and Management: A Methodological Primer and an Application. In: Young, S. (ed.) Int. Perspectives on Global Environmental Change, pp. 978–953. InTech (2012), ISBN: 978-953-307-815-1, doi:10.5772/29375
6. Malena, C.: Strategic Partnership: Challenges and Best Practices in the Management and Governance of Multi-Stakeholder Partnerships Involving UN and Civil Society Actors. Background paper prepared by for the Multi-Stakeholder Workshop on Partnerships and UN-Civil Society Relations, Pocantico, New York (2004)
7. Isak, K.G.Q., Wildenberg, M., Adamescu, M., Skov, F., De Blust, G., Varjopuro, R.: A Long-Term Biodiversity, Ecosystem and Awareness Research Network Manual for Applying Fuzzy Cognitive Mapping – Experiences from ALTER-Net. Project no. GOCE-CT-2003-505298, ALTER-Net Deliverable type: Report, WPR6-2009-02 - Deliverable 4.R6.D2 (2009)
8. Buruzs, A., Pozna, R.C., Kóczy, L.T.: Developing Fuzzy Cognitive Maps for Modeling Regional Waste Management Systems. In: Tsompanakis, Y. (ed.) Proceedings of the Third Int. Conference on Soft Computing Technology in Civil, Structural and Environmental Engineering, Paper 19. Civil-Comp Press, Stirlingshire (2013)

9. Buruzs, A., Hatwágner, M.F., Pozna, R.C., Kóczy, L.T.: Advanced Learning of Fuzzy Cognitive Maps of Waste Management by Bacterial Algorithm. In: Proceedings of IFSA World Congress and NAFIPS Annual Meeting, pp. 890–895. IEEE (2013)
10. Carvalho, J.P.: On the Semantics and the Use of Fuzzy Cognitive Maps in Social Sciences. In: WCCI 2010 IEEE World Congress on Computational Intelligence, CCIB, Barcelona, Spain (2010)
11. Buruzs, A., Kóczy, T.L., Pozna, C.R.: Developing fuzzy cognitive maps for modeling regional waste management systems. In: Tsompanakis, Y. (ed.) Proceedings of the Third International Conference on Soft Computing Technology in Civil, Structural and Environmental Engineering, Paper 19, Cagliari, Italy, Civil-Comp Press, Stirling (2013), doi:10.4203/ccp, ISBN:978-1-905088-58-4
12. Buruzs, A., Kóczy, T.L., Hatwágner, F.M., Pozna, C.R.: Advanced learning of fuzzy cognitive maps of waste management by bacterial algorithm. In: Reformat, M.Z., Pedrycz, W. (eds.) Proceedings of the 2013 Joint Papers Author Index IFSA World Congress NAFIPS Annual Meeting, pp. 890–895. IEEE, Edmonton (2013) ISBN:978-1-4799-0347-4
13. Buruzs, A., Hatwagner, M.F., Koczy, T.L.: Modeling integrated sustainable waste management systems by fuzzy cognitive maps and the system of systems concept. Technical Transactions series Automatic Control (accepted for publication, 2014)
14. Klir, G.J., Folger, T.A.: Fuzzy Sets, Uncertainty and Information. Prentice Hall (1987)

An ELM Based Multi Agent Systems
Using Certified Belief in Strength

Chong Tak Yaw[1], Keem Siah Yap[2], Hwa Jen Yap[3],
and Ungku Anisa Ungku Amirulddin[4]

[1]Department of Electronics and Communication Engineering,
Universiti Tenaga Nasional, Malaysia
takhehe@yahoo.com
[2]Department of Electronics and Communication Engineering,
Universiti Tenaga Nasional, Malaysia
yapkeem@uniten.edu.my
[3]Faculty of Engineering, University of Malaya, Malaysia
hjyap737@um.edu.my
[4]Department of Electrical Power Engineering, Universiti Tenaga Nasional, Malaysia
anisa@uniten.edu.my

Abstract. A trust measurement method called certified belief in strength (CBS) for Extreme Learning Machine (ELM) Multi Agent Systems (MAS) is proposed in this paper. The CBS method is used to improve the performance of the individual agents of the MAS, i.e., ELM neural network. Then, trust measurement is achieved based on reputation and strength of the individual agents. In addition, trust is assemble from strong elements that are associated with the CBS which let the ELM to improve the performance of the MAS. The efficiency of the ELM-MAS-CBS model is verified with several activation function using benchmark datasets which are Pima Indians Diabetes (PID), Iris and Wine. The results show that the proposed ELM-MAS-CBS model is able to achieve better accuracy as compared with other approaches.

Keywords: Certified Belief in Strength, Extreme Learning Machine Neural Network, Multi Agent System, Pattern Classification.

1   Introduction

Over the past years, the Extreme Learning Machine (ELM) has proven as an efficient learning algorithm compared to traditional learning methods in terms of generalization and learning speed [1-6]. It is important to point out that the ELM is capable of making universal approximation with random input weights and biases [7].

In order to improve the performance of ELM, many research focus on ensemble model to combine individual prediction of multiple ELMs to give a final output [8-12]. This strategy is also adopted in a Multi Agent System (MAS) [13]. Multi agent systems (MAS) have attracted a lot of attention in the past decade, whereby researchers have successfully applied them to tackle problems in various fields. This is evidenced by a widespread application of MAS to different domains including

C.K. Loo et al. (Eds.): ICONIP 2014, Part III, LNCS 8836, pp. 458–465, 2014.

e-Commerce [14], healthcare [15], military support [16], decision support [17], knowledge management [18], as well as control systems [19]. When the constituent agents in an MAS model consist of classifiers, where each agent has the ability to carry out a specific task and to make decisions.

A trust measurement method that based on the recognition and rejection accuracy rates has been proposed [13]. In the model, two teams were used where the first consists of three modified Fuzzy min-max (FMM) agents and the second team consists of three modified Fuzzy ARTMAP (FAM) agents. The model was presented with better performances as compared with other approaches mentioned in [13]. Another trust measurement strategy based on Bayesian formalism with FMM MAS was proposed in [20]. In this model, the FMM is used as a learning agent in MAS and then combined with Bayesian formalism to obtain trust measurement. As the results, the model is able to yield the better performances as compared with other approaches mentioned in [20].

In the recent development of MAS model for trust measurement, a method namely Certified Belief in Strength (CBS), which based on strength and reputation of individual FMM based agent [21]. During the training process, trust is strong elements that are related with the FMM agents which let the CBS method to improve the performance of the MAS. As the results, the CBS improved the performance of the MAS model by improving the accuracy rates of the individual agents [21].

This paper proposes an extended CBS method using Extreme Learning Machine based Multi Agent System (hereafter denoted as ELM-MAS-CBS). The difference is that MACS CBS used FMM which consist of multiple hyperboxes while the proposed approach employ "team" concept which consist of individual ELM-based agents.

This paper is organized as follows. In section 2, the algorithms of ELM-MAS-CBS is explained. Section 3 present the results and discussion of the benchmarking datasets. Lastly, conclusion is presented in section 4.

2   The Algorithms of ELM-MAS-CBS

In this paper, the ELM-MAS-CBS model has three layers shown in Fig.1, i.e., the first layer consists of several individual ELM-based agents; the second layer consists of several teams of ELM-based. The CBS is implemented in the individual ELM-based agents. Then, Manager selects the team with the highest CBS as final decision output. In this paper, the number of teams is set as 3 ($T = 3$). The number of agent used in a team is set as 5 ($K = 5$). An ELM-based agent is denoted as ELM^{tk} (for $t = 1, ..., T$, for $k = 1, ..., K$).
The step-by-step training procedures are given as following.

Step 1: Randomly assign the input weights $\mathbf{a}_i^{tk}$ and b_i^{tk}. For all steps/equations of training process, variables run for $i = 1,... L$ (number of hidden neuron of ELM), for $t = 1, .., T$, and $k = 1, 2, ...K$.

Step 2: Calculate the hidden layer output matrix for ELM^{tk}, $\mathbf{H}^{tk}$, as follows, where N is the number of training samples, G is the activation functions and $\mathbf{x}_j$ is the input vector.

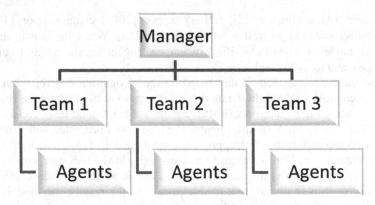

Fig. 1. Overview of ELM-MAS-CBS model

$$\mathbf{H}^{tk} = \begin{bmatrix} G(\mathbf{a}_1^{tk}, b_1^{tk}, \mathbf{x}_1) & \cdots & G(\mathbf{a}_L^{tk}, b_L^{tk}, \mathbf{x}_1) \\ \vdots & \cdots & \vdots \\ G(\mathbf{a}_1^{tk}, b_1^{tk}, \mathbf{x}_N) & \cdots & G(\mathbf{a}_L^{tk}, b_L^{tk}, \mathbf{x}_N) \end{bmatrix}_{N \times L} \tag{1}$$

$$G(\mathbf{a}_i^{tk}, b_i^{tk}, \mathbf{x}_j) = \frac{1}{1 + \exp\{-(\mathbf{a}_i^{tk} \cdot \mathbf{x}_j + b_i^{tk})\}} \quad \text{(Sigmoid)} \tag{2}$$

$$G(\mathbf{a}_i^{tk}, b_i^{tk}, \mathbf{x}_j) = \exp\{-b_i^{tk} \left\| \mathbf{x}_j - \mathbf{a}_i^{tk} \right\|^2\} \quad \text{(RBF)} \tag{3}$$

Step 3: Compute the output weights of ELM^{tk}, $\boldsymbol{\beta}^{tk}$ by following equation,

$$\boldsymbol{\beta}^{tk} = \left((\mathbf{H}^{tk})^T (\mathbf{H}^{tk})\right)^{-1} (\mathbf{H}^{tk})^T \mathbf{T}, \tag{4}$$

where $\mathbf{T} = [t_1, ..., t_N]^T$ is the respective targeted output vectors.

Step 4: Once output weights of ELM^{tk} are computed, the training samples are used to compute the outputs of ELM^{tk}, i.e.,

$$\mathbf{y}^{tk} = ELM^{tk}(\mathbf{x}_j) = \sum_{i=1}^{L} \beta_i^{tk} G(\mathbf{a}_i^{tk}, b_i^{tk}, \mathbf{x}_j) \quad \text{for } j = 1, ..., N \tag{5}$$

Step 5: Calculate the accuracy rates of the ELM^{tk}.

$$A^{tk} = \frac{N^{tk}}{N} \times 100\% \tag{6}$$

where N^{tk} and A^{tk} are number correctly classified samples and accuracy rate of ELM^{tk}.

Step 6: Given the validation samples, the output of ELM^{tk} is calculate based on Eq. (5).

Step 7: Given an initial strength of CBS for all team is 100 ($\mathbf{S} = [100\ 100\ 100]$) [21] and initial bid coefficient (C_{bid}) is 0.01 [21]. The initial team bid is in proportion to strength as follows [23],

$$B^t = C_{bid} S^t \tag{7}$$

Step 8: With the validation samples, calculate the trust element, C^t, where the equation is shown in Eq. (8). In order to calculate the C^k, use the Eq. (6) to find the accuracy rate of agents in each team. Then choose the ELM that with the highest accuracy rate (denoted as ELM^{tw} where w is the winner of the team) to represent its team and apply in Eq. (8) and summit to the manager.

$$C^t = C_{bid}(S^t + A^{tw})$$ (8)

Step 9: In the proposed paper [21], the Eq. (7) is further modified as the reward and penalty to update the strength shown in Eq. (9), where P is penalty and R is reward. When an agent generates a correct prediction, $P = 0$ while $R = B^t$; otherwise, $P = B^t$ while $R = 0$.

$$S^t(new) = S^t(now) - P + R$$ (9)

Step 10: Since S^t is updated, hence both the A^{tk} and the B^t are updated based on equation (6) and (7) respectively.

Once all the samples are trained using Step 1 to Step 10, the ELM-MAS-CBS can be used for prediction of a newly arrived and unknown input vector $\mathbf{z}$. The step-by-step prediction procedures are given as following.

Step 11: Load a_i^{tk}, b_i^{tk}, β^{tk}, A^{tk}, S^t, and C^t from completed training process. For all steps/equations of predictions procedures, variables are run for $i = 1,\ldots L$, for $t = 1, \ldots, T$, and $k = 1, 2, \ldots K$.

Step 12: Calculate the hidden layer output matrix for ELM^{tk}, $\mathbf{h}^{tk}$, as follows.

$$\mathbf{h}^{tk} = [G(a_1^{tk}, b_1^{tk}, \mathbf{z}) \quad \ldots \quad G(a_L^{tk}, b_L^{tk}, \mathbf{z})]_{1 \times L}$$ (10)

Step 13: Calculate the outputs of ELM^{tk},

$$\mathbf{y}^{tk} = \mathbf{h}^{tk}\beta^{tk}$$ (11)

Step 14: Select the highest value of accuracy rates from each team (denoted as A^{tU}), and apply in Eq. (13) to calculate the trust elements of teams.

$$A^{tU} = \arg\max_k \left(A^{tk}\right)$$ (12)

$$C^t = C_{bid}(S^t + A^{tU})$$ (13)

Step 15: Select the highest value of the C^t from all teams (denoted as C^V), where V is the winner from all teams, i.e.,

$$C^V = \arg\max_t \left(C^t\right)$$ (14)

Step 16: The final output of ELM-MAS-CBS can be found based on the Eq. (11), where $t = V$ (winning team) and $k = U$ (winning agent of the winning team).

3   Results on the Benchmarking Data

In this paper, three benchmark datasets (e.g. Pima Indians Diabetes (PID), Iris and Wine) were used to test the performance of ELM-MACS-CBS. For all experiments, the number of teams has been set as 3 ($T = 3$). Each team has 5 agents (N_1) based on

ELM. Both Sigmoid and RBF activation functions are used in the experiments. The specifications of the datasets are shown in Table 1 [20]. All experiments were run in MATLAB (ver.2010) on a personal computer equipped with Intel(R) Core(TM) i7 2.9 GHz CPU and 8 G RAM.

Table 1. Specification of Benchmark Datasets [20]

Dataset	# Attributes	# Classes	# Data Samples
PID	8	2	768
Iris	4	3	150
Wine	13	3	178

In the experiment, the train-validation-test method was adopted to evaluate three benchmark datasets. The 60% of the PID samples were used for training while the 20% were used to determine the most appropriate number of neurons (i.e., L) through a validation process. In the case of Iris, 100 % of the data samples were used for training (90 % for training and 10 % for validation) and 100 % for testing. All the experiments were repeated 10 times. The tenfold cross-validation method was used to evaluate the Wine. Each Wine data set was divided into 10 subsets, 8 for training and 1 for validation and the remaining for testing.

There are two types of activation function which are Radial Basis Function (RBFun) and Sigmoid activation function (SigAct) used in each benchmark datasets. Table 2 shows the test accuracy rates based on sigmoid activation function for PID, Iris and Wine. In addition, Table 3 also shows the test accuracy rates based on radial basis activation function for the three benchmark datasets. From the Table 2 and Table 3, the number of hidden neurons, L with the best test accuracy rate is selected for evaluating the performance of ELM-MAS-CBS.

Table 4 summarizes the results for using ELM-MAS-CBS in terms of the test accuracy and the number of hidden neurons for two types of activation function in the benchmark datasets. The proposed ELM-MAS-CBS is also compared with other ELM. As observed from Table 5, the test accuracy rates of ELM-MAS-CBS are comparable (if not superior) with MACS-TNC [13] and MACS-CBS [20].

Table 2. Testing Accuracy Rates of ELM-MAS-CBS Using Sigmoid Activation Function

# Hidden Neurons, L	PID Test Accuracy (%)	Iris Test Accuracy (%)	Wine Test Accuracy (%)
5	74.18	94.53	95.56
10	76.73	97.73	98.33
15	76.67	98.60	98.33
20	**77.19**	98.67	**98.89**
25	76.54	98.60	97.78
30	77.12	**98.73**	98.89
35	76.73	98.60	98.89
40	76.67	98.53	98.89
45	76.54	98.53	98.33
50	76.54	98.40	98.33

Table 3. Testing Accuracy Rates of ELM-MAS-CBS Using Radial Basis Activation Function

# Hidden Neurons, L	PID Test Accuracy (%)	Iris Test Accuracy (%)	Wine Test Accuracy (%)
5	71.70	96.27	91.11
10	75.03	98.07	98.33
15	75.88	98.53	97.78
20	76.67	98.53	97.78
25	76.34	**98.67**	**100**
30	**77.12**	98.60	98.33
35	76.14	98.60	97.78
40	76.21	98.67	99.44
45	76.14	98.40	98.89
50	75.49	98.40	99.44

Table 4. Summary for Test Accuracy Rates of ELM-MAS-CBS

Activation Function	PID # Hidden Neurons, L	Test Accuracy (%)	Iris # Hidden Neurons, L	Test Accuracy (%)	Wine # Hidden Neurons, L	Test Accuracy (%)
RBFun	27	**77.52**	23	**98.87**	25	**100**
SigAct	20	77.19	34	**98.87**	22	99.44

Table 5. Comparison with other approaches

Algorithm	PID Test Accuracy (%)	Iris Test Accuracy (%)	Wine Test Accuracy (%)
FMM [20]	72.46	94.65	97.15
MACS-TNC [13]	75.82	-	97.32
MACS-CBS [20]	76.58	**99.33**	97.67
ELM-MAS-CBS (RBFun)	**77.52**	98.87	**100**
ELM-MAS-CBS (SigAct)	77.19	98.87	99.44

4   Conclusion

In this paper, a trust measurement method called CBS is implemented to the ELM-MAS model which has been developed. This new approach named as ELM-MAS-CBS. ELM-MAS-CBS has been tested with two activation functions (Radial Basis Function and Sigmoid activation function) using benchmark datasets, i.e, Pima Indians Diabetes (PID), Iris and Wine. The experiment results shown that proposed model is comparable (if not superior) to other approaches.

Although the results obtained from the benchmark studies are encouraging. But more studies using datasets from various application domains are required to validate the applicability of ELM-MAS-CBS in real world application.

Acknowledgement. This work was supported by the University of Malaya Research Collaborative Grant Scheme (PRP-UM-UNITEN), under Grant Number: CG026-2013.

References

1. Huang, G.B., Zhu, Q.Y., Siew, C.K.: Extreme Learning Machine: a new learning scheme of feedforward neural networks. In: Proceedings of the IEEE International Joint Conference on Neural Networks (IJCNN 2004), Budapest, Hungary, July 25-29, pp. 985–990 (2004)
2. Huang, G.B., Zhu, Q.Y., Mao, K., Siew, C.K., Saratchandran, P., Sundararajan, N.: Can threshold networks be trained directly? IEEE Trans. Circuits Syst II 53(3), 187–191 (2006)
3. Huang, G.B., Zhu, Q.Y., Siew, C.K.: Extreme Learning Machine: theory and applications. Neurocomputing 70(1), 489–501 (2006)
4. Huang, G.B., Zhou, H., Ding, X., Zhang, R.: Extreme Learning Machine for regression and multiclass classification. IEEE Trans. Systems, Man, and Cybernetics, Part B: Cybernetics 42(2), 513–529 (2012)
5. Huang, G.B., Chen, L.: Convex incremental Extreme Learning Machine. Neurocomputing 70(168), 3056–3062 (2007)
6. Huang, G.B., Chen, L.: Enhanced random search based incremental Extreme Learning Machine. Neurocomputing 71(16), 3460–3468 (2008)
7. Yap, K.S., Yap, H.J.: Daily Maximum Load Forecasting of Consecutive National Holidays using OSELM-Based Multi-Agents System with Average Strategy. Neurocomputing 81, 108–112 (2012)
8. Zhao, G., Shen, Z., Miao, C., Gay, R.: Enhanced Extreme Learning Machine with stacked generalization. In: Proceedings of the IEEE International Joint Conference on Neural Networks, pp. 1191–1198 (2008)
9. Sun, Z.L., Choi, T.M., Au, K.F., Yu, Y.: Sales forecasting using Extreme Learning Machine with applications in fashion retailing. Decision Support Systems 46(1), 411–419 (2008)
10. Lan, Y., Soh, Y.C., Huang, G.B.: Ensemble of online sequential Extreme Learning Machine. Neurocomputing 72(135), 3391–3395 (2009)
11. van Heeswijk, M., Miche, Y., Lindh-Knuutila, T., Hilbers, P.A.J., Honkela, T., Oja, E., Lendasse, A.: Adaptive ensemble models of extreme learning machines for time series prediction. In: Alippi, C., Polycarpou, M., Panayiotou, C., Ellinas, G. (eds.) ICANN 2009, Part II. LNCS, vol. 5769, pp. 305–314. Springer, Heidelberg (2009)
12. Heeswijk, M.V., Miche, Y., Oja, E., Lendasse, A.: GPU-accelerated and parallelized ELM ensembles for large-scale regression. Neurocomputing 74(16), 2430–2437 (2011)
13. Quteishet, A., Lim, C.P., Tweedale, J., Jain, L.C.: A Neural Network-based Multi-agent Classifier System. Neurocomputing 72, 1639–1647 (2009)
14. Gwebu, K., Wang, J., Troutt, M.D.: Constructing a Multi-Agent System: An Architecture for a Virtual Marketplace. In: Phillips-Wren, G., Jain, L. (eds.) Intelligent Decision Support Systems in Agent-Mediated Environments. IOS Press (2005)
15. Hudson, D.L., Cohen, M.E.: Use of Intelligent Agents in the Diagnosis of Cardiac Disorders. Computers in Cardiology, 633–636 (2002)

16. Tolk, A.: An Agent-Based Decision Support System Architecture for the Military Domain. In: Phillips-Wren, G., Jain, L. (eds.) Intelligent Decision Support Systems in Agent-Mediated Environments. IOS Press (2005)
17. Ossowski, S., Fernandez, A., Serrano, J.M., Hernandez, J.Z., Garcia-Serrano, A.M., Perez-de-la-Cruz, J.L., Belmonte, M.V., Maseda, J.M.: Designing Multi agent Decision Support System the Case of Transportation Management. In: The 3rd International Joint Conference on Autonomous Agents and Multi agent Systems, pp. 1470–1471 (2004)
18. Singh, R., Salam, A., Lyer, L.: Using Agents and XML for Knowledge Representation and Exchange: An Intelligent Distributed Decision Support Architecture. In: The 9th Americans Conference on Information Systems, pp. 1854–1863 (2003)
19. Ossowski, S., Hernandez, J.Z., Iglesias, C.A., Ferndndez, A.: Engineering Agent Systems for Decision Support. In: The 3rd International Workshop Engineering Societies in the Agents World, pp. 184–198 (2002)
20. Quteishet, A., Lim, C.P., Saleh, J.M., Tweedale, J., Jain, L.C.: A Neural Network-based Multi-agent Classifier System with a Bayesian Formalism for Trust Measurement. Soft. Compt. 15(2), 221–231 (2001)
21. Mohammed, M.F., Lim, C.P., Quteishat, A.: A Novel Trust Measurement Method Based on Certified in Strength for a Multi-Agent Classifier System. Springer, London (2012)

Constrained–Optimization-Based Bayesian Posterior Probability Extreme Learning Machine for Pattern Classification

Shen Yuong Wong[1], Keem Siah Yap[2], and Hwa Jen Yap[3]

[1]Department of Electronics and Communication Engineering,
Universiti Tenaga Nasional, Malaysia
joeywsy77@yahoo.com
[2]Department of Electronics and Communication Engineering,
Universiti Tenaga Nasional, Malaysia
yapkeem@uniten.edu.my
[3]Faculty of Engineering, University of Malaya, Malaysia
hjyap737@um.edu.my

Abstract. Extreme Learning Machine (ELM) has drawn overwhelming attention from various fields notably in neural network researches for being an efficient algorithm. Using random computational hidden neurons, ELM shows faster learning speed over the traditional learning algorithms. Furthermore, it is stated that many types of hidden neurons which may not be neuron alike can be used in ELM as long as they are piecewise nonlinear. In this paper, we proposed a Constrained-Optimization-based ELM network structure implementing Bayesian framework in its hidden layer for learning and inference in a general form (denoted as C-BPP-ELM). Several benchmark data sets have been used to empirically evaluate the performance of the proposed model in pattern classification. The achieved results demonstrate that C-BPP-ELM outperforms the conventional ELM and the Constrained-Optimization-based ELM, and this in turn has validated the capability of ELM for being able to operate in a wide range of activation functions.

Keywords: Bayesian Posterior Probability, Extreme Learning Machines, Pattern Classification.

1 Introduction

Extreme Learning Machine (ELM) is a new learning algorithm [1] for single hidden layer feedforward neural networks (SLFNs) with addictive hidden neurons or radial basis function (RBF) hidden neuron that guarantees the universal approximation capability. ELM is prominent for providing good solutions for a large class of natural and artificial phenomena that are difficult to handle using classical parametric techniques [1]. The strength of ELM lies in the random determination of the network parameter such as the input weights (linking the input layer to the hidden layer) and hidden biases. It is in contrast to many learning algorithms which still require iterative

C.K. Loo et al. (Eds.): ICONIP 2014, Part III, LNCS 8836, pp. 466–473, 2014.

adjustment of parameters that eventually result in the dependence between different layers of network parameters for single hidden layer feedforward neural network. For example, error back propagation is one of the well-known methods used by the traditional gradient-descent based neural networks to adjust the network weights from the output layer to the input layer. This kind of feedforward neural networks may converge slowly to the solution of the given problem if the learning rate is adequately small. Yet they can be unstable or turn divergent if the learning rate is set too huge. [1, 2]. Recent works confirm the advantages of ELM over earlier approaches for ANN [1, 2, 3, 4].

Huang et al. later [5] extended the preliminary ELM from the SLFN with addictive or RBF hidden neurons to "generalized" SLFN with a wide variety of hidden neurons. These hidden neurons can be any type of piecewise continuous nonlinear hidden neurons, including the addictive or RBF type of neurons, multiplicative neurons, fuzzy rules, fully complex neurons, hinging functions, high-order neurons, ridge polynomials, wavelets, and Fourier terms, etc. . The universal approximation capability is proved [6, 7] with such random neurons using almost any type of piecewise continuous nonlinear functions.

In the most recent publication, Huang et. al. [6] has highlighted the relationship and advantages of ELM as compared with the least square support vector machine (LS-SVM) and proximal support vector machine (PSVM) for handling both regression and multi-class classification tasks. In addition, a detail survey of ELM-based neural network and applications can be found in [7]. Note that it is important to point out that the ELM-based neural networks are capable of universal approximation with random hidden nodes. This has been discussed and justified in [5, 8].

To the best of our knowledge, there is no work of applying Bayesian activation function in Constrained-Optimization-based ELM for pattern classification thus far. Therefore, we proposed a Constrained-Optimization-based ELM network structure implementing Bayesian framework in its hidden layer for learning and inference in a general form, which is still new in ELM ideology for pattern classification. A Bayesian network is a well-established tool for representing complex probabilistic knowledge. It is popular in many neural network researches [9, 10, 11]. The wide acceptance of Bayesian framework is mainly due to its great expressive power, allowing the simultaneous analysis of complex relationships between many variables. Bayesian model computes the posterior probability and thus establishes a generative and flexible model. The connections between Bayesian and neural network models motivate further exploration of the relation between the two. Here, we prove Bayesian Posterior Probability in the Constrained-Optimization-based ELM can become a viable alternative to the two popular nonlinear functions (i.e., Addictive Sigmoidal and Radial Basis Function) in approximating any continuous functions. Our work is different from [12] in which the authors invoked many iterative procedures for linear regression that has defeated the original purpose of ELM to avoid time-consuming iterative tuning steps. Many parameters have to be tuned to optimize the weights of the output layer. This approach seems to be more ad hoc in view of the iterative process is stopped only when various thresholds are met.

The organization of the paper is as follows. Section 2 provides a brief introduction of ELM and Constrained-Optimization-based ELM. Section 3 presented the proposed C-BPP-ELM algorithm in detail. In Section 4, the performance of C-BPP-ELM is evaluated by comparing it with other ELM-based learning algorithm. Finally, Section 5 offers some concluding remarks.

2   ELM

In a perfect case, the output of this ELM respectively to $\mathbf{x}_j$ should be

$$f(\mathbf{x}_j) = \sum_{i=1}^{L} \beta_i G(\mathbf{a}_i, b_i, \mathbf{x}_j) = t_j \quad \text{for } j = 1, \ldots, N, \tag{1}$$

where $\mathbf{a}_i$ and b_i are the input weights (linking the input layer to the first hidden layer) and bias (learning parameters) of the hidden nodes, β_i is the output weights (linking the hidden layer to output layer), and $G(\mathbf{a}_i, b_i, \mathbf{x}_j)$ is the output of the ith hidden neuron respectively to the input vector $\mathbf{x}_j$. Two popular nonlinear functions for addictive sigmoidal hidden neurons and RBF hidden neurons are shown in (2) and (3), i.e.,

$$G(\mathbf{a}_i, b_i, \mathbf{x}_j) = \frac{1}{1 + \exp\{-(\mathbf{a}_i \cdot \mathbf{x}_j + b_i)\}}, \quad b_i \in R, \tag{2}$$

$$G(\mathbf{a}_i, b_i, \mathbf{x}_j) = \exp\{-b_i \|\mathbf{x}_j - \mathbf{a}_i\|^2\}, \quad b_i \in R^+. \tag{3}$$

Note that the (1) can also be written as

$$\mathbf{H}(\mathbf{a}, \mathbf{b}, \mathbf{x})\boldsymbol{\beta} = \mathbf{T} \tag{4}$$

where

$$\mathbf{H}(\mathbf{a}, \mathbf{b}, \mathbf{x}) = \begin{bmatrix} G(\mathbf{a}_1, b_1, \mathbf{x}_1) & \cdots & G(\mathbf{a}_{\overline{N}}, b_{\overline{N}}, \mathbf{x}_1) \\ \vdots & \cdots & \vdots \\ G(\mathbf{a}_1, b_1, \mathbf{x}_N) & \cdots & G(\mathbf{a}_{\overline{N}}, b_{\overline{N}}, \mathbf{x}_N) \end{bmatrix}, \tag{5}$$

and

$$\boldsymbol{\beta} = [\beta_1 \quad \cdots \quad \beta_{\overline{N}}]^T, \quad \mathbf{T} = [t_1 \quad \cdots \quad t_N]^T. \tag{6}$$

As named by Huang *et al.* [1], $\mathbf{H}(\mathbf{a}, \mathbf{b}, \mathbf{x})$ is hidden layer output matrix of the neural network; the ith column of H is the ith hidden node output with respect to inputs x_1, x_2, ..., x_N. For simplicity, it can be denoted as "H". The output weights of ELM can be found by pseudo inverse matrix of (4), i.e.

$$\boldsymbol{\beta} = (\mathbf{H}^T \mathbf{H})^{-1} \mathbf{H}^T \mathbf{T}. \tag{7}$$

When an unlabeled samples $\mathbf{z}$ arrived and required classification, the output function of ELM is

$$f(\mathbf{z}) = \mathbf{H}(\mathbf{a}, \mathbf{b}, \mathbf{z})\boldsymbol{\beta}. \tag{8}$$

In the case that slack variable is introduced to ELM [6], then the learning of output weights that given in (7) ois updated with a new user selectable constrained parameter as given in (9).

$$\beta = (\frac{I}{C} + H^T H)^{-1} H^T T \tag{9}$$

3   The Proposed C-BPP-ELM Algorithm

Our work is novel in its precise and complete implementation of a Bayesian framework in the Constrained-Optimization based ELM. The proposed network takes observations as inputs and computes posterior probability for each hidden neuron. In other words, ELM with BPP hidden neurons provides an insight of the probability of which hidden neuron the new input belongs to. In general, ELM can work with a wide range of activation function. Such activation functions include the sigmoidal functions as well as the radial basis, sine, cosine, exponential, and many other nonregular functions as shown in Huang and Babri [6]. The proposed Bayesian activation function conforms to the function paradigm to become a universal approximator. In other words, ELM with Bayesian activation function in the hidden layer is capable of approximating any continuous functions.

The main advantages of implementing Bayesian posterior probability in the hidden layer of the Constrained-Optimization-based ELM are:
  (i)   Possibility to use prior information.
  (ii)   Predictive distribution for all hidden neurons.
  (iii) One-pass learning algorithm.
  (iv)   Suitably exploited to become a universal approximator.

Consider a set of N training data samples (with an input vector and respectively target output vector), $(\mathbf{x}_k, \mathbf{t}_k) \in \mathbf{R}^M \times \mathbf{R}^Q$ are used to train C-BPP-ELM. The training algorithms of proposed C-BPP-ELM are presented in following steps.

Step 1:   Define and initialize the parameters of C- BPP- ELM, i.e., Select the number of hidden neurons (L) to 1000 as suggested by Huang et al. [6]. Then Randomly assign training parameter of C-BPP-ELM, i.e., center $\mu \in \mathbf{R}^L x \mathbf{R}^M$, standard deviation of likelihood function $\sigma \in \mathbf{R}^L x \mathbf{R}^M$, and prior probability $\mathbf{n} \in \mathbf{R}^L$. Finally, define C, where C is a user-specified parameter which is chosen from the range $C \in \{2^{-24}, 2^{-23}, \ldots, 2^{24}, 2^{25}\}$ as suggested by Huang et al.in [6].

Step 2:   Using the training data samples to compute hidden layer output matrix **H**.

$$\mathbf{H} = \begin{bmatrix} P(j=1|x_1) & P(j=2|x_1) & \cdots & P(j=L|x_1) \\ \vdots & \cdots & & \vdots \\ P(j=1|x_N) & P(j=2|x_N) & \cdots & P(j=L|x_N) \end{bmatrix} \tag{10}$$

Note $P(j \mid x_k)$ is the output of jth hidden neuron with Bayesian Posterior Probability activation function respectively to the input vector $\mathbf{x}_k$ as equation below:

$$P(j \mid \mathbf{x}_k) = \frac{P(\mathbf{x}_k \mid j)P(j)}{\sum_{i=1}^{L} P(\mathbf{x}_k \mid i)P(i)} \tag{11}$$

where $P(\mathbf{x}_k \mid j)$ is a Laplacian likelihood function used to measure the similarity between $\mathbf{x}_k$ and hidden neuron-j, and is defined as

$$P(\mathbf{x}_k \mid j) = \frac{1}{2^M \prod_{i=1}^{M} \sigma_{ji}} \exp\left(-\sum_{i=1}^{M} \frac{1}{\sigma_{ji}} \left| \mu_{ji} - x_{ki} \right| \right) \tag{12}$$

where $i = 1$ to attributes M. And for the prior probability, $P(j)$, it is expressed as

$$P(j) = \frac{n_j}{\sum_{i=1}^{L} n_i} \tag{13}$$

for $i = 1$ to L.

Step 3: Use Constrained- Optimization based ELM [6] to analytically calculate the output weights as shown in (9).

4   Experimental and Results

In this section, the applicability of C-BPP-ELM is evaluated using three benchmark problems which include Satellite image, Image segmentation and DNA [13]. In the experiments, all the inputs (attributes) have been normalized into the range [0, 1]. Each experiment is conducted for 50 trials. The details of benchmark problems are evaluated. Table 1 details the specifications of these problems.

The satellite image problem comprises dataset generated from the Landsat multispectral scanner. There are four digital images of the same scene in four different spectral bands in one frame of the Landsat multispectral scanner imagery .The database is a tiny sub-area of a scene, consisting of 82 x 100 pixels. Each sample in the data set corresponds to a region of 3x3 pixels. The aim is recognize the central pixel of a region into six categories, namely red soil, cotton crop, grey soil, damp grey soil, soil with vegetation stubble, and very damp grey soil given the multispectral value for each region. The training and test sets contain 4435 samples and 2000 samples, respectively.

The image segmentation data set consists of 2310 regions of 3x3 pixels, which were randomly drawn from seven outdoor images. The objective narrates the classification of each region into one of the 7 categories, namely brick facing, sky, foliage, cement, window, path, and grass, using 19 attributes extracted from each square region.

Table 1. Specification of benchmark datasets for experiments [13]

Datasets	# Attributes	# Class	# Training Samples	# Testing Samples
Image Segmentation	18	7	1,500	810
Satellite Image	36	6	4,435	2,000
DNA	180	3	2,000	1,186

The data set of "Primate splice-junction gene sequences with associated imperfect domain theory" is known as the DNA data set [13]. Splice junctions are points on a DNA sequence at which "superfluous" DNA is removed during the process of protein creation in higher organisms. The aim of the DNA data set is to recognize the boundaries between exons (the parts of the DNA sequence retained after splicing) and introns (the parts of the DNA sequence that are spliced out) for a given sequence of DNA. This consists of three sub-tasks: recognizing exon/intron boundaries (referred to as EI sites), intron/exon boundaries (IE sites), and neither (sites). A given sequence of DNA consists of 60 elements (called "nucleotides" or "base-pairs"). Every symbolic variable representing nucleotides is coded as three binary indicator variables, thus resulting in 180 binary attributes.

Table 2 presents the classification performance of C-BPP-ELM when it is benchmarked against other learning algorithms such as the conventional ELM and Constrained-Optimization-based ELM (C-ELM). As shown in Table 4, C-BPP-ELM delivers the best performance for Satellite image data set, with average testing accuracy of 91.38%, a significant improvement as compared to ELM and C-ELM. For Image segmentation, C-BPP-ELM manages to achieve good accuracy rate of 97.06%, which is far superior to ELM and C-ELM with either Sigmoid or RBF activation function. Furthermore, C-BPP-ELM improves the testing accuracy dramatically when compared to ELM and C-ELM for DNA data set. It can be observed that C-BPP-ELM outperforms ELM with RBF activation function by 2.72% and C-ELM with Sigmoid activation function by 1.24 %. The result implies the ability of the proposed C-BPP-ELM in undertaking huge number of binary-valued input attributes. Ankerst et al. [14] suggested using only 60 out of 180 attributes for DNA experiment in the description of the training data, because many neural network models are depriving of such competency in handling high dimensional binary- valued classification problems. In short, when comes to performance evaluation, C-BPP-ELM topped the list of the classification testing accuracy.

5   Summary

In this paper, we proposed a Constrained-Optimization-based ELM network structure implementing Bayesian framework in its hidden layer for learning and inference in a general form. The key principle of Bayesian approach in the Constrained-Optimization-based

ELM is to construct the posterior probability for all the hidden neuron given an input sample. It uses prior information to produce predictive distribution for each hidden neuron. With no doubt, Bayesian model often provide powerful analytical tool across a wide range of problems in neural network. Its competencies in computing likelihoods and priors make it succeed in learning its tasks. The advantages of C-BPP-ELM have been put to evidence. The experimental results showed that C-BPP-ELM outperforms the conventional ELM and the Constrained-optimization-based ELM.

Table 2. Comparison of C-BPP-ELM and other ELM-based learning algorithms on classification tasks

Data Sets	Algorithms	Accuracy Rate (%)	
		Training	Testing
Satellite image	C-BPP-ELM	95.59	91.38
	ELM (Sigmoid) [15]	91.95	88.97
	ELM (RBF) [15]	92.94	89.03
	C-ELM (Sigmoid) [15]	n/a	89.80
	C-ELM (RBF) [15]	n/a	89.06
Image segmentation	C-BPP-ELM	98.69	97.06
	ELM (Sigmoid) [15]	96.75	95.07
	ELM (RBF) [15]	96.22	94.91
	C-ELM (Sigmoid) [15]	n/a	96.07
	C-ELM (RBF) [15]	n/a	95.54
DNA	C-BPP-ELM	97.07	95.05
	ELM (Sigmoid) [15]	96.90	94.30
	ELM (RBF) [15]	95.87	92.33
	C-ELM (Sigmoid) [15]	n/a	93.81
	C-ELM (RBF) [15]	n/a	94.81

Acknowledgement. This work was supported by the University of Malaya Research Collaborative Grant Scheme (PRP-UM-UNITEN), under Grant Number: CG026-2013.

References

1. Huang, G.-B., Zhu, Q.-Y., Siew, C.-K.: Extreme Learning Machine: Theory and Applications. Neurocomputing 70, 489–501 (2006)
2. Zhu, Q.-Y., Qin, A.K., Suganthan, P.N., Huang, G.-B.: Evolutionary extreme learning machine. Pattern Recognition 38(10), 1759–1763 (2005), Sun, Z., Au, K., Choi, T.: A Neuro-fuzzy inference system through integration of fuzzy logic and Extreme Learning Machine. IEEE Trans. on Syst., Man, and Cybern.–Part B: Cybern. 37(5) (2007)

3. Wang, Y., Cao, F., Yuan, Y.: A study on the effectiveness of extreme learning machine. Neurocomputing 74, 2483–2490 (2011)
4. Huang, G.-B., Chen, L.: Convex Incremental Extreme Learning Machine. Neurocomputing 70, 3056–3062 (2007)
5. Huang, G.-B., Zhou, H., Ding, X., Zhang, R.: Extreme Learning Machine for Regression and Multi-Class Classification. IEEE Trans. on Syst., Man, and Cybern. –Part B: Cybern 42(2), 513–529 (2012)
6. Huang, G.B., Chen, Y., Babri, H.A.: Classification ability of a single hidden layer feedforward neural networks. IEEE Trans. Neural Network 11(3), 799–801 (2000)
7. Huang, G.-B., Chen, L., Siew, C.-K.: Universal Approximation Using Incremental Constructive Feedforward Networks with Random Hidden Nodes. IEEE Transactions on Neural Networks 17(4), 879–892 (2006)
8. Doya, K., Ishii, S., Pouget, A., Rao, R.: Bayesian brain: probabilistic approaches to neural coding. MIT Press, Cambridge (2007)
9. Nikovski, D.: Constructing Bayesian networks for medical diagnosis from incomplete and partially correct statistics. IEEE Trans. Knowl. Data Eng. 12(4), 509–516 (2000)
10. Vigdor, B., Lerner, B.: Bayesian ARTMAP. IEEE Trans. on Neural Networks 18(6) (November 2007)
11. Emilia, S.-O., Juan, G.-S., Martin, J.D., Joan, V.-F., Martinez, M., Magdalena, J.R., Serrano, A.J.: Bayesian extreme learning machine. IEEE Trans. on Neural Networks 22(3) (March 2011)
12. Huang, G.B., Zhu, Q.Y., Siew, C.K.: Extreme Learning Machine: a new learning scheme of feedforward neural networks. In: Proceedings of the IEEE International Joint Conference on Neural Networks (IJCNN 2004), Budapest, Hungary, July 25-29, vol. 2, pp. 985–990 (2004)
13. Blake, C., Merz, C.: UCI repository of machine learning databases, Dept. Inf. Comput. Sci., Univ. California, Irvine, CA (1998), http://www.ics.uci.edu/~mlearn/MLRepository.html
14. Ankerst, M., Ester, M.,, H.: P Kriegel.: Towards an effective cooperation of the user and the computer for classification. In: Proceeding of 6th ACM SIGKDD Int. Conf. on Knowledge Discovery & Data Mining (KDD-2000), pp. 179–189 (2000)
15. Liang, N.Y., Huang, G.-B., Saratchandran, P., Sundararajan, N.: "A fast and accurate online sequential learning algorithms forfeedforward network. IEEE Trans. Neural Networks 17(6), 1411–1423 (2006)

Adaptive Translational Cueing Motion Algorithm Using Fuzzy Based Tilt Coordination

Houshyar Asadi[1], Arash Mohammadi[2], Shady Mohamed[3], and Saeid Nahavandi[4]

[1] Researcher, Centre for Intelligent Systems Research, Deakin University, Australia
hasadi@deakin.edu.au
[2] Researcher, Electrical Department, Engineering Faculty, University of Malaya, Malaysia
arash_7mh@siswa.um.edu.my
[3] Senior Research Fellow, Centre for Intelligent Systems Research,
Deakin University, Australia
shady.mohamed@deakin.edu.au
[4] Director – Centre for Intelligent Systems Research, Deakin University, Australia
saeid.nahavandi@deakin.edu.au

Abstract. Driving simulators have become useful research tools for the institution and laboratories which are studying in different fields of vehicular and transport design to increase road safety. Although classical washout filters are broadly used because of their short processing time, simplicity and ease of adjust, they have some disadvantages such as generation of wrong sensation of motions, false cue motions, and also their tuning process which is focused on the worst case situations leading to a poor usage of the workspace. The aim of this study is to propose a new motion cueing algorithm that can accurately transform vehicle specific force into simulator platform motions at high fidelity within the simulator's physical limitations. This method is proposed to compensate wrong cueing motion caused by saturation of tilt coordination rate limit using an adaptive correcting signal based on added fuzzy logic into translational channel to minimize the human sensation error and exploit the platform more efficiently.

Keywords: cueing motion algorithm, high pass filter, tilt rate limit, human sensation, Fuzzy logic.

1 Introduction

A simulator motion platform cannot exactly reproduce the accelerations and angular rates experienced in a real vehicle since it is constrained by its physical limits. In order to reproduce and replicate the appropriate motion cue of specific force and angular velocity by a motion platform within its physical limitation a motion drive algorithm is needed, thus washout filter algorithm is recommended [1]. The design of cueing algorithms for driving simulation is a complex task and it depends on simulator architecture and the kind of regenerated manoeuvre [1-4]. The first washout algorithm is known as the classical algorithm. Conrad and Schmidt [5, 6] have presented the basic setup for the classical method and the main idea behind this algorithm is remained the same. The classical washout algorithm is generally provided as a linear

C.K. Loo et al. (Eds.): ICONIP 2014, Part III, LNCS 8836, pp. 474–482, 2014.
© Springer International Publishing Switzerland 2014

cueing algorithm by Reid and Nahon [7-9] and it is mostly used in commercial simulators because of its relative simplicity. It is the basic solution broadly used in dynamic simulators. The second algorithm assessed in this research is the Adaptive Washout Algorithm that was proposed by Parrish et al. [10] to overcome the drawback of the constant classical filter parameters. The variation of the Coordinated Adaptive Washout Algorithm was developed by L. Reid and M. Nahon [7, 8]. The fourth algorithm that is Optimal algorithm was provided at MIT by Sivan, et al. [11] and it has been implemented in Reid and Nahon's study [7, 8]. This technique takes the human perception feature into account by minimizing the error of the perception between the dynamics ones of the simulated vehicle and of the real vehicle. In optimal washout filters, the weight coefficients and filter parameters are fixed that will not make efficient use of the workspace for producing the best motion sensation when the maneuvers are less severe and the motion will be conservative. Since the tuning process most commonly uses white noise in the approach based on the optimal control theory, optimal transfer function that is produced is optimal for a set of different input scenarios regenerated by white noise and it may not give good results for all circumstances.

Even though classical washout filter has very simple form, it has very good result and matching with the reference curve. The classical washout filter extensively used in commercial simulators because of their main advantages such as simplicity, easy to adjust, short processing time and stable performance. However, the classical washout filter has some disadvantages such as first, the phase and magnitude in the summation of the high and low frequency specific force are distorted and falsified since the high and low pass filters utilized in tilt coordination are not ideal and perfect. Second, it is tough to reproduce continuous motion in the simulator owing to the phase delay produced by the low-pass filter and tilt coordination that leads to wrong sensation of motions. This lapse is one of the reasons of the simulator sickness and this problem is more serious in reckless-driving, and also car accident simulations. In addition, classical washout filters come with the fact that the filters parameters are constant during its application that makes the resulting simulator fail to suit all circumstances. In addition, their tuning process focused on the worst case situations, which leads to a poor usage of both the workspace and the platform displacements and capabilities.

You et al. [12] proposed a washout algorithm that used new tilt coordination technique in its structure to give more accurate and real sensations to drivers within limitation of motional range. This new tilt coordination algorithm has been proposed for compensating the falsified specific force and wrong sensation that is produced by classical washout filter tilt coordination that can reappear specific force more completely. To eliminate the phase delay of the low pass filter, the low pass filter in the tilt coordination system of the classic washout algorithm is removed. Their new tilt coordination method applied the difference between translational acceleration and translational one passing through high-pass filter in order to resolve the mentioned problem of applying low pass-filter on tilt coordination in the classical washout algorithm. According to their results [12], however it presented the driver with the improved translational specific force, this proposed new tilt coordination algorithm made the roll and pitch motion sensation worse, and the quick return component made the driver feel an artificial motion sensation. Thus, their method produced somewhat extra rotational sensation motion of pitch and roll direction and an artificial motion sensation comparing with classical tilt coordination that can lead simulator sickness.

There is a problem that is produced on generating pitch and roll sensation as utilizing only high-pass filter in classical washout structure. Therefore, the low pass filter is remained in this paper to stop producing extra rotational sensation motion of pitch and roll direction and make it worse compared with the previous classical washout structure. For decreasing wrong cueing motion caused by tilt coordination and also utilizing workspace efficiently, new technique for minimizing the human perception error between real and simulator driver is provided. In this paper, we use the new translational channel technique based on tilt rate limit adaptive error compensator for providing better translational human perception compared with previous model proposed by [12] and classical washout filter without affecting roll and pitch motion. We vary the gain of subtracting low pass signals and rate limited signals adaptively by fuzzy logic system and add the compensator signal to translational channel throughout the workspace such that motion system states are less limited when they are far from the limits. This can improve human sensation more without adding artificial sensation on rotational channel and exploit the work space more efficiently. This paper is organised as follows. Section 2 introduces vestibular system and its mathematical equations. The proposed cueing motion method and its concepts are provided in details in section 3. Finally, section 4 is given to prove the ability of this method over previous methods.

2   Vestibular System

Different kinds of motions that are applied to drivers can be perceived by driver's vestibular system that is in the inner ear as signals. The vestibular system includes semicircular canals and otoliths which are the main sensors mammals use to perceive rotational and linear motion. The specific force perceived in otolith is attained by subtracting acceleration of gravity $\vec{g}$ from translational acceleration $\vec{a}$ as below

$$\vec{F} = \vec{a} - \vec{g} \tag{1}$$

Otolith organs cannot distinguish translational acceleration from gravity or tilt. Specific force sensation model transfer function that links sensed specific force to the real input specific force [13, 14] is shown in

$$\frac{\hat{f}}{f} = \frac{K\,(1+\tau_a s)}{(\tau_L s+1)(\tau_s s+1)} \tag{2}$$

Rotational motion sensation model transfer function is given in (3) according to Zacharias [14] and based on his survey on the previous research [15-17]. The sensed rotational motion is related to the input by the semicircular canals mode transfer function as equation below

$$\frac{\hat{\omega}}{\omega} = \frac{\tau_a\,s}{(\tau_a s+1)} \cdot \frac{\tau_1 s}{(\tau_1 s+1)(\tau_2 s+1)} \tag{3}$$

3   Proposed Cueing Motion Algorithm Method

The classical algorithm is composed of high and low pass filters for the translational and the rotational degrees of freedom with a crossover path to provide the steady state or gravity alignment cues.   Filtering plays an essential role in the classical washout filter because of the physical limitations of motion platforms and to generate the closest possible translational movements.   Therefore, it is needed to decrease the low frequency accelerations that tend to drive motion platforms to their workspace limits. High-pass filtering is used in the translational and rotational channel to remove low frequency components of accelerations and angular velocity (4) and (5) from the vehicle dynamic model for limiting translational movement and ensures platform displacement to return into its origin point. A low-pass filter can transfer the sustained component of the translational acceleration signal to the angular dynamic by tilt coordinating under the driver's perception threshold that uses gravity as a deceptive sustained acceleration as it is shown in Fig. 1. The used low-pass filter in classical is given in equation (6) where ς is a damping ratio and ω_n is cut off frequency.

$$HP_{translational} = \frac{s^2}{s^2 + 2\varsigma\omega_n s + \omega_n^2} \tag{4}$$

$$HP_{Rotational} = \frac{s}{s + \omega_b} \tag{5}$$

$$LP_{tilt-coordination} = \frac{\omega_n^2}{s^2 + 2\varsigma\omega_n s + \omega_n^2} \tag{6}$$

To avoid wrong rotational sensation from translational command by vestibular system, the rate of tilt angle should be kept below human perception threshold limit. Thus a rate limiter is placed after tilt coordination to limit angular velocity of rotational channel generated by sustained acceleration to have both translational sensation and avoid wrong rotational cueing motion feeling for vehicle driver. Even though, rate limiter keeps tilt angular velocity within allowed borders, it results in making unrealistic acceleration sensation for driver. Tilt angle before and after rate limit is shown in Fig. 2. The objectives of our new proposed system are compensation of error due to difference between desired tilt and rate limited tilt and improving the translational human sensation comparing to previous method [12] without disturbing the rotational sensation.

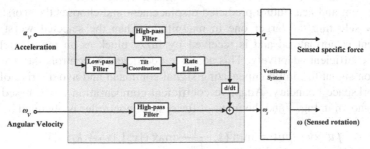

Fig. 1.   Classical washout filter schematic

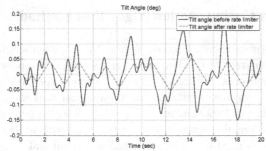

Fig. 2. Tilt angle before and after rate limiter

3.1 New Washout Filter Based on Tilt Rate Limit Error Compensator

In new proposed method, the error caused by saturation of rate limiter is obtained by the considering the difference between output and input of tilt rate limiter and multiplying it by gravity of earth. The new proposed tilt rate limit error compensator is shown in Fig. 3. The range of coefficient K is from 0 to 1 and ideally it is supposed to be 1 to fully compensate the error and when K is 0 the proposed system acts similar to the classical washout filter. The displacement filter is an LTI (linear time-invariant) transfer function with two zeros in nominator to ensure return of platform displacement to its original point. It is used to transform the produced signal to translational channel and add it to translational acceleration. The classical washout filter acts conservative in term of platform displacement. Therefore, our proposed method uses the rest of platform space in order to exploit the platform more efficiently and fulfil compensation of tilt cueing motion error. The error caused by rate limit of tilt channel is corrected by inserting compensator signal into translational channel online. The adaptive variation of the coefficient K is provided in order to exploit the platform workspace more efficiently when it is far from the limits and control it when it is close to physical boundary.

3.2 Adaptive Translational Channel Algorithm by Integrating Fuzzy Logic

In applications with uncertain relationship between inputs and output, fuzzy logic provides a reliable solution. The compensation coefficient K is controlled by a fuzzy logic block that is shown in Fig. 3. This adaptive coefficient block considers both displacement and near future predicted displacement and chooses the worst case scenario by selecting the larger one in magnitude. Then the selected worst case displacement is normalized and is received by fuzzy block as an input to change the suitable coefficient adaptively. This adaptive coefficient K controls the magnitude of correction signal to limit compensating signal command and avoid strike of platform with workspace boundary. Adaptive coefficient compensation block based on fuzzy logic is shown in Fig. 4 and the input of fuzzy logic controller is obtained by

$$fuzzy\ input = \min\left(1, \frac{1}{x_{d,max}} \max\left(|x_d|, |x_d + k_i \dot{x}_d|\right)\right) \qquad (7)$$

where k_i is set to 0.1 sec. The range of fuzzy input is from 0 to 1 on which centre, middle and boundary membership functions are applied as follows

$$\mu_{centre}(x; 0, 0.4) = \max\left(\min\left(\frac{0.4-x}{0.4}, 1\right), 0\right) \tag{8}$$

$$\mu_{middle}(x; 0.1, 0.5, 0.9) = \max\left(\min\left(\frac{x-0.1}{0.4}, \frac{0.9-x}{0.4}\right), 0\right) \tag{9}$$

$$\mu_{boundary}(x; 0.6, 1) = \max\left(\min\left(\frac{x-0.6}{0.4}, 1\right), 0\right) \tag{10}$$

Membership functions can be represented in different forms such as triangular, trapezoidal, Gaussian, sigmoid and generalized bell. However, for the sake of light computation, triangular membership functions are chosen for this proposed system as shown in Fig. 5 (a).

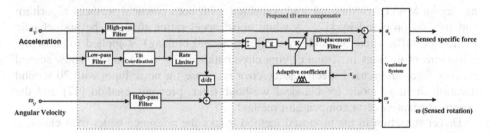

Fig. 3. New proposed tilt rate limit error compensator

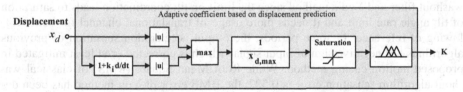

Fig. 4. Fuzzy control scheme of compensation coefficient

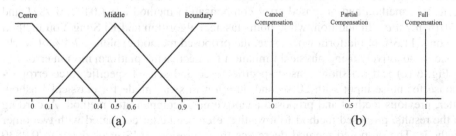

Fig. 5. (a) Input Membership Function of adaptive coefficient fuzzy controller (b) Output Membership function

The rules are implemented based on Sugeno-type fuzzy inference because of good smoothness properties. Three rules are applied for proposed fuzzy controller

If *displacement is centre* Then *gain is full compensation* $(K = 1)$
If *displacement is middle* Then *gain is partial compensation* $(K = 0.5)$
If *displacement is boundary* Then *gain is cancel compensation* $(K = 0)$

This Sugeno inference follows

$$\begin{cases} \Psi: X \rightarrow Y \\ x \mapsto \dfrac{\sum_{m=1}^{3} \mu_{A_j}(x).f_i}{\sum_{m=1}^{3} \mu_{A_j}} \end{cases} \tag{11}$$

where A_j is subset of input space.

4   Results

The simulation of proposed compensation of tilt coordination wrong cueing is implemented in MATLAB/Simulink and compared with both classical washout algorithm and method proposed by [12]. A white noise acceleration that can be a stochastic motion in traffic situation (sudden acceleration and braking) is applied to show the sensations of motions in virtual driving circumstance. Fig. 6 (a) and (b) show sensed specific force and sensed specific force error response for noise input with 20 second duration in surge mode for classical washout filter, previous method [12] and the proposed adaptive error compensator method.

Driver sensation in the proposed method tracks the reference better than classical motion cueing algorithm and previous method and they are closer to the reference sensed specific force. Sudden falls and rises cannot be correctly tracked by classical washout filter and You's method since the limit on tilt coordination leads to saturation of tilt angle rate limit and it needs more usage of translational channel for better following of reference. In some periods, the output acceleration sensation of previous algorithms has negative sign of input signal which is corrected or at least mitigated in proposed motion cueing method. While Root Mean Square (RMS) of classical washout algorithm sensation error is 0.522, the RMS error of You method has been decreased to 0.4869 and our proposed method decreases the sensation RMS error down to 0.3279 m/sec^2 . The head force RMS error for classical washout algorithm, Ki Sung You method and proposed error compensation method are 1.0402, 0.9821 and 0.7015 m/sec^2. In addition, while both classical algorithm and Ki Sung You method use only 11.3% of platform workspace, the proposed method exploits 47.4% of workspace by adaptively using physical limitation for decreasing platform motion error.

Fig. 6 (c) and (d) show sensed specific force and sensed specific force error response for noise input with 20 second duration in sway mode for classical washout filter, previous method and proposed adaptive error compensator method. According to the results, proposed method follows the reference better compared with two other methods. The proposed method decreases the sensation RMS error down to 0.2830 m/sec^2 while RMS of classical washout algorithm sensation error is 0.5005 and the RMS error of You's method has been reduced to 0.4732. Moreover, while both classical algorithm and Ki Sung You method use only 6.46% of platform workspace and they are very conservative, the proposed method exploits 52.25% of workspace by adaptively using physical limitation for decreasing platform motion error.

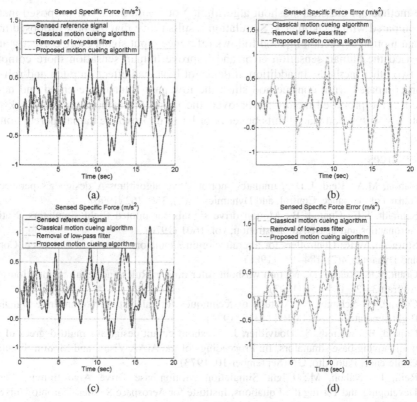

Fig. 6. Simulated results for noise input (a) Surge sensed specific force (b) Surge sensed specific force error (c) Sway sensed specific force (d) Sway sensed specific force error

5   Conclusion

The classical washout filters have some problems such as generation of wrong sensation of motions, false cue motion, and tuning process focused on the worst case situations, which leads to a poor usage of the workspace displacements. The saturation of tilt rate limit causes wrong cueing motion which can result in significant sensation error and leads to simulator sickness. This motion cueing error needs to be fixed. This error correction requires a signal in tilt coordination which is out of capacity of tilt coordination channel without generation wrong rotation sensation therefore motion correction through translational channel would be the alternative which increases platform workspace usage. In our proposed method, the difference between input of tilt coordination and its rate limited form are calculated and transformed from tilt angle to translational acceleration. An adaptive fuzzy block controls percentage of compensation from 100% when we platform is at the centre of workspace to 0% when platform is close to workspace boundaries. The result generated by fuzzy logic is needed to be filtered in order to washout platform displacement to its original point. Through this technique motion system states are less limited when they are far from the physical limits to exploit work space more efficiently. Using MATLAB/Simulink three methods of classical washout algorithm, You method and our proposed method are compared with each other's. Simulation results prove that the new proposed translational motion correction method follows reference input signal more correctly and it decreased the human sensation error and improve human sensation more compared with two other methods. In addition, it does not have bad effect on pitch and roll sensation compared with You method since the low pass filter is preserved and not removed in the proposed method. Moreover, the proposed method exploits workspace adaptively within boundary limits better in order to improve the human perception.

References

1. Nahon, M.A., Reid, L.D.: Simulator motion-drive algorithms-A designer's perspective. Journal of Guidance, Control, and Dynamics 13(2), 356–362 (1990)
2. Schmidt, S.F., Conrad, B.: Motion drive signals for piloted flight simulators. National Aeronautics and Space Administration, vol. 1601 (1970)
3. Sturgeon, W.R.: Controllers for aircraft motion simulators. Journal of Guidance, Control, and Dynamics 4(2), 184–191 (1981)
4. Grant, P.R., Reid, L.D.: Motion washout filter tuning: Rules and requirements. Journal of Aircraft 34(2), 145–151 (1997)
5. Conrad, B., Schmidt, S.: A study of techniques for calculating motion drive signals for flight simulators. NASA CR-114345 (1971)
6. Conrad, B., Schmidt, S., Douvillier, J.: Washout circuit design for multi-degrees of freedom moving base simulators. In: Proceedings of the AiAA Visual and Motion Simulation Conference, Palo Alto, CA (September 10, 1973)
7. Reid, L., Nahon, M.: Flight Simulation Motion-base Drive Algorithimns: Part 1-Developing and Testing the Equations. Institute for Aerospace Studies, Toronto University (1985)
8. Reid, L., Nahon, M.A.: Flight Simulation Motion-Base Drive Algorithms.: Part 2, Selecting the System Parameters. Utias Report, N307 (1986)
9. Reid, L., Nahon, M.: Flight simulation motion-base drive algorithms. Part 3: Pilot evaluations (1986)
10. Parrish, R.V., Dieudonne, J.E., Martin Jr, D.J.: Coordinated adaptive washout for motion simulators. Journal of Aircraft 12(1), 44–50 (1975)
11. Sivan, R., Ish-Shalom, J., Huang, J.-K.: An Optimal Control Approach to the Design of Moving Flight Simulators. IEEE Transactions on Systems, Man and Cybernetics 12(6), 818–827 (1982)
12. You, K.S., et al.: Development of a washout algorithm for a vehicle driving simulator using new tilt coordination and return mode. Journal of Mechanical Science and Technology 19(1), 272–282 (2005)
13. Meiry, J., Young, L.: A revised dynamic otolith model (1968)
14. Zacharias, G.L.: Motion cue models for pilot-vehicle analysis. DTIC Document (1978)
15. Peters, R.A.: Dynamics of the vestibular system and their relation to motion perception, spatial disorientation, and illusions (1969)
16. Young, L., Oman, C.: Model for vestibular adaptation to horizontal rotation. Aerospace Medicine 40(10), 1076–1080 (1969)
17. Jones, G.M., Barry, W., Kowalsky, N.: Dynamics of the Semicircular Canals Compared In Yaw, Pitch and Roll, vol. 35, p. 984 (1964)

Adaptive Washout Algorithm Based Fuzzy Tuning for Improving Human Perception

Houshyar Asadi[1], Arash Mohammadi[2], Shady Mohamed[3], Delpak Rahim Zadeh[4], and Saeid Nahavandi[5]

[1] Centre for Intelligent Systems Research, Deakin University, Australia
hasadi@deakin.edu.au
[2] Electrical Department, Engineering Faculty, University of Malaya, Malaysia
arash_7mh@siswa.um.edu.my
[3] Centre for Intelligent Systems Research, Deakin University, Australia
shady.mohamed@deakin.edu.au
[4] Geelong, Victoria, Australia
drahimzade@gmail.com
[5] Centre for Intelligent Systems Research, Deakin University, Australia
saeid.nahavandi@deakin.edu.au

Abstract. The aim of this paper is to provide a washout filter that can accurately produce vehicle motions in the simulator platform at high fidelity, within the simulators physical limitations. This is to present the driver with a realistic virtual driving experience to minimize the human sensation error between the real driving and simulated driving situation. To successfully achieve this goal, an adaptive washout filter based on fuzzy logic online tuning is proposed to overcome the shortcomings of fixed parameters, lack of human perception and conservative motion features in the classical washout filters. The cutoff frequencies of high-pass, low-pass filters are tuned according to the displacement information of platform, workspace limitation and human sensation in real time based on fuzzy logic system. The fuzzy based scaling method is proposed to let the platform uses the workspace whenever is far from its margins. The proposed motion cueing algorithm is implemented in MATLAB/Simulink software packages and provided results show the capability of this method due to its better performance, improved human sensation and exploiting the platform more efficiently without reaching the motion limitation.

Keywords: Motion Cueing Algorithm, Human perception, Washout filter, Fuzzy Logic.

1 Introduction

Driving simulators deliver a realistic driving feeling with the replication of several driving cues which are most necessary for perceiving by human. More technically, the use of virtual reality techniques to be as close as a real situation in the domain of driving simulation would be essential to equip the simulator with more multimodal cues. These cues can be categorized in three categories: visual, hearing and inertial cue. Inertial cues

C.K. Loo et al. (Eds.): ICONIP 2014, Part III, LNCS 8836, pp. 483–492, 2014.

are produced by the Motion Cueing Algorithm in the platform for transforming vehicle accelerations and angular velocities in to the simulator platforms within the physical limitation to present to the simulator driver the similar sensation of driving in a real vehicle.

The first washout algorithm is known as the classical algorithm and has been provided by Conrad and Schmidt [1, 2]. The classical washout algorithm is generally provided as a linear cueing algorithm by Reid and Nahon [3-5] and it is largely used in commercial simulators because of its relative simplicity. The second algorithm is the Adaptive Washout Algorithm that was provided by Parrish et al. [6] to solve the problem of fixed parameters in classical washout filter. The variation of the Coordinated Adaptive Washout Algorithm was developed by L. Reid and M. Nahon at the University of Toronto [3, 4]. For overcoming the shortcoming of the lack of human perception the fourth algorithm that is optimal algorithm was provided by Sivan, et al. [7] and it has been implemented in Reid and Nahon's study [3, 4]. This method minimize the error of the human perception between the simulated and the real vehicle driver sensation by taking the human perception feature into account. Mainly, optimal washout filter is optimal for a set of different input scenarios regenerated by white noise and it may not give good results for all of the circumstances. In addition, it is still conservative and cannot exploit the workspace efficiently.

Classical washout filters are widely used in different simulators because of their simplicity, short processing time and easy to adjust [8]. One of the main problems of the classical washout filters is that the filter parameters are fixed during its application and it is attuned by worst case scenario tuning method that is the major reason of a poor usage of the available motion platform. This leads inflexibility of the structure and makes the resulting simulator fail to suit all circumstances. Moreover, the classical does not take human perception into account for producing motion cue.

Song [9] proposed a classical washout filter with fuzzy-based tuning for a motion simulator to overcome the problems of fixed parameters in classical washout filters and problems with adaptation to different signals. Even though this method presented better performance than fixed parameter classical washout filters, the feedback of human perception error between real and simulator driver has no role in parameter tuning of his proposed method. His technique was still oriented towards minimizing the state error between the vehicle and simulator rather than the human perceptual error. Although the previous technique provides acceptable result but it cannot produce accurate sensation as implemented for vehicle washout filter in traffic situation includes brutal and sudden accelerations and braking.

The aim of this research is to provide an adaptive washout filter based on fuzzy logic tuning to produce more accurate signal and minimize the human sensation error between the simulator and real vehicle driver in simulated traffic situation while using the workspace better. The cutoff frequencies of high-pass and low-pass filters in translational, tilt coordination and rotational channels are adjusted in the real time according to the displacement and angle of platform, workspace limitation and human sensation online based on fuzzy logic system to overcome the shortcomings of fixed parameters and lack of human perception in the algorithm. In addition, an intelligent fuzzy based scaling method is proposed in this paper to guarantee the efficient usage of workspace.

This paper is organized as follows. Section 2 introduces the classical washout filter that is the base of this research and section 3 presents human perception model. The proposed adaptive washout algorithm based Fuzzy tuning is proposed in section 4 and the results of this research are presented in section 5.

2 Classical Washout Algorithm

The basic idea of the classical washout filters is generating specific forces and rotations at the driver's perception in the simulator similar to those they would experience in a real car. Classical washout algorithm is composed of high and low pass filters for translational, rotational and coordination channel as shown in Fig. 1. High-pass filtering is utilized in the translational and rotational channel to eliminate low frequency components of accelerations $\vec{a}_v$ and angular velocity $\vec{\omega}_v$ as given in (1) and (2) from the vehicle dynamic model for limiting translational motion and ensuring to return platform into its origin point.

$$HP_{translational} = \frac{s^2}{s^2 + 2\zeta\omega_n s + \omega_n^2} \tag{1}$$

$$HP_{rotational} = \frac{s}{s + \omega_b} \tag{2}$$

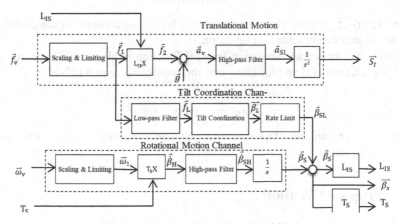

Fig. 1. Classical Washout Filter algorithm

A low-pass filter can transfer the sustained component of the translational acceleration to the angular motion by tilt coordinating under the driver's perception threshold that uses gravity as a deceptive sustained acceleration. The low-pass filter is given as

$$LP_{tilt-coordination} = \frac{\omega_n^2}{s^2 + 2\zeta\omega_n s + \omega_n^2} \tag{3}$$

where ζ is a damping ratio and ω_n is cutoff frequency.

3   Human Perception Model

Human perceive specific force and rotational motion by otolith organs and semicircular system which are located inside the inner ear. The perceived specific force by otolith is attained by subtracting acceleration of gravity $\vec{g}$ from translational acceleration $\vec{a}$ as below

$$\vec{sF} = \vec{a} - \vec{g} \tag{4}$$

This transfer function can be transformed to the following equation in the coordinative system at the center of the seat where $\vec{a}_T$ is the translational acceleration and $R(\alpha, \beta)\vec{g}$ is the acceleration of gravity in a driver seat coordinate system.

$$\vec{F}_T = \vec{a}_T - R(\alpha, \beta)\vec{g} = \begin{bmatrix} \vec{a}_{Tx} - \vec{g}\sin\beta \\ \vec{a}_{Ty} - \vec{g}\sin\alpha\sin\beta \\ \vec{a}_{Tz} - \vec{g}\cos\alpha\cos\beta \end{bmatrix} \tag{5}$$

Young and Meiry [10] provided the mathematical model of otolith and the same set of parameters are used for all three axes and they are summarized in Zacharias research [11]. The transfer function of specific force sensation model that links sensed specific force to the real input specific force is given

$$\frac{\hat{f}}{f} = \frac{K(1+\tau_a s)}{(\tau_L s+1)(\tau_s s+1)} \tag{6}$$

where τ_s, τ_L, τ_a are the short time constant, long time constant of the orolith organs, and neural processing term lead operator.

Rotational mathematic sensation transfer function is given in (7) according to Zacharias's research [11] and his survey on the previous research [12-14].

$$\frac{\hat{\omega}}{\omega} = \frac{\tau_a s}{(\tau_a s+1)} \cdot \frac{\tau_l s}{(\tau_1 s+1)(\tau_2 s+1)} \tag{7}$$

where τ_a, τ_l, τ_2 are the adaptation time constant, long time constant, short time constant respectively.

4   Proposed Method

The main drawbacks of classical washout filters are fixed parameters which are attuned by worst case scenario and lack of human perception in the algorithm. In previous methods, tuning washout filters was done while human sensation error as a control parameter was neglected. In this research, cutoff frequencies of all washout filters are tuned according to displacement of platform, workspace limitation and output human sensation error in order to overcome the drawbacks of previous washout filters. In systems with extreme stochastic behavior and variable types of input, fuzzy logic is one of the best suggestions. Fuzzy control rules of motions cueing algorithm are applied online in order to improve drawbacks of fixed parameter washout filter and help it perform better for various circumstances. The fuzzy blocks will generate suitable cutoff frequencies in order to decrease sensation error regarding physical margins. Tilt coordination based motion cueing algorithm with fuzzy tilt rate and scale factor tuning proposed by Song et al. [9] improves motion sensation more than fixed parameters ones nevertheless it still has human sensation error in brutal input cases such as traffic situations. In addition, in previous method the range of allowed tilt rate is chosen between 1 to 5 deg/sec however since human perception threshold for surge mode is 3.6 deg/sec, we avoid allowing rate limit in tilt channel to pass angular velocity rates above that angular velocity to prevent false cueing motion in human perception.

The proposed fuzzy-tuning washout filters structure based on displacement and human sensation is shown in Fig. 2. In this scheme, the control block for tuning cutoff frequency ω_n in translational high-pass washout filter receives two input signals online, sensed specific force error and platform displacement. Sensed specific force error is the difference between real vehicle and simulator driver sensed specific forces

$$\hat{f}_{error} = \hat{f}_{vehicle} - \hat{f}_{simulator} \tag{8}$$

And platform displacement limit in real time is defined as normalized remained distance to the nearest workspace limit as given

$$\text{Displacement Limit} = 1 - \frac{|x_d|}{x_{d,max}} \tag{9}$$

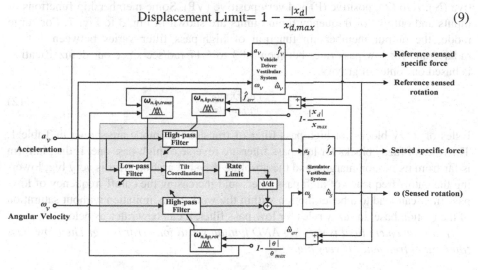

Fig. 2. Proposed fuzzy-tuning washout filters based on displacement, angle and human sensation

The fuzzy logic block is provided to control ω_n according to fuzzy membership functions for both displacement and human sensation and rules. When displacement limit is higher and less strict and when sensed acceleration error is higher, the fuzzy unit decreases cutoff frequency to let the platform use all of the physical limitation and decrease the error. The fuzzy controller unit of low-pass washout filter in tilt channel receives sensed specific force error and angle limit for adjusting cutoff frequency ω_n. Angular limit platform in real time is defined as normalized remained angle to the nearest angular physical limitation is given

$$\text{Angular Limit} = 1 - \frac{|\theta|}{\theta_{max}} \tag{10}$$

While angular limit is higher and sensed acceleration error is big, low-pass fuzzy unit increases ω_n for exploiting the workspace efficiently. The third fuzzy unit is proposed in the rotational channel to tune high-pass filter cutoff frequency according to angular limit of platform motion and perceived angular velocity error defined as

$$\widehat{\omega}_{error} = \widehat{\omega}_{vehicle} - \widehat{\omega}_{simulator} \tag{11}$$

4.1 Fuzzy Logic Controller Blocks

In this research, fuzzy blocks are chosen Mamdani-type and membership functions are chosen by trial and error in order to decrease output sensed error. The input area for distance limit is divided between five membership functions which are very near (VN), near (N), middle (M), far (F) and very far (VF). Since the human vestibular system is not able to recognize the quantity of the motion precisely and it is limited to detection of large, medium, or small motion in each direction, in our fuzzy logic design, the motion sensation of each direction is categorized within five sensation quality. Membership functions of sensed specific force error are very negative (VN), negative (N), zero (Z), positive (P) and very positive (VP). Some membership functions of inputs and outputs of frequency tuning units are shown in Fig. 3 to Fig. 7. For surge mode, the output membership function of high-pass filter varies between 1 to 3 rad/sec and for low-pass filter between 4.83 to 6.17 rad/sec since our defuzzification is based on center of gravity

$$z^* = \frac{\int \mu_i(z).z \, dz}{\int \mu_i(z).dz} \tag{12}$$

Rules of fuzzy blocks for high-pass filter of translational mode are shown in Table 1. The rules of fuzzy blocks for low-pass filter are reverse of high-pass one. If the platform is far from its motion margins and the absolute driver sensation error is very big, lowering the cutoff frequency of high-pass filters and increasing the cutoff frequency of low-pass filter can lead to a better motion within the workspace limitation without saturation of the motion base. In fuzzy rules for low-pass filter, it can be written as below

IF displacement limit is very big AND human sensation error is big, Then low-pass filter cutoff frequency is very big.

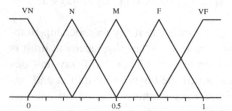

Fig. 3. Input membership of Distance Limit

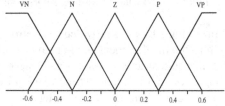

Fig. 4. Input membership functions of sensed specific force error

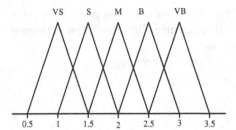

Fig. 5. Output membership of ω_n for surge high-pass filter

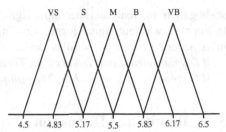

Fig. 6. Output membership of ω_n for surge tilt low-pass filter

Table 1. Fuzzy rules for translational high-pass filter cutoff frequency

		Sensed error			
		VN	N	Z	P	VP
Distance Limit	VN	B	VB	VB	VB	VB
	N	M	B	B	B	M
	M	S	M	M	M	S
	F	VS	S	S	S	VS
	VF	VS	VS	VS	VS	VS

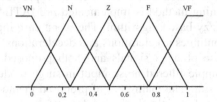

Fig. 7. Input membership of Angular Limit

4.2   Fuzzy Scaling Method

Drivers can not differentiate the real experienced forces during driving and the slightly decreased ones in the simulator since difference is not very large. Thus, scaling and limiting is used in order to decrease the amplitude of the motion signals in washout filter algorithms and this can decrease and attenuate the effects of the workspace limitations in the simulator capabilities and improve reality sensation in movement since limitations are respected, however in some cases it can lead poor usage of workspace.

The scaling and limiting of vehicle accelerations and angular rates is necessary for producing motion platform commands that are proportional to those of the simulated vehicle while it does not exceed the physical limitations of the motion platform. The scaling procedure can also be defined as a gain function it is shown in (13). Moreover, the limiting process can be deduced as a conditional statement as stated in (14) where L_i is the limiting value and the slope K_i is the scaling value.

$$Input*K_i = Scaled\ Input \tag{13}$$

$$Limited\ Input = \begin{cases} Input & If\ |Input| < L_i \\ L_i \left(\frac{Input}{|Input|} \right) & Else \end{cases} \tag{14}$$

In this paper, a scaling fuzzy based method is proposed to guarantee the efficient usage of workspace. This scaling method is using displacement information online to change the slope of scaling adaptively. It can allow the washout filter to produce more accurate signals as far as the platform is far from limitations by giving appropriate scale gain to the real reference input signal. This intelligent scaling applies fuzzy rules to improve washout signal form according to displacement limit to follow reference more accurate. Some of the rules can be written as

If Displacement Limit is Very Far, Then output scale is Very Big.
If Displacement Limit is Far, Then output scale is Big.

5 Results and Discussion

In previous method, fuzzy logics tried to decrease output error while they worked independent from human vestibular system model. When human perception error is low it is not necessary for tuning block to change frequency for purpose of decreasing error which was a trivial effort in previous methods. The new proposed fuzzy based tuning method is implemented in MATLAB/Simulink to be compared with previous fuzzy-based algorithm. The worst case input is in traffic modes which involves with continues accelerations and decelerations. Filtered white noise input can be the worst case of input signals and both proposed and previous methods are compared with sample filtered noise input in surge mode with duration of 20 seconds as shown in Fig. 8. The main goal of proposed method is better following of reference sensation signal in real vehicle and decreasing human perception error between real and simulator driver. Fig. 9 illustrates that output result of proposed method has been closer to reference signal. According to the figure, many spikes had been mitigated and the reference human sensation can be followed more accurate compared with previous fuzzy based tuning method since human vestibular sensation is taken into account in filters tuning procedure. Longitudinal perceived specific force error between the real case and simulator driver is shown in Fig. 10. It has been shown that the human sensation error has been decreased compared to previous method. The improved human sensation can eliminate simulator sickness for the simulator driver in this case. The sensed specific force below human threshold (which is $0.17\ m/s^2$ for surge mode) cannot be sensed by human vestibular system, thus any error under human threshold does not affect human perception. The output specific force sensed sensation is calculated according to human sensation model and performance of fuzzy blocks is not independent from vestibular model. When output sensed error increases, as far as displacement is far from limits, the proposed method allows high-pass filter cutoff frequency to decrease and as far as angular position is far from limits, the cutoff

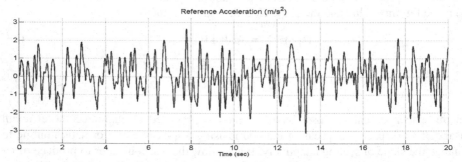

Fig. 8. Noise reference acceleration for surge mode with duration of 20 seconds

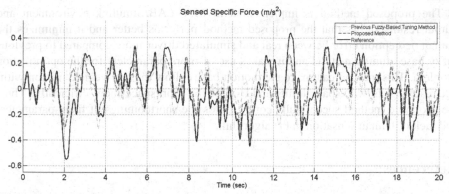

Fig. 9. Longitudinal sensed specific force of reference, proposed method and previous fuzzy-based method

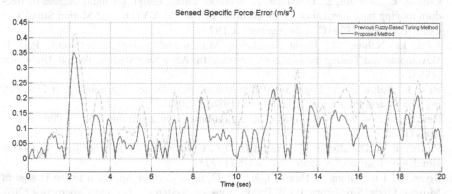

Fig. 10. Longitudinal sensed specific force error comparison of proposed method and previous fuzzy-based method

frequency for low-pass filter is increased in order to generate more realistic performance of motion cueing algorithm. Even though, consideration of online displacement and angular limit alone in tuning the cutoff frequencies improves output results, taking human perception and fuzzy scale into account changes behavior of controllers and makes them less conservative. Therefore, they would have better usage of workspace in order to decrease human perception error. This new method has caused the RMS (Root Mean Square) error of sensed specific force declines from 0.1550 m/s2 to 0.1122 m/s2.

6   Conclusion

Tuning washout filters online is one of the methods for improving performance and human sensation in cueing motion algorithm. The cutoff frequencies of high-pass, low-pass filters are adjusted according to the corresponding displacement information of platform, workspace limitation and human sensation in real time based on fuzzy logic system. A fuzzy scaling has been applied in order to use workspace displacement more efficiently by changing the scale factor adaptively.

The proposed method is implemented in MATLAB/Simulink environment and output results shows that the proposed method performs better and it minimizes the human perception error between real and simulated driver more compared to previous fuzzy based tuning method and it makes use of more parameters for the optimization problem. This method can make the motion base follow the real vehicle motion without reaching its workspace limits and exploit the workspace more efficiently. Thus, the proposed method successfully compensates the shortcomings of fixed parameters and lack of human sensation in the algorithm.

References

1. Conrad, B., Schmidt, S.: A study of techniques for calculating motion drive signals for flight simulators. NASA CR-114345 (1971)
2. Conrad, B., Schmidt, S., Douvillier, J.: Washout circuit design for multi-degrees of freedom moving base simulators. In: Proceedings of the AiAA Visual and Motion Simulation Conference, Palo Alto (CA) (September 10, 1973)
3. Reid, L., Nahon, M.: Flight Simulation Motion-base Drive Algorithimns: Part 1- Developing and Testing the Equations. Institute for Aerospace Studies, Toronto University (1985)
4. Reid, L., Nahon, M.A.: Flight Simulation Motion-Base Drive Algorithms.: Part 2, Selecting The System Parameters. Utias Report (1986) (N307)
5. Reid, L., Nahon, M.: Flight simulation motion-base drive algorithms. Part 3: Pilot evaluations (1986)
6. Parrish, R.V., Dieudonne, J.E., Martin Jr, D.J.: Coordinated adaptive washout for motion simulators. Journal of Aircraft 12(1), 44–50 (1975)
7. Sivan, R., Ish-Shalom, J., Huang, J.-K.: An Optimal Control Approach to the Design of Moving Flight Simulators. IEEE Transactions on Systems, Man and Cybernetics 12(6), 818–827 (1982)
8. Nehaoua, L., et al.: Restitution movement for a low cost driving simulator. In: American Control Conference. IEEE (2006)
9. Song, J.-B., Jung, U.-J., Ko, H.-D.: Washout algorithm with fuzzy-based tuning for a motion simulator. KSME International Journal 17(2), 221–229 (2003)
10. Meiry, J., Young, L.: A revised dynamic otolith model (1968)
11. Zacharias, G.L.: Motion cue models for pilot-vehicle analysis. DTIC Document (1978)
12. Peters, R.A.: Dynamics of the vestibular system and their relation to motion perception, spatial disorientation, and illusions (1969)
13. Young, L., Oman, C.: Model for vestibular adaptation to horizontal rotation. Aerospace Medicine 40(10), 1076–1080 (1969)
14. Jones, G.M., Barry, W., Kowalsky, N.: Dynamics of the Semicircular Canals Compared in Yaw, Pitch and Roll, vol. 35, p. 984 (1964)

Neurophysiology of Insects Using Microelectrode Arrays: Current Trends and Future Prospects

Julie Gaburro[1,2], Jean-Bernard Duchemin[2], Asim Bhatti[3], Peter Walker[2], and Saeid Nahavandi[3]

[1] Centre for Integrative Ecology, School of Life & Environmental Sciences,
Deakin University, Waurn Ponds VIC 3216, Australia
Julie.gaburro@deakin.edu.au
[2] Australian Animal Health Laboratory, Geelong, Vic 3220, Australia
{Julie.gaburro,Jean-Bernard.Duchemin,Peter.walker}@csiro.au
[3] Center for Intelligent Systems Research, Deakin University, Vic 3217, Australia
{asim.bhatti,saeid.nahavandi}@deakin.edu.au

Abstract. Simple to complex behaviors are directed by the brain, which possess nervous cells, called neurons. Mammals have billions of neurons, organized in networks, making their study difficult. Although methods have well evolved since the last century, studying a simpler model is the key to resolving neuronal communication. In this review, we demonstrate that insects are an excellent model and tool to understand neural mechanisms. Moreover, new technology, such as Microelectrodes Arrays (MEAs), is an innovative method which opens the possibility to study neuron clusters, rather than individual cells. A combined method of an insect model and MEAs technology may lead to great discoveries in neurophysiology, advancing progress in pharmacology, infectious and neurodegenerative diseases, agriculture maintenance and robotics.

Keywords: Neuron networks, MEAs, insect model, complex behavior.

1 Introduction

Neurophysiology, or the study of physical and chemical process of the nervous system, is a well-established field in the biological sciences. The study focuses on animal excitable cells, i.e., neurons, which are the functional cellular units of the central nervous system (CNS). Neurons in animals' brains have hundreds to thousands connections with surrounding cells in the body. These connections act as communication channels to pass information from one end of the body to the other. The collective response of multiple cellular networks results in a behavioral response of the organism. The processing of information through these intercellular communication channels is extremely complicated as well as incredibly robust.

Up to the early 19th century, insects were regarded as brainless, until the first dissection of the grasshopper head was conducted leading to the discovery of the CNS. Scientists began to have more knowledge in entomology only later in the century, starting with Fabre who studied insect anatomy and behavior. Considered as

C.K. Loo et al. (Eds.): ICONIP 2014, Part III, LNCS 8836, pp. 493–500, 2014.
© Springer International Publishing Switzerland 2014

non-evolved organisms for a long time, insects are actually endowed of complex behaviors. Von Frisch [1] described honeybees' methods of orientation and their remarkable ability to communicate among each other by using body language. Since then, many other complex spatio-temporal patterns in social insects have been studied [2], demonstrating high level behaviors in many insects species. In this perspective, insects can be considered close to mammals and even humans. Indeed, both nervous systems have anatomical similarities, sharing the same embryonic origins. On the cellular level, for both nervous systems, the basic unit is the neuron, which delivers signals thanks to similar hormones and neurotransmitters. These properties make insects a good model, since results can be extrapolated to humans.

Organisms such as mammals possess billions of neurons in the CNS in contrast to insects possessing around hundreds of thousands (~250 000 for fruit flies), making the study of neuronal networks technically very difficult. Nowadays, next-generation technology, such as Multi-Electrodes Arrays (MEAs), mark the first step in the comprehension of neuron networks. MEAs consist of a grid of many electrodes that could record signals from a network of excitable cells cultured on a glass substrate. Combining modern technology and adequate models, such as insects, will improve our understanding of the organization of the brain as a whole.

This review presents firstly in detail why insects are a good model in neurophysiology: they possess complex behaviors, despite a relatively simple anatomy and also share common patterns with mammals at different levels, such as cells, histology, anatomy, and physiology. Secondly, the review states new perspectives of the combination of new technology, such as MEAs, with insect model for a better understanding of neurophysiology. This demonstrates the importance of a multidisciplinary approach in this field of research.

2   Insects as Adequate Model for Neurophysiology Studies

The study of insects has been very helpful in understanding biological mechanisms in mammals. Table 1 provides the history of anatomy, physiology, pharmacology and neurophysiology research, of vertebrates (including mammals) and invertebrates (including insects). The comparison highlights that there has been a gap in the application of the MEAs technique to insects since its inception.

2.1   Simple Brain Anatomy but a Repertoire of Complex Behaviors

In comparison to mammals and humans, insect anatomy is rather simple. Indeed, the insect nervous system is composed of the CNS and a peripheral nervous system. The CNS consists of a chain of ganglia, whereas the peripheral nervous system includes a stomatogastric ganglion and sensory and motor nerves. The insect brain can be divided into three large regions: protocerebrum, deutocerebrum, and tritocerebrum, all composed of neurons and glial cells. Glial cells are considered as structural components, but they also have important functional roles, which are still poorly understood in this class of invertebrates. Neurons play the major role in information processing in the brain.

Table 1. History of discoveries and comparison of methods used in studies of vertebrate and invertebrate neurology

Biology specialisation	Discovery	Technique	Vertebrates	Invertebrates
Anatomy	Anatomical brain description	Dissection	17th century Thomas Willis (*Anatomy of the brain*, 1664) Humans	1850 Dujardin (*Mémoire sur le système nerveux des insectes*) Insects
Physiology	Organization of the nervous system / Basic biological functions of insects	Improvement of Golgi's method Microscopy	19th century Cajal (*Histology of the Nervous System of Man and Vertebrates*) Humans	1939 Wigglesworth (*The Principles of Insect Physiology*) Insects
Neurophysiology	Record of electrical discharge of single nerve fibres under stimulus	Electrodes	1928 Adrien (*The Basis of Sensation*) Toad	1955 Vowles (*The structure and connexions of the corpora pedunculata in bees and ants*) Insect
Pharmacology	Action of synthetic drugs on neurons communication	Drugs synthesis	1952 Chlorpromazine used in psychiatric clinic Charpentier et al. (*Recherches sur les diméthylaminopropyl-N phénothiazines substituées*) Humans	1939 Dichlorodiphenyltrichloroethane, or DDT, used as insecticide Müller (*World of Anatomy and Physiology*) Mosquito
Electrophysiology	Voltage mechanisms of action potentials Record of single channel currents across a membrane	Microelectrodes (Voltage-clamp/Patch-clamp)	1978 Neher, Sakmann & Steinbach (*The extracellular patch clamp: a method for resolving currents through individual open channels in biological membranes*) Frog	1952 Hodgkin & Huxley (*Measurement of current-voltage relations in the membrane of the giant axon of Loligo*) Squid
Electrophysiology	Signal record from neuron networks Cells responses after drugs stimulations	MEAs (*in vitro*)	1972 Thomas et al. (*A miniature microelectrode array to monitor the bioelectric activity of cultured cells*) Today Stett et al. (*Biological application of microelectrode arrays in drug discovery and basic research*) Mammals and Birds	Nothing

These structures process sensory input and control motor output to generate basic behaviors such as regulation of breathing, walking, flying and digesting. In a review, Howse summarized results of early electrophysiological recordings and stimulation of insect brains, showing basic patterns [3]. For example, locomotion and feeding movements were triggered by electrical stimulation of the locust brain. On the other hand, honeybee studies showed that stimulation of the mushroom bodies, in the protocerebrum, gave rise to the same behaviors, but also cleaning and aggressive behaviors.

Nonetheless, insects are capable of more evolved and complex behaviors. Von Frisch, who was a pioneer in insect intelligence, demonstrated that honeybees employ body language communication to find food [1]. This remarkable communication system is nowadays studied and serves as an important model system to investigate the mechanisms and evolution of complex behaviors [4]. Since then, other communication systems have been by then discovered in other insects species. One of these is vibrational communication, which is used commonly in social and ecological interactions. Most species use substrate vibrations alone or with other forms of mechanical signaling such as water surface ripples [5]. Some communication mechanisms are even invisible to humans, such as the courtship behavior of mosquitoes using flight as a mating signal for their partners [6]. Pheromones are also a strong communication system of social insects. These hormonal chemicals are produced and released in order to relay messages to other members of the same species. Ants and bees are two example of pheromone usage by insects; they are both social species with incredible capability of colony organization. They can release pheromones for different purposes such as mate attraction, danger signalization, and direction location. For instance, to locate food an ant walks to and from this source and deposits on the ground a pheromone that other individuals of the same species detect and follow paths to where the pheromone concentration is higher. Through this mechanism, ants are able to transport food to their nest in a remarkably effective way. This behavior even inspires a number of methods and techniques for optimization. One of the best known and studied is called ant colony optimization, which designs algorithms for optimization of solutions to logistical problems [7].

The best evidence that insects are endowed with complex patterns is their capability for learning and memorization. Learning and memory are respectively defined as the acquisition and retention of neuronal representations of new information. Recent research indicates that a variety of insects rely extensively on learning for major life activities including feeding, predator avoidance, aggression, social interactions, and sexual behavior [8]. More precisely, it has been shown that bees can be trained to recognize visual stimuli such as colors, shapes and patterns, depth, and motion contrast. Indeed, learning local cues is essential for bees to characterize places of interest, essentially for food sources [9].

2.2   Common Properties with Mammals at Different Levels

The basic structure and function of neurons is conserved across animal species in all groups. Simple nervous systems first emerged in Cnidaria, such as sea anemone and jellyfish, and evolved from loose nerves to compact centers. These centers, called ganglia, evolved by fusion and increased in complexity, leading to the development of brains. From an evolutionary perspective, insects are represented in a branch different from the vertebrates, to which humans and mammals belong. Nevertheless, their CNS have many properties in common.

The basic structural unit of the nervous system is the nerve cell, or neuron, which is a specialized cell with several dendrites and an axon along which electric impulses pass. There is evidence that stem cells in vertebrates and invertebrates are very similar. For example, the Drosophila CNS has served as a key model to study asymmetric division of stem cells and, more recently, the link between unregulated stem cell division and production of tumor [10]. More recently, similarities between fundamental aspects of neural stem cell biology in Drosophila and the mammalian cerebral cortex have been recognized [10]. As the mammalian cerebral cortex is the most highly evolved region of the CNS, use of an insect model would allow resolution of neuronal communication at a high level.

Mammalian and insect brains are also anatomically comparable. Indeed, the arthropod CNS and vertebrate basal ganglia derive from the same embryonic basal forebrain lineages. These lineages are specified by an evolutionarily conserved genetic program of the forebrain-midbrain boundary region. Thus, the CNS of vertebrates and arthropods share comparable embryological derivation and topography [11]. Also, neurotransmitters, small molecules carrying information across synapses from a nerve cell to its neighboring cells are the same in both groups. Indeed, network connectivity and neuronal activity in the substructures of each CNS are mediated by the same neurotransmitters: for instance, inhibitory (GABAergic) and modulatory (dopaminergic) transmitters, which facilitate the regulation and release of adaptive behaviors. In each case, CNS dysfunctions result in behavioral defects including motor abnormalities, impaired memory formation, attention deficits, affective disorders, and sleep disturbances. The observed multitude of similarities suggests deep homology of the arthropod CNS and vertebrate basal ganglia underlying the selection and maintenance of behavioral actions [11].

3   Evolution of Insects' Studies with MEAs: Current Trends and Future Prospects

3.1   Pharmacology and Human Health

In medical research for over a century, most of the insect model studies have been conducted in Drosophila. *Drosophila melanogaster* populations are easy to maintain under laboratory conditions, inexpensive and have fast growth and reproduction rates. Flies have been used to understand many biological functions such as vision, reproduction, feeding behavior, etc. Many neurodegenerative diseases have now their own mutant fly model, allowing reproducible testing to be conducted more diversely and in shorter times than in rodent models. The highly diverse and accessible Drosophila mutant resources have facilitated understanding of the genetics underlying neurophysiology processes. Kohler suggests [12] that experimental organisms such as *Drosophila melanogaster* have themselves become a part of technology. Like laboratory rats or mice, the flies have been designed for a particular use and thus are more than simply representative organisms: they are tools. Experiments in laboratories with insects present many advantages such as inexpensive, rapid and easy handling. This is why now, MEAs experiments could add another prospect for the application of this model to improve our global understanding of neuron communication and its implications in human and animal nervous diseases. Since the end of the 20th century, neurodegenerative diseases have increased rapidly in developed countries. Despite a huge recent boost in Alzheimer's and Parkinson's diseases research, there is no current cure. Experiments with MEAs in mouse cells culture [13] have allowed the identification of early steps in generation of Alzheimer's disease lesions in the hippocampus. Moreover, this technique has brought to light some electrophysiological properties of the sub-thalamic nucleus [14], an area involved in Parkinson's disease. The use of insect brain cell cultures could mark another step in improving our understanding and the development of a cure for these neurodegenerative diseases.

Besides neurodegenerative diseases, transmissible diseases remain a major public health problem, particularly in developing countries. Insects, such as mosquitoes, can be vectors of numerous pathogens responsible for diseases. Pathogens are transmitted mainly by bite allowing them to pass from animal to animal through a totally biological process. Insect-transmitted virus are called arboviruses (<u>ar</u>thropod-<u>bor</u>ne-<u>viruses</u>). In contrast to vertebrate hosts in which some arbovirus may trigger haemorrhagic fever or neurological symptoms, infection of insect vectors causes no or little pathology. However, pathogen fitness is critically dependent upon efficient transmission to a new host. To improve fitness, pathogens can modify vector behavior [15, 16] such as feeding behavior which can affect host range and/or the transmission rate. Such strategies are known for some pathogens transmitted by mosquitoes: West Nile virus in *Culex* mosquitoes, dengue and chikungunya viruses in *Aedes* mosquitoes, and *Plasmodium* malaria parasite in *Anopheles* mosquitoes. The driver of behavioral changes is not known but the involvement of neurons seems obvious. Viral infection of insect neurons is poorly studied and the mechanisms underlying neuronal functional changes are unknown. MEAs method represents a good potential to understand these processes.

The MEA system allows advanced research in neurotoxicity and pharmacology, because it enables recordings in real time in cell culture, and so provides robust measures of network activity. The dynamic systems can record physical, chemical, and pharmacological perturbations, which are reflect in the tissue responses. MEAs also allow testing different chemicals on cells or tissue cultures, providing a classification of the effects of these substances [17-20]. However, all studies to date have been conducted in vertebrate cells cultures, highlighting the gap for use of MEAs in insects despite their medical importance and potential suitability as a model system. However, the scope for this method is not limited to biology.

3.2   Engineering and Robotics

MEA technology has revolutionized scientific investigations in neurophysiology. Robotics could also directly benefit from biological studies in insects using MEAs technology. Indeed, simple animal brain function is very interesting and helpful to understanding of neurobiological responses to environmental stimuli. For example, integration of a light stimulus by the eye: the light signal is detected by the retina, then the signal is transmitted through excitable cells to the brain where the message is integrated. This knowledge of environment stimulus transmission to the brain can be used and extrapolated to build efficient bio-inspired sensing-perceiving-acting machines. Much work has already been reported in designing bio-inspired robots using visual systems [21] or walking behavior [22, 23] of insects. However, although the CNS plays a major role in controlling movements, scientists have used only physiological and anatomical data (e.g., nervous system anatomy, muscle coordination, and sensory system anatomy) for these studies. By way of illustration, Bässler and Wegner showed that the deafferented (without any afferent or sensory nerve fibers) CNS of a stick insect could generate rhythmic motor patterns in leg muscles [24]. The submechanism of neurons communication remains unknown and no electric data from cell clusters have yet been recorded. The panorama opened by the MEAs technique is huge in this field. However, given the various specialties involved in this field, bridged and trans-disciplinary collaborations seem to be mandatory.

4   Conclusion: Our Perspective

Figure 1 shows the research prospects by combining insects and MEAs system.

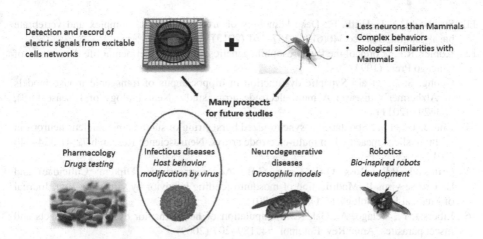

Fig. 1. A summary diagram illustrating the research perspectives of insect studies with MEAs

This review highlights a new idea that insect model is more interesting in term of easiness compared to mammals, for their brain infrastructure, costs, replicates, and especially for infections and neuropathogens. Our team aims to study virus modification to neurons communication in infected mosquitos by arboviruses. The goal of the study, circled in red in Figure 1, is to correlate a defined phenotype (infected or uninfected mosquitoes) with neuronal functions. This innovative research will allow us to characterize which neuronal modifications a virus can trigger, resulting in a phenotypic modification: the behavior of the mosquito. This research would be the first investigation using the model of insects with MEAs technology, and the first to define a characterized phenotype linked to nerves cell dysfunctions.

References

1. Von Frisch, K.: The dance language and orientation of bees (1967)
2. Wilson, E.O.: The insect societies. The insect societies (1971)
3. Howse, P.: Brain structure and behavior in insects. Annual Review of Entomology 20(1), 359–379 (1975)
4. Dyer, F.C.: The biology of the dance language. Annual review of Entomology 47(1), 917–949 (2002)
5. Cocroft, R.B., Rodríguez, R.L.: The behavioral ecology of insect vibrational communication. Bioscience 55(4), 323–334 (2005)
6. Cator, L.J., et al.: Harmonic convergence in the love songs of the dengue vector mosquito. Science 323(5917), 1077–1079 (2009)
7. Dorigo, M., Birattari, M., Stutzle, T.: Ant colony optimization. IEEE Computational Intelligence Magazine 1(4), 28–39 (2006)
8. Dukas, R.: Evolutionary biology of insect learning (2007)
9. Giurfa, M.: Cognitive neuroethology: dissecting non-elemental learning in a honeybee brain. Current Opinion in Neurobiology 13(6), 726–735 (2003)
10. Brand, A.H., Livesey, F.J.: Neural stem cell biology in vertebrates and invertebrates: more alike than different? Neuron 70(4), 719–729 (2011)

11. Strausfeld, N.J., Hirth, F.: Deep homology of arthropod central complex and vertebrate basal ganglia. Science 340(6129), 157–161 (2013)
12. Kohler, R.E.: Lords of the fly: Drosophila genetics and the experimental life. University of Chicago Press (1994)
13. Chong, S.-A., et al.: Synaptic dysfunction in hippocampus of transgenic mouse models of Alzheimer's disease: A multi-electrode array study. Neurobiology of Disease 44(3), 284–291 (2011)
14. Chu, J.-U., et al.: Spontaneous synchronized burst firing of subthalamic nucleus neurons in rat brain slices measured on multi-electrode arrays. Neuroscience Research 72(4), 324–340 (2012)
15. Grimstad, P.R., Ross, Q.E., Craig, G.B.: Aedes triseriatus (Diptera: Culicidae) and La Crosse virus II. Modification of mosquito feeding behavior by virus infection. Journal of Medical Entomology 17(1), 1–7 (1980)
16. Libersat, F., Delago, A., Gal, R.: Manipulation of host behavior by parasitic insects and insect parasites. Annu Rev. Entomol. 54, 189–207 (2009)
17. Johnstone, A.F., et al.: Microelectrode arrays: a physiologically based neurotoxicity testing platform for the 21st century. Neurotoxicology 31(4), 331–350 (2010)
18. Natarajan, A., et al.: Microelectrode array recordings of cardiac action potentials as a high throughput method to evaluate pesticide toxicity. Toxicology in Vitro 20(3), 375–381 (2006)
19. McConnell, E.R., et al.: Evaluation of multi-well microelectrode arrays for neurotoxicity screening using a chemical training set. Neurotoxicology 33(5), 1048–1057 (2012)
20. Mack, C.M., et al.: Burst and principal components analyses of MEA data for 16 chemicals describe at least three effects classes. Neurotoxicology 40, 75–85 (2014)
21. Franceschini, N., et al.: From insect vision to robot vision [and discussion]. Philosophical Transactions of The Royal Society of London. Series B: Biological Sciences 337(1281), 283–294 (1992)
22. Beer, R.D., et al.: Biologically inspired approaches to robotics: What can we learn from insects? Communications of the ACM 40(3), 30–38 (1997)
23. Delcomyn, F.: Insect walking and robotics. Annual Reviews in Entomology 49(1), 51–70 (2004)
24. Bässler, U., Wegner, U.: Motor output of the denervated thoracic ventral nerve cord in the stick insect Carausius morosus. Journal of experimental Biology 105(1), 127–145 (1983)

Neuron's Spikes Noise Level Classification Using Hidden Markov Models

Sherif Haggag, Shady Mohamed, Asim Bhatti, Hussein Haggag,
and Saeid Nahavandi

Centre for Intelligent Systems Research
Deakin University, Australia
{shaggag,shady.mohamed,asim.bhatti,hhaggag,saeid.nahavandi}@deakin.edu.au

Abstract. Considering that the uncertainty noise produced the decline in the quality of collected neural signal, this paper proposes a signal quality assessment method for neural signal. The method makes an automated measure to detect the noise levels in neural signal. Hidden Markov Models were used to build a classification model that classifies the neural spikes based on the noise level associated with the signal. This neural quality assessment measure will help doctors and researchers to focus on the patterns in the signal that have high signal to noise ratio and carry more information.

Keywords: Hidden Markov Model, Mel-Frequency Cepstrum Coefficient, Multichannel systems, neural signal.

1 Introduction

Neurons communicate with each others using electrical spikes[1]. These spikes hold all the information needed to do a specific activity. There are many ways to record neural signals; one of the most famous neural signals recording is the Encephalogram[2][3] which is known by EEG. EEG is a mainly the spikes or the electrical activity recorded in the brain. The recording of this signal is done by placing some electrodes on the human scalp[4]. Billions of brains neurons communicate with each other forming a huge complicated neural network. Another way to record neural signal is the Electromyography[5] [6]which is famously known as EMG, it is the recording of the skeletal muscles electrical activity. Also EMG records the data of millions of neurons which forms also a huge complex neural network. Recently there is a new device which is can record a specific number of neurons using specific chips. This device is called Multichannel systems[7]. Noise is the main problem which occurs while recoding any kind of the neural signals which are stated above, noise affects the recording and the noisier is the signal the more difficult to use the data[8]. It is very hard process to detect the signal to noise ratio in a neural signal[9]. Noise is undesired signal which is convoluted with the main recording of EEG, EMG or multichannel systems[10].

Noise can be divided into two categories. First, Biological noise which is commonly produced by moving limbs, eyes and head activities. The second category

C.K. Loo et al. (Eds.): ICONIP 2014, Part III, LNCS 8836, pp. 501–508, 2014.

of noise is the external noise; it is mainly generated due to technical factors[11]. Mainly the electrodes records the signal which is generated by specific number of neurons but the problem here you can not control the number of neurons so an unneeded spikes were recorded which increase the signal to noise ratio[12]. It is not possible to record the neural signals without the biological noise but it is an easy job to detect them and remove them after recording and analyzing the neural signal. On the other hand, the effect of the technical factors can be hugely minimized by knowing the amount of signal to noise ratio[9, 13, 3, 11, 14].

It is not a difficult job to know when the biological noise happens and there are ways to remove this kind of noise. On the other hand, it is very ambiguous to remove the technical noise as the amount of noise is unknown and the level of noise is very hard to know[15].

The most challenging point while removing any type of noise is to differentiate between the noise and the signal, and to put in mind not to make any distortion for the signal while removing the noise. One of the ways to remove the noise is using the wavelet transform proposed in [3], and in [16]. The main problem in this method is that it mainly depends on the assumption that the signal magnitudes dominate the magnitudes of the noise in a wavelet representation. So our main target in this paper is to give a more accurate measure for the amount of noise in an EEG signal.

2   Methodology

Our main idea is to extract the most significant features of a pure signal and the noise from a noisy signal. Then we cluster the signal into different groups based on the extracted features. The features are extracted based on the Mel-Frequency Cepstrum Coefficient(MFCC)[17, 18]. MFCC represents the short-term power spectrum of a signal, it is calculated depending on the linear cosine transform of a log power spectrum on a nonlinear mel scale of frequency. MFCC is widely used in speech and voice recognition, the reason for using MFCC is that it gives more weights to the frequencies that the human can hear which is close to the neural spikes frequencies, this frequency warping can allow for better representation of signal.

First of all a neural signal was recorded using an acquisition device (multi-channel systems), then we calculate the MFCC of the signal using the following steps which are shown in figure 1.

1. Each spike has it's own data file that can be read into an array $S(n)$, where n ranges from 0, 1, ... N-1 where N is the number of samples.
2. Split the spike into distinct frames. Each frame is 20-40 ms (25ms is standard). This means the frame length for a 16kHz signal is 0.025*16000 = 400 samples. Frame step is usually something like 10ms (160 samples), which allows some overlap to the frames. The first 400 sample frame starts at sample 0, the next 400 sample frame starts at sample 160 etc. until the end of the file is reached. If the file does not divide into an even number of frames, pad it with zeros so that it does.

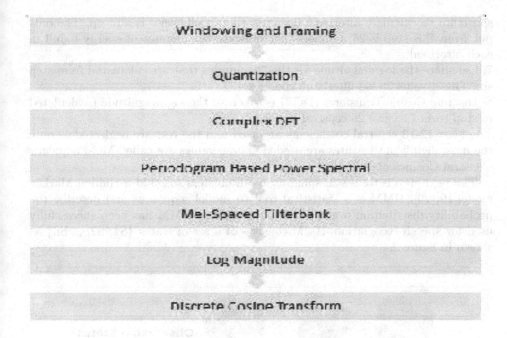

Fig. 1. MFCC calculation steps

3. Once it is framed we have $S_i(n)$ where n ranges over 1-frame size and i ranges over the number of frames. This is a form of quantization.

4. When we calculate the complex DFT, we get $S_i(k)$ - where i the denotes the frame number corresponding to the time-domain frame. To take the Discrete Fourier Transform of the frame, perform the following:

$$S_i(k) = \sum_{n=1}^{N} s_i(n)h(n)e^{-j\frac{2\pi}{N}kn}, 1 < k < K \tag{1}$$

where h(n) is an N sample long analysis window, and K is the length of the DFT.

5. Then we calculate $P_i(k)$ which is the periodogram based power spectral of frame i is given by:

$$P_i(k) = \frac{1}{N}|S_i(n)|^2 \tag{2}$$

This is called the Periodogram estimate of the power spectrum. The absolute value of the complex fourier transform is taken, and square the result. We would generally perform the FFT. 6. Compute the Mel-spaced filterbank. It is a group of 20-40 (the standard value is 26) triangular filters that is practiced to the periodogram power spectral estimate from step number five. The filterbank is represented in 26 vectors. Most values in each vector is zeros except for a particular section of the spectrum. Each filterbank is multiplied with the power spectrum to calculate filterbank energies, then coefficients is add up. The output from this step is 26 numbers that indicate the amount of energy found in each filterbank.

7. Calculate the log magnitude for the 26 energies that are calculated from step six. This give us 26 log filterbank energies.

8. Discrete Cosine Transform (DCT) is taken to the log magnitude (calculated in step seven) to give 26 cepstral coefficents.

The least 12-13 cepstral coefficients are used and the rest are neglected as only the most significant features are needed. These values are called Mel-Frequency Cepstral Coefficients.

The next step is to use 20 spikes from all clusters to build a Hidden Markov model[19, 20]. HMM is a statistical tool to model sequences and describe the probability distribution over a set of observations. HMM has been successfully used for speech recognition. HMM consists of a set of states {S1, S2, ... Sn} as shown in figure 2.

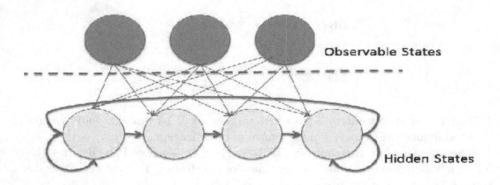

Fig. 2. Hidden Markov Model

Process moves from one state to another generating a sequence of states: $\{S_{i1}, S_{i2}, ..., S_{ik}, ...\}$ and the Markov chain property of each subsequent state depends only on what was the previous state and can be defined as: $P(S_{ik}|S_{i1}, S_{i2}, ... S_{ik-1}) = P(S_{ik}|S_{ik-1})$. Also there are states which are not visible, but each state randomly generates one of M observations (or visible states) {V1, V2, ... Vn}.

HMM was utilized to model spikes in order to understand its underlying states over a sequence of observations. For a spike, the significant shape states are silence, going up, peak and going down. Each of them is assigned to a state (from s1 to s4) of the HMM. The main problem in other clustering algorithms is that they rely on shape and distance measurement such as comparing height, width, and peak-to-peak amplitude of spikes.

These HMM models are used to classify the testing data which is more than 500 spikes per cluster. The HMM which was developed contains four states for the spike and it iterated five times and the frame size is 0.025 and the frame shift is 0.01 and it is forward backward direction and matrices are set to identity matrix. Also the data which was used in the experiments are neural signals recorded from three different neurons each has its unique shape, the same signal is used but with different noise levels and the HMM model cluster the same signal into different clusters based on the noise level.

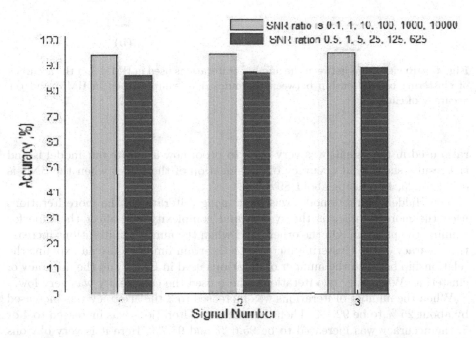

Fig. 3. Clustering Accuracy using different Signal to Noise Ratio(SNR)

3   Results

A signal was used with different signal to noise ratio. 20 spikes were used to train the Hidden Markov Model and 500 spikes were using for testing. The neural signal which was used in the experiments is the same but with different noise levels and the HMM clusters the same signal into different clusters based on the noise level.

Our model was able to differentiate between different noise levels with accuracy 95% and these noise levels (SNR) are 0.1, 1, 10, 100, 1000, 10000. Also the noise level was meant to be very close to each other and Our model was able to differentiate between them with accuracy 89% and these noise levels (SNR) is 0.5, 1, 5, 25, 125, 625.

Figure 3 shows the accuracy of classification of the spikes when the SNR is 0.1, 1, 10, 100, 1000, 10000 using three different signals. The accuracy of signal to noise ration detection is about 95 % which will help us later to estimate the amount of information in the signal. Also in Figure 3 the level of signal to noise

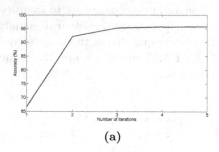 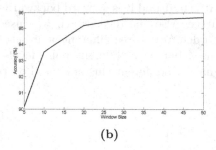

(a)   (b)

Fig. 4. a)Relationship between the number of iterations used in HMM and the accuracy of clustering b)Relationship between the number of samples used in HMM and the accuracy of clustering

ratio used in the signals was very close to proof how accurate our model is and the results shows that accuracy of classification of the spikes when the SNR is 0.5, 1, 5, 25, 125, 625 is about 89%.

The Hidden Markov model was built using 3 iterations, the more iterations used the more complex is the system and complexity will affect the time for running the program. On the other hand when the number of iterations increase the accuracy of the clustering increase to a certain limit. Figure 4a explains the relationship between the number of iterations used in HMM and the accuracy of clustering. When only two iterations where used the complexity was very low.

When the number of iterations was increased to 3 the accuracy was increased by about 25 % to be 92.1%. Then the number of iterations was increased to 4 or 5, the accuracy was increased to be 95.6 % and 95.7%. Here it is very obvious that the number of iterations should not exceed 5 iterations as the accuracy doesn't significantly change. So the HHM was build using 3 iterations and the reason behind that is to reduce the complexity and increase the accuracy of the clustering method.

One major advantage in Our model is the number of training samples. Our model only uses 20 spike for training and test the noise level for 500 spike. This reduce the complexity of the model and it is useless to use more samples to train as the model was build using specific number of iterations and states which can be easily trained by a small number of training samples. Figure 4b shows the relation between the number of training samples and the clustering accuracy. Using only 5 samples can give about 90% accuracy which is very good when there are no enough training samples. The accuracy remains nearly the same as shown when the number of training samples exceeds 20 samples, so the best number of samples to use with respect to efficiency and accuracy is 20.

The number of states used in our HMM was which was developed contains four states which basically represents the spikes. These states are: silence, going up, peak and going down. And if the number of states was changed as shown in Figure 5 the accuracy of the classification will be changed a lot due to the fact that the spike shape can be represented in 3 or 4 states maximum.

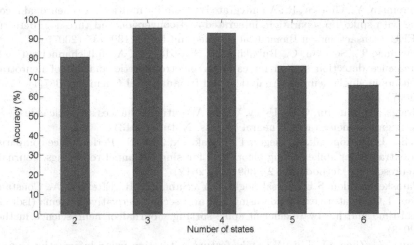

Fig. 5. Relationship between the number of states used in HMM and the accuracy of clustering

4 Conclusion and Future Work

This paper proposed a signal quality assessment method for neural signal. The method generates an automated measure to detect the noise levels in neural signal. Hidden Markov model was used to build a classification model that classifies the neural spikes based on the spike noise level. This is considered the first quality assessment measure of neural signal as we will be use this measure as a flag beside other flags (which will be developed by us) to achieve a complete quality measure for neural signal in the future.

References

[1] Moravec, H.: When will computer hardware match the human brain. Journal of Evolution and Technology 1(1), 10 (1998)
[2] Davidoff, L.M., Dyke, C.G.: The normal encephalogram. Journal of Neuropathology & Experimental Neurology 11(3), 310–339 (1952)
[3] Ramachandran, N., Chellappa, A.: Feature extraction from eeg using wavelets: Spike detection algorithm. In: International Symposium on Modern Computing (JVA 2006), pp. 120–124 (2006)
[4] Seeck, M., Michel, C.M., Mainwaring, N., Cosgrove, R., Blume, H., Ives, J., Landis, T., Schomer, D.L.: Evidence for rapid face recognition from human scalp and intracranial electrodes. Neuroreport 8(12), 2749–2754 (1997)
[5] Hermens, H.J., Freriks, B., Merletti, R., Stegeman, D., Blok, J., Rau, G., Disselhorst-Klug, C., Hägg, G.: European recommendations for surface electromyography. Roessingh Research and Development The Netherlands (1999)
[6] Da Luca, C.J.: The use of surface electromyography in biomechanics. Journal of applied biomechanics 13, 135–163 (1997)

[7] Perelman, Y., Ginosar, R.: An integrated system for multichannel neuronal recording with spike/lfp separation, integrated a/d conversion and threshold detection. IEEE Transactions on Biomedical Engineering 54(1), 130–137 (2007)

[8] Lapainis, T., Scanlan, C., Rubakhin, S., Sweedler, J.: A multichannel native fluorescence detection system for capillary electrophoretic analysis of neurotransmitters in single neurons. Analytical and Bioanalytical Chemistry 387(1), 97–105 (2007)

[9] Briggs, F., Mangun, G.R., Usrey, W.M.: Attention enhances synaptic efficacy and the signal-to-noise ratio in neural circuits. Nature (2013)

[10] Wild, J., Prekopcsak, Z., Sieger, T., Novak, D., Jech, R.: Performance comparison of extracellular spike sorting algorithms for single-channel recordings. Journal of Neuroscience Methods 203(2), 369–376 (2012)

[11] Paraskevopoulou, S.E., Barsakcioglu, D.Y., Saberi, M.R., Eftekhar, A., Constandinou, T.G.: Feature extraction using first and second derivative extrema (fsde), for real-time and hardware-efficient spike sorting. Journal of neuroscience methods (2013)

[12] Yang, Z., Zhao, Q., Liu, W.: Spike feature extraction using informative samples. Advances in Neural Information Processing Systems 21, 1865–1872 (2009)

[13] Machart, P., Ralaivola, L.: Confusion matrix stability bounds for multiclass classification. arXiv preprint arXiv:1202.6221 (2012)

[14] Bielat, V.E., Levi, A.: Open access textbook access, quality, use (2012)

[15] Khatwani, P., Tiwari, A.: A survey on different noise removal techniques of eeg signals. International Journal of Advanced Research in Computer and Communication 2(2) (2013)

[16] Quiroga, R.Q., Nadasdy, Z., Ben-Shaul, Y.: Unsupervised spike detection and sorting with wavelets and superparamagnetic clustering (2004)

[17] Zhou, X., Garcia-Romero, D., Duraiswami, R., Espy-Wilson, C., Shamma, S.: Linear versus mel frequency cepstral coefficients for speaker recognition. In: 2011 IEEE Workshop on Automatic Speech Recognition and Understanding (ASRU), pp. 559–564 (2011)

[18] Haggag, S., Mohamed, S., Bhatti, A., Gu, N., Zhou, H., Nahavandi, S.: Cepstrum based unsupervised spike classification. In: 2013 IEEE International Conference on Systems, Man, and Cybernetics (SMC), pp. 3716–3720. IEEE (2013)

[19] Ephraim, Y.: Hidden markov models. Encyclopedia of Operations Research and Management Science, 704–708 (2013)

[20] Zhou, H., Mohamed, S., Bhatti, A., Lim, C.P., Gu, N., Haggag, S., Nahavandi, S.: Spike sorting using hidden markov models. In: Lee, M., Hirose, A., Hou, Z.-G., Kil, R.M. (eds.) ICONIP 2013. LNCS, vol. 8226, pp. 553–560. Springer, Heidelberg (2013)

Improved Robust Kalman Filtering
for Uncertain Systems with Missing Measurements

Hossein Rezaei[1], Shady Mohamed[2], Reza Mahboobi Esfanjani[1],
and Saeid Nahavandi[2]

[1] Electrical Engineering Department, Sahand University of Technology, Tabriz, Iran
{h_rezaei,mahboobi}@ sut.ac.ir
[2] Centre for Integrative Systems Research, Deakin University,
Waurn Ponds VIC 3216, Australia
{shady.mohamed,saeid.nahavandi}@deakin.edu.au

Abstract. In this paper, a novel robust finite-horizon Kalman filter is developed
for discrete linear time-varying systems with missing measurements and norm-
bounded parameter uncertainties. The missing measurements are modelled by a
Bernoulli distributed sequence and the system parameter uncertainties are in the
state and output matrices. A two stage recursive structure is considered for the
Kalman filter and its parameters are determined guaranteeing that the covariances of the state estimation errorsare not more than the known upper bound.
Finally, simulation results are presented to illustrate the outperformance of the
proposed robust estimator compared with the previous results in the literature.

Keywords: robust Kalman filter, miss measurement, state estimation, norm-
bounded parameter uncertainties.

1   Introduction

The Kalman filter which is based on the minimization of the filtering error covariance
is the popular tool for the state estimation through the noisy observations. The key
assumptions in the standard Kalman filtering are that the perfect model of the underlying system is priory known and all the measurements are available[1]. However, in
the many real-world applications, for instance in the networked control systems, unre-
liability of the communication channels together with modeling uncertainties imposes
significant challenges in the optimal state estimation [2-4].

The initial work on the filter design problem with missing measurements can be
traced back to [5], and [6], where a binary sequence specified by a probability distribution were utilized todescribe the missing data. On the other hand, robust Kalman
filter with a guaranteed bound on the filtering error covariancefor systems with time-
varying norm-bounded uncertainties in the state and output matrices were proposed in
[7,8] and [9], for discreet and continues time systems; respectively.

Only a few papers have considered the common case wherein the problem of missing observations is combined with the norm-bounded modeling uncertainties. The
infinite-horizon optimal filter was derived in [10], for discrete-time systems with

C.K. Loo et al. (Eds.): ICONIP 2014, Part III, LNCS 8836, pp. 509–518, 2014.
© Springer International Publishing Switzerland 2014

stochastic missing measurements and parameter uncertainties. However, finite-horizon filters leads to better transient performance if the noise inputs are notstationary. Then, for linear discrete time-varying systems with time-varying norm-bounded uncertainty in the state matrix and missing measurements a robust finite-horizon Kalman filter was introduced in [11_ENREF_15]. In[12], robust finite- horizon Kalman filter was developed for the more comprehensive system withnorm-bounded uncertainty in the state and output matricessuffering from missing measurements. In [13], within the different framework, robust state estimator was suggested for the systems with missing measurements based on minimizing the sensitivity of the estimation errors to the parameter variations.

In this paper, robust finite-horizon filtering problem is derived for uncertain time-varying linear system with intermittent measurements. The state and output matrices of the system model are subject to norm bounded uncertainty and missing data are described by Bernoulli distributed random sequence. Unlike [11] and [12], a two stage recursive structure is adopted for the robust Kalman filter and furthermore, a different augmented state space model is utilized to extract a procedure to determine filter parameters. Finally, simulation results are presented to illustrate that the introduced estimator leads to the remarkably improved performance compared to the previously developed approach in [12].

The rest of the paper is organized as follows: The estimation problem is formulated in the section II. In section III, the optimal estimator is derived for the uncertain system with missing observations. In section IV, numerical benchmark examples are presented to illustrate the outperformance of the proposed approach. Section V concludes this note.

Notations: $\Re$ denotes real numbers set. $\mathrm{Prob}\{\}$ represents the probability of the stochastic variable. $\mathrm{E}\{\}$ is the mathematical expectation. The superscript ^{T} stands for the matrix transposition.

2   Problem Setup

Consider the following class of the uncertain linear discrete-time stochastic systems:
$$x(t+1) = (A_t + \Delta A_t)x(t) + B_t w(t) \tag{1}$$
with the measurement equation
$$y(t) = \gamma_t (C_t + \Delta C_t)x(t) + v(t) \tag{2}$$
where, $x(t) \in \Re^n$ is the state vector, $y(t) \in \Re^m$ is the measured output, $w(t) \in \Re^n$ and $v(t) \in \Re^m$ are the process and measurement noise, respectively. It's assumed that $w(t)$ and $v(t)$ are uncorrelated white noises with zero means and variances Q and R. A_t, B_t and C_t are known real time-varying matrices with appropriate dimensions. ΔA_t is a real-valued uncertain matrix satisfying:
$$\begin{bmatrix} \Delta A_t \\ \Delta C_t \end{bmatrix} = \begin{bmatrix} H_{1,t} \\ H_{2,t} \end{bmatrix} F_t E_t, \quad F_t F_t^T \le I$$

Here, $H_{1,t}, H_{2,t}$ and E_t are known time-varying matrices of appropriate dimensions and F_t represents time-varying uncertainties. The output sequence is defined in (2). Some measurement data may be lost. The stochastic variable γ_t which takes the values 0 and 1, is a Bernoulli distributed variable with mean μ. It is assumed that γ_t is uncorrelated with $w(t), v(t)$ and initial state x_0. From the properties of the Bernoulli distribution, the following relations hold:

$$\text{Prob}\{\gamma_t = 1\} = E\{\gamma_t\} = \mu, \text{Prob}\{\gamma_t = 0\} = 1 - E\{\gamma_t\} = 1 - \mu$$

$E\{(\gamma_t - \mu)^2\} = (1 - \mu)\mu$. Also, it is assumed that: $E\{F_t\} = 0$ and $E\{F_t F_j\} = I\delta_{ij}$.

The aim of this note is to design finite-horizon robust Kalman filter for discrete-time systems with parameters uncertainty and missing observations. The structure of the proposed robust Kalman filter is given in (3) and (4). The estimation of the state is computed by the following recursive equations:

$$\hat{x}(t|t) = \hat{x}(t|t-1) + K_t(y(t) - \mu C_t \hat{x}(t|t-1)) \tag{3}$$

$$\hat{x}(t+1|t) = \hat{A}_t(t)\hat{x}(t|t-1) + L_t(y(t) - \mu C_t \hat{x}(t|t-1)) \tag{4}$$

where $\hat{x}(t)$ is the estimate of the state $x(t)$, and $\hat{A}_t, L_t$ and K_t are time-varying filter parameters are determined such that filtering error $e(t) = x(t) - \hat{x}(t|t)$, and prediction error $\tilde{e}(t) = x(t) - \hat{x}(t|t-1)$ variances be smaller than positive-definite matrices $\bar{\Theta}(t)$ and $\bar{\Sigma}(t), (0 < t \leq N)$, respectively:

$$E\{(x(t) - \hat{x}(t|t))(x(t) - \hat{x}(t|t))^T\} \leq \bar{\Theta}(t) \tag{5}$$

$$E\{(x(t) - \hat{x}(t|t-1))(x(t) - \hat{x}(t|t-1))^T\} \leq \bar{\Sigma}(t) \tag{6}$$

3   Filer Design

In this section a procedure is developed to obtain the parameters of the two stage Kalman filter defined in (3) and (4). First, the upper bounds of the filtering and prediction covariance matrices presented in (5) and (6) are determined in the form of discrete time Riccarti-like difference equation.

3.1   Preliminaries

In this subsection, some preliminaries are introduced which will be used in derivation of the main results. First, new augmented state vectors   and   are defined as follows:

$$\zeta(t) = \begin{bmatrix} e(t) \\ \hat{x}(t|t) \end{bmatrix}, \tilde{\zeta}(t) = \begin{bmatrix} \tilde{e}(t) \\ \hat{x}(t|t-1) \end{bmatrix} \tag{7}$$

Then, by combination of (1)-(4), the augmented system equations can be written as follows:

$$\zeta(t+1) = (A_{c1} + H_{c1}F_k E_{c1})\tilde{\zeta}(t) + A_{e1}\tilde{\zeta}(t)$$

$$+ \tilde{A}_{e1}\tilde{\zeta}(t) + G_{v1}v(t) \tag{8}$$

$$\tilde{\zeta}(t+1) = (A_{c2} + H_{c2}F_k E_{c2})\tilde{\zeta}(t) + A_{e2}\tilde{\zeta}(t)$$

$$+ \tilde{A}_{e2}\tilde{\zeta}(t) + G_{v2}v(t) + G_{w2}w(t) \tag{9}$$

where

$$A_{c1} = \begin{bmatrix} I - \mu K_t C_t & 0 \\ \mu K_t C_t & I \end{bmatrix}, A_{c2} = \begin{bmatrix} A_t - \mu L_t C_t & A_t \hat{A}_t \\ \mu L_t C_t & \hat{A}_t \end{bmatrix}, G_{v1} = \begin{bmatrix} -K_t \\ K_t \end{bmatrix}, G_{v2} = \begin{bmatrix} -L_t \\ L_t \end{bmatrix}$$

$$A_{e1} = \begin{bmatrix} -K_t \tilde{C}_t & -K_t \tilde{C}_t \\ K_t \tilde{C}_t & K_t \tilde{C}_t \end{bmatrix}, E_{c1} = E_{c2} = \begin{bmatrix} E_t & E_t \end{bmatrix}, G_{w2} = \begin{bmatrix} B_t \\ 0 \end{bmatrix}$$

$$A_{e2} = \begin{bmatrix} -L_t \tilde{C}_t & -L_t \tilde{C}_t \\ L_t \tilde{C}_t & L_t \tilde{C}_t \end{bmatrix}, H_{c1} = \begin{bmatrix} -\mu K_t H_{2,t} \\ \mu K_t H_{2,t} \end{bmatrix}, H_{c2} = \begin{bmatrix} H_{1,t} -\mu K_t H_{2,t} \\ \mu K_t H_{2,t} \end{bmatrix}$$

$$\tilde{A}_{e1} = \begin{bmatrix} -\eta K_t \Delta C_t & -\eta K_t \Delta C_t \\ \eta K_t \Delta C & \eta K_t \Delta C \end{bmatrix}, \tilde{A}_{e2} = \begin{bmatrix} -\eta L_t \Delta C_t & -\eta L_t \Delta C_t \\ \eta L_t \Delta C & \eta L_t \Delta C \end{bmatrix} \text{ In}$$

which $A_{e1}, \tilde{A}_{e1}, A_{e2}$ and $\tilde{A}_{e2}$ are stochastic matrix sequences with the zero mean and $\eta = (\gamma_t - \mu), \tilde{C} = \eta C_t$. The covariance matrices of the augmented state vector in (8) and (9) are represented as:

$$\tilde{\Theta}(t+1) = E\{\zeta(t)\zeta^T(t)\} \tag{10}$$

$$\tilde{\Sigma}(t+1) = E\{\tilde{\zeta}(t)\tilde{\zeta}^T(t)\} \tag{11}$$

According to the equations (8) and (10) the Lyapunov equations for the filtering covariance matrix can be obtained as the following:

$$\tilde{\Theta}(t+1) = (A_{c1} + H_{c1}F_t E_{c1})\tilde{\Sigma}(t)(A_{c1} + H_{c1}F_t E_{c1})^T$$

$$+ G_{v1}RG_{v1}^T + \psi_t + \tilde{\psi}_t \tag{12}$$

Similarly, regarding to the equations (9) and (11) the Lyapunov equations for the prediction covariance matrix can be attained as follows:

$$\tilde{\Sigma}(t+1) = (A_{c2} + H_{c2}F_t E_{c2})\tilde{\Sigma}(t)(A_{c2} + H_{c2}F_t E_{c2})^T$$

$$+ G_{v2}RG_{v1}^T + \varphi_t + \tilde{\varphi}_t + G_{w2}QG_{w2}^T \tag{13}$$

where:

$$\psi_t = E\{A_{e1}\tilde{\Sigma}(t)A_{e1}^T\}$$

$$= \delta \begin{bmatrix} -K_t C_t & -K_t C_t \\ K_t C_t & K_t C_t \end{bmatrix} \tilde{\Sigma}(t) \begin{bmatrix} -K_t C_t & -K_t C_t \\ K_t C_t & K_t C_t \end{bmatrix}^T$$

$$\tilde{\psi}_t = E\{\tilde{A}_{e1}\tilde{\Sigma}(t)\tilde{A}_{e1}^T\} =$$

$$\delta \begin{bmatrix} -K_t H_{2,t}E_t & -K_t H_{2,t}E_t \\ K_t H_{2,t}E_t & K_t H_{2,t}E_t \end{bmatrix} \tilde{\Sigma}(t) \begin{bmatrix} -K_t H_{2,t}E_t & -K_t H_{2,t}E_t \\ K_t H_{2,t}E_t & K_t H_{2,t}E_t \end{bmatrix}^T$$

$$\varphi_t = E\left\{A_{e2}\tilde{\Sigma}(t)A_{e2}^T\right\} =$$

$$\delta\begin{bmatrix} -L_tC_t & -L_tC_t \\ L_tC_t & L_tC_t \end{bmatrix}\tilde{\Sigma}(t)\begin{bmatrix} -L_tC_t & -L_tC_t \\ L_tC_t & L_tC_t \end{bmatrix}^T$$

$$\tilde{\varphi}_t = E\left\{\tilde{A}_{e2}\tilde{\Sigma}(t)\tilde{A}_{e2}^T\right\} =$$

$$\delta\begin{bmatrix} -L_tH_{2,t}E_t & -L_tH_{2,t}E_t \\ L_tH_{2,t}E_t & L_tH_{2,t}E_t \end{bmatrix}\tilde{\Sigma}(t)\begin{bmatrix} -L_tH_{2,t}E_t & -L_tH_{2,t}E_t \\ L_tH_{2,t}E_t & L_tH_{2,t}E_t \end{bmatrix}^T$$

where $\delta = \mu(1-\mu)$. The following theorem which introduces two RDEs is obtained for equations (12) and (13).

Theorem 1: If there exist positive scalar a_t such that $a_t^{-1}I - E_{c2}\Sigma(t)E_{c2}^T > 0$, where $\Sigma(t)$ is symmetric positive-definite matrix, then

$$\Theta(t+1) = A_{c1}\Sigma(t)A_{c1}^T + A_{c1}\Sigma(t)E_{c1}^T(a^{-1}I - E_{c1}\Sigma(t)E_{c1}^T)^{-1}$$
$$\times E_{c1}\Sigma(t)A_{c1}^T + \psi_t + \tilde{\psi}_t + G_{v1}RG_{v1}^T + a^{-1}H_{c1}H_{c1}^T \quad (14)$$

$$\Sigma(t+1) = A_{c2}\Sigma(t)A_{c2}^T + A_{c2}\Sigma(t)E_{c2}^T(a^{-1}I - E_{c2}\Sigma(t)E_{c2}^T)^{-1}$$
$$\times E_{c2}\Sigma(t)A_{c2}^T + \varphi_t + \tilde{\varphi}_t + a^{-1}H_{c2}H_{c2}^T + G_{v2}RG_{v2}^T + G_{w2}QG_{w2}^T \quad (15)$$

and $\tilde{\Theta}(t) \le \Theta(t)$ and $\tilde{\Sigma}(t) \le \Sigma(t)$, where $\tilde{\Theta}(t)$ and $\tilde{\Sigma}(t)$ satisfy (12) and (13), respectively.

Proof: The proof can be done along the lines of [2] and [11].

Briefly, the upper bounds of the prediction and filtering covariance matrices are written as follows:

$$E\left\{\zeta(t)\zeta^T(t)\right\} \le [I \ \ 0]\Theta(t)[I \ \ 0]^T = \bar{\Theta}(t) \quad (16)$$

$$E\left\{\tilde{\zeta}(t)\tilde{\zeta}^T(t)\right\} \le [I \ \ 0]\Sigma(t)[I \ \ 0]^T = \bar{\Sigma}(t) \quad (17)$$

3.2   Design of Robust Kalman Filter Parameters

In this subsection, the upper bounds of the filtering and prediction covariances are computed in the form of Riccati-type equation. Then, the optimal values of the proposed Kalman filter parameters in (3) and (4), $\hat{A}_t$, L_t and K_t, are determined such that minimize $tr(\bar{\Sigma}(t))$ and $tr(\bar{\Theta}(t))$.

Theorem 2: Suppose a_t be a sequence of positive scalars. Let $\bar{\Sigma}(t)$ and $P(t)$ are the positive-definite solutions of the following recursive equations:

$$\bar{\Sigma}(t+1) = B_tQB_t^T + a_t^{-1}H_{1,t}H_{1,t}^T + A_t\bar{\Sigma}(t)\left(I + E_t^TM_t^{-1}E_t\bar{\Sigma}(t)\right)A_t^T$$
$$- (\mu A_t\bar{\Sigma}(t)(I + E_t^TM_t^{-1}E_t\Sigma(t))C_t^T + a_t^{-1}\mu H_{1,t}H_{2,t}^T)\Lambda_t^{-1} \quad (18)$$
$$\times (\mu A_t\bar{\Sigma}(t)(I + E_t^TM_t^{-1}E_t\Sigma(t))C_t^T + a_t^{-1}\mu H_{1,t}H_{2,t}^T)^T$$

$$\Lambda_t = \mu^2 C_t \bar{\Sigma}(t)\left(I + E_t^T M_t^{-1} E_t \bar{\Sigma}(t)\right) C_t^T + (1-\mu)\mu H_{2,t} E_t P(t) E_t^T H_{2,t}^T)$$
$$+ (1-\mu)\mu C_t P_t C_t^T + R + a_t^{-1}\mu^2 H_{2,t}^T H_{2,t}^T$$

$$P(t+1) = A_t \left(P^{-1}(t) - a_t E_t^T E_t\right)^{-1} A_t^T + a_t H_{1,t} H_{1,t}^T + B_t Q B_t^T \quad (19)$$

wherein $M_t = a_t^{-1} I - E_t \bar{\Sigma}(t) E_t^T > 0$ and $P^{-1}(t) - a_t E_t^T E_t > 0$. The Kalman filter parameters in (3) and (4) are as follows:

$$\hat{A}_t = A_t + a_t (A_t - \mu L_t C_t) \bar{\Sigma}(t) E_t^T M_t^{-1} E_t \quad (20)$$

$$L_t = (\mu A_t \bar{\Sigma}(t)(I + E_t^T M_t^{-1} E_t \bar{\Sigma}(t)) C_t^T + a_t^{-1}\mu H_{1,t} H_{2,t}^T)\Lambda_t^{-1} \quad (21)$$

where

$$K_t = \mu \bar{\Sigma}(t)\left(I + E_t^T \tilde{M}_t^{-1} E_t \bar{\Sigma}(t)\right) C_t^T \Xi_t^{-1} \quad (22)$$

in which

$$\Xi_t = \mu C_t \left(I + \bar{\Sigma}(t) E_t^T \tilde{M}^{-1} E_t\right)\bar{\Sigma}(t) C_t^T + (1-\mu)\mu H_{2,t} E_t P(t) E_t^T H_{2,t}^T$$
$$+ (1-\mu)\mu C_t P_t C_t^T + R + a_t^{-1}\mu^2 H_{2,t}^T H_{2,t}^T$$
$$\tilde{M}_t = a_t^{-1} I - E_t P(t) E_t^T$$

Proof: Regarding (13) and (15), the $\Sigma(t)$ can be rewritten as follows [8]:

$$\Sigma(t) = \begin{bmatrix} \Sigma_{11}(t) & \Sigma_{12}(t) \\ \Sigma_{21}(t) & \Sigma_{22}(t) \end{bmatrix} = \begin{bmatrix} \bar{\Sigma}(t) & 0 \\ 0 & P(t)\text{-}\bar{\Sigma}(t) \end{bmatrix}$$

wherein $\bar{\Sigma}(t)$ and $P(t)$ are defined in (18) and (19), respectively. In order to determine K_t that minimizes $\bar{\Theta}(t)$, its first variation is computed as follows:

$$\frac{\partial \bar{\Theta}(t+1)}{\partial K_t} = (1-\mu)\mu(K_t C_t P_t C_t^T + K_t H_{2,t} E_t P(t) E_t^T H_{2,t}^T)$$
$$+ (I - \mu K_t C_t)\bar{\Sigma}(t) E_t^T \tilde{M}_t^{-1} E_t \bar{\Sigma}(t)(-\mu C_t)^T \quad (23)$$
$$+ K_t R + (I - \mu K_t C_t)\bar{\Sigma}(t)(-\mu C_t)^T = 0$$

Then, the K_t in (22) is achieved by straightforward manipulation of (23). On the other hand, considering the equations (15), (17), we have:

$$\Pi(t+1) = \begin{bmatrix} I & 0 \end{bmatrix} \Sigma(t) \begin{bmatrix} I & 0 \end{bmatrix}^T$$
$$= (A_t - \hat{A}_t)(\bar{\Sigma}(t) - P(t))(A_t - \hat{A}_t)^T + (A_t - \mu L_t C)\bar{\Sigma}(t)(A_t - \mu L_t C_t)^T$$
$$+ (1-\mu)\mu(L_t C_t P(t)(L_t C_t)^T + L_t H_{2,t} E_t P(t)(L_t H_{2,t} E_t)^T) + L_t R L_t^T \quad (24)$$
$$+ a_t^{-1}(H_{1,t} - \mu L_t H_{2,t})(H_{1,t} - \mu L_t H_{2,t})^T + B_t Q B_t^T$$
$$+ (A_t P(t) - \mu L_t C_t \bar{\Sigma}(t) + \hat{A}_t(\bar{\Sigma}(t) - P(t))) E_t^T \tilde{M}_t^{-1} E_t$$
$$\times (A_t P(t) - \mu L_t C_t \bar{\Sigma}(t) + \hat{A}_t(\bar{\Sigma}(t) - P(t)))^T$$

To determine the $\hat{A}_t$, the first variation of Π is set to be zero:

$$\frac{\partial \Pi(t+1)}{\partial \hat{A}_t} = (A_t - \hat{A}_t)(\bar{\Sigma}(t) - P(t))(I)$$

$$+ (A_t P(t) - \mu L_t C_t \bar{\Sigma}(t) + \hat{A}_t(\bar{\Sigma}(t) - P(t))) \qquad (25)$$

$$\times E_t^T \tilde{M}_t^{-1} E_t((\bar{\Sigma}(t) - P(t)))^T = 0$$

Rearranging the (25) leads to:

$$\hat{A}_t = (A_t(I + P(t)E_t^T \tilde{M}_t^{-1} E_t) - \mu L_t C_t \bar{\Sigma}(t)E_t^T \tilde{M}_t^{-1} E_t)$$

$$\times (I - (\bar{\Sigma}(t) - P(t))E_t^T \tilde{M}_t^{-1} E_t)^{-1} \qquad (26)$$

Adding and subtracting of $\bar{\Sigma}(t)E_t^T \tilde{M}_t^{-1} E_t$ in (26) yields:

$$\hat{A}_t = A_t + (A_t - \mu L_t C_t)\bar{\Sigma}(t)E_t^T \tilde{M}_t^{-1} E_t$$

$$\times (I - (\bar{\Sigma}(t) - P(t))E_t^T \tilde{M}_t^{-1} E_t)^{-1} \qquad (27)$$

On the other side, the following relation is true [8]:

$$E_t^T \tilde{M}_t^{-1} E_t = E_t^T M_t^{-1} E_t [I + (\bar{\Sigma}(t) - P(t))E_t^T M_t^{-1} E_t]^{-1}$$

$$= [I + E_t^T M_t^{-1} E_t(\bar{\Sigma}(t) - P(t))]^{-1} E_t^T M_t^{-1} E_t \qquad (28)$$

$$I - (\bar{\Sigma}(t) - P(t))E_t^T \tilde{M}_t^{-1} E = [I + (\bar{\Sigma}(t) - P(t))E_t^T M_t^{-1} E_t]^{-1} \qquad (29)$$

Combining (27)-(29) the equation (20) is obtained. Substituting (20) into (24), we have:

$$\tilde{\Pi}(t+1) = (A_t - \mu L_t C_t)\bar{\Sigma}(t)(I + E_t^T M_t^{-1} E_t \bar{\Sigma}(t))(A_t - \mu L_t C_t)^T$$

$$+ (1-\mu)\mu(L_t C_t P(t)(L_t C_t)^T + L_t H_{2,t} E_t P(t)(L_t H_{2,t} E_t)^T) \qquad (30)$$

$$+ a_t^{-1}(H_{1,t} - \mu L_t H_{2,t})(H_{1,t} - \mu L_t H_{2,t})^T + B_t Q B_t^T + L_t R L_t^T$$

The matrix L_t is computed by taking the first variation of $\tilde{\Pi}$ in (30) as follows:

$$\frac{\partial \tilde{\Pi}(k+1)}{\partial L_t} = L_t R_t + a_t^{-1}(H_{1,t} - \mu L_t H_{2,t})(\mu H_{2,t})^T$$

$$+ (A_t - L_t C_t)\bar{\Sigma}(t)(I + E_t^T M_t^{-1} E_t \bar{\Sigma}(t))(-\mu C_t)^T \qquad (31)$$

$$+ (1-\mu)\mu(L_t C_t P(t)C_t^T + L_t H_t E_t P(t)(H_t E_t)^T) = 0$$

The matrix L_t in (21) is easily derived from (31). Substituting (21) into (30) leads to (18). The covariance matrix of the state is as follows:

$$P(t+1) = E\{x(t+1)x^T(t+1)\}$$

$$= E\{((A_t + \Delta A_t)x(t) + B_t w_t)((A_t + \Delta A_t)x(t) + B_t w_t)^T\} \qquad (32)$$

$$= (A_t + H_t F_t E_t)P(t)(A_t + H_t F_t E_t)^T + B_t Q B_t^T$$

Relation (32) can be transformed to (19) [12].

4   Simulation Example

we consider the following uncertain discrete-time systems with missing measurements [12]:

$$x(t+1) = \left(\begin{bmatrix} 0 & 0.1\sin(6t) \\ 0.2 & 0.3 \end{bmatrix} + \begin{bmatrix} 0.5 \\ 1 \end{bmatrix} F_t \begin{bmatrix} 0.2 & 0.1 \end{bmatrix} \right) x(t) + \begin{bmatrix} 1 \\ 0.5 \end{bmatrix} w(t)$$

$$y(t) = \gamma_t (\begin{bmatrix} 0.5 + 0.3\sin(6t) & 1 \end{bmatrix} + 4F_t \begin{bmatrix} 0.2 & 0.1 \end{bmatrix}) x(t) + v(t) \qquad \text{The noise signals}$$

$$F_t = \sin(0.6t)$$

$w(t)$ and $v(t)$ are uncorrelated with zero-mean and unity covariances. The scalar binary stochastic variable γ_t isBernoulli distributed. Figures1 and 2compare the error variances of the results obtained by the proposed method and the one in [12]

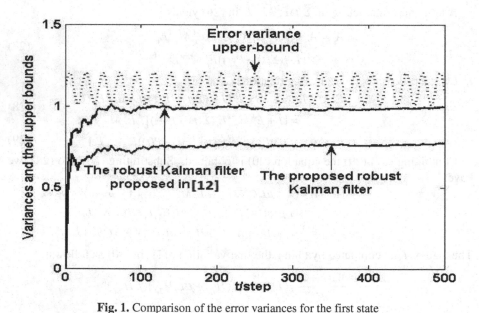

Fig. 1. Comparison of the error variances for the first state

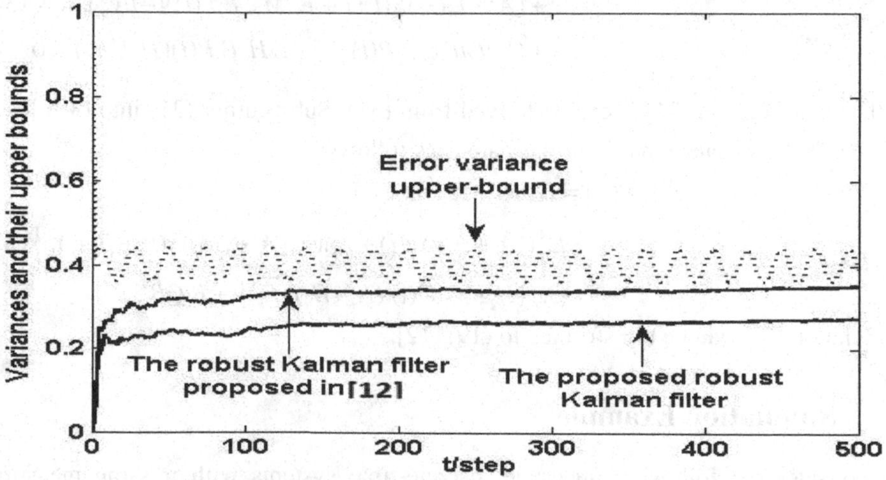

Fig. 2. Comparison of the error variances for the second state

by 100 times Monte-Carlo test, with $\mu = 0.8$ and $a_t = 3$. The outperformance of the introduced procedure is evident.

5   Conclusions

In this paper, a novel approach has been developed to design a finite-horizon robust Kalman filter for uncertain linear discrete time-varying systems subject to intermittentobservations and time-varying norm-bounded uncertainties in the state and output matrices. Filter parameters are determined such that the upper bound on the estimation error covariance matrix be minimal. The illustrative examples verified the advantages of the proposed filter.

References

1. Simon, D.: Optimal state estimation: Kalman, H infinity, and nonlinear approaches. Wiley.com (2006)
2. Yang, F., Wang, Z., Feng, G., Liu, X.: Robust filtering with randomly varying sensor delay: the finite-horizon case. IEEE Transactions on Circuits and Systems I: Regular Papers 56, 664–672 (2009)
3. Lu, X., Xie, L., Zhang, H., Wang, W.: Robust Kalman filtering for discrete-time systems with measurement delay. IEEE Transactions on Circuits and Systems II: Express Briefs 54, 522–526 (2007), Dyer, F.C.: The biology of the dance language. Annual review of Entomology 47(1), 917–949 (2002)
4. Wang, Z., Ho, D.W., Liu, X.: Robust filtering under randomly varying sensor delay with variance constraints. IEEE Transactions on Circuits and Systems II: Express Briefs 51, 320–326 (2004)
5. Nahi, N.E.: Optimal recursive estimation with uncertain observation. IEEE Transactions on Information Theory 15, 457–462 (1969)
6. Hadidi, M., Schwartz, S.: Linear recursive state estimators under uncertain observations. IEEE Transactions on Automatic Control 24, 944–948 (1979)
7. Souto, R.F., Ishihara, J.Y.: Comments on "Finite-Horizon Robust Kalman Filtering for Uncertain Discrete Time-Varying Systems With Uncertain-Covariance White Noises". IEEE Signal Processing Letters 17, 213–216 (2010), Giurfa, M.: Cognitive neuroethology: dissecting non-elemental learning in a honeybee brain. Current Opinion in Neurobiology 13(6), 726–735 (2003)
8. Zhu, X., Soh, Y.C., Xie, L.: *Design and analysis of discrete-time robust Kalman filters*. Automatica 38, 1069–1077 (2002)
9. Shaked, U., De Souza, C.E.: Robust minimum variance filtering. IEEE Transactions on Signal Processing 43, 2474–2483 (1995)
10. Wang, Z., Ho, D.W.C., Xiaohui, L.: Variance-constrained filtering for uncertain stochastic systems with missing measurements. IEEE Transactions on Automatic Control 48, 1254–1258 (2003)
11. Wang, Z., Fuwen, Y., Ho, D.W.C., Xiaohui, L.: Robust finite-horizon filtering for stochastic systems with missing measurements. IEEE Signal Processing Letters 12, 437–440 (2005)

12. Mohamed, S.M., Nahavandi, S.: Robust finite-horizon Kalman filtering for uncertain discrete-time systems. IEEE Transactions on Automatic Control 57, 1548–1552 (2012)
13. Liang, H., Zhou, T.: Robust state estimation for uncertain discrete-time stochastic systems with missing measurements. Automatica 47, 1520–1524 (2011)
14. Chu, J.-U., et al.: Spontaneous synchronized burst firing of subthalamic nucleus neurons in rat brain slices measured on multi-electrode arrays. Neuroscience Research 72(4), 324–340 (2012)

Motor Imagery Data Classification
for BCI Application Using Wavelet Packet
Feature Extraction

Imali Thanuja Hettiarachchi, Thanh Thi Nguyen, and Saeid Nahavandi

Centre for Intelligent Systems Research, Deakin University, Australia
imali.hettiarachchi@deakin.edu.au

Abstract. The noninvasive brain imaging modalities have provided us an extraordinary means for monitoring the working brain. Among these modalities, Electroencephalography (EEG) is the most widely used technique for measuring the brain signals under different tasks, due to its mobility, low cost, and high temporal resolution. In this paper we investigate the use of EEG signals in brain-computer interface (BCI) systems. We present a novel method of wavelet packet-based feature extraction and classification of motor imagery BCI data. The prominent discriminant features from a redundant wavelet feature set is selected using the receiver operating characteristic (ROC) curve and fisher distance criterion. The BCI competition 2003 data set Ib is used to evaluate a number of classification algorithms. The results indicate that ROC is able to produce better classification accuracy as compared with that from the fisher distance criterion.

Keywords: Brain-computer interface, Motor imagery data, Wavelet packet decomposition, Fisher distance criterion, Receiver operating characteristic curve.

1 Introduction

A brain-computer interface (BCI) is a system, which translates the brain's mental activity into a computer control signal [1]. Most of the BCI systems are based on the electroencephalogram (EEG) signals, owing to its noninvasive nature and its affordable price; therefore a low cost recording equipment for real-time operations. In particular, motor imagery (MI) BCI has gained significant attention among the bioengineering community in the last decade [1]. MI is the mental rehearsal of a motor action without an actual movement of limbs, fingers or tongue. One application of MI-based BCI systems is to restore sensory and motor functions in patients who have severe motor disabilities, there by improving their quality of life.

Analysis of single-trial event related potentials (ERP) for MI-based BCI systems rely on successful discrimination of the signal into different MI classes. In a BCI system, each MI task is considered to belong to a *class* of data category. A current challenge in the biomedical research is how to classify time-varying EEG

C.K. Loo et al. (Eds.): ICONIP 2014, Part III, LNCS 8836, pp. 519–526, 2014.
© Springer International Publishing Switzerland 2014

signals as accurately as possible. Extracting the most important features of EEG is performed during the stage of *feature extraction*. Then, the *classification* stage uses the extracted features to determine the class of the signal using a *classifier*.

In MI BCI applications, different methods have been utilised for the feature extraction from the EEG signals. EEG signals can be characterised with specific frequency bands such as delta (< 4Hz), theta ($4 - 7$Hz), alpha ($8 - 12$Hz), beta ($16 - 31$Hz) and gamma(> 32Hz). Frequency specific changes in EEG activity comprise one important feature in BCI systems. They utilise the decrease or increase of power in certain frequency bands (band power). A variety of other methods have been proposed for extraction of features that reliably describe several distinctive brain states. These methods include Hjorth parameters, fractal dimensions, fast fourier transform (FFT), autoregressive (AR) modelling, independent component analysis (ICA), common spatial patterns (CSP) and wavelet transform (WT). Among these methods FFT, AR and WT are frequency domain feature extraction methods.

However, the frequency domain measures such as FFT and AR models do not suit the analysis of EEG signals due to the inherent non-stationary nature of the data. In order to handle nonstationarities short-time segmentation approaches have been utilised, which results in time-frequency domain feature extraction. Short time fourier transform (STFT) is an alternative to FFT. However, FFT performs very poorly with short data segments. Adaptive AR (AAR) modelling of signals is an alternate to AR modelling of the signals, which has been proposed and used in EEG-based BCI's [4].

WT is a popular time-frequency domain feature extraction method used in BCI applications. WT is able to represent any general function as an infinite series of wavelets. The discrete wavelet transform (DWT) decomposes the signal into a series of coefficients comprising coarse approximation and detail information. Then DWT analyses the signal at different frequency bands with different resolutions. Since the DWT is powerful in selecting multi resolution features, it is an efficient and structured approach to EEG representation. In this regards, wavelet packet decomposition (WPD) is a WT in which the discrete-time signal is passed through more filters than the DWT. WPD has been applied to MI BCI [5] and mental task classification [6].

In this paper we use WPD as the preferred feature extraction tool due to its efficiency in the BCI signal analysis. After applying WPD for each channel and trial we select the sub-band coefficient average and sub-band energy to form the initial feature sets. A similar approach was used in [5]. However, unlike the work in [5], we use a merged feature set and a receiver operating characteristic (ROC) curve-based feature selection approach in this study. The details are presented in section 3.

Once the features are extracted from EEG signals, the next step is to determine to which class the features belong to. A number of widely used classifiers are Linear discriminant analysis (LDA), K-nearest-neighbor (kNN) algorithms, Support vector machine (SVM), Decision trees, Naive Bayes (NB) classifier and Neural networks (NN)[8]. Classifiers can be combined to reduce the variance;

therefore increasing classification accuracy. This is referred to as ensemble learning, which is based on the principle that a group of weak learners can be combined to form a strong learner. The classifier combination strategies such as boosting, bagging, and stacking has been used in BCI applications [8].

The winning entry for the BCI competition II Ib data set [7] used a feature extraction algorithm based on continuous WT and student's t-statistic, and reported a classification accuracy of 54.4% using a LDA classifier. The work reported in [5] analysed this data set using WPD and fisher distance feature extraction, a classification accuracy of 59.1% was produced using the probabilistic neural network (PNN) classifier. This paper aims to address classification of MI based EEG data in BCI systems using a WPD-based feature extraction and an ROC-based feature selection method. In this study we evaluate the proposed method with four widely used classifiers, namely kNN, LDA, SVM and NB.

The organisation of this paper is as follows. In Section 2, the proposed feature extraction and feature selection method is described. A series of experimental studies using the BCI competition II Ib data set is presented in Section 3. The results and discussion are in Section 4. Section 5 concludes the paper.

2 Methodology

2.1 Feature Extraction by Wavelet Packet Decomposition

Fig. 1 illustrates the filter bank structure of signal $(S[n])$ decomposition using a level 3 decomposition technique. The number of decomposition is chosen based on the level of frequency components of the signal. The frequency range of the subspace U_j^n is $[\frac{nf_s}{2^{j+1}}, \frac{(n+1)f_s}{2^{j+1}}]$, where j represents the jth level of the decomposition and n represents the node number. In a WPD there are $n = 0, 1, \ldots, (2^j - 1)$ nodes at the jth level of decomposition. The kth coefficient of WPD of the nth node at the jth level $(d_j^n(k)$) is given by,

$$d_j^n(k) = \begin{cases} \sum_m h_0(m - 2k) d_{j-1}^{n/2}(m) & \text{if } n \text{ is even} \\ \sum_m h_1(m - 2k) d_{j-1}^{(n-1)/2}(m) & \text{if } n \text{ is odd} \end{cases} \tag{1}$$

where, $h_0(i)$ and $h_1(i) = (-1)^{1-i} h_0(1 - i)$ are quadruple mirror filters. It can be seen from equation (1) that the coefficients at level j can be calculated from coefficients at level $(j - 1)$. Thus all the coefficient can be calculated recursively.

2.2 Initial Feature Set Formulation

The extracted wavelet coefficients contain information of the distribution of the EEG signal in the time-frequency domain. After applying WPD for each trial and channel separately, different discriminative features can be used to describe the signal in terms of the extracted coefficients. These discriminative features can be formed using the sub-band wavelet coefficients, transformation of wavelet coefficients by the principal component analysis or singular value decomposition,

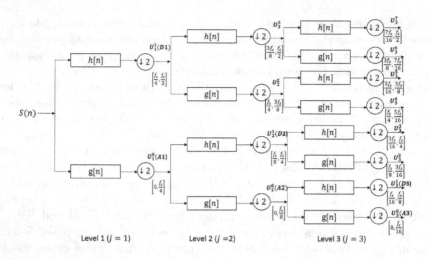

Fig. 1. A level 3 WPT decomposition on a signal S(n). U_j^n is the sub space of the nth node at the jth level of decomposition. $h[n]$ is a high pass filter and $g[n]$ is a low pass filter. f_s is the sampling frequency of the original signal $S[n]$. The sub-tree containing nodes $D1, D2, D3, D4$ and $A4$, which correspond to a level 3 DWT decomposition.

statistical features of sub-band wavelet coefficients such as average, standard deviation or skewness or sub-band energies. We have selected the sub-band coefficient average and sub-band energy, contained in the last decomposition level of the WPD as the initial features of the EEG signal. The sub-band coefficient average of node n at the jth decomposition level for channel l ($m_{j,l}^n$) is calculated as $m_{j,l}^n = \frac{2^N}{2^j} \sum_k d_{j,l}^n(k)$. The sub-band energy of node n at the jth decomposition level for channel l ($e_{j,l}^n$) is calculated as, $e_{j,l}^n = \sum_k (d_{j,l}^n(k))^2$.

The feature set for sub-band coefficient average is constructed as $M = [m_{j,1}^0, m_{j,1}^1, m_{j,1}^2, \ldots, m_{j,2}^0, m_{j,2}^1, m_{j,2}^2, \ldots, \ldots, \ldots, m_{j,L}^0, m_{j,L}^1, m_{j,L}^2, \ldots]$ while $E = [e_{j,1}^0, e_{j,1}^1, e_{j,1}^2, \ldots, e_{j,2}^0, e_{j,2}^1, e_{j,2}^2, \ldots, \ldots, \ldots, e_{j,L}^0, e_{j,L}^1, e_{j,L}^2, \ldots]$ is the obtained the sub-band energy feature set. Finally M and N are merged to obtain the initial feature set. The initial feature set formation procedure is shown in Fig. 2.

Once the initial feature set is constructed the dimensionality of the feature set is high, which include redundant features. In order to avoid the *curse of dimensionality*, we have to select the most prominent features. The next section discusses two feature selection methods utilised in this paper, namely Fisher distance criterion (FDC) and receiver operating characteristic (ROC) curve.

2.3   Fisher Distance Criterion

The FDC evaluates the separability of classes based on the class means, which can be give as, $F = \frac{S_b}{S_w}$, where $S_b = \sum_{i=1}^c n_i (m_i - m)(m_i - m)^T$ is the inter class distance, $S_w = \sum_{i=1}^c \sum_{j=1}^{n_i} (x_j - m_i)(x_j - m_i)^T$ is intra class distance and c is the number of classes. Here $m = \frac{1}{n} \sum_{k=1}^n x_k$ is the average value of samples,

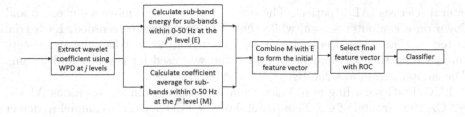

Fig. 2. Final feature set formation of the proposed method. The M and E initial features are formed using each trial and each channel of the EEG signal.

$m_i = \frac{1}{n_i} \sum_{k=1}^{n_i} x_i$ average of class i, n is the number of samples in the initial feature set and n_i is the number of samples in class i. The F value is calculated for each feature of the initial feature set. The features with larger F values are chosen as they are considered to be more discriminative features for classification.

2.4 Receiver Operating Characteristic Curve

Fig. 3(a) shows the probability density functions (pdfs) characterising the distribution of a feature in two classes, along with a threshold. If both distributions completely overlap, then for any position of the threshold it is seen that $\alpha = 1-\beta$. This can be explained with the straight line in Fig. 3(b). As the two distributions move apart, the corresponding curve moves off from the straight line. That is the more the classes of a given feature are distinct, the larger the area between the curve and the straight line (area under the curve - AUC) [9]. In practice, the ROC curve can be easily generated by sweeping the threshold and calculating the percentages of correct and incorrect classifications. The AUC, therefore, can also be computed easily. Then the features with the greatest AUC are chosen as the discriminative feature set for classification.

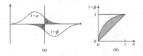

Fig. 3. Example of (a) pdfs characterising the same feature in two classes (one pdf has been inverted) (b) the resulting ROC curve

3 Motor Imagery Data Analysis

The motor imagery data analysed in this paper are from the publicly available BCI competition 2003 data sets. Here we provide a brief description of the data set. The detailed description can be found in `https://www.bbci.de/competition/ii/tuebingen_desc_ii.html` . The Ib data set provided by the University of Tüebingen, is recorded from an artificially respirated amyotrophic lateral sclerosis (ALS) patient. The subject was asked to move a cursor up and down on a computer screen, while the EEG signals were recorded. Each trial lasted 8s. The visual feedback was presented from 2s to 6.5s. The signal captured within this 4.5s interval of every trial was used for training and testing. The sampling rate was 256 Hz.

EEG/EOG recording were taken from 7 channels from the positions A1-Cz, A2-Cz, 2 cm frontal of C3, 2 cm parietal of C3, vEOG artifact channel to detect vertical eye movements, 2 cm frontal of C4 and 2 cm parietal of C4. The training data set combined 200 trials, with 100 trials belonging to each class. The class labels, '0' and '1', corresponded to left and right, respectively. The test data set contained 180 trials, each belonging to either class 0 or class 1.

3.1   Feature Set Formulation

The Ib data set of the 2003 BCI competition comprised 7 channels. However, in this study, we only used the 4 EEG channels, namely, 2 cm frontal of C3, 2 cm parietal of C3, 2 cm frontal of C4 and 2 cm parietal of C4. The mastoid referenced Cz channels and vEOG artifact channel were removed from the analysis.

We chose a level 6 WPD and the Daubechies wavelet of order 4 for wavelet coefficient extraction in this study. The resulting frequency ranges for the subspaces at the last level (6th) of the decomposition comprised [0 2], [2,4], [4,6], ..., [124 126], [126 128]. As the EEG signals only contain useful information in the $0 - 50$Hz frequency band, the sub-bands of the 6th level of decomposition less than 50Hz's was selected from each M, N initial feature sets. Therefore for one channel pertaining to one feature set, a total of 25 sub-bands belonging to $0 - 50$ Hz were available. With two initial feature sets (M, N) and four channels, the initial merged feature set comprised of 200 features. Once the initial feature set was constructed, the feature selection process was carried out. Fig. 4 shows the separability of the classes for each feature when using ROC. Based on this result, the highest 2.5% of the features ranked by ROC were

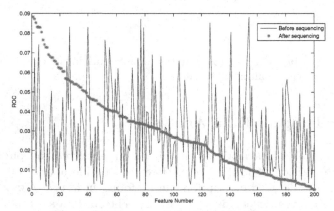

Fig. 4. ROC curve of the initial feature set

selected for classification. According to the graph in Fig. 4, the final feature set based on ROC were $Features_R = [154, 77, 126, 27, 79]$. Similarly according to the Fisher distance based method, we selected the final feature set comprised $Features_F = [77, 154, 5, 79, 52]$.

It is obvious that there is a difference between feature sets of two methods. It is this difference that leads to the dominance of the ROC method against the Fisher distance criterion, which is showcased in the next section.

4   Results and Discussion

After feature extraction and selection, the normalised feature sets were provided to LDA, SVM, kNN, and NB for classification. The classification accuracy rates for the test set is shown in Table 1. Note that LDA and SVM are most useful techniques which separate data representing different classes by using hyperplanes [8]. However, LDA and SVM classifiers have some limitations. When dealing with the nonlinear EEG data, linearity is the main limitation of LDA, which can cause poor outcomes [8]. SVM has a low speed of execution but has good generalisation properties [8]. On the other hand kNN assigns an unseen data sample to the dominant class among its k nearest neighbors formed using the training set. As stated in [8] kNN has failed in several BCI experiments due to its sensitivity to the *curse of dimensionality*. However, it performs efficiently with low-dimensional feature sets [8]. NB classifier is based on Bayes' theorem, with a strong assumption of independence of the features. The main advantage of NB is that it only requires a small number of training data to estimate the parameters. Therefore, this is an advantage in MI data classification tasks, which generally are limited to small number of trials. From the results, it is obvious that WPD-based feature extraction with ROC feature selection has given the highest accuracy (59.44%) with the NB classifier, for this two-class motor imagery classification problem.

Feature extraction is an important issue in BCI research, which can substantially affects classification accuracy. The classification performance can be increased by using an effective feature extraction method. In this study, we attempt to tackle this task by using an efficient feature extraction and selection method. The FDC operates on finding the mean distance between data samples of classes, while the ROC performs on the principle of finding the difference between the distribution of the data samples in the classes of a given feature. From the results, ROC appears to be a better method for feature selection, as compared with FDC, which only uses one statistical measure (mean) of the distribution. Therefore, better classification accuracy rates are produced by using the ROC-based feature selection method.

Table 1. Classification accuracy of the test data for BCI competition II data set Ib

Feature set formulation	LDA	SVM	kNN	NB
WPD + FDC	55	52.78	55.56	53.89
WPD + ROC	55	55.56	51.67	**59.44**

5   Conclusion

This paper has presented an effective feature extraction and selection method based on WPD and ROC criteria, respectively. From the experiments using the BCI competition II Ib data set, an accuracy of 59.44% has been achieved, which is greater than the maximum accuracy of 54.44% produced by the competition. This high accuracy rate justifies the effectiveness of the proposed method in BCI applications. Future work will focus on evaluating other data sets with the proposed method in order to fully ascertain its robustness.

References

1. Pfurtscheller, G., Neuper, C., Guger, C., Harkam, W., Ramoser, R., Schlögl, A., Obermaier, B., Pregenzer, M.: Current trends in Graz brain-computer inter-face(BCI) research. IEEE Trans. Rehabil. Eng. 8(2), 216–219 (2000)
2. Pfurtscheller, G., Lopes da Silva, F.H.: Event-related EEG/MEG synchronization and desynchronization:basic principles. Clin. Neurophysiol. 110, 1842–1857 (1999)
3. Pfurtscheller, G., Neuper, C., Flotzinger, D., Pregenzer, M.: EEG-based discrimi-nation between imagination of right and left hand movement. Electroenceph. Clin. Neurophysiol. 10, 642–651 (1997)
4. Schlöegl, A., Lugger, K., Pfurtscheller, G.: Using Adaptive Autoregressive Parame-ters for a Brain-Computer-Interface Experiment. In: Proceedings of the 19th Annual International Conference if the IEEE Engineering in Medicine and Biology Society, vol. 19, pp. 1533–1535 (1997)
5. Ting, W., Guo-Zheng, Y., Bang-Hua, Y., Hong, S.: EEG feature extraction based on wavelet packet decomposition for brain computer interface. Measurement 41, 618–625 (2008)
6. Xue, J.Z., Zhang, H., Zheng, C.X.: Wavelet packet transform for feature extraction of EEG during mental tasks. In: Proceedings of the second International Conference on Machine Learning and Cybernectics, Xi'an (2003)
7. Bostanov, V.: BCI competition 2003-data sets Ib and IIb: feature extraction from event-related brain potentials with the continuous wavelet transform and the t-value scalogram. IEEE Transactions on Biomedical Engineering 51(6), 1057–1061 (2004)
8. Lotte, F., Congedo, M., Lécuyer, A., Lamarche, F., Arnaldi, B.: A review of clas-sification algorithms for EEG-based brain-computer interfaces. Journal of Neural Engineering 4, R1–R13 (2007)
9. Theodoridis, S., Koutroumbas, K.: Pattern Recognition, 4th edn. Academic Press (2009)

Adaptive-Multi-Reference
Least Means Squares Filter

Luke Nyhof, Imali Hettiarachchi, Shady Mohammed, and Saeid Nahavandi

Deakin University - Centre for Intelligent Systems Research
Waurn Ponds, VIC, 3217, Australia
luke.nyhof@research.deakin.edu.au,
{imali.hettiarachchi,shady.mohamed,saeid.nahavandi}@deakin.edu.au
https://www.deakin.edu.au/research/cisr

Abstract. Adaptive filters are now becoming increasingly studied for their suitability in application to complex and non-stationary signals. Many adaptive filters utilise a reference input, that is used to form an estimate of the noise in the target signal. In this paper we discuss the application of adaptive filters for high electromyography contaminated electroencephalography data. We propose the use of multiple referential inputs instead of the traditional single input. These references are formed using multiple EMG sensors during an EEG experiment, each reference input is processed and ordered through firstly determining the Pearson's r-squared correlation coefficient, from this a weighting metric is determined and used to scale and order the reference channels according to the paradigm shown in this paper. This paper presents the use and application of the Adaptive-Multi-Reference (AMR) Least Means Square adaptive filter in the domain of electroencephalograph signal acquisition.

Keywords: Adaptive Multi-Reference, electroencephalograph (EEG), signal filter, biopotential, artefact filter.

Non-invasive electroencephalography *(EEG)* data acquisition is performed under tightly controlled clinical conditions, this however is not the case for everyday life. Real world scenarios can contain heavy contamination of surface EEG signals via external and internal electrical sources, these include mains line noise, electromagnetic interference and muscle noise. Myoelectric noise is generated through the activation of nerves which control muscle fiber contraction, the artefact signal propagates from the innervated muscle fibers through the tissue and contaminates the surface EEG electrode measurements [1,2].

Traditional methods of signal filtering may employ a variety of techniques to attenuate or eliminate EMG artefact contamination, including simple methods like non-linear, low & high-pass linear filters [2,3,4], notch filtering mains line noise and rejection of segments which exceed a predetermined threshold; to more advanced methods such as principle & independent component analysis (PCA & ICA)[5], canonical correlation analysis (CCA)[6], wavelet analysis and regression analysis [7,8]. While useful for off-line signal processing and static acquisitions,

C.K. Loo et al. (Eds.): ICONIP 2014, Part III, LNCS 8836, pp. 527–534, 2014.
© Springer International Publishing Switzerland 2014

these filters have limited ability to increase the signal-to-noise (SNR) during dynamic motion.

Adaptive techniques such as Least Means Squares (LMS) filters and Kalman state space estimators are useful tools for filtering highly noisy non-linear signals, especially in the case where the noise model is not known *a priori* and the signal may be under the influence of several unknown artefact sources.

Many adaptive EMG noise cancellation techniques require an external reference input which is used to minimise the output error of the filter and attenuate the noise in the signal [9,10,11] generated by the muscular contraction and extension which occurs naturally as the body attempts to maintain an upright or stationary posture[12]. However, a reference EMG reading taken from a single site looses the spatial properties of the artefact.

To increase the spatial characteristics of the reference inputs in relation to the filter, the reference channels are weighted and mixed on a channel-by-channel basis. This novel technique is demonstrated and shown to give superior results to single-input least means squares filters.

This paper takes the study out of a controlled static situation and towards a real-world environment that is under dynamic motion, presenting a method of adaptively filtering muscle noise from EEG signals using a group of EMG reference signals based on the coherence between the individual EMG and EEG sensors. The technique is shown to improve the signal-to-noise ratio (SNR) of the sensor level signal on simulated data, epoched Event Related Potential (ERP) study results are shown. The results show how multipoint referencing can be used to provide the optimal reference signal input for adaptive filter paradigms so that they can perform in noisy uncertain environments.

The work presented here is organised as follows: Methods, Adaptive Filters Theory, Simulation Data, Results & Discussion and conclusion.

1   Methods

1.1   Adaptive Filtering

This work presents new EMG noise artefact reference Adaptive Multi Reference Cascaded LMS filter, a implementation of the LMS algorithm using the adaptive noise canceller configuration where the signal output from the first filter is fed into the input of the second filter, the reference for each successive filter is adapted from the previous cascade, where the order of referential inputs, their associated filter and the order of are determined by the level of contamination.

The LMS filter, by definition, estimates the filter coefficients that relate to producing the least mean squares of the error signal. Modification to the classical filter design is shown in Figure 1 with the inclusion of multiple reference sensors that are used to acquire the muscle noise from four locations surrounding the cervical spine and neck.

In Figure 1, the EMG channels assigned to reference inputs $n1$, $n2$, $n3$, $n4$ and ECG are determined by the level of signal correlation between the EMG and EEG signals. $S(n)$ is the measured surface EEG signal that contains a mixture of

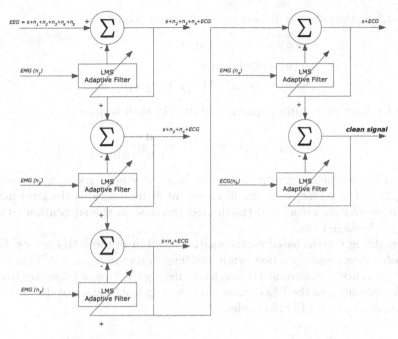

Fig. 1. Cascaded LMS Adaptive Noise Canceller configuration

the true EEG signal $x(n)$ and the noise $z(n)$. We use a weighted reference signal $R(n)$, which is correlated in someway to the noise $z(n)$. Here $h(n)$ represents the finite impulse response (FIR) filter of length p. The desired output of the noise canceller $\hat{x}(n)$ is the corrected EEG signal, the clean EEG signal,

$$\hat{x}(n) = s(n) - \hat{R}(n) \tag{1}$$

where the adaptive filter output $\hat{R}(n)$ is,

$$\hat{R}(n) = \sum_{i=0}^{p-1} h_i(n)R(n+1-i)$$

$$\hat{R}(n) = h^T(n)X(n)$$

The weight vector of the FIR filter can be defined as $h(n) = [h_0 h_1 \ldots h_{p-1}]^T$ and $X(n) = [x(n), x(n-1), \ldots, x(n-p+1)]]^T$ is the input signal vector. while Based on the LMS criterion the weights of the adaptive filter is updated according to the rule,

$$h(n+1) = h(n) + \frac{\mu\hat{x}^T(n)X(n)}{X^T(n)X(n)} \tag{2}$$

where μ is the step size controlling the speed of convergence, $i = 1, 2, \ldots, p$ and, T denotes the transpose of the matrix.

The LMS Algorithm is defined by the following equations:

$$y(n) = \mathbf{w}^T(n-1)\mathbf{u}(n)$$
$$e(n) = d(n) - y(n) \tag{3}$$
$$\mathbf{w}(n) = \alpha\mathbf{w}(n-1) + f(\mathbf{u}(n), e(n), \mu)$$

The LMS adaptive filter implemented in this work is defined as:

$$f(\mathbf{u}(n), e(n), \mu) = \mu e(n)\frac{\mathbf{u}*(n)}{\varepsilon + \mathbf{u}^H(n)\mathbf{u}(n)} \tag{4}$$

Where n is the time index, $\mathbf{u}(n)$ is a vector input at step n, $\mathbf{u}*(n)$ is the complex conjugate of the vector, $\mathbf{w}(n)$ the filter weight estimates, $y(n)$ the filter output, $e(n)$ the estimation error, $d(n)$ the desired response, μ the adaptation step size and α the leakage factor.

A weighting system based on the spatial location of the EMG sensor, EMG-EEG coherence, muscle position, origin and insertion were developed. This weighting system is firstly determined through calculating the level of coherence between the reference site and the EEG sensor site; then by the location of the muscle position relative to the EEG electrodes.

1.2   Simulation Study

Dynamic motion in environments such as driving and flying where participants are subjected to high levels movement consequently results in the neck muscle activity representative of the data set used in this study. To accommodate for the additive noise generated by muscle activation, eight individual EMG electrodes were placed over the major muscle groups which control the attitude of the head. These electrodes are configured as bipolar referential inputs.

Due to no ground truth available on experimental data we use a simulated study to validate the proposed techniques in the paper. To preserve the real nature of the simulation study, firstly the noiseless EEG is generated using BESA[13], then experimentally acquired EMG noise is added to this synthetically generated signals to generate the final contaminated EEG signal.

In order to acquire the EMG noise, electrodes were connected to the bipolar inputs of a SynAmps2[14] EEG system, with ground and active reference input points, the surface EMG electrodes were positioned on suitably prepared skin over the main central mass of the sternocliedomastoid muscle, which extends from the mastoid insertion point behind the ear at the base of the skull and attaches to the clavicular origin; and the trapezius muscle, whose origin is the occipital bone at the rear of the skull and extends longitudinally to the lower thoracic vertebrae and laterally to the scapula [15]. Placement of the electrodes was compliant with standard EMG practices, with a constant 20mm inter-electrode distance. These sites were chosen as they represent the two largest muscle groups which control the attitude of the head.

EEG electrodes were positioned on the scalp in accordance to the International 10/20 system [16]. The electrode sites were suitable prepared and a maximum impedance of less than 10 Ohms maintained. The electrodes were secured with collodion adhesive and EMG sites were secured with tape to minimise the possibility of signal drift and dropped leads. The subject instructed to keep their head in an upright position while they were subjected to external forces which would activate the muscle groups in the neck. A full EEG montage were recorded to maintain consistency in application, the real EEG channels replaced with the synthetic BESA n1p3 generated data [13].

The acquired data was processed and analysed using the Matlab environment. To ensure that the recorded EMG noise was correlated to the synthetic EEG sensor channels, the EMG data was processed using autoregressive techniques then added to the synthetic EEG channels. The contaminated signal then segmented into 197 equal length epochs, each containing the target ERP waveform. This was generated using BESA Sim [13] to ensure the synthetic target signal represented the most realistic clean EEG signal achievable. EMG contamination of the EEG signal as expected was greatest in the sites nearest to the muscle under contraction, the level of contraction were estimated using the proportion of variance due to the level of contraction, R^2 [17]. Using this measure an initial contamination scaling factor were determined, which is then used to normalise the individual reference signal to a suitable signal amplitude. This technique ensures that the reference matrix is at a level which would ensure the convergence of the filter coefficients.

2   Results and Discussion

An iterative approach was used to determine the optimal length of the filter, for this application it was found to be 3 and the step size, μ to be 0.001. Increasing the length of the filter beyond this resulted in the output failing to converge. The results from the simulated study, which used real EMG data overlaid on a realistic generated EEG signal, synthesised using the BESA Simulator software package *n1p3* preset [13].

Inspection of the real data reveals the level of disparity in signal level between EEG and EMG sites. Reference EMG signals can at times become orders of magnitude higher than the level at the surface EEG site. To account for this a scaling factor was determined based on the coherence of the EMG reading to the EEG electrode and applied to each EMG input of the filter, this results in optimising the amplitude level of the reference signal being used in the adaptive filter for any particular EEG signal.

The level of noise attenuation of the filter outputs can be seen in Figures 3 & 2, the figures show stacked ERP plot with the spectra of each successive epoch stacked on top of the previous epoch. It can be seen in the figure how the filter further improves the SNR of the target waveform over time as each iteration of the filter step reinforces the convergence of the adaptive coefficients. In Figure 2 a comparison of the filter outputs from both single reference input LMS against the AMR input LMS filter is given, the plot indicated in green shows the classic single input LMS filter output, the red plot representing the AMR LMS filter

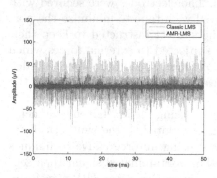

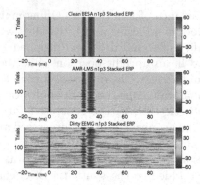

Fig. 2. Classic LMS Filter vs. Adaptive Multi-Reference LMS filter Output

Fig. 3. Stacked ERP epoch plot, showing the filter inputs and outputs

output. The level of noise attenuation is clear, additionally the overall level of the signal exhibits significantly better consistency over time.

A butterfly plot can be used to visualise the effect that a given filter has on a ERP data, the signal remains in the time domain with each successive plot overlaid on the last. The AMR filter shown in the centre effectively filters out the EMG noise whilst retaining the characteristics of the clean n1p3 waveform.

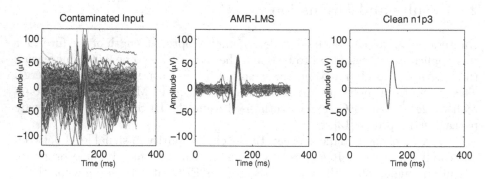

Fig. 4. Filter outputs shown as stacked butterfly overlay, each window features 200 stacked 'epochs' each representing a stimulus event

Spectral contamination of the surface EEG by EMG were analysed through study of the cross-correlation between the sensor and reference sites, from which a measure of signal similarity is determined using Pearson's R^2 technique, while simplistic this method exhibits convincing results for cases where there are two random and independent data sequences under consideration. In signal processing this tool can be used to identify the occurrence of a known signal in a continuous stream, represented by $x(n)$ and $y(n)$. In signal processing applications, as in this work, the known reference signal from the neck EMG electrode site becomes the reference $x(n)$ and the signal received by the scalp EEG electrodes becomes the target $y(n)$. The cross-correlation is calculated in this paper as Equation 5.

$$r = \frac{1}{N} \sum_{n=0}^{N-1} x_1(n)x_2(n) \tag{5}$$

However this does not consider the case where both signals are identical, in a frequency sense, and phase shifted by 180 degrees, which may indicate zero correlation between the signals or the waveforms may be 100% correlated. An important point to note is that caution must be used when utilising cross-correlation analysis with non-linear systems, there are certain situations where characteristics of the input signal can cause a system with non-linear dynamics to become blind to nonlinear effects [18]. An example of this scenario would include the situation where some moments go to zero, which mistakenly infers that the correlation between the two signals is very little when in reality it is strongly related through the underlying non-linear dynamics of the system.

The filter processes each EEG channel by cascading the output of one LMS block through to the next LMS block, changing the reference input at each stage using a reference inputs from a correlated mixture of EMG readings. The filter processed all of the five reference inputs, the four muscle and one cardiac signals were attenuated to a level which is far more conducive to mobile BCI use. Additionally, care must be taken when dealing with surface EEG recordings as a artefact free reading cannot be ascertained without the use of invasive techniques, as such the measure of success of any filter can not be made without simulated study followed by field tests.

Extensive field testing may suffer technical issues due to galvanic changes in the participant's skin due to perspiration, these issues may be alleviated by using active dry EEG electrodes [19] keeping in mind as Chi et al. suggest, that the use of adhesive with traditional wet electrodes results in maintaining the location better than dry electrode types over longer periods of time.

3   Conclusion

The proposed LMS method in this paper has shown promise in adaptively attenuating the EMG noise picked up by the surface EEG sensors, the work presented could be extended through the use of sub-band segmentation of the EMG signals before being processed through the cascaded LMS filter, this may allow a greater increase in the SNR and result in a more robust system as suggested by Lee et al. [20] in their work on improving constrained sub-band LMS convergence. A verification ERP study is planned to validate the results found in this work.

References

1. Goncharova, I., McFarland, D., Vaughan, T., Wolpaw, J.: Emg contamination of eeg: spectral and topographical characteristics. Clinical Neurophysiology 114(9), 1580–1593 (2003)

2. Luca, C.J.D., Gilmore, L.D., Kuznetsov, M., Roy, S.H.: Filtering the surface emg signal: Movement artifact and baseline noise contamination. Journal of Biomechanics 43(8), 1573–1579 (2010)
3. Jung, T., Makeig, S., Humphries, C., Lee, T., Mckeown, M., Iragui, V., Sejnowski, T.: Removing electroencephalographic artifacts by blind source separation. Psychophysiology 37(02), 163–178 (2000)
4. Barlow, J.S.: Artifact processing (rejection and minimization) in eeg data processing. In: Handbook of Electroencephalography and Clinical Neurophysiology, vol. 2, pp. 15–62 (1986)
5. Vigon, L., Saatchi, M.R., Mayhew, J.E.W., Fernandes, R.: Quantitative evaluation of techniques for ocular artefact filtering of eeg waveforms. IEE Proceedings on Science, Measurement and Technology 147(5), 219–228 (2000)
6. Junfeng, G., Pan, L., Yong, Y., Pei, W.: Online emg artifacts removal from eeg based on blind source separation. International Asia Conference on Informatics in Control, Automation and Robotics 1, 28–31 (2010)
7. Delorme, A., Sejnowski, T., Makeig, S.: Enhanced detection of artifacts in eeg data using higher-order statistics and independent component analysis. NeuroImage 34(4), 1443–1449 (2007)
8. Schlogl, A., Keinrath, C., Zimmermann, D., Scherer, R., Leeb, R., Pfurtscheller, G.: A fully automated correction method of eog artifacts in eeg recordings. Clinical Neurophysiology 118(1), 98–104 (2007)
9. Kher, R.K., Thakker, B.: Eeg signal enhancement and estimation using adaptive filtering. International Journal of Engineering 2(1) (2013)
10. He, P., Wilson, G., Russell, C.: Removal of ocular artifacts from electroencephalogram by adaptive filtering. Medical and Biological Engineering and Computing 42(3), 407–412 (2004)
11. Molla, M.K.I., Islam, R., Tanaka, T., Rutkowski, T.M., et al.: Artifact suppression from eeg signals using data adaptive time domain filtering. Neurocomputing 97, 297–308 (2012)
12. Stern, J.M.: Atlas of EEG Patterns. 2 edn. Lippincott Williams & Wilkin (2013)
13. Patrick Berg, B.G.: Besa simulator v1.0. Electronic (June 2013)
14. CompuMedics/NeuroScan: Synampsrt technical specifications (2012)
15. Gray, H., Lewis, W.H.: Anatomy of the Human Body. Philadelphia: Lea and Febiger, 1918 Bartleby, 2000 (1825-1861)
16. Jasper, H.: Report of the committee on methods of clinical examination in electroencephalography. Electroenceph. Clin. Neurophysiol. 10, 370–375 (1958)
17. Wonnacott, T.H., Wonnacott, R.J.: Introductory statistics. vol. 19690. Wiley, New York (1972)
18. Billings, S.A.: Nonlinear system identification: NARMAX methods in the time, frequency, and spatio-temporal domains. John Wiley & Sons (2013)
19. Taheri, B.A., Knight, R.T., Smith, R.L.: A dry electrode for eeg recording. Electroencephalography and Clinical Neurophysiology 90(5), 376–383 (1994)
20. Lee, K., Gan, W.: Improving convergence of the nlms algorithm using constrained subband updates. IEEE Signal Processing Letters 11(9), 736–739 (2004)

sEMG-Based Single-Joint Active Training with iLeg—A Horizontal Exoskeleton for Lower Limb Rehabilitation⋆

Jin Hu[1], Zeng-Guang Hou[1], Liang Peng[1], Long Peng[1], and Nong Gu[2]

[1] State Key Laboratory of Management and Control for Complex Systems, Institute of Automation, Chinese Academy of Sciences, Beijing, 100190, China
ustbhujin@163.com, {zengguang.hou,liang.peng,long.peng}@ia.ac.cn
[2] Centre for Intelligent Systems Research, Deakin University, 75 Pigdons Road, Waurn Ponds, VIC, 3216, Australia
nong.gu@deakin.edu.au

Abstract. In this paper, surface electromyography (sEMG) from muscles of the lower limb is acquired and processed to estimate the single-joint voluntary motion intention, based on which, two single-joint active training strategies are proposed with iLeg, a horizontal exoskeleton for lower limb rehabilitation newly developed at our laboratory. In damping active training, the joint angular velocity is proportionally controlled by the voluntary effort derived from sEMG, performing as an ideal damper, while spring active training aims to create a spring-like environment where the joint angular displacement from the constant reference is proportionally controlled by the voluntary effort. Experiments are conducted with iLeg and one healthy male subject to validate the feasibility of the two single-joint active training strategies.

Keywords: sEMG, single-joint active training, lower limb rehabilitation.

1 Introduction

Active training with rehabilitation robots turns out to be an effective method for the rehabilitation of the paralyzed patients [3,5,10] since the voluntary participation of patients is motivated. Active training usually involves the interaction between the patient and robot which therefore requires interactive signals to obtain voluntary motion intention of the patient where force signal and sEMG are usually used. Compared to force signal, sEMG directly reflects the activity of specific muscles, which contains more detailed information. In addition, sEMG has a higher sensitivity than force signal and hence has a potential for active training for paralyzed patients with weak motor function. However, it is also a challenge to use sEMG as the interactive signal, since sEMG is a electrical signal

---

⋆ This research is supported in part by the National Natural Science Foundation of China (Grants 61225017, 61175076), and the International S&T Cooperation Project of China (Grant 2011DFG13390).

C.K. Loo et al. (Eds.): ICONIP 2014, Part III, LNCS 8836, pp. 535–542, 2014.

produced by muscles [1, 8] but acquired with surface electrodes, which makes it weak, noisy, stochastic and hence difficult to be detected and processed to be an effective interactive signal.

In [9], M. Sartori *et al.* presented a biomechanical model which estimated joint torque of the lower limb from the local sEMG to predict the voluntary motion intention and furthermore to implement the active training. In [6], S. Pittacio *et al.* proposed a sEMG-based interactive control method with an ankle orthosis where the active ankle dorsiflexion was implemented based on sEMG from tibialis anterior muscle with simple threshold-based on-off control. In [11], Y. H. Yin *et al.* developed a human-machine interface between a gait exoskeleton and a hemiplegic patient which allowed patients to control the exoskeleton with sEMG from the heathy lower limb and consequently implemented active gait training of the paralyzed side.

With a horizontal exoskeleton for lower limb rehabilitation named *i*Leg that has recently been developed at our laboratory, two sEMG-based single-joint active training strategies are proposed in this paper.Fig. 1 shows the mechanical structure of one of the 3-DOF (degrees of freedom) orthoses. Joints 1, 2 and 3 correspond to the flexion-extension movement of the hip, knee and ankle joints of a lower limb respectively. The lower limb is attached along the robotic leg and the foot is strapped on the pedal during the rehabilitation training, as indicated in Fig. 2.

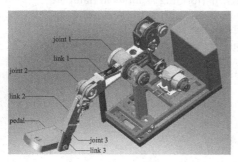

Fig. 1. The leg orthosis   **Fig. 2.** *i*Leg

The remainder of the paper is organized as follows. Section 2 introduces the acquisition and processing of sEMG. Section 3 presents the implementation of two sEMG-based single-joint active training strategies. Section 4 demonstrates the experiments and raises some discussions on the results. The last section comes up with the conclusion and the future work.

2   Acquisition and Processing of sEMG

An sEMG acquisition system is developed at our laboratory the structure of which is indicated in Fig. 3. It consists of pre-amplifier, linear isolation circuit, AD data acquisition card and post-processing software. The pre-amplifier employs differential-mode input with differential-mode amplification of 1,000, common-mode rejection ratio of 100 dB. The linear isolation circuit is implemented by linear optocoupler with amplification of 1, with pass-band higher than 1,000 Hz. The AD data acquisition card is a sophisticated commercial product with acquisition frequency of 25,000 Hz, 13-bit conversion resolution and input range of ±5 V. The post-processing software aims to filter and normalize sEMG which will be detailed in Section 2.1 and 2.2.

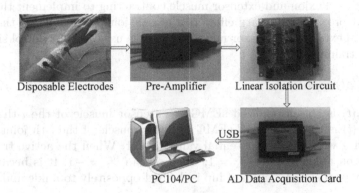

Disposable Electrodes   Pre-Amplifier   Linear Isolation Circuit

PC104/PC   AD Data Acquisition Card

Fig. 3. sEMG Acquisition System

2.1 Band-Pass and Notch Filter

The effective frequency band of sEMG is distributed within $10 \sim 500$Hz, mostly within the range of $50 \sim 150$Hz. The acquired sEMG is consecutively processed by Butterworth band-pass and notch filters which are designed using functions of "buttord" and "butter" in "Matlab". The pass-band of the band-pass filter is within $20 \sim 200$ Hz to eliminate common-mode, high- and low-frequency interference. The notch frequency is set as the power frequency in China, i.e. 50 Hz, to eliminate the power line interference.

2.2 Normalization

sEMG of the muscle under resting state and maximum voluntary contraction (MVC) needs to be acquired for normalization. In this research, the lower limbs are positioned at natural states and the subject is required to relax the muscles of the lower limbs as much as possible while resting sEMG is being acquired, filtered, rectified and averaged. The subject is required to maximally contract the muscles of the lower limbs repeatedly while sEMG is continuously recorded, filtered and rectified. Then the sEMG is grouped with unit length of 256 data. The averages are calculated within each group and the maximum average is considered as the MVC sEMG. The normalization is represented as

$$\bar{x}(t) = |x(t) - x_{\mathrm{r}}|/(x_{\mathrm{m}} - x_{\mathrm{r}}) \tag{1}$$

where: x_{r} represents the resting sEMG; x_{m} is the sEMG of the muscle under the MVC; $x(t)$ is the filtered and rectified value of sEMG at time t; $\bar{x}(t)$ is the normalized value of sEMG at time t.

The single-joint flexion-extension of lower limb is mostly controlled by the coordinated contraction of a pair of muscles, with flexor muscle contracting to implement the flexion and extensor muscle contracting to implement the extension. Therefore, the voluntary effort of the i-th joint is defined as the sEMG difference of extensor and flexor muscles, which is used as the control signal for the active training, $\tilde{x}_i(t)$:

$$\tilde{x}_i(t) = (\bar{x}_{\mathrm{e}i}(t) - \bar{x}_{\mathrm{f}i}(t))f_i \tag{2}$$

where: $\bar{x}_{\mathrm{e}i}(t)$ is the normalized sEMG of extensor muscle of the i-th joint at time t; $\bar{x}_{\mathrm{f}i}(t)$ is the normalized sEMG of flexor muscle of the i-th joint at time t; f_i is a flag variable with potential values of ± 1. When the active training is performed on ankle or knee, $f_i = 1$; when on hip, $f_i = -1$. It is because that the flexion-extension direction of hip is defined oppositely to ankle and knee.

3   Single-Joint Active Training

Two single-joint active training strategies are proposed with iLeg using the sEMG difference between the extensor and flexor muscles as the input control signal. Damping active training allows patients to control the joint velocity with their voluntary effort proportionally, while in spring active training, the joint displacement is proportionally controlled by the voluntary effort.

3.1   Damping Active Training

Damping active training is implemented by a degenerate impedance control method [2, 4, 7]—damping control where the inertia and stiffness coefficients are set as zeros. The dual closed-loop architecture of sEMG-based dampimg control is shown in Fig. 4, where the inner loop of velocity control is implemented by a sophisticated commercial system.

The outer loop of damping control modifies the reference velocity with the adjustment that is determined by the sEMG difference using damping function

$$\tilde{x}_i(t) = B_i(\dot{q}_{\mathrm{c}i} - \dot{q}_{\mathrm{r}i}) \tag{3}$$

where: B_i is the damping coefficient of the i-th joint which is a positive constant; $\dot{q}_{\mathrm{r}i}$ and $\dot{q}_{\mathrm{c}i}$ are respectively the reference and command velocities of the i-th joint.

Since the reference velocity is commonly set as zero, i.e. $\dot{q}_{r2} = 0$ (rad/s), it is can be concluded from (3) that joint angular velocity is controlled by the sEMG difference.

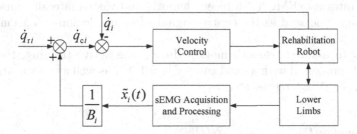

Fig. 4. Dual Closed-Loop Architecture for sEMG-Based Damping Control

3.2 Spring Active Training

Another degenerate impedance control method [2, 4, 7], stiffness control, is used to implement the spring active training where the inertia and damping coefficients are set as zeros. The dual closed-loop architecture of sEMG-based stiffness control is shown in Fig. 5, where the inner loop of position control is implemented by a sophisticated commercial system.

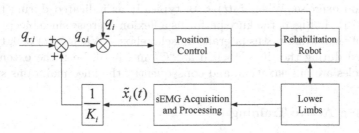

Fig. 5. Dual Closed-Loop Architecture for sEMG-Based Stiffness Control

In the outer loop of stiffness control, the reference is then modified by the adjustment which is determined by the sEMG difference using stiffness function

$$\tilde{x}_i(t) = K_i(q_{ci} - q_{ri}) \tag{4}$$

where: K_i is the stiffness coefficient of the i-th joint which is a positive constant; q_{ri} and q_{ci} are respectively the reference and command positions of the i-th joint. Since the reference position is commonly set as a constant, it is can be concluded from (4) that joint angular displacement from the constant reference is controlled by the sEMG difference.

4   Experiments

Experiments were conducted with the right knee of one healthy male subject and iLeg to validate the feasibility of the two sEMG-based single-joint active training strategies. sEMG from biceps femoris and vastus lateralis muscles was acquired and processed as the control signal of active flexion-extension motion of the knee joint.

In order to assure a smooth motion during the active training, the sEMG difference is processed with a dead zone of $[-0.1\ 0.1]$ as well a saturation zone of $[-\infty\ 0.1]$ or $[1\ \infty]$, represented as

$$\tilde{x}_2(t) = \begin{cases} 0, & \|\tilde{x}_2(t)\| < 0.1 \\ \text{sgn}(\tilde{x}_2(t)), & \|\tilde{x}_2(t)\| > 1 \\ \tilde{x}_2(t) - 0.1\text{sgn}(\tilde{x}_2(t)), \text{ otherwise} \end{cases} \tag{5}$$

4.1   Damping Active Training

The damping coefficient for the knee joint was set as 2, i.e. $B_2 = 2$. The experiment of damping active training lasted for over 350 seconds. The result in the duration of $310 \sim 350$s is shown in Fig. 6 where it is indicated that the angular velocity of the knee is proportionally controlled by the sEMG difference between the extensor and flexor muscles, performing as an ideal damper. When $\tilde{x}_2(t) > 0$, typically as indicated during the time labelled as "A" in Fig. 6, the contraction of vastus lateralis muscle plays the dominant role. As a result, the knee performs an extension exercise. When $\tilde{x}_2(t) < 0$, typically as indicated during the time labelled as "B" in Fig. 6, the knee performs a flexion exercise since biceps femoris muscle contracts while vastus lateralis muscle relaxes. When $\tilde{x}_2(t) = 0$, typically as indicated during the time labelled as "C" in Fig. 6, both the extensor and flexor muscles are in relaxation, and consequently, the knee maintains still.

4.2   Spring Active Training

The stiffness coefficient for the knee joint was set as 1, i.e. $K_2 = 1$. The reference position of the right knee was set as -0.94 (rad), i.e. $q_{r2} = -0.94$ (rad). The duration of the experiment of spring active training was approximately 300 seconds. The experiment result during $110 \sim 150$s is shown in Fig. 7 where it is indicated that the angular displacement of the knee from the reference position is proportionally controlled by the sEMG difference between the extensor and flexor muscles, performing as an ideal spring. Typically as indicated during the time labelled as "A" and "B" in Fig. 7, the displacement of the knee from the reference increases with the increase of the sEMG difference. When the sEMG difference decreases, typically as indicated during the time labelled as "C" and "D" in Fig. 7, the joint displacement decreases as well. Typically as indicated during the time labelled as "E" and "F" in Fig. 7, when $\tilde{x}_2(t) = 0$, i.e. both the extensor and flexor muscles relax, the knee moves back to the reference position and then stay still.

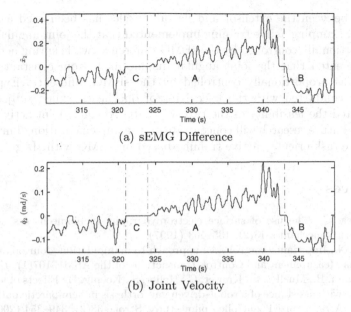

(a) sEMG Difference

(b) Joint Velocity

Fig. 6. Experiment result of single-joint damping active training

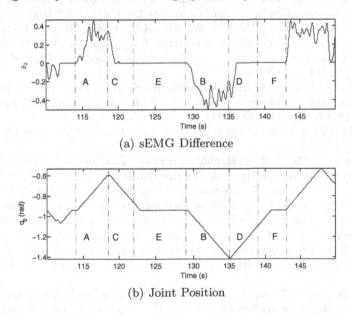

(a) sEMG Difference

(b) Joint Position

Fig. 7. Experiment result of single-joint spring active training

5   Conclusion and Future Work

Two single-joint active training strategies have been proposed with *i*Leg—a horizontal exoskeleton for lower limb rehabilitation, where the normalized sEMG difference between the extensor and flexor muscles has been used as the control signal. Damping active training implemented that the joint angular velocity was proportionally controlled by the sEMG difference, while spring active training implemented that the joint angular displacement from constant reference position was proportionally controlled by the sEMG difference. Experiments have been conducted with the knee joint of one heathy male subject, which has validated the feasibility of the two sEMG-based single-joint active training strategies. Future research will concentrate on the implementation of multi-joint-cooperative task-oriented active training based on sEMG with iLeg.

References

1. De Luca, C.: The use of surface electromyography in biomechanics. Journal of Applied Biomechanics 13(2), 135–163 (1997)
2. Hogan, N.: Impedance control - An approach to manipulation. Journal of Dynamic Systems Measurement and Control-Transactions of the ASME 107(1), 1–24 (1985)
3. Husemann, B., Mueller, F., Krewer, C., Heller, S., Koenig, E.: Effects of locomotion training with assistance of a robot-driven gait orthosis in hemiparetic patients after stroke - A randomized controlled pilot study. Stroke 38(2), 349–354 (2007)
4. Jung, S., Hsia, T.: Neural network impedance force control of robot manipulator. IEEE Transactions on Industrial Electronics 45(3), 451–461 (1998)
5. Liao, W.W., Wu, C.Y., Hsieh, Y.W., Lin, K.C., Chang, W.Y.: Effects of robot-assisted upper limb rehabilitation on daily function and real-world arm activity in patients with chronic stroke: A randomized controlled trial. Clinical Rehabilitation 26(2), 111–120 (2012)
6. Pittaccio, S., Viscuso, S.: An EMG-controlled SMA device for the rehabilitation of the ankle joint in post-acute stroke. Journal of Materials Engineering and Performance 20(4-5, SI), 666–670 (2011)
7. Pons, J.L.: Wearable robots: Biomechatronic exoskeletons. John Wiley & Sons, Ltd (2008)
8. Robertson, G., Caldwell, G., Hamill, J., Kamen, G., Whittlesey, S.: Research Methods in Biomechanics, 2E. Human Kinetics (2013)
9. Sartori, M., Reggiani, M., Mezzato, C., Pagello, E.: A lower limb EMG-driven biomechanical model for applications in rehabilitation robotics. In: Proceedings of 2009 International Conference on Advanced Robotics, pp. 905–911 (June 2009)
10. Waldman, G., Yang, C.Y., Ren, Y., Liu, L., Guo, X., Harvey, R.L., Roth, E.J., Zhang, L.Q.: Effects of robot-guided passive stretching and active movement training of ankle and mobility impairments in stroke. Neurorehabilitation 32(3), 625–634 (2013)
11. Yin, Y.H., Fan, Y.J., Xu, L.D.: EMG and EPP-integrated human-machine interface between the paralyzed and rehabilitation exoskeleton. IEEE Transactions on Information Technology in Biomedicine 16(4), 542–549 (2012)

Find Rooms for Improvement:
Towards Semi-automatic Labeling
of Occupancy Grid Maps

Sven Hellbach[1], Marian Himstedt[1], Frank Bahrmann[1], Martin Riedel[2],
Thomas Villmann[2], and Hans-Joachim Böhme[1],*

[1] University of Applied Sciences Dresden, Artificial Intelligence and Cognitive
Robotics Labs, POB 12 07 01, 01008 Dresden, Germany
{hellbach,himstedt,bahrmann,boehme}@informatik.htw-dresden.de
[2] University of Applied Sciences Mittweida, Computational Intelligence and
Technomathematics, POB 14 57, 09648 Mittweida, Germany
{riedel,thomas.villmann}@hs-mittweida.de

Abstract. Semi-automatic semantic labeling of occupancy grid maps
has numerous applications for assistance robotic. This paper proposes
an approach based on non-negative matrix factorization (NMF) to extract environment specific features from a given occupancy grid map.
NMF also computes a description about where on the map these features
need to be applied. We use this description after certain pre-processing
steps as an input for generalized learning vector quantization (GLVQ) to
achieve the classification or labeling of the grid cells. For the supervised
training of the GLVQ the assigned label is propagated to all grid cells
of a semantic unit using a simple, yet effective segmentation algorithm.
Our approach is evaluated on a standard data set from University of
Freiburg, showing very promising results.

Keywords: NMF, GLVQ, semantic labeling, occupancy grid maps.

1 Introduction

Over the last few years a number of research projects have addressed different
scenarios for the application of mobile assistance robots. However, in most cases
the scenario is defined in a way that the robot is already present within its working environment. Implicitly, this means it is assumed that a trained technician
sets up the robot, in particular runs the map building process.

In our work, we aim to find an approach that allows the robot to be set up by
any person, for example its new owner. The idea is that this person shows the
robot around – as we would do with a new colleague. In this process, the different
rooms of the environment are meant to be taught to the robot. However, with
an increasing number of similar rooms, like in an office environment, this still
can be an annoying task. We think it would be a benefit to the user, if the robot

---

* This work was supported by ESF grant number 100076162.

C.K. Loo et al. (Eds.): ICONIP 2014, Part III, LNCS 8836, pp. 543–552, 2014.

could suggest at least the type of room – or beyond this a speech recognition system could use the information to eliminate ambiguous detections.

For such a system the rooms should be classified using the available sensor information. In this paper, for the prove of concept, we limit ourselves to the usage of a laser range sensor only. For a reliable application more than a single sensor is essential. To train the classifier a semi-automatic labeling processing is necessary. Furthermore, the classifier should be able to provide a similarity measure for different classes. Both demands are fulfilled by the proposed system.

The remainder of this paper is organized as follows. After a short overview over the current state of the art in Sec. 2, the following Sec. 3 summarizes the proposed approach, as well as explains the details of NMF, GLVQ, and the data pre-processing. The experimental results are discussed in Sec. 4, while the paper concludes in Sec. 5.

2   Related Work

Achieving a semantic understanding of the environment of a mobile robot platform is an on-going topic in many research teams. A popular approach with training and testing datasets was presented by Mozos [11]. This algorithm mainly used geometrical features and an AdaBoost classifier to differentiate between three classes (room, corridor, doorway) within a metric map. The mildly noisy output was smoothed afterwards using probabilistic relaxation labeling. A variation of Mozos' solely laser-range-finder-based solution was published in [16,17]. They used $\mathcal{L}_2$-regularized logistic regression on geometrical (area of polygonal approximation of the laser scan) as well as statistical laser scan features (standard deviations of lengths of consecutive scans and of ranges). Another approach using Support Vector Machines is proposed in [18].

The concept of teaching the mobile robot semantic labels at runtime, provided by a human guide, was followed by [12]. During a tour in its new surroundings, the robot obtains place labels where each spatial region is represented by one or more Gaussians. The complete map is later classified by region growing. A similar approach is described in [2].

Assuming laser range data is insufficient to fully understand our complex environment, more sensor cues are introduced. [14] used Mozos' laser range features and combined them with visual features obtained by SIFT (Scale-invariant feature transform) and CRFH (composed receptive field histogram). Each cue produces a scoring value which are then combined by SVM-DAS (SVM-based Discriminative Accumulation Scheme) to a final class label.

Due to new inexpensive RGB-D sensors, object recognition has gained much attention among research communities. For instance,[8,1] built 3-dimensional maps via RGB-D-SLAM. Within the resulting maps, preconceived coarse (wall, ceiling, ground) and individual (printer, monitor, etc.) labels are recognized. Different kinds of rooms are inferred using an associative coupling of these lables. Conversely, it is possible to find out the most likely position of an individual label.

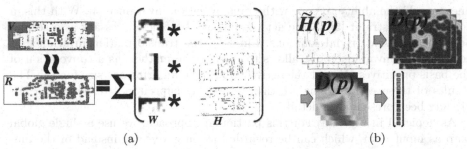

(a)   (b)

Fig. 1. (a) Application of the NMF algorithm on the global occupancy grid map. (b) Preprocessing before GLVQ. The activities H computed by the NMF for the global map are thresholded and then distance transformed. After that, local patches are cut out representing the description for the local map. Finally, the local maps are vectorized.

3   Approach

As discussed in the introduction, we propose a system that is able to classify different rooms of the current environments. As classifier we apply the well known Generalized Learning Vector Quantization (GLVQ) for prediction of the class labels for each grid cell.

The input space for the GLVQ relies on an NMF-based approach that has been introduced in [6]. Our approach takes an occupancy grid map as input. From this occupancy grid map a set of basis primitives and corresponding activities is computed using Non-negative Matrix Factorization (NMF).

For the proposed method, we continue with the activity maps and apply a distance transform. This is necessary, since the activity maps only have distinct peaks (compare Fig. 1(b)) and a slight translation within the map results in large dissimilarity. Using distance transform smoothens our representation and also encodes the distance metric within the grid map into our vector space. The combination of GLVQ and NMF is also described in [5] with another intention for the evaluation.

Since the classifier is trained during the application phase of the robot, an automatic labeling needs to take place as well. For this, we apply a method proposed in [2], which is able to segment an occupancy grid map. Hence, this method is able to automatically spread the label information for each grid cell within e.g. a room.

As discussed in [2] the method uses a simple heuristic assumption for recognizing doors, i.e. separating elements in the map, which sometimes leads to erroneous segmentations. Together with the GLVQ-classifier, which can also be trained for passages, the segmentation can be improved further.

Non-negative Matrix Factorization: Like other approaches, non-negative matrix factorization (NMF) [9] is meant to solve the source separation problem. Hence, a set of training data is decomposed into basis primitives W and their respective activations H: $V \approx W \cdot H$ For our approach we apply two different extension to the standard form of non-negative matrix factorization. For one, we add the ability to cope with transformation invariance [4]. With this, a sparseness constraint [3] on the activity matrix becomes necessary as a second adaption to avoid trivial solutions. Limiting the transformation invariance to translational invariance only, allows to rewrite the method as a convolution of the basis primitives W over the activities H with an arbitrary size for W. The results of these convolutions are then additively superimposed. This idea has already been presented in detail in [6].

As depicted in Fig. 1(a) For this particular approach, we use a single global map as input data, which can be regarded as only a vector instead of the matrix V. However, the succeeding explanation will follow the convolution based formulation, where the matrices can take almost any arbitrary structure. From the input grid map V a set of basis primitives W is generated, which can be regarded as environment specific local descriptors. Further, the corresponding activity maps H are derived using the multiplicative update rule defined in [19].

After the initial training the environment specific primitives W are kept fixed. Only the activity maps H are adapted based on the local grid map as input V. This results in an alternative description of the local environment.

Data Representation: In [6] several histogram based descriptors are evaluated. For this paper, we go back to a representation relying on the activities H computed by the NMF. As it is depicted in Fig. 1(b), the column vectors H^p of the activity matrix H with $H = (H_k^p) = (H^p)_i$ can be regarded as a map of activities with the same width and height $w \times h$ of the training map used in V. We define this map for basis primitive W_p as:

$$\tilde{H}(p) = \left(\tilde{H}(p)_j^i\right) \text{ with } \tilde{H}(p)_j^i = H_k^p \text{ with } k = j \cdot w + i \qquad (1)$$

These maps are then thresholded by θ to gain a binary activity map, which we define as a set:

$$\mathcal{O} = \left\{(i,j) \in \Omega \,\middle|\, \tilde{H}(p)_j^i > \theta\right\}. \qquad (2)$$

with $\Omega = \{1, \ldots, w\} \times \{1, \ldots, h\}$ being the set of grid cells in the map. Consequently, the grid cells where the corresponding basis primitive is to be placed belong to the set $\mathcal{O}$.

Subsequently, the binary map $\mathcal{O}$ undergoes a Euclidian distance transform [13]. This results in a map D where each grid cell contains the Euclidian distance to the nearest grid position, where a basis primitive is activated:

$$D(p) = \left(D_j^i\right) \text{ with } D_j^i = \min\left\{d((i,j),q) \,|\, q \in \mathcal{O}\right\} \qquad (3)$$

For this, $d(\cdot, \cdot)$ stands for a metric defined over Ω. We simply use the Euclidian distance here.

For the practical application these two steps need to be computed on the local maps during run time. To eliminate errors coming from the construction of these local maps, we decided to cut out patches of size $u \times v$ from the distance transformed activities $D(\text{p})$ of the global map.

$$\tilde{D}(p) = \left(D_j^i\right) \text{ with } i = k - \frac{u}{2}, \ldots, k + \frac{u}{2}, j = l - \frac{v}{2}, \ldots, l + \frac{v}{2} \qquad (4)$$

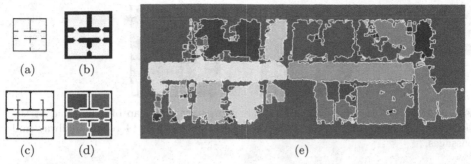

Fig. 2. (a) The original occupancy grid map (white indicates free and black indicates occupied cells). (b) Map after dilatation / erosion to close gaps and filter smaller smaller occupied regions. (c) The thinned skeleton (blue) with starting and intersection points (purple) and door hypotheses (red). (d) The resulting areas. (e) Applying the map segmentation algorithm to the occupancy grid map of building 79 at Freiburg University leads to some segmentation errors. Some rooms are oversegmented due to the heuristic used to recognize doors.

Otherwise, the experimental evaluation would also consider errors not caused by the proposed method.

The next step would be to transform the distance transformed activities into a vector space used by GLVQ. For this, each distance transformed activity is traversed row-wise. The resulting vectors are then simply concatenated to a single vector.

Before the vectorization takes place the patches or local maps respectively can be subsampled with step size s.

$$\tilde{D}'(p) = \left(D_j^i\right) \text{ with } i = k - \frac{u}{2}, \, k - \frac{u}{2} + s, \dots, k + \frac{u}{2}, \tag{5}$$

$$j = l - \frac{v}{2}, \, l - \frac{v}{2} + s, \dots, l + \frac{v}{2}$$

This reduces the number of input dimensions for the classifier. With the reduced number of dimensions it becomes possible to train the classifier with fewer training samples. This brings a tremendous advantage for the practical application, since we aim to learn characteristics from the few sensor reading of a single room.

Generalized Learning Vector Quantization: Prototype based classification, like Learning Vector Quantization (LVQ) [7], is frequently used by practitioners for different reasons: (a) the learning rule as well as the classification model are very fast; (b) due to that learning rule, adjusting an existing model by new data could be done very efficiently, and (c) the resulting classifier is interpretable since it represents the model in terms of typical prototypes. These prototypes can be treated in the same way as data. Considering popular alternatives – such as SVM – the GLVQ model is able to deal with an arbitrary number of classes very easily.

Since LVQ is a heuristically motivated approach to minimize the classification error, several modifications were applied to it to improve classification perfor-

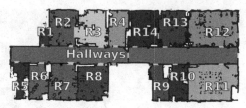

Fig. 3. Data set used for evaluation. The occupancy grid map of a building at the University of Freiburg. Sixteen different classes are labeled: 14 rooms, hallway and passages.

mance. A very famous extension of LVQ was introduced by Sato and Yamada known as Generalized LVQ (GLVQ) [15]. Based on distance evaluations a mathematical cost function is minimized during the learning process. This cost function approximates the classification error in a suited fashion:

$$E = \sum_j \Phi\left(\mu\left(x_j\right)\right) \text{ with } \mu\left(x_j\right) = \frac{d^+\left(x_j\right) - d^-\left(x_j\right)}{d^+\left(x_j\right) + d^-\left(x_j\right)}. \tag{6}$$

where $d^+(x_j)$ denotes the distance of a datum x_j to the closest prototype with the same label as x_j, and $d^-(x_j)$ refers to the best matching prototype with a class label different to the label of x_j. The transfer function Φ transform the output of the classifier function $\mu(x_j)$ to $[0,1]$ if a sigmoidal function is used. Since $\mu(x_j)$ becomes negative if a datum x_j is classified correctly and positive otherwise, E approximately counts the amount of misclassifications. Learning prototypes uses stochastic gradient descent on the GLVQ cost function E.

The application of GLVQ allow to simply add addition classes without retraining the entire classifier. In that way, the training process can take place while building the map of the environment. Classification results can already be taken into account during at this time.

Map Segmentation for Automatic Labeling: In this step a simple and computationally inexpensive assumption of areas and doorways is created. To achieve this we took an occupancy grid map which encodes the probability of each cell (representing a square tile of the environment) to be occupied (Fig. 2(a)) and applied the following steps. Initially the occupancy grid map is binarized, so cells are either free or occupied. Occupied cells will be dilated and afterwards eroded by the same value (possible gaps in unobserved wall segments are closed). Remaining smaller occupied cell groups will then be removed as they refer to chair or table legs. This preprocessing step is completed by dilating the map by the robot's radius (Fig. 2(b)).

The proposed method in [20] is used to compute a one-pixel wide skeleton of the map. A predefined set of intersection and start/end point templates provides a reliable way to convert the skeleton into a fully connected graph-like structure. A local normal is then approximated for each skeleton cell providing a side clearance distance. By interpreting this distance it becomes possible to create door hypotheses (cf. Fig. 2(c)). Occupied cells and door hypotheses result in a

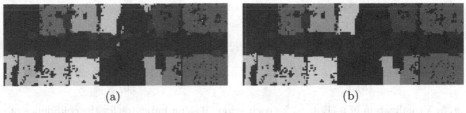

(a)   (b)

Fig. 4. Experimental results with a local map size of 50×50 (a) an additional dilation of class *passage* for (a) one and (b) five LVQ-prototypes per class

partitioned space where each group of connected free cells is declared as a room hypothesis (cf. Fig. 2(d)).

This methods robustness depends heavily on the mapping quality. If there is a multitude of distortions by small objects the naive door hypotheses assumption fails, resulting in an erroneous number of rooms (shown in Fig. 2(e)). A full description of this method is proposed in [2].

4   Experiments

The goal of experimental evaluation in this paper is to show that the proposed method is able to fulfill the demands already formulated throughout the paper. Firstly, the approach has to be able to discriminate between all semantic units, like different rooms and the hallway. Secondly, the passages (e.g. doors) need to be recognized correctly. And finally, a distance measure should be part of the algorithms for which semantically similar structure are close together. For this, we concentrate our evaluation onto the combination of NMF and GLVQ. The map segmentation has already been discussed in [2]. The evaluation of the integration of both approaches will be part of future work.

We use an online available data set[1] for evaluation, which has already been used in [10]. This data set consists of a previously computed 700×289 grid map, for which three classes of building structures have been labeled: room, hallway and passage. We extended the labeling by giving an individual label for each room as shown in Fig. 3. For computational reasons we subsample the given grid map by taking only each 4th pixel in both directions into account.

To understand the classification performance two different parameter settings for GLVQ with one and five prototypes per class respectively were evaluated. The parameters for NMF, like the basis primitive size, are identical to [5].

Since, the goal of our experiments is to understand how well different rooms can be distinguished, we used the entire map for training. Hence, we didn't use a test set for evaluation. However, the generalization of our approach has already been addressed in [5].

As mentioned, our first criterion is the accuracy of classification. In Figure 4 the classified grid cells are colored according to their predicted class label. No

---

[1] Data sets corresponding to the semantic classifications of places under http://webpages.lincoln.ac.uk/omozos/

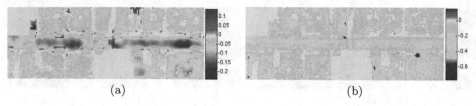

(a)   (b)

Fig. 5. Visualization of μ (Eqn. 6) for each grid cell as an indicator for the confidence of the classification. Positive values indicate a misclassification. Values below zero stand for a correct classification results. Values close to zero show that the data point is close to the decision boundary.

smoothing or majority voting has been performed here. Fig. 4(a), which shows the results for a single prototype per class, reveals some minor classification errors in particular in the corners of some rooms. In this setting we gain an accuracy of 95.64%. The results become much better using five prototypes per class as depicted in Fig. 4(b). Almost no errors are visible, which is confirmed by the accuracy of 99.87%.

To better understand the classification process, Fig. 5 shows a visualization of the value μ defined in (6). Values close to zero (yellow) stand for a representation of the grid cell is close to the class border. Colors from yellow over green to blue show the confidence of a correct classification, while reddish colors show a misclassification. In Fig. 5(a) it is clearly visible that using only a single prototype results in close decisions or even misclassifications in particular at doors and passages. This contradicts to one of our demands, to be able to support the map segmentation process.

Figure 6 show the distance of each grid cell to the class prototype. The plot shows 4×4 maps. One for each class. This visualization is meant to reveal the similarity between different environment structures. The distances for only one prototype in Fig. 6(a) again does not fulfill one of our demands. However, with five prototypes in Fig. 6(a) it can clearly be seen that the office rooms are similar to each other. The hallway, the passages and the small room labeled R1 are less similar to each other and to the other rooms.

Summarizing, the use of five prototypes for GLVQ fulfills all three demands needed for the application of our proposed method.

5 Conclusion

In this paper we proposed an approach for semi-automatic semantic labeling of occupancy grid maps. We rely on a data representation that codes the distance to environment specific features for each grid cell. The environment-specific features are trained by applying non-negative matrix factorization. A Generalized Learning Vector Quantization is used to classify different structural elements of the environment. Since GLVQ needs labeled data for each grid cell a map segmentation is performed to distribute a user given label. The evaluation was focused on the classification process. It revealed that GLVQ together with the

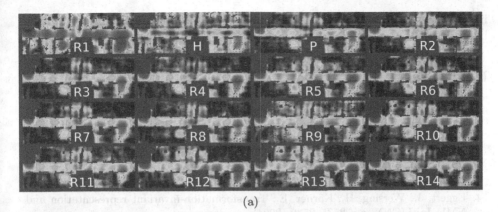

(a)

(b)

Fig. 6. Both plots for (a) one and (b) five LVQ-prototypes per class show 16 maps corresponding to the 16 classes. For each cell of such map the distance to the respective class prototype has been computed.

NMF-based data representation is able to distinguish between different rooms. Furthermore, the prototype representation allows to derive a distance measure, which provides information about the similarity of different rooms.

While in this paper only the classification process has been addressed, future work will concentrate on the evaluation of the entire system. Furthermore, it would be interesting to figure out how well the approach is able to deal with environmental changes, in particular dynamic obstacles like e.g. chairs.

Despite the already good results, it would be interesting to try different types of LVQ instead of the used GLVQ. Our results confirm that it seems to be a difficult problem to classify passages. Hence, we aim to try Generalized Matrix LVQ (GMLVQ) or the Kernel version of LVQ. Furthermore, it would be interesting to include the knowledge about the relation between different dimension of the LVQ input space. Some dimensions depend on the same basis primitive or share the same spatial location within the grid map.

Beside the introducing scenario, our approach would furthermore allow a coarse to fine localization. Within a large environment, our approach could confine the regions, where a e.g. particle filter based approach is allowed to draw hypotheses.

References

1. Anand, A., Koppula, H.S., Joachims, T., Saxena, A.: Contextually guided semantic labeling and search for three-dimensional point clouds. Int. J. Robot. Res. 32(1), 19–34 (2013)
2. Bahrmann, F., Hellbach, S., Böhme, H.-J.: Please tell me where I am: A fundament for a semantic labeling approach. In: KI, pp. 120–124 (2012)
3. Eggert, J., Körner, E.: Sparse Coding and NMF. In: IJCNN, pp. 2529–2533 (2004)
4. Eggert, J., Wersing, H., Körner, E.: Transformation-invariant representation and NMF. In: IJCNN, pp. 2535–2539 (2004)
5. Hellbach, S., Himstedt, M., Bahrmann, F., Riedel, M., Villmann, T., Böhme, H.J.: Some room for GLVQ: Semantic Labeling of occupancy grid maps. In: WSOM (in press, 2014)
6. Hellbach, S., Himstedt, M., Boehme, H.J.: Towards Non-negative Matrix Factorization based Localization. In: ECMR (2013)
7. Kohonen, T.: The self-organizing map. Proc. of the IEEE 78(9), 1464–1480 (1990)
8. Koppula, H.S., Anand, A., Joachims, T., Saxena, A.: Semantic labeling of 3d point clouds for indoor scenes. In: NIPS, pp. 244–252 (2011)
9. Lee, D.D., Seung, H.S.: Algorithms for non-negative matrix factorization. Adv. Neural Inf. Process. Syst. 13, 556–562 (2001)
10. Mozos, O.M.: Semantic Place Labeling with Mobile Robots. Ph.D. thesis, Dept. of Computer Science, University of Freiburg (July 2008)
11. Mozos, O.M., Triebel, R., Jensfelt, P., Rottmann, A., Burgard, W.: Supervised semantic labeling of places using information extracted from sensor data. RAS 55(5), 391–402 (2007)
12. Nieto-Granda, C., Rogers, J.G., Trevor, A.J., Christensen, H.I.: Semantic map partitioning in indoor environments using regional analysis. In: IROS, pp. 1451–1456. IEEE (2010)
13. Paglieroni, D.W.: Distance transforms: properties and machine vision applications. CVGIP: Graph. Models Image Process. 54(1), 56–74 (1992)
14. Pronobis, A., Mozos, O.M., Caputo, B., Jensfelt, P.: Multi-modal semantic place classification. Int J Robot Res 29(2-3), 298–320 (2010)
15. Sato, A., Yamada, K.: Generalized learning vector quantization. In: NIPS, pp. 423–429. MIT Press, Cambridge (1996)
16. Shi, L., Kodagoda, S., Dissanayake, G.: Laser range data based semantic labeling of places. In: IROS, pp. 5941–5946 (2010)
17. Shi, L., Kodagoda, S., Dissanayake, G.: Multi-class classification for semantic labeling of places. In: ICARCV, pp. 2307–2312. IEEE (2010)
18. Sousa, P., Araujo, R., Nunes, U.: Real-Time Labeling of Places using Support Vector Machines. In: ISIE, pp. 2022–2027 (2007)
19. Vollmer, C., Hellbach, S., Eggert, J., Gross, H.M.: Sparse coding of human motion trajectories with non-negative matrix factorization. Neurocomp. (2013)
20. Zhang, T.Y., Suen, C.Y.: A fast parallel algorithm for thinning digital patterns. Communications of the ACM 27(3), 236–239 (1984)

Understanding Dynamic Environments with Fuzzy Perception

Frank Bahrmann, Sven Hellbach, Sabrina Keil, and Hans-Joachim Böhme*

HTW Dresden, Fakultät Informatik/Mathematik, Dresden, Germany
{bahrmann,hellbach,keil,boehme}@htw-dresden.de

Abstract. This paper addresses the problem of dealing with different kinds of dynamic obstacles influencing a place recognition task. We improve an existing approach using independent Marcov chain grid maps (iMac). Furthermore, we add a fuzzy classification to exploit the iMac estimation to refine the likelihood field estimation. We can show that the proposed method increases the performance of place recognition, while still being a compact, interpretable framework.

Keywords: iMac, fuzzy classifier, cognitive robotics, place recognition.

1 Introduction

One of the still challenging tasks in cognitive robotics is the navigation in crowded environments. To solve this problem, an important step is to know the robot's position. For this step, which is referred to as localization or place recognition, occupancy grid maps proposed by Moravec and Elfes [16] are often used to represent the environment. The environment is divided into cells. It is assumed that the environment is static and the cells are independent from each other. Each cell contains the probability of being occupied by an obstacle. However, conventional occupancy grid maps are able to model static environments only. Especially crowded environment contains a lot of dynamic obstacles that cannot be handled correctly. To cope with this, there are already some approaches that identify or model the dynamic nature of the environment. Most of these approaches come with additional computational or storage effort.

Biber et al. [4] propose a dynamic map that adapts continuously over time storing different time scales. The current sensor data is then compared to all timescales in the chosen local maps. The best suited timescale from each local map is selected. Similar [2] extend occupancy grids to temporal occupancy grids (TOGs). A TOG consists of several occupancy grid maps, each representing a different period of time. Mitsou et al. [15] extend the occupancy grid map along the time axis instead of only a period of time. For each cell the history of the occupancy probability is stored through an index structure.

Some approaches filter the scan measurements and use a static map for robot localization. Measurements from dynamic obstacles are identified or filtered out.

---

* This work was supported by ESF grant number 100076162.

C.K. Loo et al. (Eds.): ICONIP 2014, Part III, LNCS 8836, pp. 553–562, 2014.

Algorithm 1. Cell initialization and update according to observation

```
initialization:
occ2free=free2occ=1 & obsFree=obsOcc=2 & // changed initialization
observations = 0 & cellState = UNKNOWN & η = 2000/2001

cell observed free:                        cell observed occupied:
    observations = observations + 1            observations = observations + 1
    if (cellState equals OCCUPIED)             if (cellState equals FREE)
        occ2free = occ2free + 1                    free2occ = free2occ + 1
    cellState = FREE                           cellState = OCCUPIED
    obsFree = obsFree + 1                       obsOcc = obsOcc + 1
    ## forgetting exit event                   ## forgetting entry event
    occ2free = 1 + (occ2free - 1) * η          free2occ = 1 + (free2occ - 1) * η

    obsOcc = 2 + (obsOcc - 2) * η // changed update    obsFree = 2 + (obsFree - 2) * η // changed update
```

Fox et al. [8] apply an entropy filter to measure the uncertainty of the measurements. Hähnel et al. [10] use an expectation-maximization algorithm to learn which measurements correspond to static objects.

Meyer-Delius et al. [14] came up with the idea to improve the localization with the measurements caused by semi-static objects. They use a combination of a static map and temporary maps. The temporary maps are generated when the current measurement can not be explained by the static map.

Anguelov et al. [1] and Biswas et al. [5] model shapes of dynamic objects, while there are other approaches which explicitly model the changing environment and use the resulting map for localization. Meyer-Delius et al. [13], [20] characterize the changes of the environment by the state transition probabilities of a Hidden Markov Model. A similar approach is used by Saarinen et al. [18]. In contrast to Delius et al. learning of the state transition probabilities is simplified.

Kucner [12] propose a grid-based representation that is called Conditional Transition Map (CTMap) for learning motion patterns of dynamic environments. It is assumed that cells are not independent of each other. Accordingly, the change of occupancy affects neighboring cells. Also Krajník et al. [11] model the spatio-temporal dynamics of the environment relying on its frequency spectrum.

After this short introduction, the remainder of this paper is organized as follows: Section 2 introduces our proposed methods by explaining the extended iMac (Sect. 2.1), the Fuzzy classification (Sect. 2.2), and the integration into robot localization (Sect. 2.3). The experimental evaluation is discussed in Sect. 3, while the paper is concluded in Sect. 4.

2   Approach

The proposed method is based upon the work done by [18]. To achieve better numerical stability, some improvements are suggested, which will be addressed in Sect. 2.1. They also provided some means for further applications of their approach, indicating the potential to use a Fuzzy system to classify specific types of dynamic behavior within the environment. However, they only presented a method to extract to be expected occupancy probability values. Since that is a loss of information which may be crucial for potential subsequent algorithms, we pursued the approach to extract classes. For this, the world will be divided in rectangular grid cells. Each cell consists of the dynamic model and the extracted class.

Algorithm 2. Fuzzy Classifier in FCL

```
FUNCTION_BLOCK markovAnalysis
VAR_INPUT
    lambdaEntry : REAL;
    lambdaExit : REAL;
END_VAR
VAR_OUTPUT
    cellBehavior: REAL;
END_VAR

FUZZIFY lambdaEntry
    TERM toZero := (0, 1) (0.001, 0); // triangular
    TERM low := (0, 0) (0.005, 1) (0.2, 1) (0.4, 0); //trapezoidal
    TERM high := (0.05, 0) (0.3, 1 ) (1, 1);
END_FUZZIFY
FUZZIFY lambdaExit
    TERM toZero := (0, 1) (0.001, 0);
    TERM low := (0, 0) (0.005, 1) (0.2, 1) (0.4, 0);
    TERM high := (0.05, 0) (0.3, 1 ) (1, 1);

END_FUZZIFY

DEFUZZIFY cellBehavior
    TERM staticOccupied := GAUSS 2 1; // FUNCTION MEAN VARIANCE
    TERM semiStaticOccupied := GAUSS 4 1;
    TERM semiStatic := GAUSS 6 1;
    TERM dynamic := GAUSS 8 2;
    TERM staticFree := GAUSS 10 1;
METHOD : COG;//Center of Gravity;
DEFAULT := 0;

END_DEFUZZIFY


RULEBLOCK no1
    AND : MIN;
    ACT : MIN; // activation method
    ACCU : MAX; // accumulation method
    RULE 1 : IF (lambdaExit IS toZero) AND (lambdaEntry IS high) THEN cellBehavior
        IS staticOccupied;
    RULE 2 : IF (lambdaExit IS high) AND (lambdaEntry IS toZero) THEN cellBehavior
        IS staticFree;
    RULE 3 : IF ((lambdaExit IS low) AND (lambdaEntry IS low)) OR ((lambdaExit IS toZero)
        AND (lambdaEntry IS toZero)) OR ((lambdaExit IS toZero) AND (lambdaEntry IS low))
        OR ((lambdaExit IS low) AND (lambdaEntry IS toZero)) THEN cellBehavior IS semiStatic;
    RULE 4 : IF ((lambdaExit IS high) AND (lambdaEntry IS low) OR ((lambdaExit IS high)
        AND (lambdaEntry IS high)) THEN cellBehavior IS dynamic;
    RULE 5 : IF (lambdaExit IS low) AND (lambdaEntry IS high) THEN cellBehavior IS
        semiStaticOccupied;
END_RULEBLOCK
END_FUNCTION_BLOCK
```

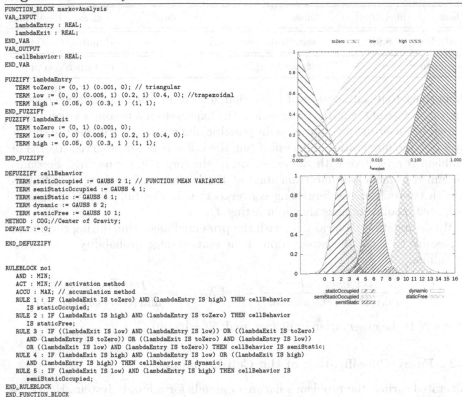

2.1 Modeling the State Change Probability

Saarinen [18] modeled an estimation of the conditional probability that a state change (free to occupied or vice versa) might occur by using the expectation of the Gamma distribution:

$$\hat{\lambda} = E\left[\lambda\right] = \frac{\alpha}{\beta} = \frac{\#positive\,events + 1}{\#observations + 1}.$$

That is

$$\lambda_{exit} = \frac{\alpha_{exit}}{\beta_{exit}} = \frac{\#occupiedToFree + 1}{\#observeOccupied + 1}$$

for the event a cell changes from occupied to free and its reverse

$$\lambda_{entry} = \frac{\alpha_{entry}}{\beta_{entry}} = \frac{\#freeToOccupied + 1}{\#observeFree + 1}.$$

Since it is not possible to change this model during runtime, [18] extended the first approach with a forgetting capability with the goal to converge unobserved

Table 1. Likelihood to observe a cell in occupied state given its specific class

class	likelihood	class	likelihood	class	likelihood
staticFree	0.01	semiStaticOccupied	0.9	dynamic	0.5
semiStatic	0.5	staticOccupied	0.99	unknown	0.35

events back to its initial value of 1. We changed this for the following two reasons: Firstly, in case of the initial observation, the expectation λ becomes larger than 1 caused by an occurring event without possible observations regarding the specific case (i.e. the observation is occupied but the cell was never observed free). This problem is repetitive with the extension of the forgetting capability. Secondly, we aimed to model an uncertain state at the beginning. For this, we initialized λ with 0.5 and let the forgetting converge to 0.5. The full algorithm with all proposed modifications is shown in listing 1.

While updating the grid-map with the presented algorithm during runtime, it is possible to query the expectation of the state change probability at any time by calling:

$$\lambda_{exit} = \frac{occupiedToFree}{\min\left(observeOccupied, N\right)} \text{ and } \lambda_{entry} = \frac{freeToOccupied}{\min\left(observeFree, N\right)},$$

where N is the observation limit (we used $N = 10000$).

2.2 Fuzzy Classification of the Cell Dynamic

As stated earlier, the problem's nature demands for a Fuzzy System. Hence, we used a fuzzy rule learner based on [3,9] to confirm this assumption. Although the results were promising, the learned fuzzy set was rather large (280 Rules). To achieve better extensibility and interpretability, we decided to build a hand-crafted fuzzy system. After analyzing testbed data, membership functions for the terms of the input variables $\lambda_{entry,exit}$ and for the terms of the output variable modeling the cell behavior (see Algorithm 2) were constructed. The terms of both linguistic variables were adapted from [18]. To understand the origin and meaning of the classes as well as the rules we would like to refer to Table 1 and Fig. 5 in [18]. The arrangement of the cell behaviors terms was chosen to achieve a semantically reasonable state transition order. The conspicuous higher variance of the dynamic term aims to include borderline values which would otherwise fall in the semi static class. The full program realized with FuzzyControlLanguage and the use of jFuzzyLogic [6,7] is shown in Algorithm 2. Further information about the classification performance of our fuzzy system as well as an analysis of the temporal characteristics is described in the experiments section.

2.3 Enhancing Robot Localization with Cell-Behavior

A common approach in range-scanner-based robot localization is the extraction of likelihood fields from pre-trained occupancy grid-maps [19]. The idea is

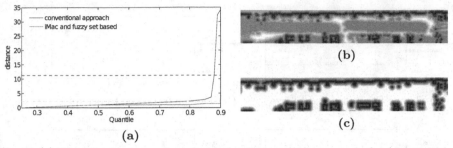

Fig. 1. (a) Statistical information about the distance between actual robot pose and the estimated one. The dashed line indicates the mean over all time steps, while the solid line shows different quantiles. (b) and (c) show the likelihood field for the proposed method and the conventional approach respectively.

to smoothen the probability of an observation which takes place in proximity of occupied cells. Simply meaning the closer a range-scan ends to an occupied area the higher its likelihood gets. Hence, the smaller the variance the higher the peakedness of the resulting probability density function (PDF). This PDF reflects the so called sensor model. Although there are works regarding the automatic adjusting of the peakedness [17], we used a static variance for the sake of simplicity. As introduced earlier, the likelihood field is extracted from a static map. Hence, it is impossible to correctly model dynamic parts of the map leading to incorrect low sensor weighting. We used the extracted classes from subsection 2.2 to propagate potential changes in the environment to the robots localization (Table 1). The results are shown in Fig. 1 (b) and (c).

3   Experiments

The goal of our experiments is twofold. On the one hand, we want to show that the expert knowledge based fuzzy sets reaches the performance of trained fuzzy sets. On the other hand, we want to understand the benefit of the combination of the iMac and fuzzy methods for future application. In particular, the experiments are following the processing for a place recognition system.

Firstly, we take a look at the behavior of the cell model. This is done by synthetically generating observation sequences, meaning static observations as well as alternating and oscillating. From this knowledge the expert knowledge based fuzzy system has been created. The next step includes examining the resulting classification stability of the models temporal characteristics. Fig. 2 presents several interesting examples of this examination to mediate the functionality of our system.

The data sets we use for the experiments consist of a recorded sequence produced by our simulator. The sequence contains 7031 time steps, which corresponds to a time of 703.1 seconds. In this sequence, four dynamic obstacles move freely in the area adjacent to the robot. Furthermore, a large element in the map was shifted to simulate a semi-static object.

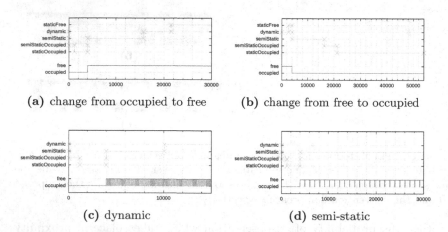

Fig. 2. Typical (a) and (b) as well as artificial (c) and (d) classifications (green) done by the expert knowledge approach according to changing observations (red)

Our simulation uses a previously recorded map from a museum environment. We have chosen a long corridor-like structure with reappearing features for the robot to patrol. These features lead to a lot of ambiguity for the place recognition process. For real world applications this situation is of highest interest. It often happens, that dynamic or semi-static obstacles occlude features of the environment that are needed to distinguish between several similar places. This results in localization errors, which may result in the robot losing track of its position entirely.

The trained map consists of 498×168 grid cells with a size of $10\text{cm} \times 10\text{cm}$ each. For our experimental evaluation we focus on the part of the map containing the relevant data meaning a sub-map of 302×48. This step is necessary since we could only train places in the part of the map which was observed by the robot. All other places share the model with the conventional approach.

To compute the data for training, the laser range scan of the robot was recorded together with label information extracted from the simulation for each time step. While the robot could observe the states free, occupied and not-observable, the label held additional information: free, occupied, unknown, semi-static obstacle, dynamic obstacle, static noise and dynamic noise. With this information our improved iMac model has been trained, additionally using a pre-trained static map for 2000 iterations as initialization. Both fuzzy models were then adapted to the trained iMac as described in Sect. 2.1 and 2.2.

The trained model reaches an accuracy of 97.88% within 2.53s on a Intel i5-3570. This is only slightly better than the expert knowledge based approach, which reaches 97.02% in 1.13s. However, the number of fuzzy rules is tremendously smaller: 5 for the expert knowledge based approach vs. 280 for the trained approach. For the final application, this results in a much shorter processing time. Furthermore, breaking down the number of fuzzy rules makes it easier to debug and leads to a better interpretability.

Table 2. Confusion matrices for both fuzzy classifiers

trained approach	STATIC_OCC	DYNAMIC	STATIC_FREE	SEMI_STATIC	expert knowledge based	STATIC_OCC	DYNAMIC	STATIC_FREE	SEMI_STATIC
STATIC_OCC	88	0	24	5	STATIC_OCC	91	0	0	30
DYNAMIC	0	1534	2	8	DYNAMIC	0	1524	20	2
STATIC_FREE	33	1	2964	8	STATIC_FREE	38	1	2939	30
SEMI_STATIC	2	12	5	34	SEMI_STATIC	1	19	0	34

(a) iMac and fuzzy set based

(b) conventional approach

Fig. 3. Plot of the weight field of the estimated robot pose summarized over time. A high weight is drawn in red, while lower weights range from yellow over green to blue. Both colormaps are scaled accordingly.

Table 2 shows the confusion matrices for both approaches. The true class is given row-wise, while the predicted class is given column-wise. It has to be noted that the label SEMI_STATIC_OCCUPIED had to be left out, since we could not provide a labeled example for this state. It is united with the STATIC_OCCUPIED class instead. However, the expert knowledge based system offers this possible intermediate state as it is defined in [18]. Furthermore, the high number of dynamic data is partly due to noise, which has no separate label yet.

It can clearly be seen that the expert knowledge based approach is better suited for the practical application. Hence, we continue all further experiments with the expert knowledge based approach. As a next experiment we want to figure out how suitable the proposed method works for place recognition. For this, we compute a likelihood field as described in Sect. 2.3 based on the proposed fuzzy classifier as well as for the conventional approach [19]. For both a radial linear kernel with a radius of 6 is applied.

Next, for each time step and each grid cell the probability $p(z_t|x)$ of the robot being at the respective grid cell x with the corresponding observation z_t is computed. This basically is the weight that a single particle of a particle filter based approach would estimate. In Fig. 3 we calculated the maximum weight over time

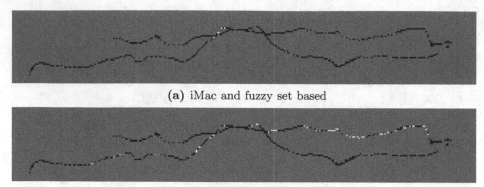

(a) iMac and fuzzy set based

(b) conventional approach

Fig. 4. Plot of actual robot trajectory. For each position the distance between the actual and the estimated position is color coded. Dark blue colors correspond to a perfect hit, while colors towards red stand for a larger distance.

Fig. 5. Plot of the actual robot trajectory as a green line against the estimated trajectories for the iMac and fuzzy set based approach (red stars) and the conventional approach (blue squares)

for each grid cell, for the sake of a better visualization. Both plots reveal the already recognizable path of the robot. However, for the conventional approach in Fig. 3(b) at some positions the path is disconnected. This disconnected part occurs at a position, where the robot mainly observes semi-static obstacles. Furthermore, the two approaches show a different level of smoothness. The proposed method exhibits a minor tendency for peakedness, which supports the stability after resampling of a particle filter.

To estimate the robot position, the maximum is taken from the time series of weights. For the final application, this corresponds to the convergence process of the particle filter. For the experiment discussed here, we decided against using the entire particle filter framework to avoid evaluation errors not originating from our approach.

From the estimated poses the Euclidian distance to the true robot position is computed. Fig 1 shows some statistical information of these computed distances.

The mean distance demonstrates the clear advantage of the proposed method. Looking at the several quantiles clearly reveals some tremendous outliers applying the conventional approach. To emphasize this, Fig. 4 and Fig. 5 show two different kinds of visualizing these outliers (the colormap was thresholded). Figure 4 draws the actual robot trajectory with the distance to the estimated robot pose for both approaches respectively. It can clearly be seen that the outliers mainly occur at the position where the semi-static obstacle is observed, as already stated. Furthermore, in the lower left part of the trajectory, the outliers result from occluding dynamic obstacles. To understand how dislocated the estimation really is, Fig. 5 plots the estimated position within the map.

4 Conclusion and Future Work

In this publication we presented an enhancement to an existing approach for modeling the dynamics of an environment. The model was extracted by designing a fuzzy system based on expert knowledge which is able to compete and outperform other algorithms in terms of extensibility and interpretability. Furthermore, the classified environment was exemplary applied to successfully improve the sensor model used for robot localization in terms of the resulting trajectory's accuracy. All presented experiments show promising results to deploy the proposed method to a real world application. This leads to the necessity to perform long time experiments with an actual particle filter algorithm. Due to the approach's adaptability, another utilization in future works will be to use the modeled understanding of the dynamic of the environment to enrich path planners by adding dynamic traversal costs or to design robot systems with adaptive behavior.

References

1. Anguelov, D., Biswas, R., Koller, D., Limketkai, B., Thrun, S.: Learning hierarchical object maps of non-stationary environments with mobile robots. In: UAI, pp. 10–17 (2002)
2. Arbuckle, D., Howard, A., Mataric, M.: Temporal occupancy grids: a method for classifying the spatio-temporal properties of the environment. In: IROS, pp. 409–414 (2002)
3. Berthold, M.R.: Mixed fuzzy rule formation. Int. J. Appr. Reas. 32(2), 67–84 (2003)
4. Biber, P., Duckett, T.: Dynamic maps for long-term operation of mobile service robots. In: RSS, pp. 17–24 (2005)
5. Biswas, R., Limketkai, B., Sanner, S., Thrun, S.: Towards object mapping in dynamic environments with mobile robots. In: IROS, pp. 1014–1019 (2002)
6. Cingolani, P., Alcalá-Fdez, J.: jfuzzylogic: a robust and flexible fuzzy-logic inference system language implementation. In: FUZZ-IEEE, pp. 1–8 (2012)
7. Cingolani, P., Alcalá-Fdez, J.: jfuzzylogic: a java library to design fuzzy logic controllers according to the standard for fuzzy control programming. Int. J. CI Sys. 6(sup1), 61–75 (2013)
8. Fox, D., Burgard, W., Thrun, S.: Markov localization for mobile robots in dynamic environments. Journal of Artificial Intelligence Research 11, 391–427 (1999)
9. Gabriel, T.R., Berthold, M.R.: Influence of fuzzy norms and other heuristics on mixed fuzzy rule formation. Int. J. Appr. Reas. 35(2), 195–202 (2004)
10. Hähnel, D., Triebel, R., Burgard, W., Thrun, S.: Map building with mobile robots in dynamic environments. In: ICRA, pp. 1557–1563 (2003)
11. Krajník, T., Fentanes, J.P., Cielniak, G., Dondrup, C., Duckett, T.: Spectral analysis for long-term robotic mapping. In: ICRA (in press, 2014)

12. Kucner, T., Saarinen, J., Magnusson, M., Lilienthal, A.J.: Conditional transition maps: Learning motion patterns in dynamic environments. In: IROS, pp. 1196–1201 (2013)
13. Meyer-Delius, D., Beinhofer, M., Burgard, W.: Occupancy grid models for robot mapping in changing environments. In: AAAI, pp. 2024–2030 (2012)
14. Meyer-Delius, D., Hess, J., Grisetti, G., Burgard, W.: Temporary maps for robust localization in semi-static environments. In: IROS, pp. 5750–5755 (2010)
15. Mitsou, N.C., Tzafestas, C.S.: Temporal occupancy grid for mobile robot dynamic environment mapping. In: MED, pp. 1–8 (2007)
16. Moravec, H., Elfes, A.: High resolution maps from wide angle sonar. In: ICRA, pp. 116–121 (1985)
17. Pfaff, P., Plagemann, C., Burgard, W.: Improved likelihood models for probabilistic localization based on range scans. In: IROS, pp. 2192–2197. IEEE (2007)
18. Saarinen, J., Andreasson, H., Lilienthal, A.J.: Independent markov chain occupancy grid maps for representation of dynamic environment. In: IROS, pp. 3489–3495 (2012)
19. Thrun, S., Burgard, W., Fox, D.: Probabilistic robotics. MIT Press (2005)
20. Tipaldi, G.D., Meyer-Delius, D., Beinhofer, M., Burgard, W.: Lifelong localization and dynamic map estimation in changing environments. In: ICSC (2012)